Dictionary of
Tropical American Crops
and Their Diseases

by

Frederick L. Wellman

*"Si se conoce el nombre de la enfermedad,
la cura esta amitad de camino."*

The Scarecrow Press, Inc.
Metuchen, N.J. & London
1977

On the title page: an old Spanish saying, freely
translated, "If you know the name of the disease it
is half way to being cured."

Other works by the author:

Coffee Botany, Cultivation, and Utilization (1961)

Plant Diseases: An Introduction for the Layman (1971)

Tropical American Plant Disease (1972)

Library of Congress Cataloging in Publication Data

Wellman, Frederick Lovejoy, 1897-
   Dictionary of tropical American crops and their
diseases.

   Includes bibliographies and index.
   1. Tropical crops--Diseases and pests--Dictionaries.
2. Tropical crops--Latin America--Dictionaries.
I. Title.
SB608.T8W44   631'.098   77-8558
ISBN 0-8108-1071-9

to Claud L. Horn the broadly experienced, highly motivated and superb horticulturist, my erstwhile exacting administrator in the tropics, whose masterly understanding and thoughtfulness has helped me in many ways to do all that I could over a long period of time, and who has remained a fine and eternal personal friend;

to Don E. Ellis the dedicated and intense builder of his group of professionals, who invited me to be a willing and enthusiastic part of the great Plant Pathology Department he headed; and

to Robert Aycock his able successor, who stimulated me to pick up and continue on my way at about the point where I had decided to retire for the fourth time,

this book is dedicated

CONTENTS

A Memorial

CARLOS LUIS SPEGAZZINI (1858-1926)

The searching, active, helping mycologist-pathologist
dominant in tropical America for half a century

Born in Italy he prepared himself by meticulous studies in
Padua under the great Pier Andrea Saccardo, and was stimulated to
make a classic contribution through work in the New World.   He
went to Argentina in 1879, taught plant pathology and inevitably became clinician-adviser to innumerable worried crop growers and disease specialists both in and away from the tropics.   He carried on
expeditions to warm-country and cool-country, untouched veritable
Edens of disease fungus life mostly in Argentina, Chile, Uruguay,
Brazil, and Paraguay but elsewhere too.

He was especially interested in his adopted country.   At the
outset only 39 Argentine fungi were known; in his lifetime some 5000
more species were found in the broad area of his activity, were determined, described, many drawn, and published in about 130 papers.
He was Director General of Studies and Senior Professor, University
of La Plata, Buenos Aires, Argentina.   In 1964 that University established in it the famous Spegazzini Museum.

To have met him (he was well known in the United States of
America, the West Indies, Central America, South America, and
Europe) must have been an unforgettable experience.   He was a man
of quickly apparent erudition and much energy and enthusiasm.   He
had fine keen eyes, a feeling of world responsibility, and enjoyed
what he did in his profession.   As I have read his writings, I recognize that his reputation as a mycologist must come first.   However,
to pathologists it is known that he worked hard and successfully on
disease problems and that his mycological preeminence does not in
the least diminish the help he gave in such groups of plant diseases

vii

Dr. Carlos Luis Spegazzini (1858-1926)
of Argentina and Tropical American Plant Pathology
(photograph courtesy of John A. Stevenson)

as tropical mildews, rusts, grass ergots, smuts, destructive leaf-spots, and others, including hyperparasites.

For an estimate of the Spegazzini contribution all a patholo-gist needs is to study lists of tropical American plant disease fungi. The largest is "Indice de fungos da América do sul" by A. P. Vié-gas (see "Speg." in Viégas, page 116). As a random sampling I opened it to about every tenth page of the fungus list; from 66 such sample pages there were 61 on which occurred fungus binomials fol-lowed by the authority "Speg." On some pages there were over twenty names with that "Speg." attached to it.

Ainsworth, G. C.   1961.   Ainsworth & Bisby's Dictionary of the
    Fungi, 5th ed.   Commonw. Myc. Inst. , Kew (England), 547p
    (see p378).

Farr, M. L.   1973.   An Annotated List of Spegazzini's Fungus
    Taxa.   Bibl. Mycol. , J. Cramer. Bd. 35, 1, v1:1-823; Bd.
    35, 2, v2:824-1661.

Fernandez Valiela, M. C.   1952.   Introducción a la fitopatología,
    2d ed.   Gadola, Buenos Aires, 872p (see p20-23).

Marchionatto, J. B.   1928.   Doctor Carlos Spegazzini, en el se-
    gundo anniversario de su fallecimento.   Rev. Fac. Agr. La
    Plata (Argentina), 18:16-20.

Molfino, J. F.   1929.   Carlos Spegazzini, su vida y su obra.   Im-
    prenta "Coni," Buenos Aires, leaflet 7p.

Toro, R. A.   1926.   Carlos Spegazzini.   Mycologia, 18:284.

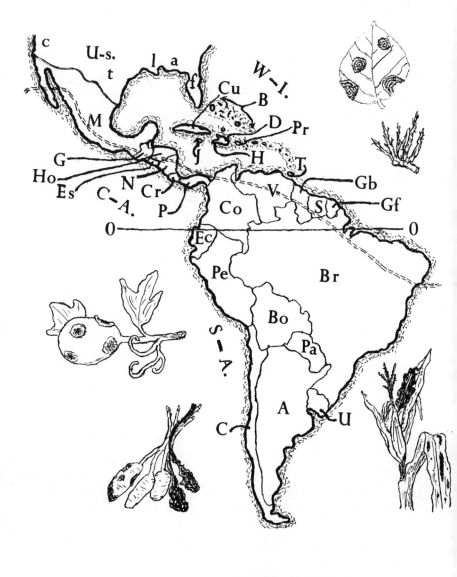

c U-s. l a
       t   f
                       W–I.
              Cu   B
       M   D Pr
G   H   T
Ho   Gb
Es   N Cr   V
C–A. P   Co   S   Gf
0 ————————— 0
          Ec
          Pe   Br
          Bo
                Pa
       S–A. A   U
       C

x

facing page:

SKETCH MAP OF THE AMERICAN TROPICS
(or Neotropics)

Legend

Straight line, 0—0, geographical equator
Undulating double line, ══, thermal equator

A	Argentina	J	Jamaica
B	Bahamas	M	Mexico
Bo	Bolivia	N	Nicaragua
Br	Brazil	P	Panama
C	Chile	Pa	Paraguay
C-A.	Central America	Pe	Peru
	(region of)	Pr	Puerto Rico
Co	Colombia	S	Surinam
Cr	Costa Rica	S-A.	South America
Cu	Cuba		(region of)
D	Dominican Republic	T	Trinidad & Tobago
E	Ecuador	U	Uruguay
Es	El Salvador	U-s.	Subtropical USA
G	Guatemala		(a. Alabama, c. Cal-
Gb	Guiana (British)		ifornia, f. Florida, l.
Gf	Guiana (French)		Louisiana, t. Texas)
Gi	Galápagos Island	V	Venezuela
H	Haiti	W-I.	West Indies (region
Ho	Honduras		of)

Sketches of disease symptoms by the author, from rough field notes, decorate the map clockwise from upper left: Chayote, Sechium, fruit with abnormally sunken spots showing raised rims; Guatemala. Lleren, Calathea, tubers attached to plant with two tubers severely diseased with black rot which, it can be seen, originated in foliage stems; also, a tuber bitten by an insect and the openings invaded by soil fungi; Puerto Rico. Camote, or sweet potato, Ipomoea; tubers rotting in the market; Virgin Islands. Maiz, or corn, Zea, Indian grown with smut of upper ear and leaf spots on lower leaf; Colombia. Cacao, Theobroma, wild seedling branch in uncultivated jungle, showing witches' broom; Ecuador. Tomatillo, Physalis, leaf with large unusual target board spots; Venezuela and Honduras.

INTRODUCTION

This book has some personal implications. I needed it in 1929 when as a professional I first went to the Neotropics. Now I hope it will help others. It has been prepared in as simple a manner as possible, a one-man activity, and presenting only the most succinct material. I do not feel it is as complete as it should be, but it has been in the process for years. I had nearly a half million "cards" of information, a great deal of which I found to be unnecessary when I began discarding and organizing for this book.

A card (a half sheet of typewriter-size paper, sometimes secondhand, sometimes purchased, sometimes purloined) was prepared, always with a Latin crop name at the top. Cards were made for each host disease, for the many names of them in local languages, field descriptives, and countries of occurrence. On consolidation, much was eliminated. After the Latin binomial of a parasite is a simple English disease descriptive that is easily translatable, and where the various countries include a whole region sometimes only that regional name is used.

In some future greater work by several coauthors, the state, and province, and even terrain and area should be reported for disease occurrences in every country, large or small. These were not always indicated from information I used, and while I kept some area reports for Mexico, Colombia, Peru, Brazil, Chile, and Argentina, they were not always adequate and such details were rejected. It is when pathologists are plentiful in a land that sufficient notices are published so that areas and states where disease occurrences are can be pinpointed. On the other hand, reference is made in the present dictionary to certain state occurrences (southern edges) in subtropical parts of the United States of America. Com-

paratively speaking, crop disease workers practically swarm in those
states and their findings make lists that are, in some cases, signifi-
cant in their relationship to disease occurrences or absences further
south in the Neotropics.

As soon as a list of a large region appears, as this proposes
to be, imperfect as it is and even if it were absolutely complete in
all respects and up to the minute, it becomes out of date when published. New things are being found and reported every month. Both
hosts and parasites are living, varying organisms, mutating, cross-
ing, and changes are being watched with interest by concerned pathologists. Corrections are always in progress. However, it is my
profound hope that this dictionary of crops with their diseases may
still be a usable tool.

I want to acknowledge here that in addition to those to whom
I dedicate this book I am also deeply indebted to Ralph H. Allee,
John A. Stevenson, and Miss May Coult. Some very helpful consultations have come from Gerald R. Thorne, Guy V. Gooding, Fran-
cis O. Holmes, Joe N. Sasser, and Mariano Montealegre. Again,
without the affection and protective measures of my wife and son, I
could never have undertaken the drudgery of manifold details such a
dictionary as this requires. Some, not all, of those in the tropics
itself who have done some special things for me are Claud L. Horn,
John S. Niederhauser, Angelina Martínez, H. David Thurston, Eu-
genio Schieber, Carlos Bianchini, Carlos Arnaldo Krug, Albert S.
Muller, Consuelo Bazan de Segura, Gino Malaguti, Eddie Echandi,
José A. B. Nolla, Julio Bird, Julian Acuña, Felix Choussy, Joan
Hayes, and Jaime Díaz Moreno. I also appreciate especially the
critical help from Larry F. Grand.

In preparing the manuscript no one could have been more
helpful and pleasant than Joyce Johnson, for which I thank her; the
late Dr. Ralph Shaw helped me a great deal, for which I am very
appreciative.

LISTS OF NEOTROPICAL PLANT DISEASES

I have experienced in the tropics inevitable limitations as to help and funds. With more facilities there would definitely be a more ideal means to gather together a universally acceptable detailed listing of all plant diseases in a great region like the Neotropics. As I reviewed what I would wish to do in that line, I found that at first I had planned more than one person, who was also involved in other things, could accomplish.

It was incumbent on me to work only on crop-plants and mostly on those that I knew. I had kept thousands of personal observations and determinations. In some instances friendly pathologists assisted me with new discoveries. This was further, and basically, influenced by consultations and clinical visits of my own with all kinds of troubled growers in their fields, with questions from regular students and graduate trainees, and with discussions among agricultural science specialists in both the tropics and the temperate zone.

To this was added study of valuable plant disease lists of various countries. It is not pretended that all details of the best lists are given here nor do uncertain diseases have much space in this book before you. However, citations of lists are not always available and the following "list of lists" is part of the source for future workers. The Latin American literature of plant pathology is progressing; some disease determinations that are old are being changed and will be changed, and much that is still unseen is certain to be published.

The citation list I give hereunder, is a more or less parenthetical inclusion in this book. References to purely mycological papers, even if they deal with a great many disease-producing organisms, are in a different category and for the professional to

secure for himself. I do not include books of diseases on the major crops, for these are readily found in libraries.

Alandia Borda, S. , and F. R. Bell.   1957.   Diseases of warm cli-
    mate crops in Bolivia.   FAO.   Plant Prot. Bull. , 5:172-173.

Alvarez-García, L. A. , and C. Chupp.   1954.   Enfermedades de
    las hortalizas en Puerto Rico y medios de combatirlas.
    Typescript, 164p.

Ancalmo, O.   1959.   Lista preliminar de enfermedades parasitarias
    en las plantas de El Salvador.   Serv. Coop. Agr. Salvadoreno
    Amer. , Bol. Tech. 22, 29p.

Baker, R. E. D.   1943.   Notes on some diseases of field crops,
    vegetables and fruits at the Imperial College of Tropical Ag-
    riculture.   Trop. Agr. (Trinidad), 20:28-32.

Batista, A. C.   1946-47.   Principias doenças das plantas em o
    Nordeste Pernambuco.   Sec. Agr. Inc. (Pernambuco), Com.
    B 13:195-252; 14:5-46.

Bazan de Segura, C.   1965.   Enfermedades de cultivos tropicales
    y subtropicales.   Editorial Juridicia, Lima (Peru), 439p.
    (A recent new edition has been published.)

Bertossi, E. O.   1961.   1962.   Fitopatología fungosa Valdiviana.
    Contrib. 1, 2, 3.   Rev. Univ. Santiago (Chile), 46:55-65,
    205-212; 47:43-55.

Bitancourt, A. A.   1941.   Brasil: plant diseases observed in the
    state of São Paulo in 1939 and 1940.   Int. Bul. Plant Prot. ,
    15:222M-223M.

Ciferri, R.   1961.   Mycoflora Domingensis integrata.   Pavia Univ.
    Bot. Lab. Crittogamico, Quad. 19, 539p.

Cralley, E. M.   1954.   Las enfermedades de las plantas en Pan-
    ama, 2d ed.   Mimeographed, 24p.

Del Prado, F. A. , and N. J. Van Suchtelen.   1955.   Lijst van
    ziekten van landen tuinbouwgewassen in Suriname.   Mimeo-
    graphed, 25p.

Echandi, E. , J. K. Knoke, E. L. Nigh, Jr. , M. Shenk, and G. T.
    Weekman.   1973.   Crop protection in Brazil, Uruguay, Bolivia,
    Ecuador and Dominican Republic.   US/AID Processed Special
    Report from Univ. California and North Carolina State Univ. ,
    68p.

Feldman, J. M. , and R. E. Pontis.   1960.   Enfermedades para-

sitarias de las plantas cultivadas, señaladas para la provincia de Mendoza. Rev. Argentina Agron. , 27:27-50.

Fernández Valiela, M. V. 1952. Introducción a la fitopatología, 2d ed. Gadola, Buenos Aires, 872p.

Galli, F. , H. Tokeshi, P. de C. Torres de Carvalho, E. Balmer, H. Kimati, C. O. Nogueira Cordoso, and C. Lima Salgado. 1968. Manual de fitopatologia; doenças das plantas e seu controle. Ceres, São Paulo, 640p.

García-Rada, G. 1947. Fitopatología agricola del Peru. Estac. Exp. Agr. La Molina (Peru), 423p.

_____ and J. A. Stevenson. 1942. La flora fungosa peruana. Lista preliminar de hongos que atacan a las plantas en el Peru. Estac. Exp. Agr. La Molina (Peru), 112p.

Koch de Brotos, L. , and C. Boasso. 1955. Lista de las enfermedades de los vegetales en el Uruguay. Uruguay Dir. Agron. Publ. 106, 65p.

Larter, L. N. H. , and E. G. Martyn. 1943. A preliminary list of plant diseases in Jamaica. Imp. Mycol. Inst. , Mycol. Papers 8, 108p.

Litzenberger, S. C. , and J. A. Stevenson. 1957. A preliminary list of Nicaraguan plant diseases. Plant Dis. Reptr. Suppl. 243, 19p.

Marchionatto, J. B. 1951. Los hongos parasitos de las plantas. Acme Agency, Buenos Aires, 118p.

Martyn, E. B. 1942. Diseases of plants in Jamaica. Jamaica Dep. Sci. Agr. Bull. 32, 34p.

Molestina, E. 1942. Indice preliminar de las principales enfermedades y plagas de la agricultura en el Ecuador. Ecuador Dep. Agr. Bol. 15, 25p.

Mujica R. , F. , and C. Vergara C. 1945. Flora fungosa Chilena, indice preliminar de los huespedes de los hongos chilenos y sus referencias bibliográficos. Imprenta Stanley, Santiago (Chile), 199p.

Muller, A. S. (From 1933 to 1962, 13 publications of findings in South and Central America.)

Nowell, W. 1923. Diseases of crop-plants in the Lesser Antilles. West India Committee, London, 383p.

Orjuela Navarrette, J. 1965. Indice de enfermedades de plantas cultivadas en Colombia. Editorial ABC, Bogota, 66p.

Pierre-Louis, F.   1955.   Maladies des plantes économiques d'Haiti. Bull. Agr. , 4:1-51.

Stevenson, J. A.   1975.   Fungi of Puerto Rico and the American Virgin Islands.   Reed Herbarium (Baltimore) Contrib. 23: 743p.

_____ and F. L. Wellman.   1944.   A preliminary account of the plant diseases of El Salvador.   Washington Acad. Sci. J. , 34:259-268.

Toler, R. W. , R. Cuellar, and J. B. Ferrer.   1959.   Preliminary survey of plant diseases in the Republic of Panama, 1955-1958.   Plant Dis. Reptr. , 43:1201-1203.

Viégas, A. P.   1961.   Indice de fungos da América do sul.   Seção Fitopat. Inst. Agron. Campinas. , 921p.

Wellman, F. L.   1972.   Tropical American plant disease (Neotropical phytopathology problems).   Scarecrow Press, Metuchen, N. J. , 989p.

Zevada, M. Z. , W. D. Yerkes, Jr. , and J. S. Niederhauser.   1955. Primera lists de hongos de México, arreglada por huéspedes. México Ofic. Estud. Espec. Tech. Foll. 14, 43p.

THE IMPORTANCE OF MYCOLOGISTS
TO TROPICAL PATHOLOGY

I feel it due to me and to my profession to formally acknowledge the help we obtain from mycology. Tropical pathologists have no doubt of the unquestioned need for precise identity, on standardized mycological bases, of the exact fungus causing a given disease. Everyone knows the scientific names we use for parasites are from the determinations of mycologists. And these determinations were not easily made; they are the products of highly trained and dedicated workers.

The general research pathologist soon learns how to name the major severe disease organisms. When variations or special forms develop, deeper pathological and mycological studies are instituted. If the pathologist has discovered these changes he is the one, often, who reports them, but the descriptions and the names are on the bases stipulated by mycological scientists.

Fungi have been carefully studied for about 280 years mostly in the temperate zone, and it is there that the largest amount of fungus research has been carried on and completed. However, in the present century more is being done in the tropics. Pathologists go to, and obtain, willing help from mycologists and their great wealth of materials, both in the tropics and in the temperate zone. It is in the tropics that I have found mycology to be increasingly complex and tremendously stimulating.

In comparison with the temperate zone, in the tropics there are many more fungi on many more cultivated hosts--and the large numbers of ever proliferating and changing new parasites require new names. Here in the tropics is where fungi multiply most vigorously; there are no seasons that kill or slow down activity and as fungus populations multiply in plentitude there is greater chance for

the reshuffling of genes. In our Neotropics, sometimes parasitic species' names become altered when the same fungus is found on another host; sometimes a difference is made in designation on the grounds that a species is found in a different hemisphere or even in a different ecological area.

Pathology and mycology have to progress and be subsidized hand-in-hand in the tropics. Without one the other is weak. Pathology is an insecure and crippled science, if it is divorced from close association and collaboration with mycology. Mycology depends on pathology for much of its reason for being as it helps with economic growing of crops. I wish to repeat that these two scientific specialties are bound inextricably and in the tropics they can not be independent.

At this point, I wish to state that I do not consider myself the past master in mycology, but my mycological friends have been generous in helping me. However, any mycological mistakes in the ensuing pages of this dictionary are mine and to be blamed on me. I have been the one who has made the final decisions on the parasite determinations that are in this book.

DICTIONARY OF CROPS
WITH THEIR DISEASES

(Note to those consulting this list: All scientific names
of crop plants are in all capital letters and are arranged in this book in simple alphabetical order. Latin
crop names follow what are to me reasonably accepted
standard usages from England, the United States, and
South America. A condensed description of the crop
is included, ending with common names in quotation
marks; the language involved in each case is indicated
immediately afterwards, thus: "Sp." [Spanish], "Po."
[Portuguese], "En." [English], and "Ind." [Indian].
Under each host its diseases are given, listed on the
basis of scientific names of the causal parasites.
These are arranged under the host, listed alphabetical-
ly, characterized as succinctly as possible, and something of the distribution indicated.)

ABELMOSCHUS see HIBISCUS

ABUTILON spp. Native to the American tropics, they are mostly
grown as ornamental perennial shrubs, some handsomely variegated, one is grown in Brazil for its edible flowers; "abutilon
ornamentál" Sp. , "abutiloniera" Po. , "abutilon" En. (Only a few
diseases have been determined or reported on abutilons, as they
are of lesser economic importance than so many other crops. I
have seen, but not identified, wilts, a branch disease, white mildew, a black and a creamy-brown stem canker, and black rootrot.)
    Asterina diplocarpa Cke. ; black mildew. Puerto Rico.
    Cercospora brachypoda Speg. ; leafspot. Puerto Rico.
    Meliola molleriana Wint. = Irenopsis molleriana (Wint.)
        Stevens; black mildew. Argentina, Brazil, Guatemala.
    Meloidogyne sp. ; rootknot. C. America, W. Indies.
    Peroneutypa tuyutensis Speg. ; wood rot. S. America.
    Phyllachora abutilonis Gar. ; leafspot. Colombia.
    Phyllosticta abutilonis P. Henn. ; leafspot. Brazil.
    Puccinia heterospora Berk. & Curt. = Uromyces malvicola
        Speg. ; rust. S. America. P. sherardiana Koern. ; rust.
        Bolivia.
    Sphaceloma abutilonis Bitanc. & Jenk. ; scab. Brazil.

1

ACACIA spp. [see also ALBIZZIA JULIBRISSIN]. Trees used in
shade agriculture, important for wood and charcoal, for temporary buildings and posts; "carbón," "palo acacia" Sp., "acácia,"
"carboniera" Po., "acacia," "charcoal-wood tree" En. (Note:
There are over 35 acacias used in the American tropics, all
seemingly diseased with leafrust. Effects are apparently mild,
but defoliation is much hastened by rust. On 25 acacias Spegazzini described a whole galaxy of rusts; at least 33 on known
acacia species and nine on unspecified acacia materials that had
been sent to or collected by him.)
    Armillaria mellea Vahl. = Armilariella melea (Fr.) Karst.;
      rootrot. Florida, W. Indies, C. America.
    Cassytha filiformis L.; green dodder. Puerto Rico, W. Indies.
    Cephaleuros virescens Kunze; algal leafspot. Florida, W. In-
      dies, C. America. Probably widespread.
    Cercospora alemquerensis Speg.; leafspot. Brazil.
    Clitocybe tabescens Bres. = Armillariella tabescens (Scop.)
      Sing.; rootrot. Florida.
    Coniothyrium decipiens Cke. & Harkn.; branch disease. Cali-
      fornia.
    Corticium salmonicolor Berk. & Br.; pink branchrot. W.
      Indies, C. America, S. America.
    Cuscuta americana L.; dodder. W. Indies, C. America.
    Cylindrocarpon obtusisporum Wr.; twigrot. California, Mexico.
    Diabole cubensis Arth. (on A. CAVEN Mol.); leafrust. Brazil,
      Cuba.
    Diaporthe medusaea Nits.; dieback. California.
    Diplodia sp. = Botryodiplodia theobromae Pat.; dieback. Flor-
      ida, California, W. Indies.
    Fomes sp.; woodrot. Puerto Rico, C. America.
    Fusarium sp.; dieback complex. Puerto Rico, Colombia.
    Ganoderma lucidum Karst.; basal disease. Costa Rica. G.
      pseudoferrum Ov. Ste.; basal disease. C. America.
    Phoradendron spp.; bushy mistletoes. California, Mexico,
      Texas, C. America.
    Phthirusa spp.; recumbent mistletoes. C. America.
    Phyllachora acaciae P. Henn.; tarspot. W. Indies, C. America,
      S. America.
    Phyllosticta acaciicola P. Henn.; leafspot. Brazil, Argentina.
    Phymatotrichum omnivorum (Shear) Dug.; dryland rootrot.
      Texas, Mexico.
    Physalospora fusca Stev.; foliage disease. Florida.
    Psittacanthus cuneifolius Blume; parasitizing bush on tree
      branches in Argentina, and P. linearis Macbr.; parasite
      bush. Peru.
    Ravenelia spp. all leafrusts, examples: R. australis Diet. &
      Neg. on A. CAVEN Mol., in Argentina, R. hieronymi Speg.
      on A. CAVEN in Chile, Brazil, Uruguay, Argentina, R. in-
      quirenda Arth. & Holw. with the synonym Uredo acaciae-
      bursariae Comm. on A. BURSARIO Sche., in Guatemala,
      R. monosticha Speg., leafrust, Paraguay, R. stevensii Arth.
      on A. RIPARIA HBK. in the W. Indies, and other Ravenelias
      on acacias from Mexico and the West Indies south into Ar-
      gentina.

ACALYPHA WILKESIANA Muell. -Arg. = A. TRICOLOR Seem. Important as an ornamental, has much variegated foliage; "acalifa" Sp. Po., "acalypha" En.

Appendiculella arecibensis (Stev.) Toro = Meliola arecibensis Stev.; black mildew, leaf disease. Puerto Rico.

Botryosphaeria dothidea (Moug.) Ces. & deNot.; stem disease. El Salvador, Guatemala.

Capnodium spp.; fumago. W. Indies, C. America.

Cassytha filiformis L.; dodder. Puerto Rico, St. Thomas.

Cephaleuros virescens Kunze; algalspot. W. Indies, C. America.

Cercospora acalypha Pk.; leafspot. Puerto Rico, Brazil, Costa Rica, Texas.

Clitocybe tabescens Bres. = Armilariella tabescens (Scop.) Sing.; rootrot. C. America, Florida.

Cuscuta americana L.; dodder. Puerto Rico. C. domingensis Urb.; dodder. Dominican Republic, St. Thomas.

Diplodia sp.; dieback. W. Indies, Costa Rica, El Salvador.

Peniophora cinerea Cke.; stalk disease. Bermuda.

Phyllachora acalyphae Chard.; tarspot. Brazil, Colombia.

Phyllosticta sp.; leaf and stem disease. Puerto Rico, St. Thomas.

Ramularia acalyphae Tharp; leafspot. Texas.

Rhizoctonia solani Kuehn = Thanatephorous cucumeris (Frank) Donk; basal rot. Puerto Rico.

Rosellinia bunodes Berk. & Br.; rootrot. Puerto Rico, Guatemala.

Schizophyllum commune Fr.; secondary stemrot. Puerto Rico, C. America.

Scytonemeae sp.; black alga of leaf. Puerto Rico, St. Thomas, Guatemala.

Strigula complanata Fée; white lichenspot. W. Indies, C. America.

Uredo acalyphae Speg. = Uromyces proeminens (DC.) Pass. Leafrust. Argentina. Uredo paulistana Speg.; rust. Brazil.

ACHRAS ZAPOTA L. = LUCUMA MAMMOSUM DC. Native to Central America, an important tree, source of much fruit in many out-of-the-way parts of Latin America; "níspero," "zapotero" Sp., "zapotizeiro" Po., "sapodilla" En.

Botryosphaeria dothidea (Moug.) Ces. & de Not.; branch disease. W. Indies.

Capnodium sp.; sootymold. C. America, Puerto Rico.

Cassytha filiformis L.; dodder. Guatemala.

Cephaleuros virescens Kunze; algalspot. W. Indies, C. America, S. America.

Colletotrichum gloeosporioides Penz.; anthracnose. C. America, W. Indies, Brazil.

Elsinoë lepagei Bitanc. & Jenk.; scab. Brazil.

Meliola sp.; black mildew. Florida, Puerto Rico.

Oryctanthus spp.; recumbent mistletoes. Costa Rica.

Pestalotia scirrofaciens Brown; limb tumor. Florida, Texas.

Phorandendron robustissimum Eich.; coarse mistletoe. El Salvador.

Agave 4

Phthirusa pyrifolia Eich.; slender mistletoe. Costa Rica, El
   Salvador, Guatemala.
Phyllosticta sapotae Sacc.; leafspot. Bahamas.
Phymatotrichum omnivorum (Shear) Dug.; dryland rootrot.
   Texas.
Septoria sp.; leafspot. Florida, El Salvador.
Strigula complanta Fée; white lichenspot. W. Indies, C.
   America.
Trentepohlia sp.; algal tufts. Puerto Rico.
Uredo sapotae Mains = Scopella sapotae Arth. & Johns.; rust.
   W. Indies, C. America, S. America, Florida.

AGAVE spp.   Native to dry-hot Neotropics, strong perennial plants
   with rosette of sword-like leaves.   (1) A. AMERICANA L., ornamental, produces fiber, pulque, cortisone; "agave," "maguey"
   Sp., "piteria" Po., "agave" En.   (2) A. FOURCROYDES Lam.,
   grown in plantations producing hard fiber from leaves; "cabuya,"
   "henequén" Sp.   (3) A. LETONAE Trel., produces leaf fiber;
   "cabúya," "pita," "penca," "henequén" Sp., Po.   (4) A. SISA-
   LANA Perr., leaf fiber for cordage; "corita," "sísal," "planta
   de cordál" Sp. Po., "sisal" En. (all have many local and Indian
   names).
      Aspergillus spp.; fiber-mold. General (1, 3, 4).
      Asterina agaves Ell. & Ev. and A. americana Ell. & Ev.;
         black mildews. Mexico (1).
      Basal necrosis, ? physiological. Brazil (4).
      Botryodiplodia theobromae Pat. = Diplodia theobromae (Pat.)
         Now., D. natalensis P. Evans; black to blue large leafspot,
         severe weakened fiber and discoloration. Neotropical (1,
         3, 4).
      Capnodium spp. = Antennaria spp., Fumago spp.; sootymolds.
         Neotropical (1, 2, 3, 4).
      Cladosporium (? herbarum); ragged leaf. Mexico, Puerto
         Rico (1).
      Colletotrichum gloeosporioides Penz. also reported as Gloeo-
         sporium agaves Cav. or C. agaves Cav., = asco-state is
         Glomerella cingulata (Stonem.) Sp. & Schr.; anthracnose.
         Neotropical (1, 2, 3, 4).
      Coniothyrium sp.; canker on leaf. Colombia (1,4).
      Didymaria sp.; rot of leaf. Mexico (1), Brazil (4).
      Dimerosporium agavectona Pat. & Har.; black mildew.
         Mexico (1).
      Diplodia (see Botryodiplodia)
      Dothidella paryii (Farl.) Th. & Syd.; eyespot. W. Indies,
         Mexico (1, 2, 3, 4).
      Fusarium spp.; associated with leafrots, often secondary,
         sometimes primary cause. Neotropics (1, 2, 3, 4).
      Gymnosporangium sp.; leafrust. Peru (1).
      Lembosia agaves Earle; blackmildew. Puerto Rico (1, 2, 3),
         El Salvador (2).
      Marssonina agaves Earle; leaf disease. Colombia (1), Vene-
         zuela (2, 4).
      Nectria sp.; stemrot. Puerto Rico (2).

Phytophthora agaves Vitl.; rot of stem base.   Mexico (1).
Pleospora sp.; leafspot.   Peru (1).
Septonema agaves Stevens; leafpatch disease.   W. Indies, El
    Salvador (1, 3).
Septoria megaspora Speg. = Rhabdospora megaspora (Speg.)
    Kuntze; leafspotting.   Argentina (1).
Stagonospora gigantea Heald & Wolf; leafscorch.   Mexico, El
    Salvador (1).
Thielavia sp.; black leaflesion.   Mexico (1).
Tip burn, physiological cause.   El Salvador, Puerto Rico (1).
Tubercularia agaves Pat.; associated with leafspot.   Costa
    Rica (3).
Viruses: Core rupture, unidentified.   Haiti (4).   Leafmottle
    and malformation, unidentified.   Brazil (4).

AGERATUM CONYZOIDES L.   Native of tropical America, a short
    and sturdy annual plant, with blue flowers, used as an almost
    trouble-free ornamental and in places important in soil erosion
    control and protection; "mendrasto" Sp., "herva San João" Po.,
    "ageratum" En.
        Albugo tragopogensis Pers. (synonyms: A. brasiliensis P.
            Henn., Cystopus brasiliensis Speg.); white blister.   Brazil,
            W. Indies, Colombia.
        Alternaria sp.; leafspot.   S. America.
        Cassytha filiformis L.; dodder.   Puerto Rico.
        Cercospora agerati Stev.; leafspot.   W. Indies, C. America,
            S. America.
        Cuscuta americana L.; dodder.   Puerto Rico.
        Erysiphe cichoracearum DC.; powdery mildew.   Guatemala,
            Panama, Puerto Rico, Florida.
        Puccinia conclinii Seym.; rust.   Venezuela, Bolivia, Puerto
            Rico, Cuba, Jamaica.   P. rosea (Diet. & Holw.) Arth.;
            rust.   W. Indies, C. America, S. America.
        Rhizoctonia solani Kuehn (= Thanatephorous cucumeris (Frank)
            Donk);   seedling and cutting rot.   Florida, Puerto Rico,
            Panama.
        Virus, unidentified; leaf mottle.   C. America.

AGLAONEMA SIMPLEX Blume.   Foliage plants introduced to West
    Indies from tropical islands of the Pacific, grown by florists for
    shipment to the temperate zone (has numerous as yet unidentified
    diseases other than what is listed below); "cara" Sp., "house
    evergreen" En.
        Bacteria (at least two); leafspot, stem collapse.   Puerto Rico.
        Colletotrichum gloeosporioides Penz. (asco-state is Glomerella
            cingulata (Stonem.) Sp. & Schr.); anthracnose lesions on
            stems, leaf-sector collapse.   W. Indies, Puerto Rico,
            Brazil.
        Cuscuta americana L.; dodder (by inoculation).   Puerto Rico.
        Pythium spp.; rootrot and cutting failures.   Florida, Puerto
            Rico.
        Rhizoctonia solani Kuehn is the parasitic stage of Thanate-
            phorous cucumeris (Frank) Donk which bears basidia and

spores (found by me on nearby plant debris); basalrot and
cutting bed disease. Puerto Rico, Florida.
"Rust," physiological cause. C. America.
Virus, dasheen mosaic. Florida, W. Indies.

AGROPYRON spp. Mostly from American temperate zone, found in
some cool dry tropics as used for pasture and may become more
important in certain localities; "agropiro" Sp., "grama de trigo"
Po., "wheatgrass" En. (many diseases still to be determined).
  Alternaria alternata (Fr.) Keiss.; black leaftip rot. Puerto
    Rico.
  Ascochyta sp.; leafspot. Guatemala.
  Curvularia geniculata Boed.; basalrot in clumps. Mexico, W.
    Indies.
  (Diplodia-like spores); blackened old stem. Guatemala.
  Erysiphe graminis DC.; powdery mildew (cattle did not eat
    badly mildewed tops in El Salvador). Puerto Rico, C.
    America, scattered in S. America.
  Fusarium (? roseum); plant base disease. C. America.
  Marasmius sp.; rot of very old plants. El Salvador, Guate-
    mala.
  Metasphaeria magellanica Speg.; on dead leaves and grass
    canes. Tierra del Fuego (? Chile).
  Nigrospora sp.; leaf edge infections. C. America.
  Ophiobolus graminis Sacc.; footrot. Mexico, S. America
    (Andes).
  Puccinia glumarum Erik. & P. Henn. = P. striiformis West.;
    leafrust. Chile. P. rubigo-vera Wint. = P. recondita
    Rob.; leafrust. Bolivia.
  Physoderma sp.; brownspot. Subtropical USA.
  Rhizoctonia solani Kuehn = Thanatephorous cucumeris (Frank)
    Donk; leafbanding, rootrot. Puerto Rico.
  Ustilago bullata Berk. and U. spegazzinii Hirsch.; smuts. S.
    America.

AGROSTIS spp. Pasture grasses known and used a long time in Old
World temperate zone, now being adapted to certain parts of the
New World tropics; "planta de pradera" Sp., "agróstide" Po.,
"red top" En.
  Alternaria sp.; leafrot. Costa Rica.
  Claviceps purpurea (Wallr.) Tul.; ergot. Chile, C. America.
  Ophiobolus graminis Sacc.; footrot. Mexico, Bolivia, Chile,
    Colombia.
  Phyllachora fuscens Speg. and P. graminis Fckl.; tarstreak.
    Argentina, Mexico, Chile.
  Puccinia moyanoi Speg.; rust. Argentina, Chile. P. poarum
    Niel in Chile. P. rhamni Wett. and P. sessilis Schn. in
    much of S. America.
  Scolecotrichum graminis Fckl.; leafrot. Brazil.
  Septoria triseti Speg.; leafspot. Chile.
  Spegazzinia sp.; on senescent leaf base. Costa Rica.

ALBIZZIA JULIBRISSIN Durazz. = ACACIA JULIBRISSIN Willd. and

A. LEBEK (L.) Benth. = ACACIA LEBEK Willd. Fast-growing trees of African and Australian tropics planted in yards and roadsides; in places are naturalized, used not only for ornament but as important timber, and cooking wood and charcoal fuel; "carboniera," "labek" Po., "lengua de mujer" Sp., "lebbek," "woman's tongue mimosa" En. (Diseases given below are on both species.)

Armillaria melea Vahl = Armilariella melea (Fr.) Karst.; rootrot. Puerto Rico, El Salvador.

Cassytha filiformis L.; green dodder. Puerto Rico.

Cephaleuros virescens Kunze; algal leafspot, defoliation in dry-season. W. Indies, Florida, C. America, Brazil, Peru.

Clitocybe tabescens Bres. = Armilariella tabescens (Scop.) Sing.; rootrot. Florida.

Cuscuta americana L.; dodder. Puerto Rico.

Eutypella spp.; dieback complex. Florida, W. Indies.

Fusarium perniciosum Hept. = F. oxysporum Schl. f. sp. perniciosum (Hept.) Toole; wilt. Florida, Subtropical USA, W. Indies, C. America and Mexico.

Meloidogyne sp.; rootknot. Florida.

Nectria spp.; branch cankers, dieback. N. cinnabarina Tode in W. Indies, Brazil. N. tucumanensis (Speg.) Chard. on lebek in Brazil, Argentina, W. Indies. N. coccinea Pers. also on lebek in Brazil, Subtropical USA.

Phoma lathyrina Sacc.; podspot. Virgin Islands.

Phyllosticta divergens Sacc.; leaf and pod spot. Florida, W. Indies, C. America.

Physalospora spp.; diebacks, various colored ascocarps. C. America, Florida.

Psittacanthus calyculatus (DC.) Don; bush woodrose mistletoe. El Salvador.

Ravenelia papillosa Speg.; leafrust. C. America, S. America.

Rosellinia bunodes Berk. & Br.; black rootrot. C. America. R. pepo Pat.; white rootrot. C. America.

Schizophyllum commune Fr.; secondary rot. Florida.

Sphaerophragmium sp.; rust. C. America.

Tryblidiella rufula (Spreng.) Sacc.; branchrot. W. Indies.

Valsa sp.; dieback complex. Costa Rica.

Verticillium (? albo-atrum); wilt. Chile.

ALEURITES FORDII Hemsl. A good-sized, important tree originally in Asia that produces a much valued drying oil and is grown in largest plantings in South America; "tungua" Sp. Po., "tung" En.

Bacterial wet-wood. Seen in Florida, Honduras (no isolation made).

Botryosphaeria dothidea (Moug.) Ces. & de Not.; branch canker. W. Indies, C. America. B. quercuum (Schw.) Sacc.; dieback. (See Physalospora.)

Cephaleuros virescens Kunze; algalspot. W. Indies, C. America.

Cephalosporium sp.; collar girdle. Louisiana.

Cercospora aleuritidis Miy.; leafspot (see Mycosphaerella).

Clitocybe tabescens Bres. = Armilariella tabescens (Scop.) Sing.; rootrot. Subtropical USA.

Colletotrichum gloeosporioides Penz. = Glomerella cingulata
(Stonem.) Sp. & Schr.; anthracnose. C. America, Sub-
tropical USA.
Fusarium spp.; seedling disease, dieback complex. Honduras.
Marasmius sp.; black cord leafblight. Panama.
Meloidogyne sp.; rootknot. W. Indies.
Mycena citricolor (Berk. & Curt.) Sacc.; leafspot. Panama.
Mycosphaerella aleuritidis Ov. asco state of Cercospora; leaf-
spot. Subtropical USA, W. Indies, Honduras, Trinidad,
Brazil.
Pellicularia koleroga Cke.; web-blight. Subtropical USA, W.
Indies, S. America.
Phoradendron sp.; bush mistletoe. Costa Rica.
Phthirusa sp.; vine-like mistletoe. Costa Rica.
Phymatotrichum omnivorum (Shear) Dug.; dryland rootrot.
Texas.
Physalospora rhodina Cke. = Botryophaeria quercuum (Schw.)
Sacc.; ascocarp associated with dieback. Costa Rica,
Puerto Rico, subtropical USA.
Phytophthora cinnamomi Rands; rootrot. Louisiana.
Pseudomonas aleuritidis (McCull. & Dem.) Stapp; bacterial
leafspot and collapse. Subtropical USA.
Rhizoctonia microsclerotia Matz; cobweb blight. Louisiana,
Honduras. R. solani Kuehn = Thanatephorous cucumeris
(Frank) Donk; seedling rot. Honduras.
Rosellinia sp.; rootrot. Brazil.
Sclerotium rolfsii (Curzi) Sacc. = Pellicularia rolfsii (Sacc.)
West; nursery blight. Texas, Florida.
Septobasidium pseudopedicellatum Burt; rootfelt. Louisiana.
S. saccardinum March.; rootfelt. Brazil.
Strigula complanata Fée; whitespot. Honduras, Panama,
Puerto Rico.
Virus (unidentified); roughbark. Louisiana, Mississippi.
Wagelina sp.; branch infection. Brazil.

ALLIUM spp. Flavorsome bulbous vegetables brought to the Americas after the days of Columbus: (1) A. CEPA L., generally a
single prominent bulb to a plant, Asiatic origin, medium strong
flavor; "cebolla" Sp., "cebola" Po., "onion" En.; (2) A. PORRUM
L., single bulb but not prominent, European origin, mild flavor;
"cebollín" Sp. Po., "leek" En.; (3) A. SATIVUM L., bulb made
up of cloves or segments, European origin, notably strong flavor;
"ajo" Sp., "alho" Po., "garlic" En.
Alternaria alternata (Fr.) Keiss. = A. tenuis Nees; blacktip,
black stalk. Subtropical USA, Peru, Ecuador, Puerto Rico,
Venezuela, Costa Rica, Colombia, St. Croix, Brazil, Argentina (all reports on 1). A. porri Cif. = A. allii Nolla;
purple blotch, stem lesion. Subtropical USA (1), Nicaragua
(1), Costa Rica (1, 2, 3), Brazil (1, 3), Puerto Rico (1, 2),
Honduras (1, 2, 3), Venezuela (1, 2, 3), El Salvador (1, 2),
Guatemala (1), Jamaica (1), Mexico (3).
Aspergillus niger v. Tiegh.; black bulbmold. Brazil (1, 3),

Peru (1, 3), Argentina (1), C. America (1, 3), Panama
(1), Puerto Rico (1), Chile (1).
Botrytis allii Munn; graymold neckrot, leaf tipburn. Brazil
(1, 3), Chile (1, 3), Peru (1, 3), Venezuela (1, 3), C.
America (1, 3), Subtropical USA (1, 3), Mexico (3), Puerto
Rico (1, 3). B. cinerea Pers.; graymold of bulbs and
leafblight. Subtropical USA (1), Puerto Rico (1), Panama
(1). B. squamosa Walk.; bullet neckrot. Mexico (1), Sub-
tropical USA (1), C. America (1).
Cassytha filiformis L.; green dodder. W. Indies (1).
Cercospora victorialis Th.; leafspot. Brazil (1, 3), Mexico
and C. America (1).
Chlorospora vastatrix Speg. = Peronospora destructor (Berk.)
Casp. latter name is most acceptable; leafspot and fleck,
mildew. General (1, 2, 3).
Cladosporium sp.; associated with leaf decadence. Chile and
Puerto Rico (1).
Colletotrichum dematium (Pers.) Grove f. circinans Arx = C.
circinans Vogl.; smudge. Subtropical USA, W. Indies,
Costa Rica, Guatemala, El Salvador, Venezuela (all on 1),
Brazil (1, 3). C. gloeosporioides Penz. (perfect state
Glomerella cingulata (Stonem.) Sp. & Schr.); bulbrot, damp-
ing-off. El Salvador (1, 2), Brazil (1, 2).
Cuscuta americana L.; yellow dodder. C. America and W.
Indies (1). C. corymbosa Eng.; lesser dodder. El Salva-
dor (3).
Didymella sp.; leafblight. Colombia and Brazil (1).
Diplodia (? natalensis); bulb dryrot. Puerto Rico, Subtropical
USA (1, 3).
Ditylenchus dipsaci Filipj.; root nematode. Costa Rica (1).
Erwinia carotovora (L. R. Jones) Holl.; softrot. Neotropics
(1, 2, 3).
Fusarium oxysporum Schl., f. sp. capae Syn. & Hans.; wilt,
bulb disease. Scattered (1, 2, 3). F. roseum Lk.; bulb-
rot, damping-off. General (1, 3). F. solani (Mart.) Sacc.;
storage dryrot. Widespread (1, 3).
"Fusarium pink root" (see Pyrenochaeta)
Helminthosporium allii Camp.; canker and dryrot. Mexico
(3), Subtropical USA (3).
Heterosporium allii Ell. & Mart.; typical elliptical leafspot,
leaftip collapse. California (1), Puerto Rico (1), S.
America (1, 3).
Macrophomina phaseoli (Maubl.) Ashby; charcoal bulbrot. Sub-
tropical USA (1, 3), Mexico (1, 3), Guatemala (1).
Meloidogyne sp.; rootknot nematode. Scattered throughout Neo-
tropics (1, 2, 3).
Mycena citricolor (Berk. & Curt.) Sacc. (Leafspot by laboratory
inoculation on onion in Puerto Rico.)
Nematodes reported as of various undetermined kinds; causing
root injuries. Peru (1).
Penicillium spp.; storage molds. General (1, 3).
Peronospora (see Chlorospora)

Phytophthora cinnamomi Rands; seedhead collapse. Brazil (1).
Pleospora herbarum (Pers.) Rab. = Stemphylium botryosum
    Wallr.; secondary adding to stalk rot. Widespread (1).
Pseudomonas alliicola Starr & Burkh.; systemic bacterial
    rot. Haiti (1).
Puccinia asparagi DC.; rust. Brazil (1, 3), Bermuda (1),
    Colombia (3), Uruguay (3), Argentina (3). P. porri (Sow.)
    Wint. = P. allii Rud.; rust. Brazil (1, 3), Subtropical
    USA (1, 3), Ecuador (1).
Pyrenochaeta terrestris (Hans.) Gor., Walk., Lars. (has been
    widely reported incorrectly as a Fusarium); pinkroot. Neo-
    tropics (1, 2, 3).
Pythium ultimum Trow; damping-off, rootrot. California and
    Panama (1).
Rhizoctonia sp.; chlorotic leaf lesions on garlic in Guatemala.
Sclerotinia sclerotiorum (Lib.) dBy. = Whetzeliana sclerotiorum
    (Lib.) Korf & Dumont; watery bulb and top rot. Chile (1),
    Florida (1).
Sclerotium cepivorum Berk. = Sphacelia allii Vogl.; whiterot.
    Subtropical USA (1, 3), Peru (1, 3), Bolivia (1), Chile
    (1, 3), Argentina (1, 2, 3), Panama, Brazil (1, 2, 3),
    Uruguay (1, 3), Colombia (3). S. rolfsii (Curzi) Sacc. =
    Pellicularia rolfsii (Sacc.) West; groundrot. Subtropical
    USA (1, 3), Puerto Rico (1), Brazil (1), Peru (3), El Salva-
    dor (3).
Tylenchorhynchus sp.; nematode root injury. Chile (1).
Urocystis cepulae Frost; leaf and bulbscale smut. Chile (1, 3),
    Peru (1), Bolivia (1), Brazil (1), Mexico (1), Ecuador (1).
Viruses: Mosaic mild mottle. Scattered (1, 2, 3). Yellow
    dwarf. Guatemala (1), Argentina (1), Subtropical USA (3).
    And Infection with sunflower virus, artificial inoculation.
    Argentina (1).

ALNUS spp. Native trees of the Americas growing in cool mountain
    areas, increasingly interesting as producers of good poles, firm
    grained wood, used for charcoal and in soil conservation; "aliso"
    Sp., "tropical alder" En. (Diseases in the American tropics not
    much studied.)
Anisochora chardonii Garces; leafstroma. Colombia.
Armillaria mellea Vahl = Armilariella melea (Fr.) Karst.;
    rootrot. California.
Cercospora sp.; leafspot. Costa Rica, Panama.
Cuscuta americana L.; dodder. Puerto Rico.
Dieback (Fusarium sp., Pestalotia sp., Colletotrichum sp.)
    following root injury; Guatemala-El Salvador border.
Diplocystis alnicola Sing.; wood decay. Argentina.
Fomes sp.; stumprot. Costa Rica.
Hydnum sp.; logrot. Costa Rica.
Lenzites sp.; woodrot. Costa Rica.
Melampsoridium alni Diet. = M. hiratisukanum Ito; rust.
    Mexico, Bolivia.
Phoradendron sp.; bush mistletoe. Mexico, Costa Rica.

Phytophthora (? cinnamomi); rootrot.   C. America.
Poria sp.; woodrot.   C. America.

ALSTROEMERIA AURANTIACA D. Don = A. AUREA Meyen.   An
ornamental from Chile with lily-like flowers thought of as having
good commercial future; "liuto" Sp. Po., "chilean lily" En.
  Aecidium alstroemeriae Diet. & Neg.; rust.   Chile.
  Hendersonia alstroemeriae (Speg.) Sacc. & Trott.; leafburn.
    S. America.
  Isothea chilensis Speg.; leaf collapse.   S. America.
  Oidium sp.; powdery mildew.   S. America.
  Pleospora alstroemeriae Speg.; leafspot.   S. America.
  Porterula alstroemeriae Speg.; leafrot.   S. America.
  Puccinia alstroemeria Syd.; rust.   Chile.
  Rhizoctonia violacea Tul. = Helicobasidium purpureum Pat.;
    rootrot.   S. America.
  Scolecotrichum alstroemeriae Allesch.; leafspot.   Brazil.
  Sphaerella alstroemeriae Speg.; leaf disease.   Chile.
  Uromyces alstroemeriae (Diet.) P. Henn.; rust.   S. America.
    (Note: These admired South American lilies have more
    rusts listable as affecting them, but some are probably
    synonyms; however, the natural rust flora on the several
    species of this host serve to indicate another instance of
    the richness of rust parasite occurrences in tropical
    America.)

ALTERNANTHERA BETTZICKIANA (Regel) Standl. = TELANTHERA
BETTZICKIANA Regel.   Of Brazilian origin.   This lush herb
is a colorful wine red ornamental much used in gardens and yards;
"vino," "planta de vino" Sp. Po., "wine plant" En.   (Disease
studies not much required as the host is seldom if ever sold as
plants or foliage.   There are a few wild species of the host.)
  Albugo (see Cystopus)
  Cephaleuros virescens Kunze; algalspot, strangle tip.   W.
    Indies, C. America.
  Cercospora alternantherae Ell. & Lan.; leafspot.   Puerto Rico,
    Louisiana, El Salvador.
  Cystopus bliti (Biv.-Bern.)dBy. = Albugo bliti (Biv.-Bern.)
    Kuntze; whiterust.   Grenada, St. Vincent, Dominica,
    Panama, Guatemala.
  Elsinoë uruguayensis Bitanic.; scab spot.   Uruguay.
  Guignardia cephalariae Stev. var. alternantherae Stev.; blast-
    blight.   Puerto Rico, S. America.
  Meloidogyne sp.; rootknot.   Subtropical USA, W. Indies, C.
    America.
  Mycena citricolor (Berk. & Curt.) Sacc.; naturally occurring
    leafspot.   Costa Rica.
  Puccinia mogiphanis Arth.; rust.   S. America.
  Pythium spp.; cutting disease.   S. America.
  Rhizoctonia sp. (not R. solani); cutting disease and stem-leaf
    rot.   Puerto Rico.
  Sclerotium rolfsii (Curzi) Sacc. = Pellicularia rolfsii (Sacc.)
    West; soilrot.   Scattered.

Uredo alternantherae Jacks. & Holw.; rust.   S. America.   U.
  maculans Pat. & Gai.; rust.   Panama.

ALTHAEA ROSEA Cav.   An herbaceous ornamental of Europe and
  USA, with origin in China, taken to certain parts of the American
  tropics for the tall and spectacular stems of flowers from top to
  bottom; "malva hortense" Sp., "malva-rosa" Po., "hollyhock"
  En.
    Ascochyta althaeina Sacc. & Bizz.; leaf-and-stem spot.   Vene-
      zuela.   A. parasitica Fautr.; leafspot.   Brazil.
    Cercospora althaeina Sacc.; leafspot.   Venezuela, Texas,
      Guatemala.
    Colletotrichum sp.; anthracnose.   Guatemala, Honduras.
    Meloidogyne sp.; rootknot.   Jamaica, Florida.
    Oidium sp.; mildew.   C. America.
    Phyllachora deusta (Fr.) Sacc.; leafspot.   S. America.
    Physalospora sp.; dying stalks.   Costa Rica.
    Phytophthora megasperma Drechs.; rootrot.   S. America.
    Puccinia heterogenea Lagh.; leaf and stem rust.   Ecuador.
      P. lobata Berk. & Curt.; leafrust.   California, Texas.
      P. malvacearum Bert.; rust.   Argentina, Brazil, Chile.
      P. scherardiana Koern; rust.   Texas, Puerto Rico.
    Rhizoctonia microsclerotia Matz; webblight.   W. Indies, C.
      America, S. America.
    Viruses: Abutilon mosaic.   Brazil.   Mosaic (unidentified).
      Puerto Rico, El Salvador.   Sida infectious chlorosis.
      Brazil.

AMARANTHUS QUITENSIS HBK., A. EDULIS Michx., and A.
  LECOCARPUS L.   All three indigenous to the American tropics,
  important field crops grown in often difficult terrain in the high
  mountains and greatly valued by the Indians for the heavy and
  nutritious crops of seed used for food; "trigo inca" Sp., "ama-
  ranto" Po., "grain amaranth" En., "quinua," "quihuicha" Ind.
  (Note: I do not know of a major study being made of the diseases
  of grain amaranths in the American tropics; they are said to be
  quite tolerant to diseases encountered in the high dry mountain
  fields.   I questioned three tourists who saw these crops in dry
  cool areas of Colombia and Peru; they were surprised to find
  these "pigweeds" were grown for grain.   I concluded they had
  seen mildew probably from Oidium; black head from weakly parasitic fungi; grain cast perhaps both insect and fungus problems;
  lodging from weak roots; and black lesions on stems.)
    Albugo bliti (Biv.-Bern) Kuntze; white rust.   Mexico, S.
      America.
    Glomosporium amaranthi Hirsch.; smut.   S. America.
    Oidium sp.; mildew.   Mexico, S. America.
    Sphacelia amaranthicida Speg.; twig and leaf rot.   S. America.
    Virus; mosaic mottle seen in Peru and a virus streak in
      Brazil.

AMARYLLIS sp. (sometimes given as A. BELLADONNA L.) =
  HIPPEASTRUM sp.   Spectacular ornamental that originated in the

African-Asia region, introduced into Latin America and now economically valuable for the large fragrant lily-like flowers borne
on tall stems growing out of large bulbs; "lirio," "amarillis"
Sp., "açucena" Po., "amaryllis" En.

Asterinella amaryllidis Ell. &. Ev.; black mildew. Puerto
    Rico. A. hippeastri Ryan; black mildew. Puerto Rico.
Botrytis cinerea Pers.; graymold bulbrot. Puerto Rico.
Cercospora amaryllidis Ell. &. Ev.; leafspot. Subtropical USA,
    Puerto Rico, El Salvador, Costa Rica.
Curvularia trifolii Boed. f. sp. amaryllidis Nolla; leafspot.
    Puerto Rico.
Epicoccum purpurascens Ehr.; purple spot. California.
Mycena citricolor (Berk. & Curt.) Sacc.; leafspot. Costa
    Rica.
Phyllosticta amaryllidis Bres.; black spot. Brazil.
Pythium debaryanum Hesse; basal rot. Florida, Texas, Puerto
    Rico.
Sclerotium rolfsii (Curzi) Sacc. = Pellicularia rolfsii (Sacc.)
    West; Sclerotium rot. Florida, Texas.
Soil rot, undetermined complex of fungi and nematodes.
    Puerto Rico.
Stagonospora curtisii Sacc.; red blotch. Puerto Rico.
Tylenchus penetrans Cobb; root disease. Florida.
Viruses: Common yellow-green mosaic. C. America, Florida,
    California, W. Indies, S. America. Spotted Wilt in Cali-
    fornia and Texas.

AMYGDALUS see PRUNUS (4)

ANACARDIUM OCCIDENTALE L. The medium-sized tree native to
    tropical America both planted and protected in forest edges,
    much valued for fruits used for wine and eaten, and the nutritious
    nuts; "marañon" Sp., "caqueiro" Po., "cashew" En. (local un-
    published names).

Asterina carbonacea Cke. var. anacardii Ryan; black mildew.
    Puerto Rico, Florida, El Salvador.
Balladynastrum anacardii Bat. & Vit.; leaf disease. Brazil.
Botryodiplodia (see Diplodia)
Botrytis anacardii Viégas; fruitmold. Brazil, El Salvador.
Cephaleuros virescens Kunze; algalspot. Brazil.
Ceratocystis fimbriata Hunt (f. anacardii); branch death, fruit-
    stem rot, and woodstain. C. America, S. America, Santo
    Domingo, St. Thomas.
Cercospora anacardii Mull. & Chupp; leafspot. Brazil.
Cladosporium sp.; spores from fruit. El Salvador.
Colletotrichum gloeosporioides Penz. = Glomerella cingulata
    (Stonem.) Sp. & Schr.; blossom blight, leafspot, fruit russet,
    fruitrot. Florida, W. Indies, C. America, S. America.
Dendrochium paraense Vinc.; leafspot. Brazil.
Dendryphiella cruzalmensis Bat.; leaf disease. Brazil.
Diplodia theobromae (Pat.) Now. = Botryodiplodia theobromae
    Now. (latter often considered preferable binomial); dieback,
    fruit mummy. No asco-state seen. C. America.

Fusarium sp.; dieback complex. El Salvador, Puerto Rico.
Helminthosporium bactridis P. Henn.; leafspot. S. America.
Irenopsis coronata (Speg.) Stev.; black mildew. W. Indies,
   C. America.
Loranthaceae, numerous kinds; mistletoe-like parasites.
   Scattered.
Meliola anacardii Zimm.; black mildew. S. America. M.
   mangiferae Earle; black mildew. C. America, W. Indies.
Microcallis megalospora Petr. & Cif.; scabspot. W. Indies.
Micropeltella sparsa Stev. & Mant.; leaf infection. W. Indies.
Nectria anacardii P. Henn.; stem disease and dieback com-
   plex. C. America, S. America.
Oidium anacardii Noack; mildew. C. America, S. America.
Parella sp.; leafy liverwort smother of flushing buds. El
   Salvador.
Parodiella melioloides Wint.; leafspot. Brazil.
Penicillium sp.; fruitrot. El Salvador.
Phyllosticta anacardicola Bat. & Vit.; small leafspots. Brazil.
   P. brasiliensis Speg. and P. mortoni Fair.; leafspot
   diseases. Brazil, El Salvador.
Rhizoctonia solani Kuehn = its perfect state Thanatephorous
   cucumeris (Frank) Donk; seedling constriction. Costa Rica.
Sclerotium rolfsii (Curzi) Sacc. = Pellicularia rolfsii (Sacc.)
   West; seedling basal rot. Florida.
Stomatochroon lagerheimii Palm; systemic algal leaf attack.
   Guatemala.
Trichomerium psidii Bat.; leafspot. Brazil.
Uredo anacardii Mains; leafrust. Guatemala.
Yeast (undetermined); fruit riperot. El Salvador, Guatemala.
Zukalia paraensis P. Henn.; foliage disease. S. America.

ANANAS COMOSUS (L.) Merr. = A. SATIVUS Schult. A native of the
  Americas and related to the Bromeliads, this plant is an important
  money making fruit crop lending itself to extensive plantation culture and factory processing; "piña" Sp., "abacaxi" Po., "pine-
  apple" En.
    Aphelenchus avenae Bast.; root nematode. (Note: The pine-
     apple crop is one of the most studied for nematodes; many
     eel worms are present in the root zone, and only some of
     those determined in the Americas are noted here.) Brazil.
    Aphelenchus sp.; root nematode. Puerto Rico.
    Aphelenchoides parietinus Stein.; root nematode. Brazil.
     Aphelenchoides sp.; root nematode. Puerto Rico.
    Aspergillus sp.; riperot. S. America.
    Asterinella stuhlmanni P. Henn.; leafspot. Puerto Rico, Virgin
     Islands.
    Bacterium (unidentified); "eye infection." W. Indies.
    Botryodiplodia theobromae Pat. (Botryosphaeria quercuum
     (Schw.) Sacc.; riperot. Jamaica, Puerto Rico, Virgin
     Islands.
    Calothyriella ananassae Viégas; leaf disease. Brazil.
    Ceratocystis paradoxa Mor. (= Ceratostomella, Ophiostoma,

Thielaviopsis); leaf and fruit rot.   Mexico, Jamaica, Trinidad & Tobago, Surinam, Florida, Puerto Rico, Brazil, ? general occurrence.

Colletotrichum gloeosporioides Penz.; anthracnose (often not severe).   Mexico, W. Indies, C. America, S. America.

Criconemoides sp.; ring nematode.   Puerto Rico.

Curvularia lunata Boed.; leafedge dryseason spotting.   Mexico, Guatemala, Costa Rica.

Cuscuta americana L.; dodder.   Puerto Rico.

Ditylenchus dipsaci Filip.; nematode.   Brazil, Puerto Rico.

Dorylaimus sp.; nematode.   Puerto Rico.

Erwinia ananas Serr.; bacterial brownrot.   Haiti, Santo Domingo, Mexico, C. America, Brazil.   E. carotovora (L. R. Jones) Holl.; soft stemrot.   Jamaica.

Fusarium moniliforme Sheld.; basal disease.   Brazil, Argentina.   F. oxysporum Schl.; fruit infection.   Argentina. Fusarium sp.; secondary to injury of leaves.   Puerto Rico.

Gibberella stage of Fusarium sp. in moist chamber studies on brown rot.   Central America, Puerto Rico.

Helicotylenchus nannus Stein.; spiral nematode.   Brazil, Puerto Rico.

Helminthosporium sp.; leafspot.   Colombia.

Leptonchrus granulosus Cobb; nematode, root ectoparasite. Brazil.

Mealybug wilt.   Puerto Rico, Jamaica, Florida.

Meloidogyne exigua Goeldi; rootknot.   Florida.   Meloidogyne sp.; rootknot nematode, decline.   Florida, Puerto Rico.

Monhystera stagnalis Bast.; root nematode.   Brazil.

Paratylenchus macrophallus Goo.; root nematode.   Brazil.

Penicillium sp.; "eye" disease.   Florida, W. Indies, widespread.

Phoma sp.; alligator leafspot.   Colombia.

Phyllosticta spp.; leafspot.   S. America.

Phytomonas ananas Berk.; black "eye."   Haiti, Puerto Rico.

Phytophthora cinnamomi Rands; heart rootrot.   C. America, Brazil, scattered in S. America.   P. palmivora Butl.; heart rootrot.   ? general.   P. parasitica Dast.; canker rot.   Venezuela, Jamaica, W. Indies.

Porella sp.; smother from leafy liverwort.   Guatemala.

Pratylenchus macrophallus Goo., P. pratensis Filipj.; root ectoparasitic nematodes.   Brazil.

Pythium spp.; wilt from rootrot in wetland.   W. Indies, C. America, S. America.

Rhizoctonia sp. (not solani); basal attack.   Colombia, Puerto Rico.

Rotylenchus sp.; reniform nematode at root.   Puerto Rico.

Septobasidium westonii Couch; felt.   Panama.

Trichoderma viride Pers.; dryseason riperot.   W. Indies, C. America.

Trichosphaeria sacchari Ban.; wateryrot.   W. Indies.

Tylenchus davianii Bast.; nematode.   Brazil, Puerto Rico.

Virus (undetermined wilt, proof claimed of mealybug transmission), Puerto Rico.

Xiphinema americana Cobb; root ectonema. Brazil.
Yeast (undetermined); ripening rot. Puerto Rico.

ANDIRA spp. These trees are from the American tropics and much
used for agricultural shade and wood products; "mocá" Sp.,
"angelim" Po., "bastard mahogany" En. (many Indian names).
Cephaleuros virescens Kunze; algal leafspot. Brazil.
Cercospora stevensii Young; large leafspot. C. America, W.
Indies, S. America.
Colletotrichum gloeosporioides Penz. (no Glomerella state
secured); leaf anthracnose. El Salvador, Puerto Rico.
Dimerosporium andirae P. Henn.; moldspot. S. America.
D. meliolicola P. Henn.; leafmold. S. America.
Dothidella andiricola Speg.; leaf scab-spot. Puerto Rico, C.
America, Argentina.
Fusarium (? roseum); dieback complex. Costa Rica.
Ganoderma lucidum Karst.; trunk baserot. Costa Rica.
Lenzites repanda Fr.; heartrot. W. Indies, C. America, S.
America.
Meliola andirae Earle; black mildew. W. Indies, C. America.
Mycena citricolor (Berk. & Curt.) Sacc.; luminous leafspot.
Puerto Rico.
Physalospora andira Stev.; dieback. W. Indies.
Polystigma pusillum Syd.; small angular leafspots. W. Indies,
C. America, Panama, Colombia.
Ravenelia goyazensis P. Henn.; leafrust. Brazil.
Tremella sp.; stumprot. Costa Rica.

ANDROPOGON spp. [see also SORGHUM]. A few perennial pasture
grasses from the American temperate zone found good in cool dry
mountains of the tropics; "pasto" Sp., "pastagem," "capim" Po.,
"pasture grass" En.
Balansia henningsiana Moel.; choke disease. Florida, Costa
Rica.
Cephaleuros virescens Kunze; agal leafspot. Brazil.
Cerebella andropogonis Ces.; black head. Subtropical USA.
Claviceps purpurea (Fr.) Tul.; ergot. Subtropical USA, ? W.
Indies.
Colletotrichum graminicola (Ces.) Wils. = Gloeosporium
tucumanensis (Speg.) v. Arx & Müll. (Physalospora tucu-
manensis Speg.); anthracnosis. Widespread.
Epichloe nigricans Speg. (Balansia strangulans (Mont.) Diehl);
strangle disease. S. America, C. America.
Fusarium moniliforme Sheld.; seed rot. Trinidad.
Leptostromella andropogonis Dearn. & House; on culms. Sub-
tropical USA.
Lophodermium andropogonis Tehon; culm lesion. Puerto Rico.
Meliola panici Earle; black mildew spots. Puerto Rico,
Panama.
Metasphaeria infuscans Ell. & Ev.; culm lesions. W. Indies.
Myriogenospora atramentosa (Berk. & Curt.) Diehl; tangle.
Texas, Louisiana. M. bressadoleana P. Henn.; fasciation.

Puerto Rico. M. paspali Atk.; leaf twist. Subtropical
USA, C. America.
Nectria epichloe Speg. = Cucurbitaria epichloe (Speg.) Kuntze;
leaf disease. Paraguay.
Phyllachora andropogonicola Speg.; black sheath and leafspot.
Argentina, ? other countries in S. America. P. brevifolia
Chard.; tarspot. Puerto Rico.
Popularia vinosa (Berk. & Curt.) Mason; culm disease.
Florida, W. Indies.
Puccinia andropogonis Schw.; rust. Subtropical USA, Chile.
P. ellisiana Th.; rust. Florida. P. meridensis Kern;
rust. Venezuela. P. posadensis Sacc. & Trott (synonyms:
P. kernbachii Arth., P. venustula Arth.); rust. Subtropical
USA, W. Indies, Bolivia, Brazil, C. America, Argentina.
P. varispora Arth. & Holw. = P. versicolor Diet. &
Holw.; rust. Widespread.
Sclerotium clavus DC. f. andropogonicola Speg.; flower disease.
Argentina.
Septoria andropogonicola Speg.; on dying leaves. Argentina.
Sorosporium ellisii Wint.; headsmut. Subtropical USA. S.
platensis Hirsch.; smut. S. America.
Sphacelotheca andropogonis-hirtifolii (P. Henn.) Clint.; smut.
S. America. S. culmiperda Clint.; culm smut. Brazil.
S. guaraniticum (Speg.) Zund. = Ustilago guaraniticum
Speg.; smut. Argentina, Brazil (? other nearby countries).
Telimena sp.; tarspot. Florida.
Tolyposporella brunkii Clint.; smut. Neotropics. T. semi-
berbis Kunth.; smut. Brazil.
(Uredo andropogonicola Speg. = Puccinia posadensis see above.)
Uromyces andropogonis Tracy; rust. Subtropical USA, Chile,
Bolivia. U. clignyi Pat. & Har.; rust. Guatemala. U.
occulta P. Henn.; culm smut. Brazil.
Virus symptoms are often encountered in C. America.

ANNONA CHERIMOLA Mill., and A. MURICATA L. Native to the
Neotropics, popular trees grown for fruits produced under a wide
range of ecological conditions; "corazón," "chirimoya" Sp.,
"articu" Po., "cherimoya," "soursop" En. (many names given
by Indians).
Aecidium anonae P. Henn.; yellow rust. Brazil.
Armillaria mellea Vahl = Armilariella melea (Fr.) Karst.;
rootrot. California. (Note: Two agaricaceous fungi cause
severe and similar rootrots in this and several other tree
and woody crops. The organisms are very much alike but
their differences, briefly, are as follows: C. tabescens is
not luminescent and it has freed decurrent gills, and no
annulus. A. mellea is luminescent, has gills that are not
free and has an annulus.)
Ascochyta cherimoliae Theum.; leafspot, stem lesion. Chile,
Ecuador.
Botryosphaeria dothidea (Moug.) Ces. & de Not.; branch
canker from inoculation. Costa Rica.

Botryodiplodia theobromae Pat.; mummy, dieback, leafspot.
 Neotropical.
Calothyriopeltis anonae Bat. & Galao; on leaves. Brazil.
Capnodium anonae Pat.; sooty mold. Widespread.
Cephaleuros virescens Kunze; algal leafspot. Brazil.
Cephalosporium sp.; target leafspot. El Salvador.
Cercospora anonae Mul. & Chupp; leafspot. Guatemala,
 Puerto Rico, Brazil. C. annonaceae P. Henn.; leafspot.
 Colombia, Venezuela, Guatemala. C. caracasensis Chupp
 & Mul.; leafspot. Venezuela. C. xylopiae Viég. & Chupp;
 leafspot. Brazil.
Cladosporium sp.; senescent leaf invasion. C. America.
Clitocybe tabescens Bres. = Armilariella tabescens (Scop.)
 Sing.; rootrot. Florida, Puerto Rico.
Colletotrichum gloeosporioides Penz. = C. annonicola Speg.
 (the asco-state is Glomerella cingulata (Stonem.) Sp. &
 Schr.); anthracnose of leaf, peduncle, and fruit. Studies
 noted in Colombia, Argentina, C. America, Bolivia, Brazil,
 Puerto Rico, Florida.
Coniosporium argentinense = Anthinium phaeospermum (Cda.)
 Ell.; branch rot. Argentina.
Corticium salmonicolor Berk. & Br.; leafrot and pink disease.
 C. America.
Cryptosphaerella annonae Speg.; branch infection and leaf
 disease. S. America.
(? Cylindrocarpon sp.); rootrot. Chile.
Diaporthe anonae Speg.; dieback. Argentina. D. medusae
 Nits.; fruitrot. S. America.
Diplodia natalensis P. Evans = Botryosphaeria quercuum
 (Schw.) Sacc.; fruit blackrot and mummy, dieback. W.
 Indies, C. America, S. America.
Elsinoë annonae Bitanc. & Jenk.; spot scab. Brazil, Vene-
 zuela, C. America.
Eutypa ludifunda Sacc. var. annonae-cherimoliae Speg.; foliage
 stroma. S. America.
Fomes ribis (Schum.) Fr. = F. lamaoensis Murr.; trunk base-
 rot (slow field spread). S. America, rare in C. America.
Fusarium sp.; secondary infections. Neotropics.
Halplosporella chlorostroma Speg.; on leaves. Argentina.
Hemicycliophora sp.; nematode. Peru.
Isariopsis anonarum Petr. & Cif.; leafspot. W. Indies,
 Bolivia.
Limacinia aurantii P. Henn.; sootymold. Brazil.
Meliola spp.; black mildew spots. W. Indies, S. America.
Melophia anonae Speg. = Linochora anonae Speg. = Linochora
 anonae (Speg.) Hoehn.; leaf disease. Paraguay.
Mycena citricolor (Berk. & Curt.) Sacc.; round leafspot.
 Peru, Costa Rica.
Mycosphorella sp.; large spreading leafspot. Puerto Rico.
Nectria episphaeria Tode; on twigs. Colombia. N. tucumanen-
 sis Speg. = Creonectria tucumanensis (Speg.) Chard.; asco-
 carps on diseased leaves and stems. S. America.

Peroneutypa heteracantha (Sacc.) Berl. var. anonae-cheri-
    moliae Speg.; branch and stem disease. S. America.
Pestalotia conglomerata Dres. var. folicola Cif. & Gonz.
    Frag; leaf and petiole spot. W. Indies.
Phaeobotryosphaeria yerbae Speg.; branch disease. S.
    America.
Phakopsora cherimoliae Cumm. (synonyms: Uredo cherimoliae
    Syd., Physopella cherimoliae Arth.); rust. Florida, W.
    Indies, Texas, C. America, S. America.
Phomopsis sp.; black spot. Colombia.
Phyllachora anonaceae Rehm; tarspot. Brazil, Bolivia. P.
    anonarum Bat. & Vit.; tarspot. Brazil. P. anonicola
    Chard.; black leafspot. Brazil. P. atro-maculans Syd.;
    dark sunken leafspot. Costa Rica. P. insularum Sacc.;
    tarspot. Brazil.
Phyllosticta tuiriensis Speg.; leafspot. Costa Rica.
Phyllostictina anonicola Bat. & Vit.; leafspot. Brazil.
Phymototrichum omnivorum (Shear) Dug.; dryland rootrot.
    Texas.
Pocosphaeria anonae Rang.; leaf disease. Brazil.
Pratylenchus sp.; nematode at roots. Peru, Chile.
(? Ramularia sp.); rootrot. Chile.
Rhizoctonia sp. (? solani); damping-off. Peru.
Rhizopus nigricans Ehr.; fruit riperot. Mexico, Colombia.
Septoria sp.; leafspot. Jamaica.
Strigula complanata Fée; white lichenspot. W. Indies, C.
    America, S. America.
Tryblidium guaraniticum Speg. = Tryblidiella spegazzinii
    Rehm (= ? T. rufula (Spreng.) Sacc.); log-wood and branch
    rot. S. America.
Uredo (see Phakopsora)
Xiphinema sp.; nematode. Chile.
Zignoella annonicola Speg.; branch and leaf disease. Colombia,
    Argentina.

ANTHURIUM spp. Handsome flowering plants of the American
    tropics protected in the wilds or grown in slat houses for exotic
    flowers and foliage, increasing importance to florists; "flor
    porcelano," "anturio" Sp. Po., "wax flower" En.
    Cephaleuros virescens Kunze; algal leafspot. Brazil.
    Cercospora sp.; leafspot. Colombia, Puerto Rico.
    Colletotrichum gloeosporioides Penz.; anthracnose. Colombia,
        Puerto Rico (Glomerella stage searched for in Puerto Rico,
        not found).
    Erwinia carotovora (L. R. Jones) Holl.; plant base softrot.
        Puerto Rico.
    Helicotylenchus nannus Stein.; root disease. Brazil.
    Hypocrella viridans (Berk. & Curt.) Petch; leaf attack. Trini-
        dad, Brazil, Cuba, Haiti, Ecuador, Mexico.
    Meloidogyne inornata Lorde.; root attack. Brazil.
    Micropeltis clava Toro; leafspot. W. Indies.
    Mycena citricolor (Berk. & Curt.) Sacc.; leafspot, flowerspot.
        Costa Rica, Puerto Rico.

Phyllachora englerii Speg.; tarspot. W. Indies, C. America,
S. America.
Physalospora inanis (Schw.) Sacc.; black leafdisease. Ecua-
dor, Surinam, C. America, W. Indies.
Placoasterina engleri P. Henn.; leafspot. S. America.
Scutellonema boocki Lorde.; rootrot from nematode. Brazil.
Sphaerella anthurii (Miles) Gonz. Frag. & Cif. = Mycosphae-
rella anthurii Miles; papery leafspot, shot hole. W.
Indies.
Uredo anthurii (Mar.) Sacc.; rust. Puerto Rico, Colombia.
Virus mottle and malformation in Venezuela.

ANTIRRHINUM MAJUS L. An herbaceous annual that came origi-
nally from the Mediterranean region, and where it grows well
flowers freely in the Americas and has become a good florist's
crop; "boca-de-león, " "conejita" Sp. Po., "snapdragon" En.
Botrytis cinerea Pers.; moldy blight. In moist tropics.
Cassytha filiformis L.; dodder. Puerto Rico.
Cercospora sp.; leafspot. Guatemala.
Cuscuta corymbosa Engl.; dodder. El Salvador.
Fusarium oxysporum Schl.; wilt. Puerto Rico. F. solani
(Mart.) Sacc. (the asco state is a Nectria); stalk dryrot.
Puerto Rico, Florida, C. America.
Meloidogyne sp.; rootknot. Jamaica, Florida, Texas, El
Salvador, Guatemala, Puerto Rico.
Mycena citricolor (Berk. & Curt.) Sacc.; leafspot. Costa
Rica.
Oidium sp.; powdery mildew. W. Indies, C. America.
Peronospora antirrhini Schr.; downy mildew. W. Indies,
Guatemala, Costa Rica.
Phyllosticta antirrhini Syd.; leafspot, stemcanker. Argentina,
Texas.
Phymatotrichum omnivorum (Shear) Dug.; rootrot. Texas,
? Mexico.
Phytophthora parasitica Dast.; stemcanker. Puerto Rico.
Puccinia antirrhini Diet. & Holw.; rust. Puerto Rico, Guate-
mala, California.
Pythium (? debaryanum); cuttingrot. Puerto Rico. P. ultimum
Trow.; damping-off. Puerto Rico.
Rhizoctonia solani Kuehn (perfect state is) Thanatephorous
cucumeris (Frank) Donk; damping-off, cutting disease, basal
rot. Puerto Rico.
Sclerotinia sclerotiorum (Lib.) Mass. = Whetzeliana sclero-
tiorum (Lib.) Korf & Dumont; flower rot. Chile, El
Salvador.
Sclerotium rolfsii (Curzi) Sacc. = Pellicularia rolfsii (Sacc.)
West; wet soilrot. Texas.
Virus, Cucumber Mosaic (CMV) Celery-Commelina strain
Wellman. Puerto Rico, Florida.

APIUM GRAVEOLENS L. Developed in Europe, later in the USA,
this famed salad vegetable is noted for its succulent and crisp

leaf stalks that grow under rich soil conditions in the cooler
tropics and subtropics; "apio" Sp. Po., "celery" En.

Alternaria sp.; leafspot. Florida, Brazil.

Aphelenchoides sp.; root feeding nematode. Brazil.

Belonolaimus gracilis Stein. and B. longicaudatus Rau; root
  ectoparasitic nematodes. Florida.

Cercospora apii Fres.; spot and blight. Subtropical USA,
  Trinidad and Tobago, C. America, Surinam, Peru, Venezuela, Argentina, Uruguay.

Criconemoides sp.; nematode from root zone. Peru.

Ditylenchus dipsaci Filip.; stem nematode. California.

Dolichodorus heterocephalus Cobb; redroot nematode. Florida.

Erwinia carotovora (L. R. Jones) Holl.; market softrot. W.
  Indies, C. America, S. America. Subtropical USA.

Fusarium spp.; seedling damping-off and secondary rootrot.
  Subtropical USA.

Meloidogyne incognita Chit.; rootknot. Guatemala, Costa Rica,
  Jamaica, Surinam, Florida, California, Peru.

Phyllosticta apii Halst.; leafspot. Puerto Rico.

Pseudomonas apii Jagger; leafspot. Florida.

Puccinia apii Desm.; rust. Mexico, Chile.

Pythium ultimum Trow.; rootrot, damping-off. W. Indies,
  Florida.

Rhizoctonia solani Kuehn = Thanatephorous cucumeris (Frank)
  Donk; damp-off, rootrot, stalk infection. El Salvador,
  Peru, Guatemala, Brazil, Subtropical USA, probably general.

Sclerotinia sclerotiorum (Lib.) dBy. = Whetzeliana sclerotiorum
  (Lib.) Korf & Dumont; market rot. Argentina, California,
  probably other countries.

Septoria apiicola Speg. (synonyms S. apii Chupp, S. petroselini
  Desm. f. apii (Br. & Cav.) Ches.); large spot. W. Indies,
  Subtropical USA, Costa Rica, Guatemala, Mexico, Brazil,
  Argentina. S. apii-graveolentis Dorog.; small leafspot. W.
  Indies, Costa Rica, Venezuela.

Trichodorus sp.; stubby-root nematode. Florida.

Tylenchorhynchus sp.; nematode from root zone. Peru.

Virus, Tomato Spottedwilt. Brazil. Virus, vigorous commelina
  strain of cucumber mosaic on celery. Subtropical USA, C.
  America, W. Indies, S. America. (Note: This virus which
  I consider Cucumber Mosaic (CMV) Celery-Commelina strain
  Wellman, is severe on celery in Florida and was found
  threatening several other crops. I identified it and studied
  it as to disease effects, field spreading patterns, general
  properties, aphid vector, host range. After Florida work
  I investigated it further in Cuba, Puerto Rico and Honduras
  and made occurrence observations in Guatemala, El Salvador,
  Mexico, Costa Rica, Panama, Colombia, and Ecuador; it
  has perhaps the widest host range of any plant virus.)

ARACHIS HYPOGEAE L. Native to the Neotropics, a valuable annual
  that bears underground seedpods in which are the nuts that are
  relished and are an important local food and for international

markets; "maní," "cacahuete" Sp., "amendoim" Po., "peanut, "
"groundnut" En. (many Indian names), "ngooba-ngoo," "goober, "
from African slave dialect of the West Indies.

Alternaria alternata (Fr.) Keiss = A. tenuis Nees; stalk
    disease, leafedge infections. C. America on Pacific slopes,
    ? S. America. Alternaria sp.; enlargement of insect
    punctures. Florida, rare in W. Indies.
Aphelenchoides sp.; nematode. Subtropical USA.
Ascochyta imperfecta Peck; leafspot. Brazil. A. pisi Lib.;
    leafspot. Argentina.
Aspergillus (? niger); nut mold. Costa Rica, Puerto Rico.
Bacillus (? subtilis); bitter nut. W. Indies.
Botryodiplodia theobromae Pat., also given as Diplodia theo-
    bromae (Pat.) Now.; peg disease, stem canker, nut rot.
    Subtropical USA, W. Indies, Costa Rica, Peru, Ecuador,
    perhaps in all peanut countries.
Cercospora arachidicola Hori = Mycosphaerella arachidicola
    W. A. Jenk.; large brown leafspot, 1-12 mm. diameter
    (most common on popular cultivars). W. Indies, C.
    America, S. America. C. personata Ell. & Ev. = M.
    berkeleyii W. A. Jenk.; smaller black leafspot, 1-7 mm.
    diameter (most common on wild peanuts and on parents of
    cultivars). W. Indies, C. America, S. America.
Chaetodiplodia arachidis Maubl.; black stem disease. S.
    America.
Chaetoseptoria wellmanii J. A. Stev.; leafspot. Mexico, C.
    America.
Cladosporium herbarum Pers.; ragged leaf, pod blackmold.
    W. Indies, C. America, S. America.
Colletotrichum sp.; necrotic spots on stem and leaf. Peru,
    Ecuador.
Fusarium oxysporum Schl. f. sp. aurantiacum Wr.; pod and
    stem disease, and wilt. S. America, C. America.
Leptosphaerulina arachidicola Yen, Chev & Wang; leafscorch,
    pepperspot. Argentina, Brazil, Paraguay.
Macrophomina phaseoli (Maubl.) Ashby; charcoal rot of seed,
    stem, root. W. Indies.
Mycosphaerella spp. (see Cercospora spp.)
Meloidogyne arenaria Chit.; rootknot. Florida. Meloidogyne
    spp.; rootknot. Brazil, C. America, W. Indies.
Parodiella sp.; secondary infection. S. America.
Penicillium sp.; podmold. W. Indies, C. America, S.
    America.
Phytophthora sp.; plant blight. Ecuador.
Pseudomonas solanacearum E. F. Sm.; bacterial wilt. Sub-
    tropical USA, W. Indies.
Puccinia arachidis Speg.; mature-plant rust. W. Indies, C.
    America, S. America, rarely local in Subtropical USA.
Pythium debaryanum Hesse; field dump-off, scattered. C.
    America, S. America. P. ultimum Trow.; hot season
    damping-off. El Salvador, Argentina.
Rhizoctonia spp.; basalrot. C. America, S. America.

Sclerotinia sclerotiorum (Lib.) dBy. = Whetzeliana sclero-
tiorum (Lib.) Korf & Dumont; plant softrot. Chile, Suri-
nam.
Sclerotium rolfsii (Curzi) Sacc.; groundwilt. W. Indies, C.
America, Subtropical USA, S. America.
Sphaceloma arachidis Bitanc. & Jenk.; spot anthracnose in
S. America. (Similar symptoms to spot anthracnose seen
in Costa Rica but no microscopic proof secured.)
Verticillium sp.; podrot. Argentina.
Viruses: Abutilon Mosaic infection where inoculated. Brazil.
Tobacco Streak when inoculated. Mottle Mosaic. Ringspot.
Rosette. Stunt. Brazil, Argentina, other countries.

ARAUCARIA ARAUCANA (Mol.) Koch. and A. BRASILIANA Rich.
Tall evergreen trees of tropical America used in ornamental
groups of specimen plantings; "pehuén" Sp., "pinhão" Po.,
"tropical pine" En.
Agrobacterium tumefaciens (E. F. Sm. & Towns.) Conn;
crowngall. Subtropical USA, Mexico.
Belanopsis tropicalis Rick; branchrot. S. America.
Caeoma sanctae-crusis Esp.; rust. Chile.
Calocera cornea Fr.; woodrot. S. America.
Capnodium araucariae (Th.) Speg.; sootymold. W. Indies, C.
America, S. America.
Cladosporium sp.; leafdecline. S. America.
Clavaria muscorum Karst.; woodrot. S. America, C. America.
Collybia apiahyna Speg.; trunkrot. Brazil.
Glonium parvulum (Ger.) Cke.; a woodrot fungus. S. America.
Hypoxylon suberosum Berk. & Curt.; branchrot. W. Indies,
C. America, S. America.
Karschia araucariae Rehm; on dead wood. Brazil.
Leptosphaeria californica (Cke. & Harkn.) Berl. & Vogl.;
foliage disease. California.
Loranthacae (undetermined); many parasitic phaneroganic
plants on trees. Mexico, Costa Rica, Colombia, Panama.
Melanconiopsis elzoi Speg.; logrot and branchrot. Chile.
Pestalotia micheneri Guba; disease of inflorescence. Ber-
muda. P. theae Sawa.; sunken leafspot. Colombia.
Phyllosticta brasiliensis Lind.; leafspot. Brazil.
Phymatotrichum omnivorum (Shear) Dug.; rootrot. Texas.
Phytophthora citrophthora (Sm. & Sm.) Leon.; basal disease.
S. America, C. America.
Polyporus rickianus Sacc. & Trott.; heartrot. S. America,
C. America.
Pseudomeliola brasiliensis Speg.; leafspot. Brazil.
Schizophyllum spp.; branchrot. C. America, S. America.
Stictis auraucariae Ph. & Harkn.; leaf disease. S. America.
Uleiella chilensis Diet. & Neg. and U. paradoxa Schr.; smuts.
S. America.
Verticillium albo-atrum Reinke & Berth.; wilt. Chile.

ARMENIACA see PRUNUS

ARMORACIA RUSTICANA Gaertn., Mey. & Scherb. Rather exotic
for the tropics, a pungent flavored rootcrop from the temperate
zone requiring special cold-dry growing conditions; "rábano
picante" Sp., "rabano rustico" Po., "horseradish" En.

Albugo candida (Pers.) Kuntze; whiterust. Mexico, Guatemala,
Ecuador.

Alternaria brassicae (Berk.) Sacc.; small gray leafspot.
Florida, rare in Costa Rica or Guatemala, a few spots
seen in Puerto Rico. A. oleracea Milb.; large black leafspot. Texas, Mexico.

Cercospora armoracia Sacc.; leafspot. El Salvador.

Erwinia carotovora (L. R. Jones) Holl.; common storage softrot. General.

Erysiphe polygoni DC.; powdery mildew. W. Indies, C.
America, California.

Peronospora parasitica Pers.; downy mildew. Texas, Brazil.

Phymatotrichum omnivorum (Shear) Dug.; dryland rootrot.
New Mexico, Mexico, Texas.

Rhizoctonia spp. (in part solani); basalrot. C. America.

Virus. Common horseradish mosaic. On all plants seen.

ARRACACIA XANTHORRHIZA Bancr. From the Andes, a yellow-
colored, flavorsome rootcrop, grown mostly for local consumption and farmers markets; "racacha," "virraca" Sp., "mandio-
quinha" Po., "andean parsnip" En. (many Indian names).

Aphelenchus parietinus Stein.; nematode root invasion. Brazil.

Axonchium amplicolle Cobb; root feeding nematode. Brazil.

Cercospora sp.; leafspot. Venezuela.

Colletotrichum gloeosporioides Penz. = Glomerella cingulata
(Stonem.) Sp. & Schr.; anthracnose of foliage. S. America.

Erwinia (? carotovora); market and storage rot. Colombia,
W. Indies, Ecuador.

Erysiphe polygoni DC.; powdery mildew. Puerto Rico.

Fusarium spp.; rootrot. Colombia.

Haplolaimus sp.; root nematode. Brazil.

Helicotylenchus nannus Stein.; root nematode. Brazil.

Leaf decline (undetermined multiple causes). C. America.

Meloidogyne sp.; rootknot. Brazil.

Puccinia arrachathae Lagh. & Lind.; brown leafrust. Guatemala, Ecuador. P. imperspicua Syd.; black leafrust.
Mexico. P. repentina Jacks. & Holw.; leafrust. Widespread.

Rosellinia spp.; root diseases. Colombia, W. Indies.

Septoria apii Chupp; leafspot. Venezuela.

ARTOCARPUS spp. Two large tropical food trees of Asia-Polynesia
tropics, much grown and becoming naturalized in many parts of
the warm and moist America tropics: A. COMMUNIS Forst. =
A. INCISA L. f.; "pana" Sp., "frutapãoseiro" Po., "breadfruit"
En. and A. INTEGRA Merr. = A. INTEGRIFOLIA L.; "pana
cimarrona" Sp. Po., "jackfruit" (also "jakfruit") En. (Both
species of these trees appear to be about equally susceptible to
the diseases listed here.)

Aspergillus candidus (Pers.) Lk.; fruitrot. S. America, C.
    America.
Botrytis artocarpi Viégas; inflorescence mold. Brazil.
Capnodium spp.; sootymolds. W. Indies, C. America, S.
    America.
Cephaleuros virescens Kunze; algalspot. Nicaragua, Puerto
    Rico, Brazil, Costa Rica.
Cladosporium artocarpi Cif. & Gonz. Frag.; ragged leaf. W.
    Indies, C. America. C. herbarum (Pers.) Lk. on leaves
    from Ecuador.
Colletotrichum gloeosporioides Penz. = C. artocarpi Delacr.;
    anthracnose, dieback. Neotropics.
Corticium salmonicolor Berk. & Br.; leafrot, pink disease.
    Puerto Rico, S. America.
Diploida sp.; dieback complex. General.
Fomes lignosus Kl.; friable heatrot. C. America, S. America.
    F. noxius Cor.; footrot. Costa Rica.
Fusarium (roseum-type), fruit peduncle disease. Scattered.
Ganoderma sp.; stumprot. Costa Rica.
Hymenochaeta noxia Berk.; bark disease. W. Indies.
Marasmius sp.; cordblight of leaves. S. America.
Marssonina artocarpi Bat.; leafspot. Brazil.
Meliola sp.; black mildew. Costa Rica.
Mycena citricolor (Berk. & Curt.) Sacc.; leafspot. Costa Rica,
    Guatemala, Puerto Rico.
Pestalotia (? funerea); sunken leafspot. W. Indies, S. America,
    C. America.
Phyllosticta artocarpi Speg.; white leafspot. Costa Rica,
    Argentina, Brazil. P. artocarpicola Bat. & Vit.; brown
    leafspot. Brazil.
Phytophthora palmivora Butl.; shootrot, cuttingrot. Costa
    Rica.
Porella sp.; leafyliverwort smother. Puerto Rico, Virgin
    Islands.
Rhizoctonia sp. (aerial form); dieback. Costa Rica.
Rhizopus artocarpi Rac.; ripe fruitrot. ? General.
Rhombostilbella rosea Zimm.; foliage disease. Scattered in
    C. America and S. America.
Rosellinia bunodes (Berk. & Br.) Sacc.; black rootrot. C.
    America, Cuba, Puerto Rico.
Sclerotium rolfsii (Curzi) Sacc. = Pellicularia rolfsii (Sacc.)
    West.; sclerotial rot. Puerto Rico.
Setella coracina Bat.; leafrot. Brazil.
Sphaerostilbe repens Berk. & Br.; rootstain. W. Indies.
Stomatochroon lagerheimii Palm; leaf infection by systemic
    alga. Guatemala.
Strigula complanata Fée; lichenspot. Costa Rica, probably
    widespread.
Trichosporium nigricans Sacc.; leaf disease. S. America.
Uredo artocarpi Berk. & Br.; rust. S. America, Cuba,
    Puerto Rico.
Zignoella algaphila F. L. Stev.; on dying leaves infected with
    a Cephaleuros. Puerto Rico.

ARUNDINARIA spp. Important native canes used in many ways, sometimes with and instead of bamboo; "caña brava" Sp. Po., "wild cane" En. (Work is in progress on identifying parasitic organisms but this crop is still in large part harvested from the wild where there is little thought given to disease effects.) (Note: Caña brava strips are used in house walls, fencing, baskets, and floor mats and often last for years; they become the mostly superficial support of a multitude of miscellaneous organisms: bacteria, algae, nematodes, minute insects, and lignicolous fungi. Many still challenge mycologists, and even some of those listed on these pages are possibly not altogether admissible in a disease list.)

　　Anthostoma atro-cinctum Rick; weak parasite. S. America.
　　Anthostomella stegophora (Mont.) Sacc.; crust on culm. El
　　　　Salvador, probably general.
　　Apiospora montagnei Sacc.; on dead culms. Subtropical USA.
　　Arthrobotryum rickii Syd. & P. Syd.; on culms. General.
　　Ascochyta arundinariae Tassi; leafspot. S. America, C.
　　　　America.
　　Aulographium arundinariae Cke.; black mildew on leaves and
　　　　dead culms. Subtropical USA, C. America.
　　Botryosphaeria arundinariae Earle; canker on dead culms.
　　　　Subtropical USA.
　　Ceratocystis sp., dark streaks in culm. S. America.
　　Coniosporium bambusae (Thuem. & Bolle) Sacc. (sporuliferous
　　　　stage of Papularia which see below); leaf decay. General.
　　Diatrype consobrina Mont.; culm infection. Subtropical USA.
　　Dicellomyces gloeosporus Olive; leafspot. Subtropical USA.
　　Dothidella minima Sacc. & Syd.; old-leaf-spot. Subtropical
　　　　USA.
　　Echinodothis tuberiformis (Berk. & Rav.) Atk.; on culms.
　　　　Subtropical USA.
　　Eriosphaeria calospora Speg. = Gaeumannia calospora (Speg.)
　　　　Petr.; stain. S. America.
　　Eutypa linearis Rehm; weak parasite. General.
　　Hypoderma scirpinum DC.; culm tissue disease. Scattered
　　　　occurrence.
　　Hypoxylon culmorum Cke.; on old culm. Subtropical USA,
　　　　and C. America. H. rubiginosum Pers.; on dead culms.
　　　　C. America.
　　Leptothyrium cylindrium Atk.; leafspot. Subtropical USA.
　　Lophiosphaeria schizostoma (Mont.) Trev.; weak parasite.
　　　　Widespread.
　　Meliola tenuis Berk. & Curt.; black mold leafspots. Sub-
　　　　tropical USA. Meliola sp., black mildew. C. America.
　　Metasphaeria rimulorum (Cke.) Sacc.; on culms. Florida.
　　Mycocitrus aurantium Moel.; culm fungus. S. America.
　　Mycosphaerella arundinariae (Atk.) Earle; leaf spots. Costa
　　　　Rica (asci secured in moist chamber). Subtropical USA.
　　Ophiobolus stictisporus (Cke. & Ell.) Sacc.; culmspot. Sub-
　　　　tropical USA.
　　Papularia ecuadorensis Petr. (see Coniosporium); mycelial

disease. Ecuador and S. America, C. America. P. sphaerosperma (Pers.) Hoehn.; on dead culms and leaves. General.

Phyllachora arundinariae Orton; black leafspeck. Subtropical USA, W. Indies, C. America, ? S. America.

Physalospora arundinariae Orton; canecanker. Subtropical USA, C. America.

Puccinia arundinariae Schw.; rust. Subtropical USA, ? Mexico. P. bambusarum (P. Henn.) Arth.; rust. S. America, C. America.

Pyrenopeziza arundinariae (Berk. & Curt.) Sacc.; disease of culms. Subtropical USA, C. America.

Sclerotium sacidioides Speg.; on leaves (specimens secured in Virginia, USA).

Scolecotrichum graminis Fckl.; leaf brown-stripe. Subtropical USA, probably other parts of tropics.

Trematosphaeria fusispora Rick; on leaf and culm. W. Indies.

Volutella tecticola Atk.; leafspot. Subtropical USA, ? El Salvador.

ASPARAGUS OFFICINALIS L. Grown for its succulent shoots, a temperate zone plant that is somewhat exotic for the tropics; often roots have to be replanted yearly; "esparago" Sp., "asparago" Po., "asparagus" En.

Alternaria alternata (Fr.) Keiss.; rarely associated with dieback. Florida, Guatemala, Puerto Rico.

Axonchium amplicolle Cobb; root nematode. Brazil.

Cercospora asparagi Sacc. (Synonyms: C. caulicola Wint., Cercosporina asparagicola Speg.); leaf-and-stalk spots. Subtropical USA, Mexico, Cuba, Brazil, Argentina.

Cladosporium herbarum Lk.; on senescent foliage. El Salvador.

Colletotrichum sp.; anthracnose spots and canker. Subtropical USA, Puerto Rico, Colombia, Brazil.

Fusarium oxysporum Schl. f. sp. asparagi Cohen; wilt. California, Florida, Puerto Rico, Peru. F. roseum Lk.; stem disease, mold of shoots in markets. Occasional.

Macrophomina phaseoli (Taub.) Ashby; stemblight. Mexico.

Meloidogyne sp.; rootknot. Florida, Puerto Rico.

Puccinia asparagi DC.; rust. W. Indies, Peru, Argentina.

Xiphinema americanum Cobb; ectoparasitic root nematode. Brazil.

ASPARAGUS PLUMOSIS Baker. Introduced to the American tropics for producing florists' foliage, lightweight and readily shipped by air freight; "helecho de esperigo," "ala de pájaro" Sp. Po., "asparagus-fern" En.

Ascochyta asparagina Petr.; canker and blight. Florida.

Cercospora asparagi Sacc.; leaf and branch spots. Subtropical USA.

Cladosporium herbarum Lk.; leafspot, mold and leaf drop. Subtropical USA, Puerto Rico, C. America.

Colletotrichum sp.; stem lesion. Florida.

Coniothyrium sp.; stemcanker. Brazil.
Didymosphaeria brunneola Niessl; stem injury. Florida,
    Puerto Rico.
Fusarium sp.; rootrot. Florida, Puerto Rico.
Helminthosporium sp.; branch disease. Florida.
Hendersonia asparagi Pass.; weak parasite causing stem
    disease. Florida.
Leptosphaeria asparagina Karst.; stem disease. Florida.
Meloidogyne sp.; rootknot. Florida.
Pleospora (? asparagi); leaf disease. Chile.
"Rust," following poor growth resulting in attack by weakly
    parasitic species of Alternaria, Fusarium, Curvularia,
    and others. W. Indies.

AVENA SATIVA L.   A grain crop from the temperate zone adapted
in special tropical American areas; "avena" Sp., "aveia" Po.,
"oats" En.
Alternaria alternata Keiss. = A. tenuis Nees.; black leaf-
    edge, glumespot. A few places in C. America. A. avenae
    Oud.; leafspot. Chile.
Colletotrichum spp.; anthracnose. Brazil, Guatemala, Uru-
    guay.
Erysiphe graminis DC.; powdery mildew. Peru, Chile, Ecua-
    dor, Guatemala.
Gibberella zeae Petch = Fusarium graminearum Schwabe which
    is the conidial stage; blight. Peru, Chile, Ecuador.
Helminthosporium avenae Eid. = its perfect stage is Pyreno-
    phora avenae Ito & Kur.; large spot and stripe. Colombia,
    Mexico, Brazil, Guatemala, Argentina. H. victoriae Meeh.
    & Mur., blight. Chile, Peru, Argentina. H. pedicellatum
    P. Henn.; leaf disease. Brazil.
Heterosporium avenae Oud.; spot on senescent leaf. Argentina.
Oidium sp.; white mildew. Argentina, Colombia.
Ophiobolus graminis Sacc.; bottomrot. Mexico, Bolivia, Peru,
    Colombia, Puerto Rico, Louisiana.
Phyllosticta avenophila Tehon & Dan.; halo blight. Argentina.
Pratylenchus sp.; root nematode. Peru.
Pseudomonas coronofaciens (Elliott) Stevens; bacterial halo
    blight. Argentina. P. striafaciens (Elliott) Starr & Burkh.;
    stripe. Argentina, Colombia.
Puccinia coronata Cda.; crownrust. Argentina, Peru, Brazil,
    Bolivia, Uruguay, Chile, C. America, Mexico, Colombia,
    Ecuador. P. graminis Pers. f. avenae Erik. & E. Henn.;
    stemrust. Brazil, Venezuela, Argentina, Uruguay, Chile,
    Colombia, Peru, Chiloe Islands, Mexico, Ecuador, Bolivia.
Sclerotium rolfsii (Curzi) Sacc.; seedling collapse. Argentina.
Septoria gramineum Desm.; leafspot in Argentina. S. tritici
    Rob. f. avenae Spr.; leafspot in Colombia.
Ustilago avenae (Pers.) Rostr.; loosesmut. Venezuela, Argen-
    tina, Brazil, Uruguay, Peru, Chile, Colombia, C. America,
    Ecuador, Mexico. U. kolleri Wille = U. levis (Kell. &
    Swing.) Magn.; coveredsmut. Bolivia, Peru, Uruguay,
    Brazil, Guatemala, Mexico.

Virus: Enanismo (stunt). Colombia, Ecuador, Guatemala.
Hoja Blanca of rice. Colombia, Brazil.

AXONOPUS MICAY Hitch. A tropical green-cut hay crop; "pasto
micay" Sp. and A. SCOPARIUS Hitch.; "pasto imperiál" Sp.
(Note: Further work will result in more diseases to be listed
on these two tropical soilage (green-cut hay) crops. Other
miscellaneous species of this grass are used in tropical graze-
pastures and tropical lawns; a few diseases on these are:
Angiospora compressa Mains; Balansia lineare Diehl; Cerebella
sp.; Colletotrichum sp.; Curvularia sp.; Gnomonia (Melanconium)
iliau Lyon; Helminthosporium spp.; Meloidogyne sp.; Myriogenospora bressadoleana P. Henn.; Phyllachora spp.; Sorosporium
paranense Hirsch.; Sphacelotheca inconspicua Zund.; Stagonospora
procelura Petr.; Trichostroma axonopi Tehon; and virus symptoms.)
        Botrytis sp.; mold. Colombia.
        Balansia strangulans (Mont.) Diehl; strangle. Colombia.
        Cercospora sp.; large leafspot. Colombia, Costa Rica.
        Claviceps purpurea (Fr.) Tul.; ergot. Costa Rica.
        Fusarium graminium Schwabe; moldy cut hay. Colombia.
        (? Gnomonia-Melanconium); a severe culm disease. Costa
            Rica.
        Puccinia substriata Ell. & Barth.; rust. Colombia, Puerto
            Rico.
        Rhizoctonia (of aerial type); leaf lesions. Colombia, Costa
            Rica.
        Xanthomonas axonoperis Starr & Garces; gumming. Colombia.

AZALEA   see   RHODODENDRON

BACTRIS   see   GUILIELMA

BAMBOOs (many species of giant grasses, some genera are: BAM-
BUSA, GYNERIUM, SHROPHIRA, GUADUA, CHUSQUEA, sometimes ARUNDINARIA is included, and others). Pulp is used for
paper shoots for food, large numbers of tropical houses and
shelters are constructed from them including roof, floor, sides,
walls, doors and studs, while fences, water-conducting pipes,
ladders, containers and other things come from bamboo; "bamba,"
"bambú," "guadua" Sp., "bambou," "mimbrás" Po., "bamboo"
En., numerous Indian names. (Several hundred parasites are
reported on bamboos but space is limited and I believe I have
selected and give below at least some of the interest and special
importance.)
        Actiniopsis bambusae Starb.; lichen attack (?) causing culm
            injury. S. America, C. America.
        Agaricus heteropus Speg. = Pleurotus heterotropus Speg.; on
            rotting bamboo trash. Argentina. A. succineus Speg. =
            Omphalia succinea Speg.; on old building material. Para-
            guay. Agaricus spp.; on old trash at bases of bamboo
            clumps. Puerto Rico, Costa Rica.
        Algae on culms very often in most places (genera noted:
            Nostoc, Geococcus, Trentepohlia, Spirogyra, Phycopeltis).
            Scattered.

Anthostomella chusqueicola Speg.; stroma on culm. Chile,
    Argentina. A. puiggarii Speg.; stroma on leaves. Brazil
    and adjacent countries. A. sublimata Speg.; a culm
    stroma. Chile.
Apiospora chilensis Speg.; culm disease. Chile. A. striola
    Sacc.; culm blemish. Subtropical USA, Panama.
Arthrobotryum stilboideum Ces.; mold. Brazil, Puerto Rico.
Ascopolyporus gollmerianus P. Henn.; twig disease. Brazil,
    adjacent countries. A. polychrous Moel.; twig infection.
    Brazil. A. puttemansii P. Henn. and A. villosus Moel.
    also reported. S. America.
Asteridium bambusellum Speg. (synonyms: Asterina bambusella
    Speg., Zukalia bambusella Speg., Coscinopellis argentinen-
    sis Speg.); leafspotting. S. America.
Asterina microscopica Lév.; black mold on green leaf. Chile.
Auerswaldia bambusicola Speg. (see Hypoxylon)
Auricularia spp.; on dead culms used in jungle buildings. C.
    America.
Bacteria, non-pathogenic isolated by me from internal tissues.
    Honduras.
Balansia chusqueicola P. Henn.; black crust of stem. Costa
    Rica. B. claviceps Speg.; stem disease. Chile, Brazil,
    Costa Rica. B. regularis Moel.; witches' broom. Brazil.
Birdsnest fungi, unidentified, on bamboo used as "pots" for
    growing cinchona seedlings. Guatemala.
Brefeldiella brasiliensis Speg. and B. subcuticulosa Th.; culm
    cankers. S. America.
Broomella guaranitica Speg.; weak culm-disease parasite.
    Brazil, Argentina.
Calonectria guaranitica Speg. = Malmeomyces pulchella Starb.;
    severe disease on young branches. Brazil.
Cenangella bambusicola Rick; on old branches. Brazil.
Chaetophoma pellicula Sacc. & Syd.; leaf disease. S. America.
Colletotrichum sp.; anthracnose. Widespread.
Coniosporium crustaceum Speg.; on dead stalk tips. Neo-
    tropics. C. inquinans Dur. & Mont.; dieback. S. America,
    C. America. C. shirainum Bub.; culm tiprot. Texas.
Cordella coniosporioides Speg.; leaf and culm disease. Para-
    guay.
Corticium sp., mycelium and basidia on culms but causing no
    weakening by infections. Ecuador, El Salvador.
Corynaea crassa Hook; tuberous rootparasite. Panama.
Criconemoides sp.; root nematode. Peru.
Crinipellis bambusae Pat.; cord fungus causing culm disease.
    Brazil.
Cylindrosporium bambusae Miy. & Hara; leafblight. Sub-
    tropical USA.
Didymosphaeria rhytidosperma Speg. (synonyms: Eudidyma
    rhytidosperma (Speg.) Kuntze, Clypeosphaeria rhytidosperma
    (Speg.) Rick); culm disease. Paraguay, Brazil.
Diplodia bambusae Ell. & Lang. (not D. theobromae) (? =
    Microdiplodia valdiviensis Speg.); culm tipdisease. Guate-
    mala, El Salvador, Chile, Argentina.

Eutypa hypoxantha St.; cane disease. Bolivia. E. phaselina
   Sacc.; cane disease. Trinidad.
Fusarium gigas Speg. = F. episphaeria (Tode) Sny. & Hans;
   decay of old culm. Paraguay and widespread. Fusarium
   spp.; isolated from "white plume" branchlet disease. El
   Salvador, Costa Rica (in the latter the specimens were in-
   cubated in moist chamber and there developed red and brown
   ascocarps apparently Nectria sp.).
Gilletiella chusqueae (Pat.) Sacc. & Syd.; stem disease. Ecua-
   dor.
Glaziella cyttarioides Speg.; branch infection and culmrot.
   Paraguay.
Gloniella chusqueicola P. Henn.; on foliage. S. America.
Helminthosporium caaguazense Speg. = Spiropes caaguazense
   (Speg.) Ell.; cane lesions. Paraguay. Helminthosporium
   spp.; large leafspots. Florida, W. Indies, C. America.
Hemicycliophora sp.; nematode near roots. Peru.
Hormomyces pezizoides Speg.; foliage disease. Brazil.
Hypoxylon bambusicolum Speg. = Auerswaldia bambusicola
   Speg., H. culmorum Cke.; decay of culms used in construc-
   tion. Ecuador, Argentina, Paraguay.
Laestidia sp.; anthracnose-like lesions on leaves and on culms.
   Florida, C. America.
Lentinus perpusillus Speg. = Crinipellis perpusilla (Speg.)
   Sing.; canerot. Brazil.
Leptosphaeria schneideriana Rick; leaf flyspeck. S. America.
   L. weddellii Sacc., culm flyspeck. Bolivia.
Lichens, various kinds; disfigure culms. Moist parts of C.
   America.
Melanconium bambusinum Speg. = Papularia bambusinum
   (Speg.) Petr.; culm rinddisease. S. America. M. saccha-
   rinum Penz. & Sacc.; culm rinddisease. Subtropical USA,
   Peru.
Melanographium spinulosum (Speg.) Hug.; on culms. Argentina,
   Dominican Republic, Panama.
Munkia chusqueae March.; stem disease. C. America, S.
   America. M. martyris Speg. = M. strumosa (Cke.) Mar.;
   branch disease. Paraguay and adjacent countries.
Mycosphaerella sp.; leafspot. California, Puerto Rico, Ecua-
   dor.
Nectria puberula Speg. = Cucurbitaria puberula (Speg.) Kuntze,
   and N. vagabunda Speg. = C. vagabunda (Speg.) Kuntze;
   both on dying foliage. Paraguay, Brazil.
Nigrospora oryzae (Berk. & Curt.) Petch; accompanying ragged-
   leaf condition. Neotropics.
Omphalia pergracilis Speg.; root disease. Chile.
Ophiodothis linearis Rehm; leaf disease. Brazil.
Papularia (see Melanconium)
Paranectria albo-lanata Speg. = Puttemansia albolunata (Speg.)
   Hoehn.; leaf disease. Paraguay.
Pestalotia macrochaeta (Speg.) Guba; leafspot. Argentina.
Phillisiella graminicola Hoehn.; leaf stroma. Brazil.
Phoma bambusina Speg.; leafspot. Paraguay.

Phyllachora bambusina Speg. (often associated with the
Phoma); shiny leaf blackspot. Brazil. P. bonariensis
Speg.; tarspot. Brazil. P. chusqueae P. Henn. & Lind.;
black leafspot. Chile, Panama. P. gracilis Speg.; tar-
spot. Brazil, Paraguay. P. portoricensis Orton; leaf
blackspot. Puerto Rico.
Phyllosticta bambusinifolia Bat.; spot of young leaf. Brazil,
Chile.
Physalospora sp.; diseased branch. Puerto Rico.
Pratylenchus sp.; nematode at roots. Peru.
Puccinia ignava (Henry) Arth. = Uredo ignava Henry; yellowish
rust on both sides of leaf. Florida, W. Indies, C.
America. P. melanocephala Syd.; brown rust on under-
side of leaf. Subtropical USA, W. Indies.
Pucciniospora chusqueae Speg.; rust. Brazil.
Rhabdospora bambusae Viég.; on culms. Brazil.
Rosellinia chusqueae Speg.; rootrot. Chile. R. pseudohy-
poxylon Speg.; root and base disease. Chile. R. subver-
ruculosa Rehm; rootrot. Brazil.
Schizophyllum bambusinum Speg. and S. commune Fr.; on
dead culms. Neotropics.
Sclerotium rolfsii (Curzi) Sacc. (no perfect state seen); basal
rot. Puerto Rico.
Selenophoma donacis Spr. & Johnson; culmspot. California.
Septoria bambusella Speg.; dark leafspots. Brazil.
Serella coracina Bat.; sootymold. Brazil.
Sphaerella chusqueicola Speg.; lesion on foliage. Chile.
Tomentella bambusina Viég.; root and shoot rot. Brazil.
Ustilago shiriana P. Henn.; stem smut and witches' brooming.
Subtropical USA, C. America.
Zythia lancispora Speg.; green leaf disease. Paraguay.

BAUHINIA spp. There are in the American tropics about 20 species
many with large orchid-like flowers, mostly trees, used for
agricultural shade, ornament, poles, charcoal, structural wood;
"hojas gamelas," "palo de orchidia" Sp. Po., "twin leaf tree,"
"orchid tree," "twin leaf" En.
Aphanopeltis bauhiniae Bat.; on foliage. Brazil.
Bacteria (isolated from non-fungus leafspot); leafspot by
inoculation. Puerto Rico.
Catacauma weirii Chard.; scab-spot on leaves. Bolivia.
Cephaleuros virescens Kunze; algal spot. Costa Rica.
Cercospora bauhinicola Syd. & P. Syd.; round leafspot. S.
America, C. America.
Cladosporium herbarum Lk.; on necrotic leaf edges. Neo-
tropics.
Clitocybe tabescens Bres. = Armilariella tabescens (Scop.)
Sing.; rootrot. Subtropical USA.
Colletotrichum gloeosporioides Penz. = Glomerella cingulata
(Stonem.) Sp. & Schr.; leafspot and leafdrop. Texas, W.
Indies, Costa Rica, probably General. (Glomerella state
collected on old flower petioles in Costa Rica.)

Corticium salmonicolor Berk. & Br.; pink disease of branch.
  Costa Rica, S. America, W. Indies.
Diaporthe eres Nits. Sacc. = D. eres Speg.; foliage dieback.
  S. America.
Diplodia sp. = Botryodiplodia sp.; on dead stubs after pruning.
  Puerto Rico.
Eutypa heteracantha Sacc. f. bauhinia-grandiflorae Speg.; stem
  lesions. S. America.
Ganoderma sp.; rot of old tree. Costa Rica.
Grallomyces portoricensis Stev.; superficial web growth on
  leaves. W. Indies.
Helminthosporium decacumiatum Thuem. & Pas.; blackened
  leafspot. S. America.
Loranthus sp.; large parasitizing bush. Costa Rica.
Marasmius sp.; white-thread blight. Costa Rica.
Meliola pazschkeana Gaill.; black mildew. W. Indies. M.
  perexigua Gaill.; black mildew. W. Indies.
Micropeltis bauhiniae Bat. & Gay.; black mold on living leaves.
  Brazil.
Microsphaeria diffusa Cke. & Pk.; white mildew. Florida, El
  Salvador.
Mycosphaerella sp.; leafspot. Brazil.
Myrmaciella hoehneliana Rick; leaf disease. C. America.
Nidularia sp.; on fallen branch in dieback complex. Puerto
  Rico.
Periconiella missionum Speg.; attack of living leaves. Argen-
  tina.
Phthirusa sp.; a drooping parasitic bush. Costa Rica.
Phyllachora tenuis (Berk.) Sacc.; tarspot. Mexico, Nicaragua.
  (Other Phyllachoras recorded.)
Phyllosticta bauhinicola P. Henn., P. juruana P. Henn., and
  P. missionum Speg.; all leafspots. Brazil, Argentina,
  Florida.
Phytophthora citrophthora (Sm. & Sm.) Leo.; root disease.
  Argentina.
Pilostyles blanchetii R. Br.; phanerogams, small parasitic
  plants, remarkable systemic invasion of trunk and root at
  base. Brazil. P. caulotreti Hook. are small plants (see
  above), trunk disease. Venezuela. P. globosa Hemsl. the
  small parasitic plants (see above) attacking base of tree and
  exposed roots. Mexico. Pilostyles sp., found in Costa
  Rica on tree base.
Ramularia bauhiniae Ell. & Ev.; leaf lesion covered with white
  mold. Jamaica.
Rhytisma bauhiniae Nees; twig disease. Argentina.
Scolecopeltis bauhiniae P. Henn.; raised spots on leaves and
  petioles. Ecuador.
Scopella bauhiniicola (Arth.) Cumm. = S. bauhiniicola Arth.;
  leafrust. C. America.
Stomatochroon lagerheimii Palm; systemic algal infection of
  leaf. Guatemala.
Uromyces spp. (rusts: numerous species have been named on

Bauhinias in the Neotropics and a number are given here
to not only have them on the record, but also to indicate
the notably high specialization some of them show for the
very definite parts of the host on which they develop): U.
anthemophilis Vest.; rust of only calyx. Brazil. U.
bauhiniicola P. Henn.; twig and leaf rust. Cuba, Mexico,
Brazil. U. dietelianus Paz.; leafrust. Brazil. U. fiebrigii P. Henn. & Vest.; typical leafrust. Paraguay,
Brazil. U. florialis Vest.; severe rust of flower but with
mild rust on leaf and calyx. Brazil. U. faveolatus Juel.;
typical leafrust. Brazil. U. goyazensis P. Henn.; rust
of flower and on bark of green branch. Brazil. U. guate-
malensis Vest.; the common brown leafrust. W. Indies,
C. America. U. hemmendorfii Vest.; a not commonly reported leafrust. Brazil. U. imperfectus Arth. = U.
superfixus Vest.; perhaps the most regularly encountered
leafrust. Special reports in Nicaragua, Costa Rica,
Jamaica, Brazil, Venezuela, Argentina, Bolivia. U. jamaicensis Vest.; leafrust. Mexico, Puerto Rico, Jamaica,
Trinidad. And the following, all leafrusts and all in
Brazil: U. pannosus Vest., U. parafinis Diet., U. perle-
biae Vest., U. praetextus Vest., and U. regins Vest.
Xylaria ianthino-velutino Mont., black rootrot. S. America,
C. America.

BEGONIA spp. Ornamental lush tropical American herbs, to bush-
like plants, of considerable value to florists and appreciated by
home makers for both indoor and outside growing; "begonia"
Sp. Po. En.
Aphelenchoides fragariae Chr.; leaf nematode. Florida.
Asterina filamentosa Pat.; leafmold. S. America.
Bacteria (undetermined); leafspots. Widespread in C. America.
Bartalina begoniae Bat.; leaf disease. Brazil.
Botrytis cinerea Pers.; plant blight, cuttingrot. Costa Rica,
W. Indies.
Cassytha filiformis L.; green dodder. Puerto Rico.
Cephaleuros virescens Kunze; algal spot. Brazil, Puerto
Rico, and other countries.
Coleosporium begoniae Arth.; yellow rust. Mexico.
Colletotrichum sp.; anthracnose, branch decay. Scattered
throughout C. America, Puerto Rico.
Cuscuta americana L.; yellow dodder. Puerto Rico.
Erysiphe polygoni DC.; powdery mildew. Neotropics.
Leaf eruption in aggregate spots. Cause undetermined. Not
edema. Costa Rica, Puerto Rico.
Meloidogyne sp.; rootknot, occasionally giant size. Puerto
Rico, Florida.
Mycena citricolor (Berk. & Curt.) Sacc.; leafspot both in field
and by inoculation. C. America.
Mycosphaerella begoniae Pat.; leafspot. Ecuador.
Myxosporium sordidum Tassi; spots on leaf and stem. S.
America.

Oidium begoniae Putt.; white powdery mildew.   Florida,
     Brazil.
Phyllachora begoniae Pat.; black and shiny leafspot.   Ecuador.
Phyllosticta begoniaecola Rangel.; leafspot and collapse.
     Brazil.
Pythium ultimum Trow.; warm land damping-off.   Peru,
     Panama.   P. vexans dBy.; cool season damping-off.   C.
     America, W. Indies.
Rhizoctonia solani Kuehn = Thanatephorous cucumeris (Frank)
     Donk; basé and collar-rot.   Costa Rica, Puerto Rico.
Sclerotinia sclerotiorum (Lib.) = Whetzeliana sclerotiorum
     (Lib.) Korf & Dumont; stem rot.   California, W. Indies.
Trichothyrium dubiosum (Bomm. & Rouss.) Th.; on leaves.
     S. America.
Virus, mosaic symptoms mild but common.   Costa Rica.
Xanthomonas begoniae (Buchw.) Dows.; bacterial spotting.
     Costa Rica.

BERTHOLLETIA EXCELSA Humb. & Bompl.   From the northern
     Amazon, a large forest and jungle tree protected in wild growths,
     some recently being tested for growth in plantations, important
     for producing popularly known nutritious and oily nuts for a large
     international market; "nuez de brasíl" Sp., "pau-brazil," "turwi,"
     "castanha-do-pará" Po., "brazil nut" En.   (The following list is
     based on incomplete reports.   Intensive work on this crop must
     have yielded many disease occurrences that I have not seen.)
     Actinomyces brasiliensis Spenc.; endosperm rot.   Brazil.
     Aspergillus flavus Mont., A. ochraceus Wilh., A. wentii
          Wehm.; all nut decay.   Brazil.
     Cephalosporium bertholletianum Spencer; nut whitemold.
          Brazil.
     Coemansia braziliensis Thaxt.; nutmold.   Brazil.
     Colletotrichum gloeosporioides Penz.; common foliage an-
          thracnose.   Brazil.
     Diplodia sp.; dieback on specimen tree.   Puerto Rico.
     Elsinoë bertholletia Bitanc.; foliage spot-anthracnose.   Brazil.
     Fusarium sp.; dryrot of nut.   Brazil.
     Leaf spotting (brown and discreet edges) from unidentified
          fungus, on specimen tree.   Costa Rica.
     Monilia sitophila (Mont.) Sacc.; mold.   Brazil.
     Myxosporium sp.; nut bitterrot.   Brazil.
     Pellionella macrospora Spenc.; blackcrust on nut.   Brazil.
     Penicillium aureum v. Tiegh.; nutmold.   Brazil.
     Pestalotia sp.; leafspot.   Brazil.
     Phomopsis bertholletianum Spenc.; endosperm rot.   Brazil.
     Porella sp.; leafy liverwort smother on leaves.   Puerto Rico.
     Rhizopus nigricans Ehr.; nut rot.   Brazil.

BETA VULGARIS L.   A garden rootcrop from the Old World intro-
     duced to the American tropics; "remolacha," "betterave" Sp.,
     "betterrabe" Po., "common beet" En.; B. VULGARIS L. var.
     CICLA (L.) Moq. is a selection of an agronomic crop of beet,

specially adapted to sugar accumulation in the root but botanically indistinguishable from the garden vegetable; "remolacha acrícac," "acelga" Sp., "acelga," "betterrabe sacarina" Po., "sugarbeet" En. (These two crops appear, where I have seen them, to be practically identical from the standpoint of disease susceptibility and countries of disease occurrence.)

Agrobacterium tumefaciens (E. F. Sm.) Conn.; crowngall. Argentina.

Albugo bliti (Bir.-Bern.) Kuntze; white blister. S. America (Andean areas).

Alternaria alternata (Fr.) Keiss. = A. tenuis Nees; seed mold. El Salvador. A. (? amaranthi); rare spreading spots. El Salvador.

Aphalenchus radicicolus Stein.; root nematode. W. Indies.

Aphanomyces cochlioides Drechs.; damping-off. Uruguay.

Belonolaimus longicaudatus Rau; root nematode. Florida.

Cercospora beticola Sacc.; common leafspot. W. Indies, C. America, S. America, Subtropical USA, Mexico.

Erwinia carotovora (L. R. Jones) Holl.; market softrot. General.

Erysiphe polygoni DC.; powdery mildew. El Salvador.

Fusarium oxysporum Schl.; rootrot. Argentina. F. scirpi Lam. & Fau.; on roots. Argentina. Fusarium sp.; damping-off. Uruguay.

Glomerella sp.; ascocarps on dead leaf petioles. Guatemala, El Salvador.

Heterodera schachtii Schm.; root nematode. California, Chile.

Macrophomina phaseoli (Taub.) Ashby; ashy rootrot. Uruguay.

Meloidogyne sp.; rootknot. Subtropical USA, Brazil, Chile.

"Nematodes," undetermined; root disease. Nicaragua.

Nothotylenchus acris Thorne; root nematode. Subtropical USA.

Paratylenchus sp.; root nematode. Chile.

Penicillium sp.; market rot. C. America.

Peronospora schachtii Fckl.; downy mildew. Argentina.

Phoma betae Frank; root and leaf disease. Neotropical.

Physalospora rhodina (Berk. & Curt.) Cke. = Diplodia sp.; leaf disease. Subtropical USA.

Physoderma pulposum Wallr.; brown rootspot. S. America.

Phytophthora sp.; rootrot. Uruguay.

Pleospora oligomera Sacc. & Speg.; leafspot. S. America.

Pythium aphanidermatum (Edson) Fitz.; damping-off. W. Indies, C. America. P. ultimum Trow.; warm season damping-off. Uruguay.

Ramularia betae Rostr.; leafspot. S. America.

Rhizoctonia solani Kuehn = Thanatephorous cucumeris (Frank) Donk; damping-off, market rot, dry rot, stem infection. Peru, Costa Rica, Uruguay, Mexico, Panama, Brazil.

Sclerotinia sclerotiorum (Lib.) dBy. = Whetzeliana sclerotiorum (Lib.) Korf & Dumont; rot. Uruguay.

Sclerotium rolfsii (Curzi) Sacc.; rootrot, seedling wilt.

Subtropical USA, Venezuela, Puerto Rico, C. America,
Uruguay, Bahamas, Virgin Islands.

Septoria betae West; leafspot.   Chile.

(Uredo marmoxiae Speg.; rust.   Found in Mediterranean islands, potential problem in tropical America.)

Uromyces betae Tul. = U. hemisphaericus Speg.; rust.   Chile,
Uruguay, Bolivia, Argentina, Brazil.

Urophlyctis leproides (Trab.) Magn. = Physoderma leproides
(Trab.) Karl.; root tumor.   Uruguay.

Viruses: Brazilian curly top in Brazil, Uruguay, Argentina.
Argentina Sunflower Mosaic in Argentina.   Brazilian Tobacco Streak in Brazil.   Yellow wilt (Mycoplasma causation) in Argentina.   Cucumber mosaic and its Commelina-
celery strain in Florida, Puerto Rico.

BIDENS spp.   Some species are weeds and a few are important for
food of Indians, often cultivated in mountain soils; "verdura
comestible" Sp.   (Many Indian names.)

Cassytha filiformis L.; dodder.   Puerto Rico, Florida.

Cercospora megalopotamica Speg.; large leafspot.   W. Indies,
C. America, S. America.

Cuscuta americana L. and C. umbellata HBK.; dodder.
Puerto Rico.

Cystopus brasiliensis Speg. = Albugo trogopogoni Pers.; white
blister.   C. America, S. America.

Entyloma bidentis Speg. = E. calendulae (Oud.) dBy. f.
bidentis Viegås and E. guaraniticum Speg.; blister smut.
First in S. America, W. Indies, C. America, second in
Brazil, Argentina.

Erysiphe lamprocarpa (Wallr.) Lév. = Oidium sp.; mildew in
dry season.   W. Indies, C. America, S. America.

Mycena citricolor (Berk. & Curt.) Sacc.; leafspot on shaded
plant in coffee plantation.   C. America, Puerto Rico.

Plasmopara halstedii (Farl.) Berl. & deT.; downy mildew.
Scattered.

Septoria balansiae Speg.; small leafspots.   Argentina, Brazil.

Sphaceloma bidentis Bitanc. & Jenk.; spot anthracnose.   S.
America.

Sphaerotheca humuli (DC.) Burr. f. fuliginosa Schl.; mildew.
Scattered.

Thecaphora pustulata Clint.; stemsmut.   Puerto Rico.

Uredo bidenticola Speg. = Uromyces bidenticola Arth.; rust.
W. Indies, S. America, C. America.

Viruses: Cucumber mosaic and its Commelina-celery strain.
W. Indies, Florida.   An unidentified yellowing and "crespera." Cuba.

BIGNONIA spp.   Ornamental and much planted bushes and vines;
"copa-de-oro" Sp., "bignonia" Po., "flowering creeper," "cup
of gold" En.

Calonectria ambigua Speg. = Subiculicola ambigua Speg.;
foliage disease.   Argentina, ? S. America in general.

Cephaleuros virescens Kunze; algal leafspot. W. Indies.
Cercospora bignoniecola Speg.; gray leafspot. S. America.
Cercosporella unguis-cati Speg.; purple leafspot. Argentina.
Colletotrichum gloeosporioides Penz. (asco-state not found);
    anthracnose. S. America, W. Indies, C. America, Sub-
    tropical USA.
Darluca australis Speg.; hyperparasite on rusts. General.
Dimerosporium bignoniicola Speg.; dark mold on leaf. Ar-
    gentina, probably widespread in S. America.
Ectosticta bignoniicola Speg.; on leaf (may be a hyperparasite.)
    Argentina.
Hypocrella disjuncta Seav.; leaf disease. W. Indies.
Meliola spp.; black mildews. Neotropical.
Mycosphaerella passiflorae Rehm f. bignoniae Rehm; leafspot.
    C. America, S. America.
Pestalotia versicolor Speg. var. americana Speg.; leaf and
    stem spotting. General.
Phyllachora amphigena Speg. and P. tenuis Speg.; tarspots.
    S. America.
Pistillaria tucumanensis (Speg.) Cor.; root and basal rot.
    Argentina.
Puccinia adenocalymatis (P. Henn.) Arth. & J. R. Jonst.;
    leafrust. Trinidad, Subtropical USA. P. appendiculatoides
    P. Henn.; rust. Brazil.
Seynesia paraguayensis Speg.; black moldy leaf disease.
    Paraguay.
Strigula complanata Fée; cream-spot of leaf. W. Indies.
Uredo cuticulosa Ell. & Ev.; leafrust. Nicaragua.
Uropyxis rickiana Magn. and U. reticulata Cumm.; gall rusts.
    S. America.

BIXA ORELLANA L. A small tree native of tropical America
noted for its fruit extract increasingly valuable as an export to
Europe and North America for a natural flavor and color added
to foods (it is known to add vitamins), in addition is employed
as a so-called "fugitive" dye for certain textiles. At this
writing, habitually semi-nude jungle-dwelling indigenes of tropical
America color their exposed skin surfaces with this dye for its
insect repellent character; "anato," "achiote" Sp., "urucuzeiro,"
"annato," "acafrá" Po., "annato," "roucon" En., and many
Indian names.
    Cephaleuros virescens Kunze; red alga-spot. Florida, Costa
        Rica, El Salvador, Puerto Rico, Brazil.
    Cercospora bixae All. & Noack; medium sized circular spots
        on leaves. Guatemala, Brazil, Costa Rica, Venezuela,
        Puerto Rico, El Salvador, Trinidad.
    Colletotrichum gloeosporioides Penz. (asco-state developed on
        old diseased twigs in moist chamber and identified as
        Glomerella cingulata (Stonem.) Sp. & Schr. in Puerto Rico);
        anthracnose and defoliation. W. Indies, C. America, S.
        America.
    Corticium salmonicolor Berk. & Br.; leafrot and pink branch-
        disease. Guatemala.

Meliola arachnoidea Speg. = Irenina arachnoidea (Speg.) Stev. ;
  black mildew on leaves.   C. America, S. America, W.
  Indies.
Oidium bixae Viégas; white mildew.   C. America, S. America.
Penicillium sp. ; fruit mold.   Costa Rica.
Peronospora sp. ; downy mildew.   Guatemala.
Phoma sp. ; black leafspot.   S. America.
Phthirusa spp. ; small leaved parasite bushes.   Costa Rica.
Phyllosticta bixina E. Young; leafspot.   C. America, W.
  Indies.
Rosellinia bunodes (Berk. & Br.) Sacc. ; black rootrot.   Ja-
  maica.
Sphaeropsis sp. ; fruit disease.   Brazil, ? Argentina.
Uredo bixae Arth. ; powdery leafrust.   Brazil, Puerto Rico.
Virus symptoms of distinct mottle and leaf malformation seen
  in Guatemala (jungle clearing with Indian inhabitants).

BLECHNUM OCCIDENTALE L.   An important tropical American
  fern used by florists, taken long ago to temperate zone in
  Europe and North America for pot culture as house plant and
  for greenery in flower bouquets, grown in gardens in the tropics;
  "helecho común, " "helecho" Sp. Po. , "common tropical fern"
  En.
    Cassytha filiformis L. ; dodder.   Puerto Rico.
    Cladosporium herbarum Lk. ; spores from ragged leaf.   W.
      Indies.
    Cuscuta americana L. ; dodder.   Puerto Rico.
    Elsinoë blechni Bitanc. & Jenk. ; spot anthracnose.   Brazil.
    Glomerella cingulata (Stonem.) Sp. & Schr. = Colletotrichum
      sp. ; anthracnose.   Puerto Rico.
    Infection body (an unidentified gemma-producing fungus, not
      Mycena citricolor); caused a leafspot in Costa Rica.
    Milesia australis Arth. ; leafrust.   Venezuela, Puerto Rico,
      Chile, Colombia, Ecuador, Trinidad.
    Mycena citricolor (Berk. & Curt.) Sacc. ; leafspot natural
      occurrence in garden from infection body.   Costa Rica.
    Mycosphaerella filicum (Desm.) Schr. ; leafspot.   Florida.
    (On B. tabulae Mett. , attack by M. tabularis Syd. ; leafspot.
      Chile.)
    (On Blechnum spp. , attack by Uredinopsis mayoriana Diet. ;
      rust.   Colombia.   Uredo blechni Diet. & Neger; rust.
      Chile.   U. blechnicola P. Henn. ; rust.   Brazil.)

BLIGHIA SAPIDA Koen.   A medium-sized, attractive tree that
  came from Guinea, grown in scattered locations in tropical
  America for the delicious fruits that are edible only when fully
  ripe; "seso vegetál, " "huevo vegetál" Sp. , "akee" En. Po.
    Ascochytella cupaniae Gonz. Frag. & Cif. ; leafspot.   W.
      Indies.
    Capnodium sp. ; sootymold.   Honduras.
    Colletotrichum gloeosporioides Penz. ; anthracnose.   W. Indies.
    Cuscuta americana L. ; dodder on seedling.   Honduras.

Fusarium sp.; secondary in senescent dieback. W. Indies.
Galls (undetermined cause), Jamaica.
Ganobotrys blighiae Gonz. Frag. & Cif.; leaf disease. W.
    Indies.
Loranthus sp.; parasite bush. Honduras.
Meliola sp.; black mildew. Costa Rica.
Mold, gray color (unidentified fungus). Fruitrot. Honduras.
Virus? (leafmottle and shortened twig internodes). Guate-
    mala.
Yeast (unidentified), fruitrot. Honduras.

BOEHMERIA NIVEA Gaud. A plant of the nettle family, about
    shoulder high and strong growing that produces a fine fiber that
    is wear resistant and has special weaving qualities; "ramie" Sp.
    Po. En., "grass linen" En.
    Asterina sp.; black moldspot. Puerto Rico, Florida.
    Cercospora krugiana Mull. & Chupp; leafspot. Brazil.
    Diaporthe boehmeriae Speg. = D. arctii (Losch.) Nits.; causes
        leaf and branch infections. S. America.
    Fusarium solani (Mart.) Sacc.; rootrot and stem disease.
        British Honduras, El Salvador.
    Colletotrichum gloeosporioides Penz. conidial state of Glome-
        rella cingulata (Stonem.) Sp. & Schr.; anthracnose disease.
        W. Indies, C. America, S. America.
    Meloidogyne arenaria-thamesi Chitw.; rootknot. Florida.
    Nectria sp.; rot at base. Florida.
    Ophiobolus oedistoma Speg.; root disease. Argentina.
    Phytophthora boehmeriae Saw.; rootrot. C. America.
    Puccinia boehmeriae P. Henn.; yellow leafrust. Brazil.
    Rosellinia bunodes (Berk. & Br.) Sacc.; rootrot and wilt.
        Costa Rica.
    Sclerotium rolfsii (Curzi) Sacc. (not perfect stage); seedling
        collapse. Brazil, El Salvador.
    Seynesia balansae Speg.; moldspot of leaf and stem. S.
        America.
    Struthanthus sp.; mistletoe-like vine. Brazil, Paraguay.
    Virus, moiré mottle and malformed leaf. El Salvador.
    (Note: On B. caudata Sw. is found in Brazil the spot anthrac-
        nose from Elsinöe boehmeriae Bitanc. & Jenk. In Puerto
        Rico under artificial laboratory bench conditions I induced
        infection on B. nivea Gaud. with the two dodders Cuscuta
        americana L. and Cassytha filiformis L.)

BRASSICA spp. Herbs from Eastern Hemisphere gardens, grown
    successfully in many parts of the American tropics where they
    add materially to food and economics. (A number in parentheses
    before a Brassica species, or a B. oleracca variety, designates
    the host then referred to in the disease list which follows. It is
    known in scientific agriculture that diseases of cabbage or a
    relative are infectious to most other brassicas; only rarely is
    such a disease confined to only one species or a species variety.)
    (1.) B. JUNCEA Coss, and B. NIGRA L.; "mustard" En.,

"mostarda" Po., "mostaza" Sp.   (2.) B. NAPOBRASSICAE Mill.;
"rutabaga" En. Po. Sp., "nabo-sueco" Sp.   (3.) B. OLERACEA
L. var. BOTRYTIS L.; "cauliflower" En., "coliflor, Sp.,
"couve-flor" Po.   (4.) B. OLERACEA L. var. CAPITATA L.;
"cabbage" En., "repollo," "berza" Sp., "repolho," "couve" Po.
(5.) B. OLERACEA L. var. GEMMIFERA DC.; "brussels
sprouts" En., "bretón," "botones de repollo" Sp.   (6.) B.
OLERACEA L. var. VIRIDIS L.; (= B. OLERACEA L. var.
ACEPHALA L. of some botanists); "kale" En., "repollo de hojas"
Sp.   (7.) B. PEKINENSIS Rupr.; "chinese cabbage" En., "repollo chino," "colchino" Sp., "repolho china" Po.   (8.) B. RAPA
L. = B. CAMPESTRIS L.; "turnip" En., "nabo" Sp., "yuyo" Po.

Albugo candida (Pers.) Kuntze; white rust, leafblister.   Often
    in cool-land plantings, on all brassicas.   W. Indies, C.
    America, S. America.
Alternaria alternata (Fr.) Keiss. = A. tenuis Nees; blackened
    tissue attack, ragged leafedge, (on 3, 4, 5, 7).   W. Indies,
    C. America, S. America.   A. brassicae (Berk.) Sacc.;
    mild leaf spotting, in coolest locations.   On all cultivated
    brassicas.   W. Indies, C. America, S. America.
Aphelenchoides parietinus Stein.; root nematode.   Brazil (on
    4).
Aphelenchus avenae Bast.; root nematode.   Brazil (on 4).
Botrytis cinerea Pers.; market gray mold-rot.   Brazil (on
    3 & 4).
Cassytha filiformis L.; dodder by inoculation.   Puerto Rico
    (on 4).
Cercospora brassicola P. Henn.; leafspot on all cultivated
    brassicas.   General.
Cercosporella brassicae (Fautr. & Roum.) Hoehn. and C.
    albo-maculans (Ell. & Ev.) Sacc.; white spot.   Mexico
    (on Brassica sp.), Florida (on 2, 3, 4, 8).
Choanephora cucurbitarum (Berk. & Rav.) Thaxt.; seedling
    disease.   Brazil (on 4).
Cladosporium herbarum Lk.; leaf disease.   Chile (on 8).
Colletotrichum gloeosporioides Penz. = C. higginsianum
    Sacc. (no perfect state encountered); anthracnose, on all
    cultivated brassicas.   General.
Erwinia carotovora (L. R. Jones) Holl.; market softrot.   On
    all cultivated brassicas.   General.
Erysiphe polygoni DC.; powdery mildew.   Florida, Texas,
    Chile, C. America (on 3, 4, 6, 8), ? S. America.
Fusarium oxysporum Schl. f. sp. conglutinans (Wr.) Sny. &
    Hans.; wilt, yellows.   Occurs at low elevations in friable
    soils of circum-neutral reaction (on 2, 4, 8) in subtropical
    USA, Costa Rica, Panama, Cuba, El Salvador, Brazil and
    Fusarium spp.; root dryrot (on 4, 6) in warm soils.   Peru,
    Chile, Brazil, Costa Rica, Panama.
Helicotylenchus nannus Stein.; root nematode.   Brazil (on
    3, 4).
Heterodera schachtii Schm.; root nematode.   California (on
    Brassica sp.).

Meloidogyne sp.; rootknot.   Brazil, Surinam, Venezuela,
      Subtropical USA (on 4), Brazil (on 3).
Monhystera stagnalis Bast.; root nematode.   Brazil (on 4).
Mycena citricolor (Berk. & Br.) Sacc.; leafspot from
      laboratory inoculation.   Costa Rica (on 4).
Mycosphaerella brassicola (Duby) Lind.; ring spot, mostly
      in highland plantings.   Mexico, Panama, Colombia, Suri-
      nam, Argentina, Brazil, Uruguay, Jamaica (on 4).
Oidium balsamii Mont.; white powdery mildew.   Colombia,
      Brazil, Peru (on 3, 4), Brazil (on 5).
Olpidium brassicae (Wr.) Dang. = O. brassicae Rang.; in-
      fections of internal root tissues, benign effects.   El
      Salvador, Costa Rica (on 4), S. America.
Peronospora parasitica Pers. = P. parasitica Gaum.; downy
      mildew.   Subtropical USA, C. America, W. Indies, S.
      America (mostly on 4), Brazil (on 3, 4, 6).
Phoma lingam (Tode) Desm.; blackleg, leaf and stem attack,
      in cool areas.   C. America, Puerto Rico, Colombia (on
      4), Chile (on 4, 2).
Phyllosticta brassicae (Tul.) West. = P. brassicicola McAlp.;
      leafspot.   California, Chile, Colombia (on 4), Chile (on 6).
Phymatotrichum omnivorum (Shear) Dug.; rootrot.   Texas (on
      cultivated brassicas).
Phytomonas (see Xanthomonas)
Plasmodiophora brassicae Wor.; clubroot.   Scattered occur-
      rence in moist-cool areas.   S. America.
Pratylenchus pratensis Filip.; root nematode.   C. America,
      scattered in W. Indies, S. America (on 4), Chile (on 4,
      6 and 7).
Pseudomonas maculicola (McCul.) Stev.; bacterial leafspot.
      Subtropical USA, El Salvador (on 4), Brazil (3, 4).
Psilenchus hilarulus deMan; root nematode.   Brazil (on 3).
Pythium spp.; damping-off.   General, all brassicas as tender
      seedlings.
Rhizoctonia solani Kuehn = the basidial state is Thanate-
      phorous cucumeris (Frank) Donk also called Pellicularia
      filamentosa (Pat.) Rog.; common damping-off, rot in
      button or head, bottom rot, wire stem.   Subtropical USA,
      W. Indies, C. America, S. America (most reports on 4,
      but on all cultivated brassicas).
Rhizopus stolonifer (Ehr.) Lind. and R. nigricans Ehr.;
      market blackmold rot.   Chile, C. America (on 4).
Sclerotinia sclerotiorum (Lib.) dBy. = Whetzeliana sclero-
      tiorum (Lib.) Korf & Dumont; market cottony-rot.   Costa
      Rica, Peru, Brazil, Argentina (on 4), Brazil (on 3, 8),
      Peru (on 8), Subtropical USA (on 3, 4, 6, 7).
Sclerotium rolfsii (Curzi) Sacc. (the basidial state is Pelli-
      cularia rolfsii (Sacc.) West); groundrot and wilt.   Sub-
      tropical USA, Brazil, scattered in W. Indies and C.
      America, a few reports in S. America (on 3, 4, 8).
Trichodorus sp.; root nematode.   Florida (on 4).
Tylenchus dauainii Bast.; root nematode.   Brazil (on 4).

Viruses: Argentine sunflower virus in Argentina (on 5).
Common cabbage mosaic in W. Indies, Subtropical USA,
Brazil and it may be general (on 4, more rarely on 3, 6,
8). Cauliflower mosaic in Brazil, Puerto Rico, Subtropi-
cal USA (on 3, 5, 6). Cucumber mosaic virulent strain
from Commelina in Florida (on 4), Brazil (on 6, 8).
Yellow vein necrosis virus in Brazil (on 6).
Xanthomonas campestris (Pam.) Dows. ; systemic blackrot.
General in cool areas (mostly on 4 but occurs on other
cultivated brassicas).
Xiphinema americanum Cobb and X. campinense Lor. ; large
nematodes at roots. Brazil (on 4).

BROMUS spp. A genus of several much used, some planted,
pasture grasses; "bromus" En. Sp. , "cebadilla" Sp. , "cevadilla"
Po.
Aecidium graminellum Speg. ; rust. S. America.
Claviceps purpurea (Fr.) Tul. ; ergot. Chile, and other parts
of S. America, C. America.
Colletotrichum sp. ; grass anthracnose. General, seldom
serious.
Erysiphe graminis DC. ; powdery mildew. Chile, C. America,
Puerto Rico.
Marssonina graminicola (Ell. & Ev.) Sacc. ; leafspot. S.
America.
Oidium monilioides Lk. ; mildew. Widespread and common.
Ophiobolus graminis Sacc. and O. cariceti Sacc. ; footrot.
S. America.
Phyllachora bromi Fckl. ; tarspot. Brazil, Chile, ? General.
Placosphaeria magellanica Speg. ; culm and leaf disease. S.
America.
Puccinia anamola Rostr. , P. brachypus Speg. var. bromiphila
Speg. , P. recondita Rob. , P. bromina Eriks. , P. clema-
tidis (DC.) Lagh. , P. graminis Pers. , P. montanensis
Ell. , and P. rubigo-vera (DC.) Wint. var. agropyrina Eriks.
& Arth. and several other species; all rusts. Tropical
Americas.
Scolecotrichum graminis Fckl. var. brachypoda Speg. ; stripe
and leafspot. Brazil.
Septoria bromivora Speg. ; leaf spotting. Brazil, Argentina,
Chile, ? W. Indies.
Uredo auletica Speg. ; leafsheath rust. S. America.
Uromyces bromicola Arth. & Holw. ; rust. Chile.
Ustilago bromivora Tul. = U. bullata Berk. ; smut in cold
areas. Chile, Uruguay.

BUGINVILLAEA SPECTABILIS Willd. (genus also spelled Bugain-
villea). The spectacular flowering vine of South America now
well distributed in the tropics; "trinitaria" Sp. Po. , "bougain-
villia vine" En.
Botryodiplodia (see Diplodia)
Cassytha filiformis L. ; dodder. Puerto Rico.

Cephaleuros virescens Kunze; algalspot on leaf and sepals of
    flower.   Puerto Rico.
Cercospora bougainvillea Mun.; leaf disease.   El Salvador,
    Cuba.
Cladosporium arthinoides Thu. & Bel.; leaf disease.   Texas.
    C. herbarum Lk.; ragged leafedges.   W. Indies.
Cuscuta americana L.; dodder.   Puerto Rico.
Diplodia theobromae (Pat.) Now. = Botryodiplodia theobromae
    Pat.; dieback.   Puerto Rico.
Glomerella cingulata Sp. & Schr. (conidial state is Colletotri-
    chum); anthracnose.   W. Indies, C. America, S. America,
    Subtropical USA.
Leafspots, large and clear area appearance, cause unknown.
    Puerto Rico.
Puccinia bougainvillea (Speg.) Schr.; yellow to brown leafrust.
    Argentina.
Strigula complanata Feé; lichen spot.   W. Indies.
Tryblidiella rufula (Spreng.) Sacc.; part of dieback complex.
    Puerto Rico.
Virus, distinct leaf mottle and some stunt.   Florida, W.
    Indies, C. America.   Virus, yellow leaf and severe stunt.
    Cuba.

CAESALPINIA PULCHERRIMA Sw.   A popular, warmth-loving,
    ornamental small tree much grown in dooryards, with a pro-
    fusion of poinciana-like blossoms; "flamboyán ennano," "clavel-
    hina" Sp. Po., "pride of Barbados," "peacock flower" En.
Botryodiplodia theobromae Pat. (genus often given as Diplodia);
    dieback.   Neotropics.
Botryosphaeria dothidea (Moug.) Ces. & de Not.; branch
    canker, dieback.   Florida, Costa Rica, El Salvador, Puerto
    Rico.
Cephaleuros sp.; small appressed algal spots.   Puerto Rico.
Cercospora guanicensis Young; in W. Indies and C. America.
    C. poincianae Chupp & Mul.; in Venezuela, both leafspots.
Colletotrichum (? gloeosporioides); dieback, leaf disease.
    Puerto Rico, Florida, Costa Rica, probably general.
Corticium salmonicolor Berk. & Br.; branch disease.   Guate-
    mala.
Lenzites striata Fr.; woodrot.   C. America but widespread.
Mycena citricolor (Berk. & Curt.) Sacc.; leafspot.   C.
    America.
Pellicularia koleroga Cke.; thread blight.   Costa Rica.
Pestalotia funerea Desm.; leaf and twig lesion.   Puerto Rico.
Polyporus spp.; stumprots.   W. Indies.
Puiggarina formosa Speg.; leaf disease.   S. America.
Pythium (? ultimum); seedling decay.   W. Indies.
Ravenelia humphreyana P. Henn.; mild leafrust.   Widespread.
    R. inconspicua Arth.; severe leafrust.   Mexico.
Schizophyllum commune Fr.; senescent wood decay.   W.
    Indies, Florida, C. America.
Witches' broom from unknown primary cause, probably insect,
    invaded by numerous bacteria and fungi.   Costa Rica.

CAJANUS CAJAN (L.) Millsp. = C. INDICUS Spreng. From the
Old World tropics, it has become well adapted to the American
tropics, a highly important edible-seed-bearing shrub that is
relished and keeps multitudes from starving; "gandúl," "guin-
choncho" Sp., "gandu," "ervilha pombo" Po., "pigeon pea,"
"congo bean," "jungle bean" En.
  Acanthonitschkea argentinensis Speg.; on fallen decaying
    branches. Argentina.
  Botryosphaeria xanthocephala (Syd. & Butl.) Th. & Syd.;
    stemrot. W. Indies.
  Calonectria rigidiuscula Berk. & Br.; infectious small stem-
    galls (causes "green point gall" in Theobroma). British
    Guiana, Puerto Rico, El Salvador.
  Cassytha filiformis L.; dodder. Puerto Rico.
  Cephaleuros virescens Kunze; algal leafspot. W. Indies, C.
    America.
  Ceratocystis fimbriata Ell. & Halst. f. cajani (?) (I found
    my isolations did not infect Ipomoea); black-stem and
    branch death. El Salvador, Venezuela.
  Cercospora cajani P. Henn. = Mycovellosiella cajan Rang.
    (the ascigerous stage); foliage spot and rot. General and
    C. instabilis Rang.; small leafspot. Puerto Rico.
  Chaetoseptoria wellmanii J. A. Stev.; large round grayleaf-
    spot. C. America.
  Colletotrichum cajani Rang. = C. gloeosporioides Penz.,
    see Glomerella; anthracnose. Neotropics. C. lindemu-
    thianum (Sacc. & Magn.) Scrib.; small lesion anthracnose.
    Puerto Rico, Florida, Panama.
  Corticium salmonicolor Berk. & Br.; pink disease. W.
    Indies.
  Creonectria gramnicospora (Ferd. & Winge) Seav.; on dead
    stems. Puerto Rico, W. Indies.
  Cuscuta americana L.; yellow dodder. Puerto Rico.
  Diplodia theobromae (Pat.) Now. (synonyms: D. vulgaris Lév.,
    D. cacaoicola P. Henn., Botryodiplodia theobromae Pat.)
    with Physalospora rhodina Cke. = Botryosphaeria quercuum
    (Schw.) Sacc. as the ascigerous stage; black stem, die-
    back. Florida, Cuba, Puerto Rico, Costa Rica, probably
    Neotropical.
  Erysiphe cichoracearum DC.; mildew of pods, foliage. W.
    Indies.
  Fusarium spp.; pod rot, stem lesion, branch rot. C.
    America, W. Indies, S. America.
  Glomerella cingulata (Stonem.) Sp. & Schr., ascigerous
    fruiting bodies often recovered in moist chamber studies
    of Colletotrichum cajani-diseased plants. W. Indies, C.
    America.
  Helicobasidium purpureum Pat. = Rhizoctonia crocorum DC.;
    purple and gray rootrot. W. Indies.
  Leafcurl, caused by toxin from leafhoppers feeding on leaves.
    W. Indies, C. America, S. America.
  Macrophomina phaseoli (Maubl.) Ashby; black rootrot. W.
    Indies.

Meloidogyne spp.; rootknot. Brazil.
Phoma cajani Rang.; canker, podspot. W. Indies, C.
    America.
Phyllosticta cajani Rang.; leaf and pod spots. W. Indies,
    C. America, S. America.
Phymatotrichum omnivorum (Shear) Dug.; dryland rootrot.
    Texas.
Physalospora (see Diplodia)
Pleonectria megalospora Speg. = Thyronectria megalospora
    (Speg.) Seaver & Chard.; stem disease. S. America, C.
    America.
Rhizoctonia ferruginea Matz (not R. solani or R. micro-
    sclerotia in my studies); a distinctly red seedling disease.
    Puerto Rico. R. microsclerotia Matz; tangled-web.
    Costa Rica, W. Indies. R. solani Kuehn (no basidial
    state found); damping-off. Neotropical.
Rosellinia bunodes (Berk. & Br.) Sacc.; black-root. C.
    America. R. pepo Pat.; white root. C. America.
Sclerotium rolfsii (Curzi) Sacc.; soil-line rot. C. America,
    W. Indies.
Septobasidium pseudopedicellatum Burt.; root-sheath. C.
    America (rarely found).
Thyronectria (see Pleonectria)
Uredo cajanus Syd.; rust. S. America.
Uromyces dolicholi Arth.; common rust. General.
Viruses: Mycoplasma-like organism, witches' brooming.
    Dominican Republic, El Salvador, Puerto Rico. Cowpea
    mosaic. W. Indies, ? genera. Mosaic tangletop. Nica-
    ragua, Costa Rica. Severe stunt. El Salvador, St.
    Thomas. Leaf curl. Puerto Rico. Yellow foliage, no
    stunting. St. Croix.
Xiphinema americanum Cobb and X. campinense Lorde.; both
    root nematodes, ecto parasites. Brazil.

CALADIUM BICOLOR Vent. From the American tropics, grown
    for ornamental leaves; "caladio" Sp., "colorado" Po., "elephant
    ear" En.
        Cercospora caladii Cke. synonyms: C. caladicola (P. Henn.)
            Chupp, and C. verruculosa Stev. & Solh.; a common leaf-
            spot. W. Indies, S. America.
        Cladosporium sp.; on old ragged leaf. W. Indies, Florida,
            ? General.
        Colletotrichum spp.; anthracnoses. W. Indies, C. America,
            S. America.
        Erwinia carotovora (L. R. Jones) Holl.; tuber softrot. W.
            Indies, Florida.
        Helminthosporium caladii Stev.; large leafspot, collapse. W.
            Indies.
        Macrophoma surinamensis (Berk. & Curt.) Ber. & Vogl.; on
            leaves and petioles. Surinam.
        Meloidogyne arenaria Chitw., M. incognita Chitw., and M.
            spp.; rootknot. Scattered in W. Indies, C. America, S.
            America.

Mycena citricolor (Berk. & Curt.) Sacc.; leafspot by inocula-
tion, and in garden by diseased coffee.   Costa Rica.
Phyllosticta colocasia Hoehn.; spot and streak.   W. Indies.
Rhizoctonia sp.; basal rot on leaves.   Puerto Rico.
Sclerotium rolfsii (Curzi) Sacc.; basal rot with sclerotia.
Puerto Rico, Florida.
Virus; Mild mosaic, unidentified.   Florida, Puerto Rico.

CALATHEA ALLUIA Lindl.   A perennial shade-tolerating herb,
native to the West Indies, grown for the crisp-tasting under-
ground tubers; "leren" Sp. Po., "water chestnut of the West
Indies," "topinambour" En., "lleer," "rlereens" Ind.
(Cephaleuros virescens not an infection by itself, although
abundantly present on adjacent hosts and within easy splash
distance.   But there is development of Strigula, see below.)
Puerto Rico.
Colletotrichum gloeosporioides Penz., with asco-state =
Glomerella cingulata (Stonem.) Sp. & Schr.; mostly leaf-
spots and petiole lesions.   Puerto Rico.
Epiphytic microorganisms on old plants (apparently non-para-
sitic but flourishing: surface inhabiting bacteria, algae,
small lichens, leafy liverworts, sterile fungus mycelium).
Puerto Rico.
Fusarium sp.; secondary disease after insect injury.   Puerto
Rico.
Mycena citricolor (Berk. & Curt.) Sacc.; leafspot.   Natural
infection.   Puerto Rico.   (Years before it was reported on
another species of Calathea in Costa Rica.)
Nectria sp.; stem infection.   Costa Rica.
Oidium sp.; white mildew.   Puerto Rico.
Pestalotia sp.; stem and leaf lesions.   Puerto Rico.
Phyllosticta (? marantaceae); leafspot.   W. Indies.
Puccinia cannae P. Henn.; rust.   W. Indies, C. America, S.
America.
Rhizoctonia sp. (not solani or microsclerotia); superficial
mycelium on tubers, causes small lesions on petioles.
Puerto Rico.
Rosellinia bunodes (Berk. & Br.) Sacc.; blackroot and plant
collapse.   W. Indies.
Strigula complanata Fée; lichen spotting. Puerto Rico, Cuba.

CALENDULA OFFICINALIS L.   A popular herbaceous plant grown
for flowers of much value to homeowners and florists; "flamen-
quilla," "calendúla" Sp., "mal-me-quer," "kalendula" Po., "pot
marigold," "calendula" En.
Cercospora calendulae Sacc.; leafspot.   Brazil, ? Peru, C.
America.
Coleosporium calendulae Speg. = C. senecionis Pers.; rust.
C. America, S. America.
Cuscuta americana L.; dodder.   Puerto Rico.
Entyloma calendulae (Oud.) dBy.; smut.   California, Chile,
Argentina.

Meloidogyne sp. ; rootknot.   Florida, W. Indies.
Puccinia flaveriae Jacks. ; rust.   Texas.
Sclerotium rolfsii (Curzi) Sacc. ; basal rot.   Texas.

CALLA (cultivated)   see   ZANTEDESCHIA

CALLISTEPHUS CHINENSIS Nees.   A free-flowering, handsome
  Oriental plant especially selected for the florist, that grows
  and produces well in relatively dry cool tropics, important in
  local markets and for air shipments; "aster" Po. , "astro-chína"
  Sp. , "aster, " "china aster" En.
    Alternaria sp. (large spored); leaf disease.   Florida, Guate-
      mala, El Salvador.   A. alternata (Fr.) Keiss = A. tenuis
      Nees (small spored) causing blackening of dead leaf edges
      as well as spots on flowers.   Florida, Mexico and Guate-
      mala markets, Panama.
    Basidiophora entospora Roze & Cornu; downy mildew.   Sub-
      tropical USA, C. America.
    Cassytha filiformis L. ; pungent smelling dodder.   Puerto Rico.
    Colletotrichum gloeosporioides Penz. (perfect state is Glome-
      rella cingulata (Stonem.) Sp. & Schr.); anthracnose.
      Florida, Puerto Rico, probably well spread.
    Cuscuta americana L. ; common dodder.   Puerto Rico.
    ? Erysiphe sp. ; powdery mildew.   Puerto Rico.
    Fusarium oxysporum Schl. f. sp. callistephi (Beach) Sny. &
      Hans. ; wilt.   Subtropical USA, Chile, Uruguay, Mexico,
      Puerto Rico.
    Meloidogyne sp. ; mild rootknot.   Florida, Texas.
    Phymatotrichum omnivorum (Shear) Dug. ; dryland rootrot.
      Texas.
    Phytophthora parasitica Dast. ; stem canker.   Argentina,
      Puerto Rico.
    Rhizoctonia solani Kuehn = Thanatephorous cucumeris (Frank)
      Donk; damping-off, basal rot.   General (scattered).
    Septoria callistephi Gloy. ; leafspotting.   C. America.
    Stem disease, complex of nematodes, Pythiums, Phytophthoras,
      Alternarias, algae, bacteria.   Puerto Rico, Costa Rica.
    Verticillium albo-atrum Reinke & Berth. ; wilt.   California,
      Mexico, C. America.
    Viruses: Streak in Brazil, Spotted wilt in C. America, Yel-
      low-leafage, unidentified in El Salvador.

CAMELLIA JAPONICA L.   A handsome flowering bush originally
  from the Orient successful in subtropics and cool mountain locations in the American tropics; "camélia" Sp. Po. , "camellia"
  En.
    Armillaria mellea Vahl = Armilariella mellea (Fr.) Karst. ;
      white rootrot.   California, a few reports in C. America.
    Botrytis cinerea Pers. ; rainyseason budblight.   W. Indies.
    Capnodium sp. ; sootymold.   Puerto Rico.
    Cephaleuros virescens Kunze; common algal leafspot.   In
      "frostfree" parts of subtropical USA, and in C. America,
      W. Indies, S. America.

Colleotrichum gloeosporioides Penz. (under moist-cool con-
  ditions ascosporic stage Glomerella cingulata (Stonem.)
  Sp. & Schr.); branch disease, anthracnose. Neotropics.
Cladosporium herbarum Lk. = Pleospora herbarum (Pers.)
  Rab. f. camelliae Speg.; leaf disease. S. America.
Clitocybe tabescens Bres. = Armilariella tabescens (Scop.)
  Sing.; rootrot very much like Armillaria disease. Sub-
  tropical USA.
Cuscuta americana L.; dodder. Puerto Rico.
Diplodia (see Physalospora)
Elsinoë leucospila Bitanc. & Jenk. = Sphaceloma sp.; spot
  anthracnose and scab. Brazil, Florida, and probably
  occurs in between these two areas.
Exobasidium camelliae Shir.; stem and leafgall (of note as the
  close fungus relative E. vexans Mass. is a long time
  limiting factor in tea production in the Eastern Hemisphere
  and as tea (the close relative of Camellia) is now expanding
  in plantation culture in the Occident our Western Exobasidium may be capable of attacking both hosts). Subtropical
  USA.
Fusarium sp. (= Pseudomicrocera henningsii (Koord.) Petch);
  a severe stem disease. S. America.
Marasmius sp.; cord blight. Preserved sample without sporocarp from Paramaribo was sent for determination to my
  laboratory in Costa Rica.
Nematodes, common names personally reported to me: root-
  knot (? Meloidogyne sp.), rootlesion (? Pratylenchus),
  Puerto Rico, Subtropical USA.
Pestalotia guepini Desm.; leafspot, blight. Neotropics. P.
  discolor Speg.; on fading leaves. Uruguay. P. puttemansii
  P. Henn. and P. theae Saw. also reported in S. America,
  and a Pestalotia, species unidientified, is a hyperparasite
  of Exobasidium.
Phyllosticta camelliae West.; weakly parasitic on leaves.
  Chile, Subtropical USA.
Physalospora rhodina Cke. = Botryosphaeria quercuum (Schw.)
  Sacc. (ascigerous stage of Diplodia spp.); dieback, stem
  lesions. General.
Pleospora (see Cladosporium)
Rhizoctonia solani Kuehn as non sporulating mycelium; damping-
  off. General in warm, friable soils, moist areas.
Sclerotinia camelliae Hara (introduced from Orient); flower
  blight. Subtropical USA. S. sclerotiorum (Lib.) dBy. =
  Whetzeliana sclerotiorum (Lib.) Korf & Dumont; flower
  blight. Chile, Subtropical USA.
Septobasidium sp.; felt fungus. C. America.
Sphaceloma (see Elsinöe)
Strigula complanata Fée; lichen white-spot. General, if
  Cephaleuros is present.
Virus, unidentified, an infectious variegation. Subtropical
  USA.

CAMELLIA SINENSIS see THEA

CANAVALIA ENSIFORMIS (L.) DC. A native of the West Indies,
   it is a coarse and strong growing type of bean plant with edible
   seeds but mostly used as forage and serves well in soil conser-
   vation practices where terrain is difficult; "pallár" Sp., "feijão
   de porco" Po., "jackbean" En., "orrá" Ind.
   Ascochyta pisi Lib.; leafspot. Brazil.
   Cephaleuros virescens Kunze; algal leafspot. El Salvador,
      Brazil.
   Cercospora canavaliae Syd. & P. Syd.; leafspot. Trinidad
      and Tobago, Puerto Rico, Venezuela, Brazil. C. cruenta
      Sacc.; leafspot. Trinidad and Tobago.
   Cerotelium canavaliae Arth.; brown rust on leaves. Puerto
      Rico.
   Chaetoseptoria wellmanii J. A. Stev.; leafspot. Costa Rica,
      Mexico.
   Colletotrichum gloeosporioides Penz. (I believe C. canavaliae
      Gonz. Frag. & Cif. is probably a form of C. gloeos-
      porioides more adapted to its host in the West Indies), its
      perfect state is Glomerella cingulata (Stonem.) Sp. & Schr.
      with synonyms G. canavaliae Petr., G. canavaliae Lyon.
      West Indies, C. America.
   Meliola sp.; moldy leafspot. Mexico, El Salvador.
   Mycena citricolor (Berk. & Curt.) Sacc.; leafspot by inocula-
      tion. Costa Rica.
   Rhizoctonia microsclerotia Matz; web blight. Puerto Rico,
      Cuba, Dominican Republic.
   Rosellinia pepo Pat.; white rootrot. W. Indies.
   Sclerotium rolfsii (Curzi) Sacc.; collar rot. C. America.
   Scolecotrichum sp.; rust. Colombia.
   Vermicularia (? Colletotrichum) capsici Syd.; pod spotting.
      I believe this is actually C. gloeospoioides. W. Indies.
   Viruses: Brizilian tobacco streak in Brazil. Canavalia
      mosaic remains unidentified, is severe limiting factor
      in Nicaragua, seen in El Salvador. Euphorbia mosaic
      virus in Brazil. Abutilon mosaic virus in Brazil. Cow-
      pea mosaic in W. Indies and C. America, probably in
      S. America. Leaf curl in Puerto Rico.
   Wilt, severe in C. America, unidentified cause. In El Salva-
      dor species of Fusarium, Pestalotia, Colletotrichum, and
      Phytophthora were all recovered from wilted plants.

CANNA EDULIS Ker-Gawl. An important warm-country, often
   wild-growing food tuber; and C. GENERALIS Bailey with selected
   varieties and hybrids used for ornament; in both much disease
   tolerance; "adeira," "moracha," "chisgua" Sp., "caete," "banan-
   heirinha," "achiera" Po., "canna," "arrowroot," "tous-les-moi"
   En. (Both cannas seem susceptible to same diseases.)
   Alternaria alternata (Fr.) Keiss. = A. tenuis Nees; rarely
      found on senescent leaf edge and tip blackening. Scattered.
   Anthostomella achira Speg.; leafspots often with rifted epi-
      dermis. Widespread.
   Ascochyta cannae Rang.; small leafspot. Brazil.

Cladosporium herbarum Lk.; usually on old injured leaves.
General.

Cordella argentina Speg.; on foliage.   Argentina.

Eudarluca australis Speg. = Myrmaecium cannae Dearn. &
Barth.; the hyperparasite on rust.   General.

Fusarium (? oxysporum); wilt.   Peru.

Haplographium portoricense Stev. & Dal.; foliage disease.
W. Indies.

Lesionectria cannae Speg. = Nectria cannae Sacc. & Trott.;
stem and root disease.   Scattered W. Indies, C. America,
S. America.

Myrmaecium (see Eudarluca)

Ophiobolus linosporioides Speg.; leafspot.   S. America.

Periconia pycnospora Fres.; leaf tissue collapse disease.
Occasional.   W. Indies.

Physalospora sp.; stem canker.   El Salvador.

Puccinia cannae P. Henn. = P. thaliae Diet. = Uromyces
cannae Wint.; the common rust.   Neotropical.

Rhizoctonia sp.; stemrot.   S. America, Puerto Rico.

Rosellinia sp.; rootrot, wilt.   C. America, W. Indies, S.
America.

Sclerotium rolfsii (Curzi) Sacc.; collar rot.   Texas.

Scolecotrichum sp.; rust.   S. America.

Sphaerella cannae Speg.; leaf disease.   Argentina.

Trichothecium roseum Lk.; inflorescence and leaf disease.
C. America, S. America.

Viruses, at least the following: Common canna mosaic, not
fully studied.   W. Indies, C. America, S. America.
Virulent strain of cucumber mosaic.   Florida, W. Indies.
Mosaic unstudied, symptoms of yellow streaking and necrotic spots.   Cuba.

Yellowing and stunt with browned vasculars.   Non-parasitic
bacteria recovered in isolation studies.   Cuba.

(Note: Genera of fungi on Canna edulis in disease collections, made during 15 years, from Cuba, Puerto Rico,
Costa Rica, El Salvador, and Panama, but causing
only slight blemishes, were: Pestalotia, Trichoderma,
Aspergillus, Gliocladium, Fusarium-Nectria, Curvularia,
Chaetomium, Nigrospora, Phoma-like, Valsa, non-patho-
logical Mycelia sterilis (some with, some without, clamp
connections), Verticillium, Pyricularia, Cylindrocladium,
Oidium, rarely a few large spored Alternaria, and five
types of algae, abundant benignly acting bacteria, and in
two cases canna roots seemed normal growing intermingled
with coffee roots severely diseased with Meloidogyne nematode. All microscopic examinations were made with interest
and care.   When is it proper to include or exclude some
such determinations from a disease list?)

CAPSICUM FRUTESCENS L. includes C. ANNUUM L., and C.
BACCATUM L.   The variable large and small, red and sweet,
peppers which are perennial woody plants in the tropics.

Varietal names are often given subspecific rank such as cerasiforme, conoides, fasciculatum, grossum, longum, and typicum and cultigens selected for disease resistance are beginning to find a place in Capsicum growing; "garden pepper," "sweet pepper," "cayenne pepper," "tabasco pepper" En., "pimiento," "pimentón," "ají," "ají-pimentón" Sp., "pimentão," "ají pimentão" Po. (It appears that the three species named are susceptible to all the diseases listed below.)

Acrostalagmus sp., fungus parasite on sap sucking insects. Scattered.

Aecidium capsici Kern & Whet.; rust. Colombia.

Alternaria alternata (Fr.) Keiss. = A. tenuis Nees; leaf-edge disease, black rot of fruit, seed mold. Rare but occasional. A. solani Sor.; leaf spot, rare as fruitrot. W. Indies, C. America, S. America.

Aphanomyces sp.; stemrot. Ecuador.

Ascochyta hortorum (Speg.) Smith; leaf infections. Argentina. A. phaseolorum Sacc.; leafspot. Brazil.

Asterina coriacella Speg. (= A. azarae Lév.), A. peraffinis Speg., and A. solanicola Berk. & Curt.; black mildews, mildew black spots on all peppers. Scattered in Neotropics.

Blossom-end rot, physiological cause but weak parasites infect tissues. Scattered.

Botrytis cinerea Pers.; stemrot, fruitrot in markets. From highland-grown plots. Guatemala, El Salvador.

Cassytha filiformis L.; the pungent-smelling dodder. Puerto Rico.

Cercospora capsici Heald & Wolf; a common leafspot, wet-season defoliation, fruit stemend rot, stalk canker. Tropics in general. (Hyperparasitized by Rhinotrichum griseo-roseum March. in S. America.) C. diffusa Ell. & Ev.; a more diffuse leafspot. El Salvador. C. melongena Wel.; leafspot. Costa Rica. C. nigrum Ell. & Halst.; fruit blackrot, foliage disease. Costa Rica. C. rigospora Atk.; black spot. Venezuela, Brazil (? in all S. America). C. unamunoi Cast. (= ? C. diffusa); velvety leafspot. Brazil, Argentina, El Salvador, Venezuela, Florida, California, Texas.

Choanephora cucurbitarum Thaxt.; flower disease. Florida, El Salvador.

Cladosporium herbarum Lk. = Pleospora herbarum Rab.; leaf-edge rot, fruit gray-mold. California, Texas, Puerto Rico, C. America, Panama.

Colletotrichum gloeosporioides Penz. = its asco-state is Glomerella cingulata (Stonem.) Sp. & Schr.; the anthracnose of stem, leaf, fruit. Subtropical USA, in tropical moist areas generally. (Reports of C. atramentarium, C. nigrum, C. piperatum, C. phomoides may be considered synonyms of C. gloeosporioides.)

Curvularia lunata (Wakk.) Boed.; fruit dryrot often on small fruited types of host. Mexico, C. America.

Cuscuta americana L.; most common tropical dodder. Puerto

Rico. Also at least six other species of Cuscuta in W.
Indies, C. America, S. America.
Diaporthe (? phaseolorum); fruitrot. Mexico, C. America.
Dimerina mindoensis Syd.; moldyspots. S. America.
Diplodia sp.; associated with dieback. C. America.
Erwinia carotovora (L. R. Jones) Holl.; market softrot.
General.
Erysiphe sp.; mildew. Mexico, Pacific side of C. America.
Fusarium oxysporum Schl. f. sp. vasinfectum (Atk.) Syn. &
Hans. (synonyms: F. annuum, F. scirpi, F. semitec-
tum, F. vasinfectum); wilt. Scattered reports in Texas,
Venezuela, Puerto Rico, Panama, Guatemala, Costa Rica,
Mexico, Chile, Ecuador, Brazil, Jamaica, Peru. (It
must be noted here that many times wilt from Fusarium
may be confused with Phytophthora wilt.) F. solani
(Mart.) Sacc.; collar rot, stem disease. Puerto Rico,
Uruguay, Brazil, El Salvador.
Haplocystis vexans Speg.; attack of fruit, and root infection.
Argentina.
Helminthosporium sp.; petiole and leaf spot. Ecuador,
Surinam.
Leveillula taurica Arn.; an oidium. Peru, Mexico.
Macrophomina phaseoli (Taub.) Ashby; charcoal disease of
stem and root. Subtropical USA, W. Indies.
Meliola capsicola Stev. and other Meliola spp.; black mil-
dews. W. Indies, C. America, S. America.
Meloidogyne sp.; rootknot. Argentina, Costa Rica, Subtropical
USA, Jamaica, Panama, Brazil, Peru, Guatemala.
Nematodes other than Meloidogyne in Florida and Chile.
Nematospora coryli Pegl.; pod disease. Florida, W. Indies.
Mycena citricolor (Berk. & Curt.) Sacc.; rare leafspot.
Costa Rica, Puerto Rico.
Penicillium spp.; fruit molds following sunscald. El Salvador,
Puerto Rico, Panama.
Peronospora tabacina Adam; seedbed downymildew. Sub-
tropical USA, El Salvador.
Phacidium macrocarpum Pat., severe leafspot. Scattered.
Phoma destructiva Plowr.; fruitrot and leaf disease. Sub-
tropical USA, C. America, Mexico, Brazil, Ecuador.
Phomopsis capsici (Mgngh.) Sacc.; fruitrot. Puerto Rico.
Phyllosticta capsici Speg. = ? Ascochyta hortorum (Speg.)
Smith; causes gray leafspot. Argentina, Brazil. P.
carica-papaya Alles.; leafspot. S. America.
Phymatotrichum omnivorum (Shear) Dug.; dryland rootrot.
Arizona, Texas, ? Mexico.
Phytophthora (? cactorum); fruitrot. El Salvador. P. capsici
Leon.; a mildew blight, wilt, fruitrot. Neotropics. P.
citrophthora (Sm. & Sm.) Leon.; wilt. Subtropical USA,
Peru, Argentina, probably much more widespread. P.
infestans (Mont.) dBy.; fruit infection occasional, leaf
collapse. Mexico. P. palmivora Butl.; in wilt complex.
C. America, P. terrestris Sherb.; from rootrot. C.

America. (Phytophthoras sometimes are not easily isolated and wilt from Phytophthora can be readily confused with Fusarium wilt. Furthermore in some instances isolations from "Phytophthora wilt disease" were not clearly either Phytophthora or Pythium and my concept of the morphology put some isolates in between these two genera.)

Pratylenchus sp.; root nematode. Chile.

Pseudomonas solanacearum E. F. Sm.; bacterial brown wilt. Scattered, many localities in C. America and S. America.

Puccinia capsici May. and P. capsici Av.-Sac.; leafrust. Colombia, Brazil. P. capsicola Kern & Thur.; rust. Colombia. P. gonzalezi May.; rust. Colombia. P. paulensis Rang.; leaf and twig malforming rust. Brazil, Mexico, Peru, Guatemala.

Pythium aphanidermatum Fitz.; fruitrot. Subtropical USA, Puerto Rico. P. debaryanum Hesse; damping-off and fruitrot. Subtropical USA, Ecuador. P. ultimum Trow; fruitrot and damping-off. Puerto Rico, Louisiana, Florida, California.

Ramularia sp.; leaf disease. Bolivia.

Rhizoctonia solani Kuehn = Thanatephorous cucumeris (Frank) Donk; damping-off, soil rot of stem and fruit. General in lowlands.

Rhizopus stolonifer (Ehr.) Lind; fruit blackmold. In markets of Florida, C. America.

Rosellinia bunodes Berk. & Br.; rootrot. C. America.

Sclerotinia sclerotiorum (Lib.) dBy. = Whetzeliana sclerotiorum (Lib.) Korf & Dumont; cottony-rot of stem and fruit. Chile, Brazil, Ecuador, Florida, Uruguay.

Sclerotium rolfsii (Curzi) Sacc. = Pellicularia rolfsii (Sacc.) West (spore stage); soil neckrot. W. Indies, Florida, Venezuela, Surinam, Peru, Puerto Rico, Panama.

Stemphylium botryosum Wallr. = ? S. solani Weber; gray leafspot. Florida, W. Indies, El Salvador.

Viruses: Pepper streak in Brazil. Tobacco ringspot in Florida, El Salvador, Mexico. Spotted wilt in Brazil. Cucumber mosaic virulent mottle strain in Subtropical USA, W. Indies, C. America, S. America. Etch virus in Venezuela. Common tobacco mosaic in all the Neotropics. Vein-banding in W. Indies, Brazil. Curly-top in Brazil and W. Indies. Tobacco rattle virus in Brazil. Typical common pepper mosaic virus in Puerto Rico.

Xanthomonas vesicatoria (Doidge) Dows.; bacterial fruit and foliage spot. Mexico, C. America, Puerto Rico, Brazil, probably other countries.

CARICA PAPAYA L. An important garden or huerta fruit tree, commercially exploited for market and shipment from such areas as in countries like Puerto Rico, St. John Island, Cuba, Florida, Peru, Brazil; "papaya" or "papayo," "lechoza" Sp., "mamão" or "mamoeiro" Po., "papaya," "pawpaw" En.

Alternaria alternata (Fr.) Keiss. = A. tenuis Nees has small

spores; flower blight and blackened edges on older leaves. Ecuador, Costa Rica, Puerto Rico. Alternaria sp. large spored; peduncle rot cause of premature fruit drop. Brazil, Puerto Rico.

Ascochyta caricae Pat.; small spot of leaf. Mexico, Brazil, El Salvador, Guatemala, Puerto Rico (perhaps more widespread).

Asterina caricarum Rehm; black leafmildew, linear petiole spots. Florida, Puerto Rico, Brazil, Venezuela, Trinidad.

Asterinella papayae Gonz. Frag. & Cif.; on leaves. Dominican Republic.

Botryodiplodia theobromae Pat.; its perfect state is Botryosphaeria quercuum (Schw.) Sacc. (many names for both conidial and perfect state, but these just given are well accepted in our tropics, although "Diplodia" is often used as a name); fruitrot, trunk necrosis, leaf and petiole infections. Scattered in Neotropics, some special reports of severe effects from Cuba, Ecuador, Peru, Costa Rica, Brazil.

Botrytis cinerea Pers.; wet season budblast. Puerto Rico, S. America.

Ceratocystis paradoxa Mor. (on papaya it probably deserves the subspecific name f. caricae); peduncle disease, fruit fall. Dominican Republic, Mexico, Ecuador.

Cercospora caricae = Fusicladium caricae (Speg.) Sacc. = Asperisporium caricae (Speg.) Maubl.; leafspot. Special reports from Paraguay, Brazil, and Panama. C. manaoensis Viég. & Chupp; leafspot. Brazil.

Cerotelium fici Arth.; rust. Scattered in S. America.

Colletotrichum gleosporioides Penz. = Glomerella cingulata (Stonem.) Sp. & Schr.; severe anthracnoses, sometimes mild types that are systemic without symptoms. Subtropical USA, W. Indies, C. America, S. America.

Corynespora cassiicola Wei.; stem lesion, bottle trunk, decline, leafspotting. W. Indies, C. America. (Spores easily confused as possibly Cercospora.)

Dimeriosporium guarapiense Speg.; black moldy spot. S. America.

Diplodia (see Botryodiplodia)

Fusarium (? oxysporum); wilt. Peru, Chile, Colombia, C. America, Puerto Rico. F. solani (Mart.) Sacc.; dryrot, stem disease. Brazil, Colombia, Argentina.

Leptosphaeria sp.; aureolate leafspot. Brazil.

Meliola spp.; leaf moldspots. W. Indies, Mexico, C. America.

Meloidogyne sp.; rootknot. Texas, Surinam, Chile, Florida.

Melophia superba Speg. = Linochora superba (Speg.) Hoehn.; leaf-scorch. Paraguay, Brazil.

(Nematodes, various kinds in Chile, Ecuador, Guatemala.)

Oidium caricae Noack; mildew. General seasonally in many countries, on new growths.

(Penicillium spp. along with Aspergillus, Chaetomium and

many other weak parasites on senescent leaf and petiole, and causing ripe rots. General.)

Phaeseptoria papayae Speg.; leafspot. Brazil.

Phomopsis papayae Gonz. Frag. & Cif.; fruitrot. Dominican Republic, Trinidad.

Phyllosticta caricae-papayae Alle.; shothole, leafspot. Brazil, Cuba, Puerto Rico, Ecuador.

Phytophthora cinnamomi Rands; rootrot. W. Indies, Peru. P. palmivora Butl.; fruitrot, root attack. W. Indies, C. America, Ecuador, Panama.

Pseudomonas carica-papayae Robbs; stem rot. Brazil.

Pythium aphanidermatum Fitz., P. debaryanum Hesse, P. intermedium dBy., P. irregulare Buis., and P. ultimum Trow.; all reported as attacking seedlings. In moist tropics of Americas.

Rhinotrichum gossypium Speg.; increases size of Cercospora spots. Paraguay.

Rhizoctonia papayae Cif.; rainyseason fruitrot. Dominican Republic. R. (solani-types); soil rot, damping-off. Neotropical.

Saccharomyces spp.; riperots. General.

Septoria caricae Speg.; leafspots that tend to be small. S. America, C. America.

Sphaerostilbe repens Berk. & Br.; rootstain. W. Indies.

Viruses: Cuban mosaic in Cuba, Puerto Rico. Cucumber mosaic, common type in Florida, Venezuela, Puerto Rico, El Salvador. Papaya mosaic (mottle, malformation, stunt) in Cuba, Haiti, Trinidad, Dominican Republic, Puerto Rico, El Salvador, Jamaica, Nicaragua, Colombia, Brazil. Distortion ringspot in Venezuela, Dominican Republic, Puerto Rico. Bunchytop with Mycoplasma-like bodies in tissues with leafhopper as vector in Puerto Rico, Florida, Virgin Islands, Venezuela, Colombia.

CARTHAMUS TINCTORIUS L. and C. LUNATUS L. (another old generic name: Kentrophyllum). There are plantings in countries where the crop is still being tested experimentally, but it produces a good dye and a fine edible oil; "cartamo" Sp. Po., "carthamus," "safflower" En.

Alternaria sp.; concentric-zoned leafspot. Mexico, Guatemala, Puerto Rico, Brazil.

Bacterium, unidentified; leafspot. Brazil.

Cercospora sp.; leaf injury. Argentina, Brazil, Costa Rica.

Cladosporium sp.; spores from old leaf. Puerto Rico. Reported on Carthamus lanatus from Chile.

Diaporthe kentrophylli Speg. = D. arctii (Lasch) Nits.; at base of stems on C. lanatus in S. America.

Fusarium (? oxysporum); wilt. Mexico, Ecuador.

Meloidogyne sp.; rootknot. Costa Rica, C. America.

Phoma kentrophylli Speg.; black-leaf and stem injury. On C. lanatus in Brazil.

Phytophthora (probably parasitica); branch and base rot. Mexico, S. America.

Puccinia carthami Cda.; rust.   Mexico, S. America.
Pythium oligandrum Drechs.; damping-off.   Mexico, S.
   America.
Virus; mottle type mosaic in Mexico.

CARYOPHYLLUS see EUGENIA

CASIMIROA EDULIS Lla. & Lex.   A large native fruit tree, valued
   specially by Indians and farmers; "sapote" Sp. Po. En. Ind.
   Botryodiplodia sp. (= Diplodia sp.); dieback, swollen branch
      canker.   Honduras, Puerto Rico, Costa Rica.
   Capnodium sp.; sootymold.   C. America, W. Indies.
   Cephaleuros virescens Kunze; small algal leafspots.   El
      Salvador.
   Cercospora coleroides Sacc.; leafspot.   Mexico, C. America.
   Corticium sp.; foliage blight on young tree.   Honduras.
   Diplodia theobromae (Pat.) Now.; dieback, branch canker.
      Puerto Rico, Costa Rica.
   Glomerella cingulata (Stonem.) Sp. & Schr. ascosporus stage
      of Colletotrichum sp.; stem disease, anthracnose, fruit
      spot-rot.   Honduras.
   Mycena citricolor (Berk. & Curt.) Sacc.; leafspot.   Costa
      Rica.
   Pellicularia koleroga Cke.; foliage mass blight.   Honduras.
   Strigula complanata Fée; lichen leafspot and peduncle attack.
      El Salvador.
   Tryblidiella rufula (Spreng.) Sacc.; twig disease.   Puerto
      Rico.
   Wilt, undetermined primary cause.   Sporophores of Ganoderma,
      Xylaria, Hydnum, Poria, Agaricus found on fallen branches
      and rootbases.   C. America.

CASSIA spp.   Some botanists consider these leguminous trees under
   such generic names as Chamaecrista, Ditremixa, and Emelista;
   however Cassia is often the most accepted by growers using
   these trees so important for shade, ornament, and medicinals;
   "canafistula," "casia" Sp., "fedegosa," "cassia rosada" Po.,
   "shower-of-gold," "pink cassia," "senna" En.   (Diseases among
   Cassias are apparently able to attack most of the host species,
   hence this list under one heading.)
   Asterina elaeocarpa Syd.; black mildew.   Puerto Rico.
   Botryosphaeria dothidea (Moug.) Ces. & de Not.; common
      branch canker.   General.
   Cassytha filiformis L.; green pungent smelling dodder.
      Puerto Rico.
   Cephaleuros virescens Kunze; algal leafspot.   (On several
      species, probably all, of the host.)   Brazil, Colombia,
      Ecuador.
   Cercospora cassae P. Henn.; leafspot.   Brazil.   C. occiden-
      talis Cke.; leafspot.   Barbados, Grenada (W. Indies).   C.
      paulensis P. Henn.; leafspot.   Brazil.   C. simulata Ell.
      & Ev.; leafspot.   Puerto Rico.   C. sphaeroidea Speg.;
      leafspot.   Argentina.

Clitocybe tabescens Bres. = Armilariella tabescens (Scop.)
  Sing.; rootrot. Subtropical USA.
Colletotrichum sp. (see Glomerella)
Corynespora cassiicola Wei. (easily mistaken for Cercospora);
  small leafspots. C. America.
Cuscuta americana L.; common brittle dodder. Puerto Rico.
Diplodia sp. (perfect state Botryosphaeria sp.); dieback.
  Widespread.
Erysiphe polygoni DC. (conidial state is commonly seen as
  an Oidium); white mildew. Neotropics.
Fomes spp.; brown rootrot. W. Indies, C. America.
Fusarium spp.; associated with dieback. W. Indies.
Ganoderma lucidum Karst.; rootrot. C. America scattered.
Glomerella cingulata (Stonem.) Sp. & Schr. = Colletotrichum
  sp.; dieback, anthracnose. General but not severe.
Irenina cassiaecola Bat. & Si.; black mildew. Brazil.
Irenopsis toruloidea Stev.; moldy leafspots. Trinidad,
  Panama, Ecuador, Costa Rica, Puerto Rico.
Loranthus spp.; parasitizing coarse bushes. C. America.
Meliola chaemaecristicola Stev. and other species; black
  mildews. W. Indies, C. America, S. America.
Mycena citricolor (Berk. & Curt.) Sacc.; leafspot. Costa
  Rica.
Oidium (see Erysiphe)
Phyllachora canafistulae Stev. & Dal.; tarspot. W. Indies.
  P. cassiaecola Kel. & Swi.; leafspot. Brazil.
Pratylenchus brachyurus Godf.; root nematode. Subtropical
  USA.
Rhizoctonia (solani-type); seedling damping-off, nursery tree
  root disease. El Salvador.
Rosellinia bunodes (Berk. & Br.) Sacc., and R. pepo Pat.;
  severe root diseases. C. America usually in tree-crop
  plantations.
Rotylenchulus reniformis Lin. & Olive; root nematode.
  Florida.
Rusts: Ravenelia cassiaecola Atk. in Barbados, Puerto Rico
  and S. America. R. texensis Ell. & Gall. in Subtropical
  USA. R. pazschke Diet. in Brazil. R. portoricensis
  Arth. in Puerto Rico and W. Indies. (Note, these are a
  few of the Ravenelias listed on tree hosts, several others
  are on herbaceous and bush Cassias.) Uredo cassia-rugosae
  Thur. is in Brazil and U. cubensis (Arth. & Johns.) Cumm.
  in Cuba.
Schizophyllum sp.; secondary invasion on old branches. Puerto
  Rico.
Sclerotium rolfsii (Curzi) Sacc. = Pellicularia rolfsii (Sacc.)
  West; ground sclerotial rot. Florida.
Sphaceloma cassiae Bitanc. & Jenk.; scab and spot anthrac-
  nose. Brazil.
Struthanthus spp.; small leaved, drooping bushes parasitizing
  host tree branches. W. Indies, C. America.

White rootrot caused by unidentified Basidiomycete. El
Salvador.

Xylaria sp. ; branch dieback. Costa Rica.

CASTANEA SATIVA Mill. A temperate zone nut tree grown under
special conditions in a few localities in the American tropics;
"castana" Sp. , "castanheiro" Po. , "chestnut" En.

Cercospora castaneae Mull. & Chupp; leafspot. Brazil and
other countries nearby.

Corticium sp. (believed to be different from C. salmonicolor);
branch disease. S. American locations.

Fusarium (solani-like); weak infection of dieback twig.
Colombia. F. lateritium Nees; in dying branch wood.
Argentina.

Mucor mucedo L. ; storage disease of nuts. Chile and Peru.

Nectria sp. ; twig disease. Chile.

Penicillium glaucum Lk. ; storage disease of nuts. Chile,
Peru.

Phyllosticta castanea Ell. & Ev. ; tarspot. Subtropical USA.

Physalospora obtusa (Schw.) Cke. = Botryosphaeria sp. =
Sphaeropsis malorum Pk. ; fruit-bur disease. S. America.

Phytophthora cinnamomi Rands; rootrot. Subtropical USA,
S. America.

Scleroderma sp. ; ? saprophyte. S. America.

Tubercularia endogena Speg. ; branch disease. Argentina.

Urocystis italica (Sacc. & Speg.) deT. ; tree smut. S.
America.

CASTILLOA ELASTICA Cerv. A medium-sized, fast-growing
native tree of Central America that has produced rubber-con-
taining latex for inhabitants use since prehistoric times, but
thus far presents serious commercial difficulties for extensive
growing; "goma, " "caucho" Sp. Po. "panama rubber" En.

Cephaleuros (? virescens); unusually small algal leafspots.
Costa Rica.

Cladosporium sp. ; on failing three year old leaf. C.
America.

Corticium salmonicolor Berk. & Br. ; pinkdisease of branches.
Costa Rica, Guatemala.

Cucurbitaria sp. ; disease of petioles (these are perennial
structures on this tree). Costa Rica.

Diplodia theobromae (Pat.) Now. also given as D. cacaoicola
P. Henn. = Botryodiplodia theobromae Pat. (which is the
most acceptable), perfect state is sometimes given as
Physalospora, see below.

Fomes spp. ; root and trunk base honeycomb rots, characterized
by slow field spread. C. America. F. lignosus (Kl.)
Bres. ; trunk heartrot. W. Indies, C. America.

Leaf spotting, unidentified, on old trees, causing little injury.
Costa Rica.

Marasmius (? equirinus); profuse black cord-blight. Costa
Rica.

Meliola sp.; black mildew.   Costa Rica.
Ophiodothella sp.; stromatic leafspot.   Costa Rica.
Phycopeltis (? epiphyton); algal epiphyte and weak penetration.
   Costa Rica.
Physalospora rhodina Cke. = Botryosphaeria quercuum
   (Schw.) Sacc.; dieback, extensive lesions in three to six
   year old petioles in Costa Rica.   C. America, W. Indies,
   Florida.   (? Not typical Diplodia.)
Polyporus sp.; heartrot.   W. Indies, C. America.
Rosellinia bunodes (Berk. & Br.) Sacc.; black rootrot.   W.
   Indies, C. America, S. America.
Strigula companata Fée; lichen spotting.   Costa Rica.
Struthanthus orbicularis (HBK.) Bl.; vine-like destructive
   woody vine parasite on host branches.   El Salvador.

CASUARINA CUNNINGHAMIANA Miq., and C. EQUISETIFOLIA L.
   Pine-like tropical trees native to Australia, planted in the Neo-
   tropics for their red wood, ornament, windbreak, and soil re-
   tention; "pino australiano" Sp. Po., "shee-oak," "beefwood,"
   "australian pine" En.
   Agrobacterium tumefaciens (E. F. Sm.) Conn.; crowngall.
      Peru.
   Armillaria mellea Vahl. = Armilariella mellea (Fr.) Karst.;
      rootrot.   California.
   Capnodium (see Microxyphium)
   Clitocybe tabescens Bres. = Armilariella tabascens (Scop.)
      Sing.; rootrot.   Florida.
   Dendrophoma casuarinicola Speg.; branch dieback, twig
      disease.   Argentina.
   Diaporthe casuarinae Speg., on dead twigs of C. STICTA
      Mig.   Argentina.
   Fenestella endoxantha Speg.; fallen-branch rot.   Argentina.
   Fusarium oxysporum Schl. (? f. sp. casuarinae, considered
      by me as a good forma from repeated isolations and
      culture studies and with twigs infected in the laboratory
      appearing especially well adapted to both species of the
      host).   Puerto Rico.
   Ganoderma lucidum Karst., base disease and stumprot.   C.
      America.
   Lophiostoma humilis Speg.; branch disease.   Argentina.
   Macrophomina phaseoli (Taub.) Ashby; seedling blight.   Sub-
      tropical USA.
   Marasmius brachypus Speg.; bark disease.   Argentina.
   Meloidogyne sp.; rootknot.   Florida.
   Microxyphium casuarinae Speg. is perfect state of Capnodium
      casuarinae McAlp.; sootymold.   Widespread.
   Odontia arguata Quel.; logrot.   Bermuda.
   Oidium sp., white mildew.   Peru.
   Peroneutypa heteracantha (Sacc.) Berl. f. casuraniae-sticta
      Speg. on C. sticta; (possibly saprophyte) on stems and
      branches.   Argentina.
   Phytophthora cambivora Buis.; streak in wood.   Subtropical
      USA.

Polyporus sp.; logrot.   Bermuda.
Pseudomonas solanacearum E. F. Sm.; seedling blight.   Sub-
   tropical USA, symptoms seen, but no isolations made, in
   Puerto Rico.
Sclerotium rolfsii (Curzi) Sacc., without basidial state;
   seedling base rot.   W. Indies, Florida.
Sphaerostilbe repens Berk. & Br.; root disease.   W. Indies.
Virus: Casuarina mosaic, causing leaf mottle, witches
   brooming, killing.   C. America.
Zignoella claypolensis Speg. on C. sticta; branch disease.
   Argentina.

CATHARANTHUS   see   VINCA

CECROPIA PELTATA L.   An important, fast-growing "weed" tree,
   often the first tree to occupy a jungle clearing, their hollow
   trunks used for blow guns and water conducting pipes from
   springs, ashes serve as insecticide and bark fiber used for
   jungle thread and weaving yarn, the flush growths believed to
   have medicinal properties and mature to old leaves with textures almost like sandpaper are employed as smoothing tools
   in making bows, arrows, blow gun darts, axe handles, and the
   like, even if of hard woods, a very important tree for small
   milpa farmers and sequestered jungle peoples; "guarúmo," "llag-
   rumo" Sp., "embauba" Po., "trumpet-tree" En., and many indigenous names.
   Amylis memorabilis Speg.; branch disease.   Paraguay and
      other countries of S. America.
   Anthostomella cecropiae (Rehm) Hoehn. = Auerswaldia
      cecropiae P. Henn.; leaf scab-spot.   Puerto Rico, S.
      America.
   Appendiculella tonkinensis (Karst. & Roum) Toro var. cecro-
      piae (Stev.) Hansf.; leaf mold injury.   Puerto Rico,
      Panama.
   Balladynella amazonica (Hoehn.) Th. & Syd.; leafspot.   S.
      America (Amazon basin).
   Cephaleuros virescens Kunze; algal leafspot.   Costa Rica,
      Panama, Ecuador.
   Cercospora cecropiae Muller & Chupp; leafspot.   Brazil,
      Ecuador.   (Also see Passalora).
   Clitocybe tabescens Bres. = Armilariella tabescens (Scop.)
      Sing.; rootrot.   Subtropical USA.
   Fracchiaea cucurbitarioides Speg. f. cecropiicola Speg. = F.
      heterogenea Sacc.; branch infection.   C. America, S.
      America.
   Fusicladium cecropiae (Stev.) Toro; leaf blotch disease.
      Puerto Rico.
   Hirneola auricula-judae (L.) Berk.; log decay.   C. America,
      W. Indies, S. America.
   Irene echinus Stev.; black leafspot.   Panama, Puerto Rico.
      I. obesa (Speg.) Stev., and I. tonkinensis (Karst. &
      Roum.) Stev. var. cecropiae Stev.; leafspots.   C. America.

Meloidogyne sp.; rootknot. Florida.
Mucor arcuatus Mart.; pod softrot, fiber discoloration. C.
America.
Oudemansiella orinocensis Speg.; agaric on log decay. S.
America.
Passalora cecropiae (Stev.) Toro (a 2-celled Cercospora,
some question about its separation from Cercospora); in-
distinct leaf blotch. Puerto Rico, Florida.
Periconia cecropiae Bat. & Vit.; attack on vigorous leaf.
Brazil.
Phialea cecropiae (P. Henn.) Seav.; leaf-sheath lesion.
Puerto Rico.
Phyllosticta starbaeckii Sacc. & Syd.; leaf spotting. Brazil.
Strigula complanata Fée; white lichen spot. Neotropical.
Thyridaria insueta Petr.; branch disease. S. America.
Torula spegazziniana Sacc. = T. pulvinata Speg.; on decaying
trunk. C. America, S. America.

CEDRELA MEXICANA Roem. and C. ODORATA L. The first is
called "cedro reál" Sp. and "royal cedar" En., and the second,
"cedro macho" Sp., "cedro" Po., "great or West Indian cedar"
En. These are large forest trees, relatives of mahogany,
they produce fine lumber, and are at times planted scattered
among other kinds of trees in thinned forests and jungle.
Cephaleuros virescens Kunze; algal leafspot. Brazil, Costa
Rica.
Cercospora sp.; leafspot. Cuba, Costa Rica, probably wide-
spread.
Corticium salmonicolor Berk. & Br.; branch disease. C.
America.
Doassansia sintenisii Bres.; smut. S. America.
Fomes sp.; honeycomb heartrot, slow spread in field. C.
America.
Ganoderma sp.; trunk baserot. Costa Rica.
Loranthus sp.; parasitic bush on branches. El Salvador.
Mycena citricolor (Berk. & Curt.) Sacc.; rare but occasional
leafspot, on lowhung foliage. W. Indies, C. America.
Parodiopsis melioloides (Berk. & Curt.) Maubl.; black-patch
leaf disease. Puerto Rico.
Phyllachora balansae Speg. (synonyms: Physalospora balansae
Speg., Puiggarina balansae Speg.); tarspot of mature
foliage, seedling blight. Brazil, Argentina, Paraguay,
Bolivia, Trinidad, Puerto Rico.
Phyllosticta agnostoica Speg.; leafspot. Brazil.
Poria sp.; rot of bark on log. Costa Rica.
Psittacanthus americana Mart. and P. calyculatus (DC.) Don.;
parasite bushes (swollen host tissue at point of attack forms
famed "wood roses"). El Salvador, Guatemala.
Rhizoctonia solani Kuehn = Thanatephorous cucumeris (Frank)
Donk; typical damping off in seedbeds. W. Indies, C.
America, S. America.
Rosellinia bunodes (Berk. & Br.) Sacc.; black rootrot.

Puerto Rico. R. pepo Pat.; a white rootrot. El Salvador,
C. America. (Note: Rootrots from these two fungi associated somewhat with other causes have resulted in serious
decline and death in plantations of Cedrela trees at just
about the time they reach 20-cm. diameter at 3-meter
height.)

Sphaerella cedrelae Speg.; leaf infection. Paraguay.

Struthanthus sp.; bush with parasitizing surface roots attacking
tree branches. Costa Rica, El Salvador.

"Sunscald of seedlings," sometimes associated with lack of
shade in Mexico, but seen in Guatemala under slat house
conditions.

CHAETOCHLOA see SETARIA

CHENOPODIUM spp. Herbaceous crops that develop heavy heads
of seed, harvested for grain in areas where few carbohydrate
producing grains can be cultivated, of great importance in high
mountains. C. PALLIDICAULE Ael.; "canigua," "ayara" Ind.,
and C. QUINOA Willd.; "quinoa," "hufa," "suba" Ind. (Diseases
have not been much studied, as native growers are shy and un-
communicative with visitors.)

Cercospora dubia Wint.; leafspot. S. America.

Cladosporium sp.; often on senescent leaves. Bolivia (probably widespread).

Damping-off of seedlings (causes not carefully studied). Peru.

Hendersonia chenopodiicola Speg.; a stem disease. Argentina
and nearby countries.

Heterodera rostochiensis Wor.; root nematode. Peru. (Note:
This occurrence on quinoa has suggested this may have
been the original host on which this nematode first became
a disease on an economically important crop.)

Hypochnus peronosporioides Speg. an old name for a web
blight (Rhizoctonia-Corticium-Pellicularia) and not given
much study since Spegazzini described it. S. America.

Peronospora effusa Rob.; downy mildew. Peru.

Phoma desmazieri Dur. & Mont.; blackstem and leaf lesions.
S. America.

Phyllosticta dimorphospora Speg.; abundant spotting of leaves.
Argentina.

Physoderma hemisphericum (Speg.) Karl. and P. pulposum
Wallr.; stem brownspots. S. America.

Ramularia chenopodii Speg.; a rough surfaced leafspot.
Argentina.

Sphaerella chenopodiicola Speg.; on dry and rotting stems.
Argentina.

Virus: Mosaic mottle, in several locations in S. America.
Quinoa is used as a laboratory host for studying certain
viruses in Brazil, Puerto Rico, Venezuela, Subtropical
USA.

CHLORIS GAYANA Kunth. A large and vigorous grass from Africa

that has gained well-earned popular usage among livestock men
of the American tropics; "rhodes grass" En.
Balansia epichloe (Weese) Diehl in Puerto Rico and Sub-
    tropical USA, B. henningsiana Diehl in Florida, and B.
    strangulans (Mont.) Diehl f. discoidea (P. Henn.) Diehl
    in C. America, Brazil and Peru, all cause blighting ring-
    disease of culm, and inflorescence sterility.
Cercospora cespitosa Ell. & Ev.; leafspot.   Subtropical USA.
Colletotrichum sp. (grass race of an unidentified species is
    possibly Vermiculariella poiophila Speg.); cause of plant
    base infection and no spots but leaf decay.   Puerto Rico,
    Brazil, Argentina.
Coniosporium chloridis Gonz. Frag. & Cif. and C. tripsaci
    Gonz. Frag. & Cif.; leaf and culm diseases.   W. Indies.
Curvularia sp.; leaf moldyness.   Florida.
Cyphella microthele Speg.; culm disease (? partly saprophytic).
    S. America.
Dinemasporiella poiophila Speg.; leaf disease from weak
    parasite.   Argentina.
Fusarium graminearum Schw. = Gibberella zeae (Schw.)
    Petch; pasture decline.   S. America, C. America.   F.
    roseum Lk.; secondary infections after insect feeding.
    Puerto Rico.
Gnomonia chloridis Gonz. Frag. & Cif.; leafspot and at times
    collapse.   W. Indies.
Helminthosporium sp.; large leaf spots, when severe plant
    killing.   Costa Rica.
Heterosporium chloridis Speg.; leaf disease.   Brazil, Para-
    guay, Argentina.
Meliola panici Earle; black mildew on leaf and panicle.
    Puerto Rico.
Ophiobolus graminis Sacc.; footrot.   Chile.
Phyllachora boutelouae Rehm = P. cynodontis (Sacc.) Nie.; a
    shiny tarspot.   Texas, Mexico, W. Indies, S. America, C.
    America.   P. chloridicola Speg.; black stromatic spots.
    Argentina.
Puccinia cacabata Arth. & Holw. and P. chloricidis Speg.;
    leaf rusts.   C. America, S. America.
Uredo gayana Lindq. and U. incognita Speg.; leaf rusts.
    Brazil, Argentina.
Ustilago deserticola Speg. and U. ulei P. Henn.; culm smuts.
    Brazil, Bolivia, Argentina.
Viruses.   Leaf stripe virus in El Salvador and Florida.
    Stunt with discolored vasculars like sugarcane ratoon stunt,
    studied in Costa Rica.

CHRYSANTHEMUM MORIFOLIUM Hemsl.   A flowering plant be-
    coming increasingly valuable to florists and market women in
    the American tropics; "crisantemo" Sp. Po., "chrysanthemum"
    En.   (Note: There are several other ornamentals grown that
    are of this genus, and C. COCCINEUM Wild., the source of
    commercial "pyrethrum," is a promising crop of the American

tropics, but until it is studied in greater detail by several
pathologists it is believed the list of parasites given hereunder
are on the whole applicable to either the flower crop or the
insecticide crop.)

Alternaria alternata (Fr.) Keiss = A. tenuis Nees; black
  edging of old leaves, mild leafspot. Occasional.
Botrytis cinerea Pers. ; bud disease and rainyseason foliage
  blight. Scattered places in C. America, W. Indies.
Cercospora chrysanthemi Heald & Wolf = C. chryanthemi
  Putt. ; leafspot. Brazil.
Cladosporium herbarum Lk. (probably a special form or race
  that would be f. chryanthemi); flower petal disease.
  Brazil, Puerto Rico.
Erysiphe cichoracearum DC. = Oidium sp. ; white powaery
  mildew. Widespread.
Fusarium oxysporum Schl. sp. f. callistaphi Sny. & Hans. ;
  wilt and stemrot. Florida, Chile.
Glomerella cingulata (Stonem.) Sp. & Schr. is asco state of
  Colletotrichum gloeosporioides Penz. ; stem lesions and
  leaf disease, cool land infection. Costa Rica.
Meloidogyne sp. ; mild rootknot. Subtropical USA, W. Indies.
Mycosphaerella ligulicola Bak. , Dim. & Dav. ; stem and
  flower infections. Subtropical USA, C. America, S.
  America.
Oidium chrysanthemi Rab. = Erysiphe, see above.
Phyllosticta chrysanthemi Ell. & Dear. ; a fairly common
  leafspot. General.
Phymatotrichum omnivorum (Shear) Dug. ; dryland rootrot.
  Arizona, New Mexico, and Texas in Subtropical USA.
Pseudomonas syringae v. Hall; bacterial leafspot. Scattered.
Puccinia chrysanthemi Roze; rust. Widespread if uncontrolled.
Pythium aphanidermatum Fitz. and P. debaryanum Hesse;
  cutting and root rots. Subtropical USA, probably in most
  parts of the American tropics where the crop is grown
  under intensive conditions.
Rhizoctonia solani Kuehn = the basidial state Thanatephorous
  cucumeris (Frank) Donk; cutting decay, footrot. General.
Rhizopus nigricans Ehr. ; flower rot in market. Florida.
Sclerotinia sclerotiorum (Lib.) dBy. = Whetzeliana sclerotiorum
  (Lib.) Korf.& Dumont; common cottony stemrot. Subtropi-
  cal USA.
Sclerotium rolfsii (Curzi) Sacc. = Pellicularia rolfsii (Sacc.)
  West; blight. Florida.
Septoria chrysanthemella Sacc. = S. chrysanthemi Cav. ; often
  small leafspot. Subtropical USA, C. America, Uruguay,
  Argentina. S. leucanthemi Sacc. & Speg. and Rhabdospora
  leucanthemi (Sacc. & Speg.) Kuntze; leafblotch and blight.
  Widespread.
Stemphylium sp. ; flower speck. Subtropical USA.
Viruses: Cucumber mild mosaic and occasionally with it the
  severe mottling and necrotic race (? from Commelina?),
  in C. America, Colombia, Florida. Spotted wilt in California and Argentina.

CHRYSOBALANUS ICACO L.   A widely used indigenous tropical
tree both planted and in the wilds having fruits of much impor-
tance in local fruit markets and in home consumption; "icaco"
Sp., "uajuru" Po., "cocoplum" En.
   Asterina schroeteri (Rehm) Th.; black leafmold.   W. Indies.
   Cephaleuros virescens Kunze; algal leafspot.   General, in-
      cluding Florida.
   Cercospora chrysobalani Ell. & Ev. = (perfect state) Mycos-
      phaerella chrysobalani Miles; common leafspot.   Florida,
      W. Indies, S. America.
   Helminthosporium chrysobalani P. Henn.; leafspot.   Colombia.
   Leprieurina radiata Toro; leaf lesion with pycnidia.   Puerto
      Rico.
   Leptothyrella chrysobalani P. Henn.; on foliage.   Colombia,
      Chile.
   Pestalotia funerea Desm.; leaf lesions.   Puerto Rico.

CHRYSOPHYLLUM CAINITO L.   A rather large native tree, its
fruit much in demand by Indians and small farmers; it is sold in
local markets but fruits are eaten by monkeys in tree tops, often
just before full maturity of these attractive fruits; "caimito" Sp.,
"gaucá" Po., "caimito," "star-apple" En.   (Diseases have not
been given intensive study and the list here is just a start.)
   Aschersonia turbinata Berk.; spores from fungus parasitizing
      insects on leaves.   Warm areas in Florida, W. Indies, C.
      America, S. America.
   Asterina chrysophylli P. Henn.; black moldy leafspot.   W.
      Indies, S. America.
   Capnodium sp.; sootymold.   Neotropical.
   Cephaleuros virescens Kunze; algal leafspot.   W. Indies, C.
      America.
   Diplodia sp. = Botryodiplodia theobromae Pat.; often blackened
      twigs and dieback.   Neotropical.
   Loranthus sp.; parasitizing large bush on a host branch.
      Costa Rica.
   Meliola lucumae Stev., M. ocoteicola Stev. and other species
      not described; black mildews.   General.
   Phyllachora sp.; tarspot, defoliated a weak old tree.   Costa
      Rica.
   Polyporus sp.; basal infection on old trees.   C. America.
   Strigula complanata Fée; lichen creamy whitespot.   W. Indies,
      C. America.
   Struthanthus sp.; vine-like parasitizing bush.   Costa Rica.
   Uredo amicosa Arth.; rust.   Puerto Rico.   U. chrysophylli
      Syd. and U. chrysophyllicola P. Henn.; rusts.   Brazil
      and adjacent countries.
   Zukalia chrysophylli Gonz. Frag. & Cif.; mold patch on leaf.
      W. Indies.

CICER ARIETINUM L.   From Asia, it produces well in tropical
America and the crops of edible seeds are of increasing importance; "chickpea" En., "garbanzo" Sp.

Chaetomium sp.; on old stem in moist area.   Puerto Rico.
Fusarium (oxysporum ?); wilt.   Peru, Chile.   F. solani
    (Mart.) Sacc.; stem disease.   W. Indies, Peru.
Meloidogyne sp.; rootknot.   Florida, Puerto Rico.
Mycosphaerella sp. (? imperfect state is a Cercospora);
    leafspot.   Florida, C. America.
Phyllosticta fabiei Trot.; coalescing leafspots.   Colombia.
Pythium spp.; both damping-off and rainy season plant decay.
    W. Indies, C. America, Subtropical USA.
Rhizoctonia solani Kuehn = Thanatephorous cucumeris (Frank)
    Donk; basal rot, root infection, seedling damping-off, stem
    lesion.   General.   Rhizoctonia sp.; thread blight.   Costa
    Rica.
Sclerotinia sp.; top infection.   Chile.
Toxic injury from leaf hoppers.   W. Indies.
Viruses: Pea mosaic in S. America.   Pea enation virus in
    Argentina.

CINCHONA OFFICINALIS L.   Originally from the Andes, this tree
    was taken to the East Indies, flourished and produced medicinals
    for over a century in world trade, is now being planted in the
    Americas as improved trees, and the anti-malarials from it are
    important in continuing health programs; "quina" Sp. Po., "cin-
    chona," "quinine tree" En.
    Algal growths.   Genus and species undetermined.   Infection
        found under bark as it was being removed for processing.
        Mountain plantation in Costa Rica.
    Armillaria mellea = Armilariella mellea (Fr.) Karst.; root
        disease.   Peru.
    Boletus thelephoroides Kunth; woodrot.   Peru.
    Cercospora cinchonae Ell. & Ev.; leafspot.   C. America, S.
        America.
    Cladosporium sp.; seedling root disease.   Andean S. America.
    Colletotrichum sp. (extremely variable in cultures); foliage
        injury, especially on fruiting twigs.   Guatemala, Costa
        Rica.
    Corticium salmonicolor Berk. & Br.; leaf collapse and branch
        disease.   C. America.
    Corynaea purdiei Hook; a remarkable tuber forming root para-
        site (member of Balanophoraceae).   In Andean Peru and
        Colombia at least.
    Elsinoë cinchonae Jenk.; scabspot.   S. America.
    Meloidogyne incognita Chitw.; rootknot.   Guatemala.
    Mycena citricolor (Berk. & Curt.) Sacc.; abundant leafspot
        and severe defoliation.   Costa Rica.
    Pellicularia koleroga Cke.; threadblight of large sectors of
        tree foliage.   Costa Rica, Brazil.
    Pestalotia funerea Desm.; isolated from leafspot.   Costa Rica.
    Phyllosticta cinchonae Pat.; angular leafspot.   Ecuador.
    Phytophthora cinnamomi Rands; root disease, and stripe
        canker.   Peru, C. America.   P. palmivora Butl.; stem
        canker (not stripe), dieback.   C. America, Colombia.

P. parasitica Dast.; foliage and stem disease, girdling at
times. C. America, Peru. P. quininea Cran.; root disease. Peru.

Porella sp.; leafy liverwort smothering. Guatemala, Costa
Rica.

Prillieuxina cinchonae J. A. Stev.; black mildew of leaves.
Costa Rica, Guatemala.

Rhabdospora thallicola Tassi; speck-spots on leaf and petiole.
Wild Cinchona leaf. Peru.

Rhizoctonia solani Kuehn (the mycelial and parasitic stage) =
Thanatephorous cucumeris (Frank) Donk (the basidiospore
and non-parasitic stage); associated with seed bed damping-
off and seedling girdle and leaf decay at ground level.
Widespread.

Rosellinia bunodes (Berk. & Br.) Sacc., R. pepo Pat., and
Rosellinia sp.; all three cause mature-tree root diseases.
Costa Rica, Guatemala.

Sclerotium rolfsii (Curzi) Sacc. (but no perfect state found or
reported on this crop); in warmer locations causes collar
rot. C. America, Peru, Colombia.

Septobasidium atropunctum Couch and S. purpureum Couch;
branch brown-felt. Jamaica.

Stilbum sp.; branch canker. Guatemala.

Thelephora pavonia HBK. and T. sanguinea HBK.; collected
from forest grown trees. Peru.

Valsa sp.; recovered from dead branches incubated when
wrapped in moist leaves. Costa Rica, El Salvador. In
Guatemala collected from large stake-like cuttings that
died after being plunged in sphagnum.

Xylaria sp.; at base of killed tree. Guatemala.

(Note: Much difficulty is encountered with delicate seedlings
of cinchona when the seed is sown in lowland seedbeds
where it germinates faster than in cold mountain places.
A partial list of genera of organisms found associated with
this watery damping-off is as follows: Pythium, Phytophthora, Fusarium, Botrytis, Pestalotia, Penicillium, Rhizoctonia, Cladosporium, Diplodia, Rosellinia.)

CINNAMOMUM ZEYLANICUM Nees. A tree originally from the
Pacific Islands, is grown in the American tropics for its pungent
flavored bark; "canela" Sp. Po., "cinnamon tree" En. (C.
CAMPHORA (L.) Nees & Eberm. is the "camphor-tree" En.)

Bacteria, unidentified, isolated from internal branch tissues
of specially weak tree. Costa Rica.

Botryodiplodia theobromae Pat. (synonyms: Diplodia cam-
phorae Tassi, D. natalensis P. Evans, D. cacaoicola P.
Henn.); dieback, canker, bitter bark. Florida, W. Indies,
Costa Rica, Colombia, Venezuela.

Capnodium sp.; sooty skin-mold on leaves and stems. Puerto
Rico.

Cephaleuros virescens Kunze; the common algal leafspot. W.
Indies, C. America, S. America.

Colletotrichum gloeosporioides Penz. = Glomerella cingulata
    (Stonem.) Sp. & Schr. (also by some reported as C. cin-
    namomi, C. camphorae, Gloeosporium camphorae, and G.
    ochraceum); dieback, leaf and shoot anthracnose, stem-
    canker. Costa Rica, Puerto Rico, Honduras, Peru, Pana-
    ma, Florida, Brazil.
Corticium salmonicolor Berk. & Br.; pink disease killing
    branches. Costa Rica.
Cytosporella cinnamomi Turc.; branch infection and flush
    blight. S. America.
Elachopeltis cinnamomae Bat. & Vit.; fungus growth on leaf.
    Brazil.
Fomes spp.; brown rootrot, honeycomb woodrot. Guatemala,
    Panama.
Hypoxylon sp.; weak-branch-disease. W. Indies.
Lembosia camphorae Earle; leafspot. Florida on camphor,
    in Costa Rica on cinnamon.
Melophia cinnamomae Bat. & Vit.; leafspot disease. Brazil.
Microsphaera alni DC., var. cinnamomi Olive; = Oidium sp.
    (the conidial and most common state in the tropics); powdery
    mildew. Subtropical USA, Guatemala, Honduras, Panama,
    probably widespread.
Mycena citricolor (Berk. & Curt.) Sacc.; leafspot by inocula-
    tion. Costa Rica.
Oidium sp. is Microsphaera.
Pellicularia koleroga Cke.; web-thread-blight of leaves and
    branch tips. W. Indies, C. America.
Pestalotia funerea Desm. and P. cinnamomi Petr.; twig
    lesions. Puerto Rico, St. Thomas.
Phomopsis sp.; branch decadence. W. Indies, Florida, Costa
    Rica.
Phthirusa sp.; parasitizing bush. Honduras, Costa Rica.
Phyllosticta cinnamomi (Sacc.) Lindau; leafspot. W. Indies.
Phytophthora cinnamomi Rands; rootrot. Scattered but not
    restricted to any region.
Polyporus spp.; stump rot. Honduras, Guatemala, El Salvador.
Porella sp.; leafy liverwort on leaves in moist area. El
    Salvador.
Rosellinia bunodes (Berk. & Br.) Sacc.; rootrot and wilt.
    Florida, Costa Rica, Panama.
Sphaerella cinnamonicola Gonz. Frag. & Cif.; leaf disease.
    Dominican Republic.
Strigula complanata Fée; lichen leafspot. W. Indies, C.
    America.
Struthanthus sp.; a woody bush that parasitizes host branch.
    Honduras.
Xylaria sp.; from dieback of main branch. Guatemala.

CITRULLUS VULGARIS Schrad. Originally from tropical Africa,
    these are annual herbaceous vines grown in the Americas for
    their large and sweet but perishable fruits; "sandía" Sp. "melan-
    cia" Po., "watermelon" En.

Alternaria spp.; leafspot.   In many hot-moist watermelon
    areas, irrigated deserts appear to be free.
Blossomend rot and browning, combined effect of disturbed
    growth, infections of weakened tissue by Pythium, Colleto -
    trichum, etc.   Widespread.
Cercospora citrullina Cke.; common black leafspot, often
    large.   May be scattered in field.   C. America, Puerto
    Rico, Peru, Brazil, Venezuela.
Cladosporium cucumerinum Ell. & Arth.; leafmold, fruitscab.
    Puerto Rico, El Salvador.
Colletotrichum lagenarium (Pass.) Ell. & Halst.; anthracnose
    on leaf, shoot, and rot of fruit.   Mexico, El Salvador,
    Puerto Rico, Venezuela, Cuba, Uruguay, Brazil, Argen-
    tina.
Cuscuta americana L.; dodder.   Puerto Rico.   Cuscuta spp.;
    dodders.   Argentina, and Brazil.
Cylindrocladium scoparum Morg.; leaf disease, fruit rot.
    Florida.
Diplodia natalensis P. Evans and D. theobromae (Pat.)
    Now. = Botryodiplodia theobromae Pat.; common fruit
    stemend rot as well as vine canker.   Scattered where
    crop is grown.
Erysiphe cichoracearum DC.; powdery white mildew, "oidium."
    General and seasonal.
Fusarium oxysporum Schl. f. sp. niveum (E. F. Sm.) Sny.
    & Han.; wilt.   Mexico, Florida, California, Puerto Rico,
    Cuba, Panama, Argentina, Virgin Islands, Chile.   F.
    roseum Lk.; fruit rot.   Peru.
Macrophomina phaseoli (Taub.) Ashby; stem and fruit charcoal
    rot.   Texas, Guatemala.
Meloidogyne sp.; rootknot.   Scattered in general in C. America,
    W. Indies, Brazil, probably all other watermelon growing
    countries.
Mycosphaerella citrullina (C. O. Sm.) Gross; gummy swollen
    stem, fruit rot, leafspot.   Florida, Puerto Rico, C.
    America.
Oidium (see Erysiphe)
Phyllosticta cucurbitacearum Sacc. (considered as separate
    from Mycosphaerella); leaf disease.   Subtropical USA,
    Puerto Rico.
Phytophthora spp.; fruitrots and stem attack.   Occasional.
Pseudoperonospora cubensis (Berk. & Curt.) Rost.; severe
    downy mildew.   Neotropical.
Pythium spp.; blossom-end rot, damping-off.   Mexico,
    Florida, Guatemala, Panama, Puerto Rico, Haiti.
Rhizoctonia solani Kuehn (the seldom reported perfect state
    is Thanatephorous cucumeris (Frank) Donk); damping-off
    and soil rot of fruit.   C. America, W. Indies, S. America.
Sclerotium rolfsii (Curzi) Sacc. perfect state is Pellicularia
    rolfsii (Sacc.) West; collar rot, fruitrot.   Florida, Mexico,
    Puerto Rico.
Trichoderma sp.; a basal disease of large plants (repeated
    isolations).   Puerto Rico.

Viruses: Cucumber mosaic viruses both common mottle and
virulent type strain.   Florida, Puerto Rico, El Salvador.
Squash mosaic virus.   Venezuela.   Watermelon mosaic
virus.   W. Indies, Venezuela, Argentina.
Weakspot, organism involved not determined with certainty,
possibly from infection started by a Cercospora that did
not progress in leaf tissue.   Puerto Rico.

CITRUS spp.   Over 20 species, and numerous crosses of them,
have been brought from the tropics and subtropics of the Eastern
Hemisphere to the American tropics but the two species I have
been required to give most attention to are C. AURANTIFOLIA
(Chr.) Swingle; "lime" En., "limón" Sp., "limão" Po., and C.
SINENSIS (L.) Osbeck, the common seedling and/or grafted
orange that grows in dooryards, as accidental roadside seedling
trees or as small to large plantation-groves; "sweet orange"
En., "naranjo" Sp., "laranja" Po.   Following is a list of foliage
and tree diseases; fruit diseases are given as a supplemental
list beginning at the end of this list.
    Alternaria citri Ell. & Pierce; occasional leafspot and drop.
        Subtropical USA, W. Indies, C. America, Guayana, Vene-
        zuela.
    Amylirosa aurantiorum Speg.; branch disease.   Argentina,
        Paraguay.
    Aphelenchoides parietinus Stein.; root semiparasite.   Brazil.
        A. radicicolus Stein.; root nematode.   Florida.
    Armillaria mellea Vahl = Armilariella mellea (Fr.) Karst.;
        rootrot.   California, Guatemala.
    Armilariella (see Armillaria and see Clitocybe)
    Ascochyta citri Penz.; an often mild leafspot.   Costa Rica,
        Mexico, Guatemala, Puerto Rico, Peru, Venezuela,
        Brazil.
    Aspergillopsis pulchella Speg.; fruitrot.   Argentina.
    Bacterium gummis Comes; gummyness of twig and branch.
        Mexico, Paraguay, Brazil.
    Belonolaimus gracilis Stein.; root sting nematode.   Florida.
    Botryodiplodia lecanidion (Speg.) Petr. & Syd. synonyms:
        Diplodia natalensis P. Evans, D. hesperidica Speg.,
        Botryodiplodia theobromae Pat., Pseudodiplodia aurantiacum
        Speg.; dieback, girdling gumose of seedling, rootrot, red-
        edge canker.   Neotropical.
    Botryosphaeria dothidea (see Dothiorella)
    Camarosporium pulchellum Speg.; branch decay.   Argentina.
    Capnodium citricola McAlp. = C. citri Berk. & Desm.
        (perfect state is Microxyphium citri (Berk. & Desm.)
        Speg.); sootymold growth on insect honeydew.   General in
        Neotropics but scattered in fields, often on shaded and
        protected foliage.
    Cassytha filiformis L.; pungent-smelling dodder.   Florida,
        Puerto Rico.
    Cephaleuros virescens Kunze; splitting of pedicel, algal spot,
        occasional in moist areas.   Florida, W. Indies, C. America,
        S. America.

Cephalosporium lecanii Zimm.; appears to be twig disease but is associated with insects. Bolivia, Colombia. C. omnivorum Crand. & Arrel.; apparently true branch infections. El Salvador.

Cercospora aurantia Heald & Wolf = C. penzigii Sacc.; leafspot. Texas, C. America (for C. citri-grisea see Mycosphaerella citri). C. fumosa Penz.; leafspot on lower surfaces. Argentina. C. gigantea Fish.; tarspot. Florida, Peru.

Chaetophoma citri Sacc.; in leafspot complex. Dominican Republic. C. stromaticola Speg.; on foliage. S. America.

Cladosporium spp. (C. furfuraceum McAlp., C. herbarum Lk. and others); old ragged leaf. (In unsprayed trees). W. Indies, Florida, Colombia, Mexico, Chile, Panama, Brazil.

Clasterosporium maidicum Sacc.; weak parasite. Trinidad.

Clitocybe tabescens Bres. = Armilariella tabescens (Scop.) Sing.; rootrot. Florida, Panama.

Colletotrichum gloeosporioides Penz. (synonyms: Gloeosporium citricolum Speg., G. spegazzinii Sacc.) asco-state is Glomerella cingulata (Stonem.) Sp. & Schr.; irregular sunken lesions of twigs, anthracnoses. Generally on trees in poor soils. C. limetticolum Claus.; lime withertip and anthracnose. This seems different from C. gloeosporioides and it may be a race. Many locations.

Coniochaeta pulveracea (Ehrb.) Munk = Rosellinia pulveraceae (Ehrb.) Fckl. var. platysporella Speg.; rot of denuded wood. S. America.

Coniothecium citri McAlp.; leaf disease and C. olivaceum Bon.; stem disease. Both in Mexico.

Corticium areolatum Stahel also reported as Pellicularia filamentosa (Pat.) Rog. (This may be a special race and is not Leptosphaeria.); areolate leafspot. Surinam. C. salmonicolor Berk. & Br.; leafrot and pink-arm disease. W. Indies, C. America, S. America.

Criconemoides morgense Tay.; root nematode. Brazil, Chile.

Cryptosporella aurantiicola Speg.; branch canker. S. America.

Cuscuta spp. all typical dodders: C. americana L. Puerto Rico, Dominican Republic, Mexico, Florida. C. boldinghii Urb. Florida. C. subinclusa Dur. & Hil. California. And other species. Widespread.

Daldinia concentrica (Bolt.) Ces. & DeNot.; trunk heartrot. Florida, Puerto Rico.

Dendropemom spp.; small parasitic woody bushes. W. Indies.

Deuterophoma tracheiphila Petri; rootrot, "mal secco." Colombia, Brazil, Argentina.

Diaporthe citri Wolf, its imperfect state is Phomopsis citri Faw.; raised granules on leaves and twigs. Widespread.

Didymella aurantiiphila Speg. and D. citri Noack; twig disease, bark spot. Brazil, Argentina, Paraguay.

Diplodia (see Botryodiplodia)

Dothiorella guaranitica Speg. is conidial stage of Botryosphaeria dothidea (Moug.) Ces. & de Not.; causes stem-and-twig rot and canker. Scattered in S. America, W. Indies, Trinidad.

Elsinoë australis Bitanc. & Jenk. ascospore stage of Spha-
celoma australis Bitanc. & Jenk.; anthracnose spot-scab
especially of lime and sour orange but sweet orange is
also susceptible.   Brazil, Bolivia, Argentina, Paraguay,
Uruguay.   E. fawcetti Bitanc. & Jenk. ascospore stage of
Sphaceloma fawcetti Jenk.; anthracnose spot-scab of both
lime, and sour orange, but sweet orange is practically
immune.   Subtropical USA, W. Indies, C. America, S.
America.
Ephelidium aurantiorum Speg.; twig bark disease.   Argentina,
Paraguay.
Epiphytic microorganisms are found on matured old leaves
(when epidermal tissues remaining uninvaded become
hardened), non-parasitic types of bacteria, many fungi as
slender and somewhat fuscus-colored mycelia but no spores
on hyphal tips, small colonies of algae such as Geococcus-
like, Trentepohlia sp., Phycopeltis sp., lichens such as
the white Strigula sp., gray and brown types, and micro-
liverworts such as Porella.
Erysiphe sp.; dry season mildew.   Guatemala, Panama.
Eutypa auranticola Speg. = Enzigostoma aurantiicola (Speg.)
Kuntze; rootrot.   Paraguay.
Eutypella citricola Speg. and E. pusilla Speg.; twig and
branch diseases.   W. Indies, S. America.
Favolus paraguayensis Speg.; rot of dead branch.   Paraguay.
Fomes spp.; trunk heartrot.   Florida, El Salvador, Argentina,
Panama, Honduras, Puerto Rico, Virgin Islands, Paraguay,
Brazil.
Fumago vagans Pers. (sometimes reported as Capnodium);
sooty mold.   C. America, S. America.
Fusarium spp.   1. F. lateritium Nees or F. limoni Bri. =
the ascosporus stage is Gibberella baccata Sacc.; branch
and twig disease, dieback.   2. F. orthoceras Ap. & Wr.;
twig and fruitrots.   3.   F. solani (Mart.) Sacc.; root-tip
rot, twig and flush-blight with gumming.   4. F. roseum
Lk.; secondary invasion after insect injury, Diplodia infections and other primary effects.   5. F. oxysporum Schl.
f. sp. aurantiacum Sny. & Hans.; seedling wilt and red
fruitrot.   There are no special geographical limitations on
these.
Ganoderma australe (Fr.) Pat. = G. applanatum (Pers.) Pat.
in Jamaica.   G. fornicatum Fr., in Argentina.   G. lucidum
(Leyss.) Karst., in Florida and W. Indies.   G. sessilis
Murr., in Florida, W. Indies.   Ganoderma spp. in C.
America, S. America, W. Indies.   All cause rootrots on
old trees, heart decay, and "caries."
Gibberella (see Fusarium)
Glomerella sp. (said that it is not cingulata) on orange twigs
in Colombia.
Haplosporella hesperidicola Speg. = Sphaeropsis hesperidicola
Speg.; attack on living branches.   Argentina.
Hemicycliophora sp.; nematode at roots.   Peru.

Heterosporium sp.; dry galls. Brazil.
Hymenochaete noxia Bancr.; root canker. British Guiana.
Hypoxylon guarapiense Speg. = H. rubigenosum Pers.; on old
   branches. Paraguay. H. turbinatum Berk. var. guaraniticum Speg. = Phylacia turbinata (Berk.) Dennis; disease on
   trunk. S. America.
Insect effects: 1. Aphid, and other sucking insect deposits
   of honey dew on which capnodiums, fumagos, and other
   fungi subsist. 2. Leprosis from Brevipalpus feeding.
   3. Nailhead spots on leaf and twig from Tenuipalpus feeding
   (numbers 2 & 3 can be classed together). 4. Scale skele-
   tal remains, often notable after wetseason abates. 5.
   Stigmonose or sharply raised protuberances on leaf and
   twig from insect puncture feeding. All are of more or
   less general occurrence.
Leptosphaeria bondari Bitanc. & Jenk.; secondary parasite
   exaggerating injury from other infections. Brazil.
Leptothyrium pomi (Mont. & Fr.) Sacc.; fruit and leaf fly-
   speck. C. America.
Lichens, common on citrus trees. Many types. General.
   (See Strigula.)
Limacinia aurantii P. Henn.; sootymold. S. America. L.
   citri Sacc. and L. penzigi Sacc.; sooty mold of leaves.
   Chile, Dominican Republic.
Longidorus elongatus Th. & Sev.; root nematode. Florida.
Lophidiopsis paraguayensis Speg.; weak twig parasite. S.
   America.
Loranthus spp.; a few species of strong growing parasitic
   shrubs on tree branches. Usually on old neglected trees,
   often near forest and jungle. C. America.
Macrophoma paraphysata Speg.; twig lesion. Argentina.
Marasmius (? equirinus); "horse hair," black cord blight.
   W. Indies, C. America.
Melanographium spinulosum (Speg.) Hughes; twig disease.
   Argentina, Dominican Republic, Panama, Paraguay.
Meliola citri (Briosi & Pass.) Sacc. and M. camelliae (Catt.)
   Sacc.; black mildew spots on foliage. Occasional in
   Mexico, El Salvador, Costa Rica, Panama, probably in
   S. America. M. penzigii Sacc.; black mildew. S.
   America.
Meloidogyne sp.; rootknot. Peru.
Monilia microspora Speg.; leaf lamina infection and collapse.
   Paraguay. (Spores suggestive of this organism obtained
   from leaf scorch in Panama.)
Mycena citricolor (Berk. & Curt.) Sacc.; leafspot. Tree
   near spot-diseased coffee in wet tropics. Guatemala,
   Costa Rica.
Mycosphaerella citri Whi. (conidial stage is Stenella, not
   Cercospora); greasy spot. Subtropical USA. M. loefgreni
   Noack, M. horii Hori and M. lageniformis Rehm (possibly
   same as above); on leaf and twig, large greasy spots.
   Florida, California, Guyana, British Guiana, Surinam,

Trinidad and Tobago, Nicaragua, C. America, Ecuador,
Brazil, Argentina, Dominican Republic.
Nectria cancri Ruts. f. aurantii Av.-Sac. and N. episphaeria
  Fr.; twig cankers. Brazil, Argentina, Puerto Rico,
  Colombia. N. vulgaris Speg.; dieback. Argentina.
Nothopatella lecanidium (Speg.) Sacc.; large-branch and trunk
  canker. Brazil.
Oidium tingitanium Car.; mildew. Ecuador, Mexico, Costa
  Rica, Panama, Subtropical USA, Colombia, Argentina.
Oospora citri-aurantii (Ferr.) Sacc. & Syd., fruit sourrot,
  also isolated from crushed flush leaves. Neotropics.
Oryctanthus spp. (O. botryostachys Eich., O. cordifolius
  Urb., O florulentus Urb., and O. ruficaulis Eich. some
  of these exhibit the prehensile leaf character); are para-
  sitic drooping shrubs with aerial roots that sink into host
  branches. C. America, Peru, Surinam, Trinidad.
Pellicularia koleroga Cke.; web-thread blight. Dominican
  Republic, Jamaica, Peru, Argentina, Panama, Trinidad.
Peroneutypa tuyutensis Speg. f. citri-limettae Speg.; rare
  foliage disease. Argentina.
Pestalotia spp.; sunken leaf and twig spots. Subtropical USA,
  W. Indies, C. America.
Peziza hematochlora Speg. var. microspora Speg.; bark in-
  habitant. Paraguay.
Phoma limonii Thuem. and P. puttemansii Ben.; leaf and
  young green-stem disease. Occasional in Mexico, Florida,
  Brazil. P. paraguayensis Speg. = Sphaeropsis paraguayensis
  (Speg.) Kuntze; on foliage. Paraguay.
Phomopsis citri Fawc.; on twigs. Uruguay, Brazil, Trinidad,
  Colombia.
Phorodendron spp. (P. ceibanum Trel., P. crassifolium Eich.,
  P. dimidiatum Eich., P. rensoni Trel., and P. robustis-
  simum Eich.); coarse and upright parasitic bushes on
  branches. Occasional in plantings near undisturbed jungles.
Phthirusa spp. (pyrifolia Eich., seitzii Kr. & Urb., squamu-
  losa Eich.); pendant parasitic shrubs. C. America, Suri-
  nam, W. Indies.
Phycopeltis sp.; algal plaques on leaf. Puerto Rico.
Phyllosticta spp. (P. aurantiicola Sacc., P. citricola Hori,
  P. disciformis Penz. f. brasiliensis Speg., P. hesperidearum Penz., P. longispora McAlp.); all leaf and twig
  attack. Brazil, Argentina, Paraguay, Ecuador, Peru,
  Colombia, Dominican Republic.
Physoderma citri Chi.; bark infection of benign nature.
  Florida, no doubt other areas.
Phytophthora spp. morphological identifications require specialists. Of much concern are the following: 1. P.
  cinnamomi Rands; root disease. S. America, C. America.
  2. P. citrophthora (Sm. & Sm.) Leo.; a sweet orange and
  lime footrot (the classic gomosis or mal-de-goma). Sour
  orange root is resistant to this, but is susceptible to
  tristeza virus. However, cleopatra mandarin is tristeza

resistant, and also resists gomosis, and it is now much used for a rootstock with sweet orange. There are other employable rootstocks. This most important Phytophthora also will cause shoot and leafblight. 3. P. palmivora Butl.; flush blight and gummosis. 4. P. parasitica Dast.; stem gumming, canker, and budding/grafting disease. 5. P. syringae Kleb.; leafblight. (No. 2 is most severe in moist areas, and in poor and heavy soils, but can occur in practically any place citrus is planted. No. 4, widespread, appears specially notable during latter part of rainyseason.)

Pleospora citrorum Sacc. and P. hesperidearum Catt.; usually as minor leafspots. California, dryer parts of S. America.

Polyporus hispidus Fr., P. zonalis Berk., P. squamosus (Huds.) Fr., and others; old tree-base infections, terminal rots. Scattered.

Polystictus hirsutus Fr.; on dying and dead roots. W. Indies, C. America. P. occidentalis Kl.; on old wood. S. America.

Poria albocincta Cke. & Mass. and P. heteromorpha Murr.; rots of branchwood fallen off old trees. W. Indies, Florida.

Pratylenchus pratensis Filip.; root lesion nematode. Florida, California, Brazil, Costa Rica, Peru.

Pseudohalenchus minutus Tarj.; root nematode. Florida.

Pseudomonas garcae Am., Teix., & Pinh.; leaf disease. Brazil. P. syringae Van Hall; tip blast. California, Guatemala, Argentina, Uruguay.

Psittacanthus calyculatus (DC.) Don; parasite bush. Mexico, C. America.

Pythium spp.; damping-off, rootlet rot, decline. Subtropical USA, W. Indies.

Radopholus similis Thorne; burrowing root nematode. Causes "spreading decline." Sweet orange suscept, lime resist. Widespread.

Rhabdospora flexuosa Sacc.; a bark disease. Brazil. R. hesperidica Speg.; secondary branch infection. Argentina, Paraguay.

Rhizoctonia solani Kuehn = perfect state is Thanatephorous cucumerus (Frank) Donk (sometimes given as Pellicularia filamentosa Rog.); seedling damping-off, basal disease in nursery. Neotropical.

Rosellinia bunodes (Berk. & Br.) Sacc. and R. pepo Pat. (sometimes R. necatrix (Hartig) Berl. and R. pulveracea (Ehr.) Fckl. var. platysporella Speg. are also given); rootrots. Often in special areas in newly cleared jungle land. W. Indies, C. America, S. America.

Rotylenchus sp.; root nematode. Peru.

Schizophyllum commune Fr.; branch and trunk rot. W. Indies.

Schizoxylon licheoide Speg.; branch rot. Argentina.

Sclerotinia fuckeliana Fckl. (Whetzeliana sclerotiorum (Lib.)

Korf & Dumont); twig blights.   Subtropical USA, C.
America.
Sclerotium rolfsii (Curzi) Sacc. = Pellicularia rolfsii (Sacc.)
West; young plant rot.   Florida, W. Indies. S. hesperidum
Sacc. = S. succineum Speg.; seedling and fruit rot. S.
America.
Scolecopeltis tropicalis Speg.; twig and fruit skin blemish.
Brazil, Argentina.
Septobasidium spp. (most commonly cited is S. pseudo-pedi-
cellatum Burt, but there are a dozen or more.); felt
diseases.   General.
Septoria arethusa Penz.; a major leaf disease.   Argentina,
Brazil. S. aurentiicola Speg. = Rhabdospora aurantiicola
(Speg.) Kuntze; leafspot.   Brazil. S. citri Pass.; severe
spotting on leaves.   Venezuela, Colombia, Argentina,
Brazil, Chile, Mexico, California, Bolivia.
Sirothyrium citri Bitanc.; a leaf spotting.   Brazil.
Solenoplea ceracea Viég.; on foliage.   Brazil.
Spegazzinia ornata Sacc.; on orange leaf.   Brazil, Argentina.
Sphaceloma (see Elsinoë)
Sphaeropsis tumefaciens Hed.; an orange branch-knot.   Domini-
can Republic, Cuba, Guyana, Jamaica, Florida, Peru.
Sphaerostilbe repens Berk. & Br.; redroot, associated with
insects.   W. Indies, S. America, Florida.
Stenella (see Mycosphaerella)
Stomiopeltis citri Bitanc.; black leaf disease.   Brazil.
Strigula complanata Fée; white lichen-spot.   Neotropical,
severe in old moist areas.
Struthanthus dichotrianthus Eich., S. emarginatus Blume, S.
orbicularis Blume, S. oerstedii Sta., S. syringifolius
Mar.; all are drooping parasitizing bushes (mistletoes),
some have prehensile leaves.   W. Indies, C. America,
northern countries of S. America. S. emarginatus in
Brazil.
Terana rosella (Speg.) Kuntze; attack of trunk.   Argentina.
Trentepohlia sp.; algal hairy growth.   C. America.
Tryblidiella rufula (Spreng.) Sacc. = Rhytidhysterium rufulum
(Spreng.) Speg.; branch-tip dieback.   W. Indies, C.
America, S. America.
Tylenchulus semipenetrans Cobb; slow decline.   California,
Trinidad and Tobago, Florida, Puerto Rico, Dominican
Republic, Jamaica, Argentina, Brazil, Chile, Costa Rica.
Ustulina vulgaris Tul.; rot of old roots.   Florida, C. America.
Valsa sp.; dying trees.   Colombia.
Verticillium tubercularioides Speg. (asco-state is Nectria vul-
garis Speg.); dieback complex.   S. America.
Viruses (Not always readily determined, require host range
studies, cross inoculations, etc.): 1. Exocortis or Scaly
Butt on orange in Uruguay, Bolivia, Florida, Surinam,
Trinidad and Tobago, Jamaica, Guadeloupe, Brazil. 2.
Grenning, not common, is in northern part of Neotropics.
3. Infectious chlorosis yellow vein disease in California,

Florida (I have seen symptoms, but no experimental proof
of it in C. America). 4. Leaf mottle chlorosis in Uruguay.
5. Psorosis, various strains such as A, B, concave gum,
alveolar. Symptoms are bark patches, long bark scales,
roundish bark scales, blind pocket, diamond-shaped bark
cavities, etc. Mostly these are on sweet orange, grapefruit, and lemon. In W. Indies, Subtropical USA, C.
America, S. America. 6. Psorosis crinkly leaf in Florida.
7. Stubborn disease. Believed by careful workers to be
caused by a mycoplasma. Bushy growth, bent growth,
little leaves in California. 8. Tatter Leaf virus. Scattered in W. Indies. 9. Tristeza quick decline. On sweet
orange, the top is a failure if it is grafted onto a sour
orange root, since in the presence of this virus the rootlets rot, the tree stunts, then dies. This occurs with
sweet orange tops if under them are some non-tolerant
rootstocks such as grapefruit, certain acid limes, sour
orange. With sweet orange no tristeza symptoms if
tolerant rootstocks are used such as sweet orange, common lemon, rough lemon, mandarins, tangelo. Tristeza
is found thus far in all of the American tropics except
Central America. 10. Xyloporosis virus/cachexia. Subtropical USA, Brazil.
Xanthomonas citri (Hasse) Dows.; bacterial canker. (Was, at
    one time in Florida, at great expense eradicated.) C.
    America, S. America, rare in W. Indies.
Xiphinema americanum Cobb; root nematode. W. Indies, C.
    America, Subtropical USA, Chile, Peru.

CITRUS spp. [continued]. Fruit infections list. (Where authorities
    are not indicated, refer to the list of scientific binomials for
    foliage and tree diseases, immediately above.)
    1. Alternaria citri; internal black rot, enters blossom end.
    General and scattered. 2. Amylorosa aurantiorum Speg.;
    fruit disease. S. America. 3. Aspergillus niger v. Tiegh.;
    black soft rot. General. 4. Botryosphaeria dothidea; fruit
    spot and stem-end rot. Dry parts of C. America, Subtropical
    USA. 5. Botrytis cinerea; store and windfall rot. Scattered,
    nearly general. 6. B. citricola Bri.; storage rot. Brazil.
    7. Candelospora sp.; skin lesion extending into albido.
    Florida, California. 8. Capnodium spp.; black molds that
    scrub off. General. 9. Colletotrichum spp.; anthracnose
    (at first superficial skin lesions, that later penetrate).
    General. 10. Diaporthe = Phomopsis; multiple small raised
    lesions, mud-cake and shark-skin, tear-stain, and stem-end
    rot. General. 11. Didymella aurantiiphila Speg.; fruit lesion.
    S. America. 12. Diplodia spp.; stem-end rot. Scattered.
    13. Elsinoë spp.; scab. S. America, W. Indies. 14. Er-
    wina sp.; bacteria in rot. Mexico. 15. Fusarium spp.; red
    rot. Scattered widely. 16. Guignardia citricarpa Kiely =
    Phoma citricarpa McAlp. and P. limonii Thue.; blackspots.
    Occasional W. Indies, C. America, S. America. 17. Insect-

caused apparent "disease on fruit" symptoms: irregular dark
lesions, rust-like, buckskin, silverscurf, tear-stain, scales,
nail-head, sunken spots. 18. Leptothyrium pomi Sacc.; flyspeck (small spots with entire edges). Occasional, more or
less general. 19. Nectria cancri; firm rot. Brazil. 20.
Nematospora coryli Pegl. and N. gossypii Ash. & Now.; dryrot associated with pentatomid feeding. Scattered. 21. Oospora citri-aurantii Sacc. & Syd.; yeasty sour-rot. Brazil,
Surinam, Florida, Peru, California (? widespread). 22.
Penicillium digitatum Pers.; greenmold contact-type rot.
General. 23. P. italicum Wehm.; bluemold contact-type rot.
General. 24. Physoderma citri; fungus bodies in fruit albido.
25. Phytophthora spp. especially citrophthora, and parasitica;
brownrot. In moist tropics. 26. Pleospora spp.; scabby
lesions and blackish rot. California, Uruguay, Paraguay,
Argentina, Brazil at least. 27. Pseudomonas syringae V.
Hall; blackpit, blast. California, Mexico, Guatemala, Florida,
Panama. 28. Pythium (? splendens); soft fruitrot. W. Indies.
29. Rhizoctonia solani; brown groundrot. Uncommon. 30.
Rhizopus nigricans Ehr.; blackened softrot. Scattered, probably in all citrus countries, reported mostly in S. America.
31. Septobasidium spp.; felt occurring along peduncle grows
onto fruit. Not common but seen in Brazil, C. America.
32. Sphaerostilbe aurantiicola Petch; red fungus on scale,
surface occurrence, really not a fruit attack. C. America,
Subtropical USA. 33. Sclerotinia spp.; cottonyrot. California, Chile. (for Sphaceloma, see Elsinoë.) 34. Sphaeropsis henriquessi Thuem; rot. Mexico. 35. Stomiopeltis citri
Bitanc.; flyspecks with fringes. Brazil. 36. Virus caused
markings: dark markings, zonate chlorosis, in green fruits
black spots with yellow corona, ripe fruit sunken spot deep
into rind, irregularly shaped lesions. Many different countries.
36. Xanthomonas citri; bacterial canker-forming scab. C.
America, S. America, but not frequent.

CLUSIA ROSEA Jacq. An epiphytic bush or small tree (not a parasite), valued for its ornamental effect, and is harvested in the
West Indies for its resin, it had religious connotations to indigenes of the Caribbean Region; "matapalo," "cupey" Sp. Po.,
"clusia" En.
    Amazonia clusiae Stev.; black leaf-mildew. Puerto Rico,
        Dominican Republic.
    Amphisphaeria clusiae Pat.; on twig and leaves. W. Indies.
    Asterina coriacella Speg. = A. solanicola Berk. & Curt.;
        black leaf-patches. Puerto Rico, Virgin Islands.
    Bertia clusiae Gonz. Frag. & Cif.; on leaves. W. Indies.
    Botyodiplodia theobromae Pat. = Diplodia spp.; stem canker.
        W. Indies.
    Cephaleuros virescens Kunze; algal leafspot. Puerto Rico.
    Clithris clusiae Tehon = C. platyplacum (Berk. & Curt.)
        Tehon and C. minor Tehon; leaf spots. Puerto Rico, Cuba,
        Dominican Republic, Trinidad.

Coccomyces clusiae (Lév.) Sacc.; large leafspots. W. Indies,
    C. America, Colombia, Venezuela.
Diplodia (see Botryodiplodia)
Ellisiella portoricensis Stev.; on dead leaves. Puerto Rico,
    Venezuela.
Grallomyces portoricensis Stev.; commensal development
    with black mildew fungi. W. Indies, C. America, Brazil.
Guignardia clusiae Stev.; striking leafspots. W. Indies,
    Nicaragua.
Helminthosporium clusiae Cif. & Gonz. Frag.; leaf and petiole
    disease. Dominican Republic.
Lophodermium platyplacum Sacc.; leaf disease and defoliation.
    Puerto Rico, Cuba, Trinidad and Tobago. (Clithris sp.
    may be synonymous.)
Meliola sp. = Amazonia clusiae.
Mycena citricolor (Berk. & Curt.) Sacc.; leafspot by inocula-
    tion. Puerto Rico.
Mycosphaerella clusiae Stev.; leafspot. Puerto Rico only.
    M. guttiferae Miles; a leafspot more widespread in W.
    Indies.
Parodiopsis melioloides Maubl.; black mildew. W. Indies.
Pestalotia clusiae Griff. & Maubl.; sunken leafspots. Brazil.
    P. funerea Desm.; sunken leafspots. Puerto Rico, C.
    America.
Phyllosticta clusiae Stev.; leafspots. Puerto Rico, Brazil.
    P. clusiae-rosae Gonz. Frag. & Cif.; leafspots. W.
    Indies.
Physalocria clusiae Syd.; on decaying stalk. Puerto Rico.
Physarum bitectum Lis.; epiphytic myxomycete. Puerto Rico.
Septogloeum clusiae Karst. & Har.; leaf disease. Colombia.
Sporodesmium millegrama Berk. & Curt. (genus also given
    as Sporidesmium); on leaves. S. America.
Stictis follicola Berk. & Curt.; on decayed leafspot and old
    petiole. Puerto Rico, Cuba.
Strigula (? companata); algal tomentose spots on leaves.
    Puerto Rico under cool high rainfall conditions.
Trentepohlia sp.; hairy algal growths, epiphytic. Puerto Rico.
Uredo clusiae Arth.; rust. Puerto Rico only.
Xylaria aristata Mont.; in aerial root and at base of old dying
    plant. Might have been anywhere Clusia grows. I have
    seen several old plants in Cuba and Puerto Rico with these
    much blackened bases.

COCCOLOBA UVIFERA Jacq. A native tropical American tree for
    seaside and popular ornamental plantings, salt hardy; "uva-de-
    playa" Sp.; "seagrape" En.
        Asterina coccolobae Ferd. & Winge; black leafpatches. Virgin
            Islands, Puerto Rico, Honduras.
        Cassytha filiformis L.; the pungent dodder. Puerto Rico, and
            probably all W. Indies.
        Cephaleuros virescens Kunze; algal leafspots. Occasional in
            Puerto Rico.

Clypeotrabutia portoricensis (Stev.) Seav. & Char. = Phylla-
  chora portoricensis (Stev.) Petr.; raised tarspots on leaves.
  Puerto Rico, Virgin Islands.
Dictyonella erysiphoides (Rehm) Hoehn.; a grayish sootymold.
  W. Indies, S. America.
Endothia coccolobii Viz.; branch and trunk disease.   Bermuda.
Fabraea coccolobae Speg.; brown leafspots.   Brazil.
Gnomonia pulcherrima Seav. & Wat.; leaf disease.   Bermuda,
  Puerto Rico.
Irenopsis rectangularis Stev.; one of many black mildews.
  Puerto Rico.
Lembosia spp. (coccolobae Earle, philodendri P. Henn., porto-
  ricensis Speg., tenella Lév.); black mildews.   W. Indies,
  Florida.
Melasmia coccolobae Stev.; black leafspots.   Puerto Rico.
Meliola spp. (amphitricha Fr., coccolobae Stev. & Teh.,
  praetervisae Gail.); black mildews.   W. Indies, S.
  America.
Micropeltella constricta Stev. & Mart.; moldy spots.   S.
  America.
Oryctanthus ruficaulis Eichl.; a parasitizing woody plant.
  Surinam.
Oudemansiella canari Hoehn.; agaric disease.   Florida, Puerto
  Rico, S. America.
Pestalotia coccolobae Ell. & Ev.; small lesions on leaf and
  petiole.   Mexico, Costa Rica, Florida, Puerto Rico, W.
  Indies.
Phthirusa sp.; parasitizing woody plant.   Puerto Rico.
Phyllachora coccoloba Speg., P. simplex Starb.; black
  stromata on leaves.   W. Indies, Argentina, Paraguay,
  Florida.
Phyllosticta coccoloba Rehm and P. coccoloba Rehm f. domini-
  cana Cif. & Gonz. Frag.; abundant purplish leafspots.
  Puerto Rico, Bermuda, Cuba, Hispanola.
Physalospora coccolobae Rehm; foliage disease.   Brazil.
Scolecopeltis pachyasca Speg.; intensely black mildew.   Puerto
  Rico.
Sporotrichum globuliferum Speg. = Beauveria globulifera
  (Speg.) Pic.; insect parasite which has appearance of leaf
  disease.   Neotropics.
Stictis cocolobii Seav. & Wat.; stem and leaf disease.   Ber-
  muda, Florida, C. America.
Strigula sp.; mottled gray lichen spots.   Puerto Rico.
Struthanthus sp.; vine-like parasitizing bush.   Puerto Rico.
Uredo coccolobae P. Henn. = U. uviferae Syd.; leafrust.
  Florida.
Verticicladium effusum Earle; a semi saprophyte on leaves.
  Florida.
Virus: Symptoms of mottle, shoe-string leaf malformation.
  Cuba.

COCOS NUCIFERA L.   Where jungle trees are viewed as natural

enemies, millions of planted coconut palms are friends of which
peasants say they are "living in most vigor within earshot of
man or sound of the sea." These palms furnish shade, orna-
ment, roof thatch, mat, walls, rot-defying wood, meat-like
food, oil, salad, beverages, shade, tools and fibers; "cocotero"
Sp., "coqueiro" Po., "coconut palm" En.
Agaric (not Marasmius); root decay.   W. Indies.
Aphelenchus cause of redring (see Rhadinaphelenchus)
Bacillus coli Miq.; isolated from budrot.   Honduras, Colom-
    bia.
Bacterial leafblight, observed in British W. Indies.
Botryodiplodia theobromae Pat. (Many synonyms for example:
    the asco-state Botryosphaeria quercuum (Schw.) Sacc.
    with conidial form Diplodia sp., with also the asco-states
    often given as the primary cause Physalospora rhodina Cke.
    or P. obtusa (Schw.) Cke.); spot of leaf lamina, at times
    on midrib resulting in collapse of leaflet end, occasionally
    severe on petiole resulting in its break, husk infection
    (from which I secured perfect state in moist chamber), in-
    florescence spot in weak tree, fruit drop.   Neotropics.
Bud failure, sporadic, from rats, mechanical causes, and old
    age.
Cassytha filiformis L.; dodder by inoculation in nursery.
    Puerto Rico.
Catacauma torrendiella Bat.; scabby leafspot.   Brazil.
Cephaleuros virescens Kunze; algal leafspot on young trees
    sheltered from much saltspray.   W. Indies, C. America,
    S. America.
Cephalosporium lecanii Zimm.; insect parasite spores scraped
    from spots on leaflets.   S. America.
Ceratocystis fimbriata Ell. & Halst.; canker and dark vascu-
    lars of trunk.   Guatemala.   C. paradoxa (Dade) Mor. =
    Thielaviopsis paradoxa (de Sey.) Hoehn.; bleeding stem,
    trunk blood, "pineapple" disease.   Mexico, Puerto Rico,
    Subtropical USA, El Salvador, Colombia, Venezuela, Brazil,
    Jamaica, Guayana, Dominican Republic, Trinidad.
Chaetomium orientale Cke.; weak parasite associated with
    decay of old leaf.   General.
Colletotrichum gloeosporioides Penz. = Glomerella cingulata
    (Stonem.) Sp. & Schr.; anthracnose leaf disease.   Often
    not serious.   Neotropical.
Corticium penicillatum Petch; stringy thread-blight and whitish
    spore-bearing stroma, in moist locations away from
    beaches.   W. Indies.   (See Pellicularia.)
Cytospora palmicola Berk. & Curt.; leafspot.   Mexico.
Echidnodes cocoes Syd.; mildly parasitic black mildew.
    Dominican Republic, El Salvador.
Exosporium durum Sacc. (= ? E. palmivorum Sacc.); notably
    dark colored large spores from leafspot.   W. Indies.
Fomes ligneus (Kl.) Bres.; root/trunk disease.   Friable rot,
    fast field spread.   W. Indies, C. America.   F. noxius
    Cor.; brownroot, honeycomb rot inside trunk.   Slow field

spread. W. Indies, Brazil (and other species of Fomes
are reported). Widespread.
Fusarium moniliforme Sheld.; a husk-infection of nut. Co-
lombia. Fusarium sp.; associated with budrot but not
primary cause. From isolations in El Salvador, Guate-
mala, Honduras, Panama, Florida, Puerto Rico, Virgin
Islands.
Ganoderma lucidum (Leyss.) Karst.; root disease. Scattered.
Graphiola phoenicis (Moug.) Poit. (has been called G. cocotis
and G. cocoina); false smut on leaves. General.
"Gum splits" with poor to no inflorescence formed. Thought
to be physiological but exact cause undetermined. W.
Indies.
Herpotrichia albidostroma Sacc.; felt. Puerto Rico, Sub-
tropical USA, El Salvador.
Hypoxylon quisquiliarum Mont.; bursting midribs of leaves. S.
America.
Laestadia cocophila (Cke.) Sacc.; leaf spot disease. British
Guiana.
Leaf-bitten symptoms. Numerous causes, edge infections
from: Diplodia, Colletotrichum, Phytophthora, Strategus
(rhinoceros beetle) feeding, etc. Scattered occurrence.
Leptosphaeria sp.; diseased leaflet tips. El Salvador.
Lightning strike. Often kills; sometimes at first only killed
bud shows injury, is then called "false budrot."
Macrophoma cocos Pass.; leaflet disease. Mexico.
Marasmius palmivorus Shar.; thread blight. W. Indies, C.
America, S. America but infrequent.
Memnomium palmicolum Cke.; scabby leaf spots. Scattered
occurrence.
Metasphaeria cocogena (Cke.) Sacc.; spots on leaves. S.
America.
Microcera coccophila Desm. (cf. Sphaerostilbe); root insects
diseased, cause discoloration of tissues. Trinidad,
Puerto Rico, Costa Rica (Caribbean coast.)
Neohenningsia suffulta (Berk. & Curt.) Petch; leafsheath
disease. W. Indies.
Pellicularia koleroga Cke. (some consider the Corticium,
above, as synonymous, others doubt it); typical thread-
blight, "black rot." Florida, C. America, W. Indies.
Pestalotia fuscens Cke. f. sacchari Wakk.; leafblight.
Jamaica. P. funerea Desm., P. leprogena Speg., P.
palmarum Cke.; sunken grayspots, leafbreak. Subtropical
USA, W. Indies, C. America, S. America.
Phialea leucopsis (Berk. & Curt.) Sacc.; on nut. (? Locality).
S. America.
Phoma cocoina Cke. and P. palmarum Cke.; leaflet-black-
spot. General.
Phomopsis cocoes Petch; semi-sunken fungus on nuts.
Florida, Colombia, El Salvador.
Phyllosticta sp.; leafspotting. Florida, C. America.
Phytophthora palmivora Butl. has been given as P. faberi

Maubl. (on coconut seems to be a special strain in C.
America separate from that on cacao); severe budrot, leafdrop, wilt. Not in Florida now, after costly eradication.
In practically all other tropical American coconut-growing
countries. P. parasitica Dast.; leafstalk rot, peduncle
rot. Occasional.

Pyricularia grisea (Cke.) Sacc. (perfect state Ceratosphaeria
grisea); Small gray spot on leaflet. C. America, S.
America.

Pythium sp.; trees wilted. Florida.

Rhadinaphelenchus cocophilus (Cobb) Goodey = Aphelenchus
cocophilus Cobb (beetle vector is Rhynchophorus palmarum,
but disease is also said to be carried at times by the
black vulture, Coragyps atratus foetans); redring, spectacular and fatal nematode disease. Venezuela, Brazil, Surinam, Guyana, Trinidad, Grenada, St. Vincent, El Salvador,
Puerto Rico, Dominican Republic, Haiti, Mexico.

Rosellinia sancta-cruciana Ferd. & Winge; root disease. W.
Indies, C. America (probably in S. America).

Scoptria chlorina Petr.; weak parasite on leaflets. Dominican
Republic.

Sooty mold on nursery seedlings in Cuba.

Sphaerella cocophylla Cke., S. zonata Ell. & Ev.; leaf spots.
W. Indies.

Sphaeropsis palmarum Cke.; leafspot and petiole infections.
Scattered.

Sphaerostilbe coccophila Tul; insect disease (see Microcera).

Sporotrichum foliaecolum Lk.; abundant insect disease develops symptoms that suggest leaflet infections. C.
America.

(? Strigula spp.); varied lichen leafspots. General.

Tapering stem, poor nutrition.

Trichobasis palmarum Cke.; leaflet rust. S. America.

Trichosporium palmicolum (Cke.) Sacc., fungus recovered
from palm insects. Scattered.

Ustulina spp.; root disease on old trees. C. America, S.
America.

Viruses: 1. Lethal yellowing and wilt also called "unknown
disease," Mycoplasma bodies present, in Jamaica, Cayman
Islands, Cuba, Bahamas, Dominican Republic, Haiti,
Florida, British Guiana, from symptoms possibly also in
El Salvador and Guatemala. 2. Bronze-leaf drop. Ja-
maica, Dominican Republic, Trinidad, Venezuela. 3.
Malda folha (rolling and malformed leaves). Brazil.

Valsa chlorina Pat.; root disease. S. America. V. chlorina
Pat. f. dominicana Gonz. Frag. & Cif.; root infections.
W. Indies.

Xylaria scruposa (Fr.) Berk.; root attack. W. Indies.

Yeast-like organisms causing leaf damaging infections. W.
Indies.

Yellowing and chlorotic spotting from poor root growth, sometimes apparently potash "hunger," other times denotes
drought. Scattered.

CODIAEUM VARIEGATUM Blume.   A very popular and remarkably
variable woody bush with leaves that are many-colored, may be
twisted and curiously formed that generally come true from cuttings, and have been known in the Orient for centuries and grow
well in the American tropics; "curiosodades," "crotón de jardín"
Sp. Po., "garden croton" En.

Asteroma codiaei Noack and other genera and species;   sootymolds.   W. Indies, C. America, S. America.

Cassytha filiformis L.; green pungent smelling dodder.   Puerto
 Rico, Florida.

Cephaleuros virescens Kunze; small algal leafspots, not abundant.   W. Indies, C. America, S. America.   (Seems difficult to establish this alga on uninjured croton leaves, but
 when bitten or scratched the lichen from it more readily
 attacks the epidermis.)

Colletotrichum gloeosporioides Penz. (a common reported
 synonym on this host is: Gloeosporium sorauerianum Atk.
 the asco-state is Glomerella cingulata (Stonem.) Sp. &
 Schr.); common anthracnose.   General but benign except on
 weakened plants.

Curvularia sp.; on a ragged leaf.   Honduras.

Cuscuta americana L.; yellow dodder.   W. Indies, Florida,
 C. America.

Fomes noxius Cor.; brownroot disease with typical sporocarp.
 Puerto Rico.

Ganoderma lucidum (Leyss.) Karst.; on old dead stump.
 Puerto Rico.

Mycena citricolor (Berk. & Curt.) Sacc.; leafspot (occurrence
 only near diseased coffee).   Scattered.

Nematodes, not determined.   Part of cutting disease complex.
 Panama, Cuba, El Salvador.

Pellicularia koleroga Cke.; thread blight.   Florida, Puerto
 Rico, Costa Rica, El Salvador, Trinidad, Brazil.

Phoradendron spp.; woody plant parasites.   Puerto Rico.

Phthirusa sp.; woody vine-like parasite.   Costa Rica.

Phycopeltis sp.; algal plaque.   Costa Rica, Puerto Rico.

Phytophthora sp.; cutting failure.   Puerto Rico.

Pythium sp.; cutting failure.   Puerto Rico, Cuba.

Rhizoctonia bataticola Taub. (Macrophomina phaseoli Taub.
 Ashby); charcoal root disease.   Brazil.   R. solani Kuehn;
 cutting failure.   Puerto Rico.

Rosellinia spp.; root diseases.   In moist areas.

Sphaeropsis codiaei Cif. & Gonz. Frag.; branch disease.
 W. Indies.

Strigula complanata Fée; white lichen-spotting.   More common
 than alga alone.   General.

COFFEA ARABICA L., C. CANEPHORA Pierre, and C. EXCELSA
Chev.   By far the predominant species of these three that are
grown in the Western Hemisphere tropics is C. arabica, called
"café" Sp., "cafeeiro" Po., and "arabica coffee" En.   Over 60
species are in the genus, all originated in Africa and Asia but

representative trees of most can be found in experiment stations
in Latin America. (Note: Coffee is one of the two most-liked
stimulant drinks of humanity and it has a harmless hold on people. The good effect is from the alkaloid caffeine which even
has medicinal use; it is also extracted and added in small
amounts to many "soft" drinks. Dry coffee is the most stable,
most important cash crop of Latin America and it has the pre-
mier place there, affecting international monetary exchange,
while inside its countries it is the lifeblood of much business
and labor. Vigorous coffee growers and associations of them in
the Americas make every effort to advance scientific agriculture,
including plant pathology, lending outstanding influence in the
American tropics. (The following disease items are notable, to
me, and mostly on C. arabica but a few reported saprophytes
and some that are at most very weak parasites are not listed.)
  Aithaloderma longisetum Syd.; scabby mold leafspot. S.
    America.
  Aleurodiscus moquiriarum Viégas; branch disease. Brazil.
  Antennaria sp.; a sootymold. Surinam.
  Anthostomella coffeae Del.; leafspot, stem spot and dieback.
    Mexico, Panama.
  Aphelenchoides spp.: A. bicaudatus Goodey in Guatemala, A.
    coffeae Stein. in S. America, and A. parietinus Stein. in
    Brazil are all inhabitants of rhizosphere and by some be-
    lieved questionably injurious.
  Aphelenchus avenae Bas. and A. coffeae Noack; rhizosphere
    feeders, questionably injurious. In C. America and Brazil
    respectively.
  Armillaria mellea Vahl = Armilariella mellea (Fr.) Karst.; a
    not exactly common rootrot. Colombia, Peru, Surinam,
    Dominican Republic, Cuba.
  Ascochyta coffeae P. Henn.; darkedged brown leafspot. Brazil,
    Paraguay.
  Aspergillus gigas Speg. and A. niger Ehr.; leafrot and black-
    rot of fruit and pergamino. General.
  Aspergilopsis intermedia Speg.; on decayed leaf and harvested
    fruit. Brazil, Argentina, Paraguay.
  Asterina coffeicola Ell. & Ev. and A. pelliculosa Berk.; black
    mildews. C. America, S. America.
  Atichia sp.; black mildew. In moist warm C. America, S.
    America.
  Auricularia auricula-judae (Fr.) Schr. and A. polytrichia
    (Mont.) Sacc.; on dead pruned wood. Colombia, Costa
    Rica.
  Bacteria (unidentified), aseptically isolated from root and trunk
    tissues. Costa Rica. (Bacterial decay of wood is reported
    in Guatemala.)
  Blepharospora cambivora Preti.; blue gangrene of stem cortex.
    Colombia, Guatemala.
  Boron deficiency results in death of growingtips, numerous
    palmate-formed branches, leafedge necrosis, aborted
    flowers. Scattered.

Botryodiplodia theobromae Pat. (synonyms for conidial state
are: Diplodia theobromae (Pat.) Now., D. coffaeiphila
Speg., D. natalensis P. Evans, D. assumptionis Speg.,
Lasiodiplodia theobromae (Pat.) Gr. & Maubl.), its perfect
state often given as Physalospora rhodina Berk. & Curt.
is Botryosphaeria quercuum (Schw.) Sacc. (synonyms: P.
coffeicola Speg., P. obtusa (Schw.) Cke.); associated and
actually primarily involved in dieback, often sclerotia form
on coffee stems. Neotropics.

Calcium deficiency causes paling, but not exactly whitening,
of leaf green; fruit drop. Scattered.

Capnodium coffeae Pat. is the imperfect state with synonyms:
C. brasiliense Putt. = Paracapnodium brasiliense (Putt.)
Speg. and C. trichostomum Speg. in which the perfect state
is Microxyphium coffea (Pat.) Speg. with its synonyms of
Microxyphiella trichostoma Speg. and M. trichostoma (Speg.)
Bat. & Cif.; sootymold also called hollin and fumago. Neo-
tropics.

Cassytha filiformis L.; laboratory infection. Puerto Rico.

Cephaleuros virescens Kunze also called C. coffeae Went.; the
algal leafspot, "red rust." W. Indies, C. America, S.
America.

Cephalosporium omnivorum Crand. & Guis.; branch and leaf
disease. El Salvador. C. zonatum Stahel; zonal leafspot.
Costa Rica, Peru, Surinam, Trinidad, Puerto Rico, Cuba,
El Salvador, Guatemala, Colombia.

Ceratocystis fimbriata Ell. & Halst. (there seems to me to be
a species physiologically adapted f. coffeae, distributed by
fly and raindrop); canker, woodstain. Costa Rica, Ecuador,
Colombia, Surinam, Guatemala.

Cercospora coffeicola Berk. & Cke., with synonyms: C.
coffeicola Speg., C. herrania Far. (see Mycosphaerella);
common leafspot with gray center, branch lesion, fruit
disease, peduncle infection. General.

Chaetophoma coffeicola Av.-Sac.; stemlesions. S. America.

Chaetostroma sp.; on foliage. Mexico.

Cladosporium (herbarum-like); on two and three year old
leaves. Neotropical. Bean infection on patio, affecting
flavor. Brazil.

Clusia spp.; tree-like smothering epiphytes. Brazil.

Clypeolum megalosporium Speg. = Schizothyrium rufulum
(Berk. & Curt.) Arx; a blackening leaf disease. Costa
Rica.

Colletotrichum gloeosporioides Penz. is the conidial and parasitic state with a few synonyms such as: C. coffeanum
Noack, C. coffaeophilus Speg., C. brachysporum Speg.,
Gloeosporium apiosporum Speg., G. spegazzinii Sacc.,
and its asco-state is Glomerella cingulata (Stonem.) Sp. &
Schr.; anthracnose on stem, fruit spot, angular spreading
leafspot, part of dieback complex, flower blast (I have isolated it from roots and in a few instances from seedling
tree stems showing no external disease symptoms).

Neotropics. (Note: In Central America this organism
shows it has at least one race specially adapted to coffee:
C. gloeosporioides Penz. f. coffeanum (Noack) ? Wellman.)
Coniothyrium sp.; fruitrot, leafspot. Brazil.
Corticium salmonicolor Berk. & Br.; pink disease (especially
in sun grown plants in C. America), causes fruit drop.
W. Indies, C. America, S. America.
Criconemoides sp.; nematode in root area. Peru.
Crinipellis septotrichia Sing.; wood attack with small external
mushrooms on main trunk. Guadeloupe, Cuba, Puerto Rico,
Bermuda.
Cryptosphaerella coffaephila Speg.; on branch and stem. Ar-
gentina, Paraguay.
Cucurbitaria crustosa (Speg.) Pet. & Syd.; wound infection.
Brazil.
Curvularia sp.; causes a slow spreading canker. Costa Rica.
Cuscuta americana L.; dodder in plantation of liberica
coffee in Puerto Rico, and in an arabica plantation in
Costa Rica.
Cyathus lesueurii Tul.; birds-nest fungus on coffee chaff.
Cuba.
Cylindrocarpon sp.; reddish branch canker. C. America.
Diatrypella coffaeicola Speg.; branch disease. Brazil, Para-
guay.
Dictyothyriella mucosa Syd.; a black mildew. Dominican
Republic.
Didymella coffaeicola Speg.; branch disease as well as
peduncle and fruit rot. S. America.
Didymosphaeria coffeicola Speg.; on dieback branches. Brazil,
Paraguay.
Dimerosporium coronatum Speg.; hyperparasite of Capnodium.
Scattered in C. America.
Diplodia (see Botryodiplodia)
Ditylenchus procerus Filip.; nematode at root system. Guate-
mala.
Epicoccum sp.; found on epicarp of fruits. Brazil, C.
America.
Euryachora coffeicola Av.-Sac.; cold region leafspot. Brazil,
Argentina.
Flammula aureobrunnea Berk. & Curt.; an agaric associated
with rot of old cut trunks. S. America, C. America.
Fomes spp.; in old plantations rootrot associates and trunk
rot disease. W. Indies, C. America, S. America.
Fusarium spp.; 1. F. lateritium Nees, on arabica and libe-
rica coffee at lowland locations causing occasional stem
canker from pruning. Panama, Costa Rica, Guatemala,
Colombia. 2. F. moniliforme Sheld., probably saprophytic,
isolations from moldy coffee bean shells. Costa Rica. 3.
F. oxysporum Schl. f. sp. coffeae (Alv.) Wellm., on arabica coffee at medium low elevations in slightly acid soils
causing wilt with discolored cortex vasculars. C. America,
Puerto Rico, Peru. 4. F. roseum Lk., on arabica,

liberica, and canephora coffees, peduncle weakening, fruit-
branch canker, isolated from patio dried whole cherry, and
from enlarged leafspots.   C. America, Brazil, Colombia.
5.   F. solani (Mart.) Sacc., associated with dieback of
fruiting branches after overbearing, part of general dieback
complex, from blackened feeder roots.   Costa Rica, Puerto
Rico.   All of these fusaria also have been secured from
damping-off and seedling collar-rot diseases in C. America.
Fusicoccum sp.; on mummied fruit.   Guatemala (Pacific
    slope).
Ganoderma sp.; accompanying death of old trees.   Surinam.
Gleoporus thelephoroides Bres. (? = Polyporus sp.); on old
    pruned discarded branches.   C. America.
Helicobasidium compactum Boed.; typical felt-sheath over root-
    harboring insects, result often is collarcanker.   El Salva-
    dor, Colombia, Costa Rica, Guatemala.
Helicotylenchus erythrinae Gold. and H. nannus Stein.; root
    feeding nematodes.   W. Indies, C. America, S. America
    (scattered occurrences).
Helminthosporium sp.; small black leafspot.   Nicaragua, Costa
    Rica, El Salvador.
Hemicriconemoides sp.; nematode from soil next to roots.
    Peru.
Hemileia vastatrix Berk. & Br.; the yellow leafrust.   (Acci-
    dentally arrived on diseased seedlings in Puerto Rico in
    1903 but was completely eradicated.)   Seen in 1970 in
    Brazil, has spread to Argentina, Paraguay, probably soon
    to other countries.   (In 1976 spread to Nicaragua.)
Hendersonia coffeae Del.; mild leaf disease.   Mexico.
Hirneola coffeicola Berk.; part of bark-rot complex.   C.
    America.
Hymenochaete noxius Berk.; stemcanker of young tree.   S.
    America, Guatemala.
Insect related effects: Bacterially killed aphids result in un-
    usual leaf smears.   Beauveria Sp. is a fungus parasitizing
    Stephanoderes berry borer.   Bronzing from red spider
    (actually not an insect).   Capnodium spp. and relatives,
    subsisting on insect honeydew.   Ceratocystis sp. distributed
    by fly.   Helicobasidium sp. felt growth around feeding in-
    sects.   Mildews, dark colored, believed by some to be
    insect related.   Verticillium sp. parasite of insects.   Virus
    often insect carried.
Iron deficiency, leaf chlorosis becoming whitish, with narrow
    green veins, short internodes, no fruiting.   Scattered.
Kalmusia coffeicola Speg., and K. oranensis Speg.; on dead
    branches.   Argentina.
Koleroga (see Pellicularia)
Laestadia coffeicola Speg.; whitish leafspot.   Costa Rica.
Lentinus tener Klo., L. velutinus Fr., L. villosus Klo.; decay
    of stump and of base on aged trees.   C. America, S.
    America.
Lenzites tenuis Lév.; leather-cap rot on old tree.   Scattered
    in mountains of C. America.

Leptosphaeria coffeigena (Berk. & Curt.) Sacc. = L. pusilla
(Speg.) Sacc.; leaf and fruit-branch disease. Brazil,
Guatemala, Panama, Venezuela, Mexico, Paraguay, Costa
Rica.
Leptothyrium costaricense Speg.; fruit skin flyspeck. Costa
Rica, El Salvador. L. discoideum (Cke.) Sacc., flyspeck.
S. America.
Leucothyridium crustosum Speg. = Cucurbitaria crustosa
(Speg.) Petrak & Syd.; branch disease. Argentina.
Lichens, smothering excess growth under foggy and cloud-
forest conditions (although Strigula is a lichen, not requiring
fog or excess moisture).
Limacinia coffeicola Putt. = Trichomerium coffeicola (Putt.)
Speg.; a sootymold. S. America.
Lisea tonduzii Speg.; disease of green fruit epicarp. C.
America.
Longidorus sp. (very similar to Xiphinema); root nematode.
Puerto Rico.
Lophiotrema coffeicola Speg.; rot of old bark. Argentina.
Loranthus avicularia Mart. in Venezuela, L. brasiliensis Spr.
in Brazil, L. orinocensis Spr. and L. parviflorus Sev. in
Venezuela, Loranthus spp. in El Salvador and Costa Rica.
All flower-bearing parasitic mistletoe-like bushes seen at-
tacking mature coffee trees.
Manganese deficiency; leafspotting rare in occurrence.
Marasmius bermudensis Berk. and M. equirinus Mull.; cord
or horse-hair blights. Nicaragua, W. Indies. M. viega-
sii Sing., S. America.
Massarina (see Pseudodiaporthe)
Meliola glabra Berk. & Curt. and Meliola spp.; black mildews
on leaves. In semi-moist areas, Neotropics.
Meloidogyne coffeicola Lord. & Zam., M. exigua Goel.,
Meloidogyne spp.; rootknot. Scattered in Dominican Republic, Brazil, El Salvador, the Guyanas, Guatemala,
Puerto Rico, Peru, Colombia, Barbados, probably elsewhere.
Metasphaeria bifoveolata Speg. and M. coffaeiphila Speg.; not
uncommon but almost microscopic ascocarps on small
branch, fruit and leaf lesions. Former in W. Indies and
C. America, second in S. America.
1. Micropeltis applanata Mont. with synonyms M. coffeicola
P. Henn. (= M. longispora Earle = Scolecopeltis longispora (Earle) Toro), 2. M. mulleri J. A. Stev., 3. M.
tonduzii Speg.; black mildews. Neotropics.
Microthyrium laurantii P. Henn.; black mold of stem and
leaf. Mexico.
Microxyphiella (see Capnodium)
Monilia candida Bon.; harvested fruit rot. Scattering occurrence.
Mosses, extravagant growth under certain conditions results
in smothering effect that needs control. Often in wet
mountainside plantations.

Mucor sp.; ripe stored fruitrot, moldy beans.   General.
Mycena citricolor (Berk. & Curt.) Sacc. (synonyms: Omphalia
    flavida Maubl. & Rang., Stilbum flavidum Cke., Stilbella
    flavida Lind., Sphaerostilbe flavidum Mass.). (Note: This
    gamma developing fungus has a very wide host range, as
    over 500 species are suscepts, in over 50 separate fami-
    lies.   I have even found it infecting mosses and liverworts.
    It will not produce infection bodies on certain hosts, and
    some--not many--plants appear immune from it.) The
    "imperfect" state is the gemma or infection body that is not
    a conidium, most often it is rain-drop-splashed-around, and
    the "perfect" state is a dwarf yellow agaric with a relatively
    small number of basidiospores borne on its few gills.   Field
    spread is by gemmae; causes rounded leaf and fruit spots
    with, while young, in each a straw-colored central point,
    and the defoliation is severe.   Not in subtropical USA,
    mainly where heavily shaded coffee is grown in all countries
    of W. Indies, C. America, S. America.
Mycosphaerella coffeae Noack perfect state of Cercospora;
    brown spreading leafspots.   Brazil.   M. coffeicola (Cke.)
    J. A. Stev. & Wellman; leafspots brown with narrow gray
    margins and fungus ascocarps.   Jamaica, C. America, S.
    America.
Myrothecium roridum Tode; round leafspot, stem girdle.   Costa
    Rica, Guatemala, Colombia, El Salvador, Brazil.
Myxosporium noackianum Allesch.; possibly a semi-saprophyte,
    developing on bark of dieback fruit branch, in old leafspots.
    El Salvador, Guatemala, Brazil.
Nectria anisophilum Pic.; pruning disease.   Costa Rica.   N.
    dodgei Heis.; machete-cut disease.   Guatemala, El Salva-
    dor, Costa Rica, Mexico.   N. tropica Toro; branch disease.
    Colombia, Guatemala.
Nematospora coryli Pegl.; seed, or bean, injury.   Scattered.
Nigrospora sp.; blackening fruit-rot, spores secured from
    green-fruit drop.   C. America.
(Nolanea cubensis Murr., on leaf, apparently a small agaric,
    may have been an early collection of Mycena citricolor.
    Cuba.)
Odontia coffaeina Speg. with Hydnum-like sporocarp; branch
    and trunk disease.   S. America.
Oraniella coffeicola Speg. = Massarina coffeicola (Speg.) Bose;
    common on branches.   S. America.
Oryctanthus botryostachys Eich.; a creeping mistletoe-like tree
    branch parasite.   Surinam, Costa Rica.   O. cordifolius
    Urb.; similar creeping parasite.   El Salvador, Guatemala.
Pellicularia koleroga Cke.; the typical common web-thread
    blight and rot of leaf, fruit branch and fruit.   W. Indies,
    C. America, S. America.   Pellicularia sp. (morphologically
    I found it was not koleroga); aerial threads with a thin
    stroma on which were the basidia, associated with dieback.
    Panama, El Salvador.
Penicillium gliocladioides Speg. and some other unidentified

species; one causes seedbed damping-off, fruitrot, one associated with leaf infection and dieback in El Salvador,
moldy bean results in flavor change. Neotropics.
Pestalotia funerea Desm.; associated with dieback and fruitrot.
  C. America. P. versicolor Speg. var. vagans Speg.; dieback, fruitrot. Costa Rica, Brazil, Colombia.
Phaeosaccardinula costaricensis (Speg.) Theiss.; leaf disease
  in Costa Rica. P. javanica Yam.; scabby leafspot. Jamaica.
Phoma coffaeicida Speg.; black lesion of cherry. Costa Rica,
  Guatemala. P. costarricensis Ech.; highland tip-blast of
  branches. Costa Rica, Guatemala, Panama, El Salvador.
Phomopsis sp.; fruitrot. Brazil.
Phoradendron spp.; coarse growing parasitizing bushes on
  coffee trees: P. crassifolium (Pohl.) Eich., and other
  species in Brazil, P. chryssocarpum Krug & Urban in
  Puerto Rico, P. piperoides (HBK.) Trel. in El Salvador
  and Puerto Rico, P. rensoni Trel. in El Salvador, and
  P. rubrum Gr. in Paraguay.
Phthirusa bicolor Engl. and P. pirifolia (HBK.) Eich.; vine-
  like, weeping woody parasites. C. America. Rarely in
  W. Indies.
Phycopeltis epiphyton; tomentose algal growths on old tree-
  branches. C. America, W. Indies.
Phyllosticta coffeicola Speg. (note: Spegazzini described this
  parasite in 1896; it was again described using the same
  specific name, by Delacroix in 1904; "Speg." therefore has
  priority); tan colored leafspot, shothole, cool country
  disease. Costa Rica, Panama, Brazil, Colombia, Venezuela, Guatemala, Argentina. P. usteri Speg. in Brazil
  and Costa Rica.
Physalospora spp. = Botryosphaeria spp. (the perfect or
  ascigerous states of Diplodias and possibly other such conidial state fungi): in the W. Indies is reported as P. abdita
  Stev., in S. America is reported as P. coffeae Speg., often
  in C. America is given as P. rhodina Cke. In all my observations occurrence in wetseason on dieback tissues.
Phytomonas leptovasorum Stahel (spirochaete-like, transmission by root feeding by Rhizoecus and Cerococcus that are
  sucking insects on roots. Disease found first on Coffea
  liberica in Surinam since confirmed and reconfirmed);
  phloem necrosis, red phloem coupled with wilt on C.
  liberica, C. arabica, C. canephora. Surinam, Guyana,
  Dominican Republic, Brazil, Colombia (El Salvador?).
Phytophthora palmivora Butl.; fruit brownrot. Brazil. Phytophthora sp. caused pre-emergence damping-off in seed
  bed. Costa Rica.
Pistillaria flavida Speg.; woodrot. S. America.
Pleurotus lignatilus Fr.; decay on old tree. S. America.
Polyporus spp.: P. blanchetianus Berk. & Mont., P. flavi-
  porus (Murr.) J. A. Stev., P. rubidus Berk., P. sanguineus L., P. sapurema Moell., P. tricholoma Mont.; all

root and trunk-base rots. Occasional, throughout the
American tropics, no doubt originally attacking Neotropical
jungle and forest trees.
Polystictus sobrius Berk. & Curt.; root disease. S. American
  polypore fungus, see item immediately above.
Porella spp.; leafy liverwort smother. W. Indies, C. America.
Poria sp.; rootrot. El Salvador.
Pratylenchus spp.; rhizosphere occurrences. Surinam, French
  Guyana, Surinam, Grenada, Guatemala, Dominican Republic,
  Puerto Rico, Jamaica, Peru, El Salvador, Brazil, Martinique,
  Colombia. Probably general, uncertain about severity of
  parasitism.
Pseudodiaporthe coffeae Speg. = Massarina coffeae (Speg.) Bose;
  leafspot on arabica coffee in S. America. P. major Speg.;
  dieback on liberica coffee in Argentina.
Pseudomonas garcae Am., Tei., & Pinh.; bacterial halo-blight.
  Brazil.
Psilenchus hilarulus deMan; coffee nematode. Brazil.
Pythium sp.; damping-off. El Salvador.
Pythium-Phytophthora (uncertain which genus); damping-off. El
  Salvador, Costa Rica, Guatemala, Panama.
Radopholus similis Thorne; decline. French Guyana, Trinidad
  and Tobago, Martinique, Guadeloupe, El Salvador.
Ramularia goeldiana Av.-Sac. (? = Cercospora sp.); zonate
  leafspot, stemlesion. Brazil.
Rhizoctonia solani Kuehn = perfect state is Thanatephorous cu-
  cumeris (Frank) Donk; damping-off, seedling stemlesion.
  W. Indies, C. America, S. America. R. subepigea Bertoni
  and R. lamellifera Small; root disease, subterranean sclero-
  tial plates. El Salvador, Guatemala. Rhizoctonia sp.; a
  distinct white-root disease. Surinam, Dominican Republic.
Rhizomorpha sphaerocrystalligena Speg. (? = R. corynephora
  Kunze); rhizomorphic fungus on decaying roots. Costa Rica,
  Brazil, Colombia, El Salvador (in dryer areas).
Rhombostilbella rosea Zimm.; on leaf. Brazil.
Rosellinia spp. (parasites on coffee roots often in relatively
  new plantings made in recent jungle clearings, but may be
  in long established fields, develop in sun-grown as well as
  shaded plantations): R. bunodes (Berk. & Br.) Sacc.;
  black rotting in-between layers of tissue in the root, apparently most common of these diseases and in Brazil is
  the typical "mal araraquara" or "mal dos quatro años."
  Neotropical. R. coffeicola Pat., especially reported in
  Guyana and the Lesser Antilles. R. necatrix (Hart.) Berl.,
  scattered in C. America. R. pepo Pat.; rot characterized
  by white felty pads in root tissues. W. Indies, C. America,
  possibly less common in S. America. R. quercina Hart.
  (may be a questionable designation). Mexico, ? S. America.
Rotylenchus sp.; spiral nematode in rhizosphere. Peru.
Rust-like fungus spores from round reddish leaf lesions. In
  Costa Rica seen by Tonduz in 1893, by Echandi and Wellman in 1959. (However, we germinated spores on agar

plates and were so relieved it was not a rust, we made
no further study as to pathogenicity or mycology.) (Note
of cautionary interest: Pathologists are constantly watching
for rusts on coffee in the Americas. Hemileia vastatrix
Berk. & Br. long in the Orient and Africa is now in the
Western Hemisphere, and its close relative H. coffeicola
Maubl. & Rog., also in Africa and spreading in that con-
tinent, will likely eventually appear in the Americas. Another coffee rust, Aecidium travancorium Ram., develops
in India on Coffea travancorensis Wig. & Arn. and is a
potential danger not to be overlooked.)

Saccardinula costaricensis Speg. = Limacinula costaricensis
(Speg.) Hoehn. and S. usteriana Speg.; moldy leafspots.
C. America, S. America.

Saccharomyces spp. such as S. apiculartus Ree., S. ellipsoideus Han., S. exiguus Ree. & Han., S. ludwigii Han., and
others; are yeasts, part of fruitrot complexes, and aid in
important vat fermentation for processing in "beneficios"
before drying beans on "patio."

Schizophyllum alneum (L.) Schroet., S. commune Fr.;
secondary branch decays after dieback. C. America, S.
America.

Sclerotium coffeicola Stahel; spectacular target-board leafspot.
First seen on liberica coffee but it is since found that
coffees in general are suscepts. Surinam, Guyanas, Dominican Republic, Costa Rica, Venezuela, Brazil. S. rolfsii (Curzi) Sacc. = Pellicularia rolfsii (Sacc.) West; on
several species of young coffee causes ground-rot, death.
Trinidad & Tobago, Martinique, Guadeloupe, Cuba, Haiti,
El Salvador, Brazil.

Scolecopeltis (see Micropeltis)

Septoria berkeleyi Sacc. & Trav.; spot on leaf and greenstem.
Bolivia, Brazil, Colombia.

Speira trimera Speg.; fruitrot. S. America.

Sphaerella coffeicola Cke. = Laestidia coffeicola (Cke.) Speg.;
leaf infection. Costa Rica.

Sphaerotheca coffaeicola Speg.; on branch. Paraguay.

Sporhelminthium coffeae Speg. (with synonyms: Helminthosporium coffeae Mass., Pleurophragmium capense (Thuem.)
Hughes, Spiropes capense (Thuem.) Ell.); spots of leaf and
green flush stems. Colombia, Argentina, Brazil, Costa
Rica.

Stereum coffeanum Berk. & Curt.; trunk decay in Cuba, S.
fasciatum Speg.; stump rot, decay in large branch wounded
from improper pruning in Costa Rica, S. spegazzinianum
(Speg.) Sacc. & Manc. = S. pergameneum Speg.; in piles
of rotting pruned-off twigs and small branches in El Salvador, Costa Rica, Argentina.

Strigula complanata Fée (can be found on any species of coffee
that I have examined growing where this lichen-spot occurs--
a lichen being a symbiotic development of fungus and alga--
and I have seen this lichen on leaves where no parasitic alga
was present). Widespread.

Struthanthus marginatus (Dearn.) Blume; parasitizing "bird
vine," woody drooping perennial with aerial roots over the
host branches sucking nutrients from them. Mexico, Guatemala, El Salvador, Costa Rica, Panama, Brazil, Paraguay. S. oerstedii (Oliv.) Standl.; similar to above but not
so abundant. El Salvador. S. orbicularis (HBK.) Blume;
more compact growth than the first named, near badly infested old citrus plantation. El Salvador. S. polystachys
(Ruiz & Pavón) Blume; drooping bird vine on many young
plants and seedlings near old attacked coffee trees.
Panama.

Thyridium coffeicola Speg. = Xylosphaeria coffeicola (Speg.)
Petr.; branch and trunk rot. Argentina.

Torula sphaerella Cke.; rare wood "stain" following tree
injury. C. America, S. America.

Trachythyriolum brasilianum Speg. = Chondropodium spegazzinii Petr.; on leaves. Brazil.

Trametes elegans Fr.; large-branch attack. Guatemala (on
living trees one year after much injury from volcanic
eruption).

Trentepohlia sp.; algal non-parasitic hair-like growths. C.
America.

Trichoderma (? viride); branch canker, root invasion (following exceptionally heavy fruit harvest). Costa Rica.

Trichodorus christei Allen; stubbyroot nematode in Guatemala,
Peru. T. monohystera Stein. Peru.

Triposporium gardneri Berk.; sootymold on foliage. W.
Indies.

Tubercularia sp.; reddish-brown stemcanker. Costa Rica.

Tylenchorhynchus sp.; nematode in root zone. Peru.

Valsa sp. possibly a Cytospora; dieback complex. C. America.

Venturia coffeicola Av.-Sac.; scab. Brazil.

Viruses: 1. Blister Spot (I found it to be both aphid and graft
transmitted), in C. America. 2. Bottle or viñas, symptoms of swollen stem in Guatemala. 3. Crespera in Colombia, Brazil. 4. Grease or Oil Spot studied by Bitancourt in Brazil. 5. Coffee ringspot also in Brazil. 6.
Rugose virus only on robusta coffee and I found it only
graft transmitted in Costa Rica.

Vivianella coffaeiphila Speg.; apparently from branch lesions.
S. America.

Volutella sp.; sporodochial studded spots on leaves and fruits.
Mexico (seen in Costa Rica in abandoned plantation).

Weakspot. An arrested fungus leaf-infection. Apparently
started by Cercospora sp. in some cases. Other fungi
are isolated. Abundant in Neotropics.

Witches' broom. Cause unknown. Small, compact, infrequent.
Costa Rica, El Salvador. (Similar growths seen in West
Africa and Malaya.)

Xiphinema americanum Cobb and X. radicicola Good.; large
nematodes in rhizosphere. ? non-parasitic. Costa Rica,
Peru, Guatemala.

Xylaria aristata Mont., X. polymorpha (Pers.) Grev.; tree
decline, part of complex. W. Indies, C. America, S.
America.
Zinc deficiency causes little leaf, stunt, no fruiting. Scattered.
Zukalia vagans Speg. and Z. vagans Speg. var. brachycarpa
Speg.; hyperparasites on Meliola black mildews on leaves.
Brazil.
(Concluding note: The above compilation of coffee disease
parasites is not absolutely complete. I have left out some
but the most important organisms are represented, as
well as some of perhaps no great concern. There is no
doubt that I have missed some of the newest reports. We
can be certain that in the future some present names will
be reduced to synonymy, and there are organisms losing
their positions as notable parasites and these will be discarded in lists. However, there is equal certainty that a
number of diseases at present unknown will inevitably appear on coffees in the Neotropics.)

COLEUS BLUMEI Benth. A quick growing, not permanent, popular
variecolored lush ornamental, cuttings grow easily; "coleo" Sp.
Po., "coleus" En.
Aphelenchoides fragariae (Ritz. Bos.) Chr.; leaf injuring
nematode. Florida.
Cercospora coleicola Chupp & Mull.; leafspot. C. America.
Erwinia carotovora (L. R. Jones) Holl.; cutting softrot.
General.
Meloidogyne sp.; rootknot. W. Indies.
Mycena citricolor (Berk. & Curt.) Sacc.; leafspot (spread
from adjacent diseased coffee tree). Costa Rica.
Pestalotia sp.; stem canker. Puerto Rico.
Phytophthora drechsleri Tucker, P. palmivora Butl.; wilt.
C. America.
Pythium ultimum Trow.; rot at base, rootrot. W. Indies,
C. America.
Rhizoctonia solani Kuehn (its basidial stage, found several
times, is Thanatephorous cucumeris (Frank) Donk); stem
collapse. Florida, Puerto Rico, Costa Rica.
Uredo pallidiscula Speg.; rust. S. America.

COLOCASIA ESCULENTA (L.) Schott. A crop introduced long ago
from warm Oriental islands, perennial, low woody herbs (large
spear-shaped leaves) with much eaten tuberous root; "eddoe,"
"dasheen," "taro" En. Sp., "malanga" Sp., "inhame" Po.
Cephaleuros virescens Kunze; algal leafspot. Brazil.
Ceratocystis paradoxa (Dade) Mor.; common dark-colored
rot. Jamaica, lesser Antilles.
Cercospora caladii Cke.; large leafspots. Widespread.
Cladosporium colocasiicola Saw.; irregular leafspot. Widespread in Caribbean region.
Colletotrichum gloeosporioides Penz.; anthracnose. W. Indies.

Depazeae caladii Schw.; on foliage. (? Panama).
Diplodia sp. = Botryodiplodia sp.; black stem, rot at upper
    end of tuber. Florida, Puerto Rico, Cuba, Panama,
    Nicaragua.
Erwinia carotovora (L. R. Jones) Holl. and other species;
    common market slimy softrots. General.
Fusarium solani (Mart.) Sacc.; common dry tuber-rot, wilt.
    W. Indies.
Helminthosporium caladii Stev.; black bordered leafspot. W.
    Indies.
Heterosporium colocasiae Mass.; leafspot. Jamaica, Cuba.
Macrophoma surinamensis (Berk. & Curt.) Berl. & Vogl.;
    stemrot, tuber disease with shrivelling. W. Indies, Suri-
    nam.
Meloidogyne sp.; rootknot. W. Indies. (Also a specimen
    sent to me from Venezuela.)
Paratylenchus pratensis Filip.; root nematode. W. Indies
    islands.
Pestalotia sp.; leaf and petiole lesions. Puerto Rico.
Phyllosticta colocasiae Hoehn.; leafspot. W. Indies, Panama,
    Brazil.
Phytophthora sp.; leaf collapse. Puerto Rico.
Pleosphaerulina colocasiae Rang.; plant top disease. Brazil.
Pythium ultimum Trow.; rootrot. Subtropical USA, Puerto
    Rico.
Rhizoctonia (? microsclerotia); webby leaf disease. Puerto
    Rico. R. (? solani); root disease and scurf on tubers.
    Panama.
Rosellinia bunodes (Berk. & Br.) Sacc. and R. pepo Pat.;
    rootrots, wilts (first is black fungus, second white). C.
    America, W. Indies, Trinidad.
Sclerotium rolfsii (Curzi) Sacc. = Pellicularia rolfsii (Sacc.)
    West (both stages reported); stemrot. Florida, Texas,
    Puerto Rico.
Vasculomyces xanthosomae Ashby; tuber dryrot. Jamaica.
Virus. Dasheen Mosaic (similar to Potato Y) in Florida.
    Leaf mottle in Puerto Rico.

COMMELINA ELEGANS HBK. = C. VIRGINICA L., and C. LONGI-
CAULUS Jacq. = C. NUDIFLORA L. Ground cover, two important holdover sources of virus, anthracnose, and rootknot. Soft
almost ubiquitous weeds; "cohitre" Sp. Po., "dayflower" En.
Diseases listed below seem to attack both weeds.
    Cassytha filiformis L.; strangling green angelhair dodder.
        Puerto Rico, Florida.
    Cercospora commelinicola (Kalch. & Cke.) Chupp; leafspot.
        S. America.
    Colletotrichum commelinae Ell. & Ev. = C. gloeosporioides
        Penz. but ascigerous Glomerella stage not seen or reported;
        mild leaf lesion disease. Neotropics.
    Cuscuta americana L.; strangling yellow angelhair dodder.
        Puerto Rico, Florida.

Illosporium commelinae Stev.; brown leafspot. W. Indies, C.
    America, S. America, Florida.
Mycena citricolor (Berk. & Curt.) Sacc.; leafspot (near
    diseased coffee tree). Costa Rica, Guatemala, Puerto Rico.
Mycosphaerella tetraspora Seav.; large leafspot. W. Indies,
    Honduras, Subtropical USA.
Phakopsora tecta Jacks. & Holw.; severe rust. Puerto Rico,
    St. Thomas, Cuba, Trinidad, Argentina, Bolivia, Honduras.
Phyllosticta commelinicola Young; medium sized leafspots;
    Puerto Rico, Costa Rica.
Puccinia commelinae Holw.; rust. Mexico.
Pythium (? ultimum); isolated from rotted plants. Puerto
    Rico.
Uredo ochracea Diet.; a brown rust. Brazil.
Uromyces commelinae (Speg.) Cke. (= Uredo spegazzinii deT.,
    Uredo commelinae Speg.); common rust. Probably general.
    Brazil, Argentina, Costa Rica, Honduras, Virgin Islands.
Virus(es), carriers of: Cucumber Mosaic Virus strains at
    least as follows: CMV Banana Infectious Chlorosis and
    Heartrot; CMV Banana Rayadilla; CMV Canna Mosaic;
    CMV Virulent Celery Virus; CMV Commelina "strain 4";
    CMV on orchid; CMV Tomato Narrow Leaf; Numerous
    cereal mosaics.

CORCHORUS CAPSULARIS L. Although long grown in the Eastern
tropics, recently is being studied in the Neotropics for possible
mechanization of fiber extraction; "jute" En., "yute" Sp., "juta"
Po.
    Cercospora corchori Saw.; spot of leaf and stem and capsule.
        Peru, Brazil.
    Colletotrichum sp.; anthracnose of leaf and stem. Peru.
    Corticium sp. (aerial type, basidia spores on canes); leaf
        lesions. Costa Rica.
    Fusarium spp.: F. semitectum Berk. & Rav., on capsules,
        F. solani (Mart.) Sacc. basalrot of plant, and F. monili-
        forme Sheld. from seed. All in Trinidad.
    Macrophomina phaseoli (Taub.) Maubl.; ashy stem rot. Uru-
        guay, Surinam.
    Rhizoctonia solani Kuehn = Thanatephorus cucumeris (Frank)
        Donk; damping-off. Wherever grown.
    Rosellinia sp.; rootrot. S. America.
    Sclerotium rolfsii (Curzi) Sacc. (no perfect stage reported);
        soil-line rot. Surinam.
    Verticillium sp.; wilt. Peru.

CORDYLINE TERMINALIS Kunth = DRACAENA TERMINALIS L.
An old-time Asian introduction, relatively small colorful and
palm-like, cemetery and garden/pot plants; called "chucho, "
"lengua-de-vaca" Sp., "cocomacaco" Po., "dracaena" En.
    Botryodiplodia theobromae Pat. is the conidial state which
        has synonyms: Diplodia theobromae (Pat.) Now., D.
        gossypina Cke., D. dracaenae P. Henn. and its perfect

state is Botryosphaeria quercuum (Schw.) Sacc.; stem
lesions, leaf edge rot, root tip blackening, cutting failure.
Widespread.

Capnodium spp.; sootymolds. Mostly seen in dryseasons. C.
America.

Cassytha filiformis L.; the slower growing and berrybearing
dodder. Puerto Rico.

Cephaleuros virescens Kunze; algal leafspot. In protected
locations. General.

Cercospora cordylines Speg. = C. cordylines P. Henn.; irregular shaped leafspot. Widespread in C. and S. America
(notably in more exposed locations).

Cladosporium (? herbarum); on old and ragged leaf. W.
Indies, C. America.

Colletotrichum gloeosporioides Penz. (the Glomerella cingulata
(Stonem.) Sp. & Schr. stage found in rainyseason in Costa
Rica); spot of leaf and stem, common anthracnose. W.
Indies, C. America, Brazil.

Cuscuta americana L.; dodder. Puerto Rico.

Daedalia effusa Speg.; basal rot of stem. Argentina.

Fusarium spp.; associated with leaf spots and root lesions.
El Salvador.

Helminthosporium sp. (extremely large spores); dark leafspot.
Puerto Rico.

Leptosphaeria cordylines (Speg.) Sacc. & Trott.; on leaves
causing rot and collapse. Argentina.

Macrophoma cordylines Speg., M. bakeri Syd.; stem disease.
C. America, S. America.

Meloidogyne sp.; rootknot. Florida.

Pestalotia cordylines Speg., P. funerea Desm.; recovered
from mountain growth plants with diseased leaves. S.
America, W. Indies.

Phyllosticta dracaenae P. Henn.; leafspot. Puerto Rico, C.
America. P. maculicola Halst.; leaftip disease. Puerto
Rico.

Physalospora sp. (conidial stage is a small-spored Diplodia
apparently not D. theobromae); tip blight. Puerto Rico.

Phytophthora (? palmivora); cutting disease, rot of stalk.
Puerto Rico, Costa Rica. P. parasitica Dast.; shallow
stem lesion. Puerto Rico.

Sphaerulina paulistana Speg.; on leaves. Brazil.

Strigula complanata Fée; creamy spot. Widespread in moist
tropics.

CRACCA   see   TEPHROSIA

CRESCENTIA CUJETE L.   Tree native of Mexico-Central America,
important source of vessels for drinking, used in bathing, for
food-carrying and storage, special type used in heated stone
cooking, some artistically carved and stained, and some used in
music. The tree has rough-barked, wide spread branches;
"cabaçaiero," "cujete" Po., "calabazo" Sp., "calabash tree" En.

Algae, aerial types of many kinds as benign lodgers in rough
   bark. General but especially in moist areas.
Bactridiopsis crescentiae Gonz. Frag. & Cif.; on leaf and
   small branch. W. Indies.
Bromeliads; epiphytes in moist warm country. Common.
Calonectria crescentiae Seav. & Wat.; on calabash shell.
   Bermuda.
Cassytha filiformis L.; dodder in tops of one tree. St.
   Thomas.
Colletotrichum sp.; anthracnose. El Salvador, St. Thomas,
   probably widespread.
Cuscuta spp.; dodders. W. Indies, El Salvador.
Hydnum-like fungus; on old tree, rot at base. El Salvador.
Hysterium calabash Seav.; scab-like lesion on fruit. St.
   Thomas.
Irenina arachnoides (Speg.) Stev.; scanty black mildew. W.
   Indies, S. America.
Loranthus sp.; a coarse woody bush parasitic on branch. St.
   Thomas.
? Marasmius spp.; cords causing blights. Panama, Costa
   Rica.
Meliola crescentiae Stev.; black mildew. W. Indies.
Mosses, numerous kinds, epiphytic on branches in moist
   locations, aid in establishing orchids as added epiphytes.
Orchids, small and large, adapt well to living on this tree.
   W. Indies, C. America.
Phoradendron racemosum Krug & Urban; a woody bush para-
   sitizing branch. Surinam.
Phyllachora sp.; black leafspot. W. Indies, C. America.
Phyllosticta sp.; small leaf spots, abundant on old leaves.
   C. America.
Polyporus spp.; trunk woodrot. Scattered. (Note: Trunks
   are often injured in climbing and cutting by harvesters in
   securing calabashes for sale and home use; these bark
   openings are subject to various infections.)
Rhizopus sp.; occasional on stored, poorly cured calabashes
   seen in village markets. C. America.
Strigula complanata Fée; whitish lichenspot on leaf. Scattered.
Struthanthus confertus Mart.; a scandent bush parasite. Suri-
   nam, St. Thomas. S. concinnus Mart.; small bush parasite. Surinam, Puerto Rico.
Xanthochrous pavonius Pat.; trunk rot. This very similar to
   Polyporous. Scattered.

CROTALARIA spp. Several used as ornamentals, cover crops and
   for fiber, there are species grown as cover for the protection of
   the typically fragile tropical soils and sometimes their root zones
   appear to be rootknot nematode-free whereas nearby soils retain
   the eelworms, plants used in crop rotation; "manduvira" Po.,
   "cassabelilla" Sp., "crotalaria" En.
      Aecidium crotalariicola P. Henn.; rust. Brazil.
      Alternaria sp.; leaf and pod spots. Florida, Peru.

Aphelenchoides parietinus Stein.; root semiparasite. Brazil.
Aphelenchus avenae Bast.; nematode near roots. Brazil.
Armillaria mellea Vahl = Armilariella mellea (Fr.) Karst.
  rootrot. Guatemala, Panama.
Ascochyta pisi Lib.; leafspot. S. America (on C. juncea L.
  and C. spectabilis Roth.)
Axonchium amplicolla Cobb; root nematode. Brazil (on C.
  spectabilis).
Botrytis cinerea Pers.; blight, leaf and pod mold. Florida,
  Peru, El Salvador in rainyseason.
Cassytha filiformis L.; green dodder. Puerto Rico (on C.
  spectabilis).
Ceratocystis fimbriata Ell. & Halst.; branch disease, rainy-
  season stemrot. Brazil, Trinidad, Venezuela, Argentina,
  Costa Rica, El Salvador (the fungus from coffee had
  slightly larger perithecium and thicker subiculum than that
  from Crotalaria).
Cercospora spp. all cause defoliation from leafspots: C.
  canescens Ell. & Mart. in Brazil and Subtropical USA, C.
  cotizensis Muller & Chupp in S. America, C. crotalariae
  Sacc. in C. America and W. Indies, C. cruenta Sacc. in
  Brazil and C. demetrioniana Wint. in Brazil.
Clitocybe tabescens Bres. = Armilariella tabescens (Scop.)
  Sing.; rootrot. Florida.
Colletotrichum gloeosporioides Penz. the conidial and para-
  sitic state (synonyms: C. crotalariae Speg., C. crotalariae
  Petch, C. crotalariae Mass., C. curvatum Briarct & Mar-
  tyn); anthracnose on all host species. Neotropical.
Criconemoides sp.; nematode around roots of Crotalaria spec-
  tabilis.
Cuscuta sp.; slender yellow dodder. El Salvador.
Erysiphe polygoni DC.; white powdery mildew. Florida, El
  Salvador.
Fusarium oxysporum Schl. f. sp. tracheiphilum (E. F. Sm.)
  Sny. & Han.; wilt. (in warm, friable, subacid soils).
  Subtopical USA, W. Indies, C. America, S. America. F.
  solani (Mart.) Sacc. (perfect state is Nectria haematococca
  Berk. & Br.) stemrot and collapse. Trinidad.
Helicotylenchus nannus Stein.; nematode found in root zone.
  Brazil.
Helminthosporium sp.; black pod disease. Florida, Puerto
  Rico.
Macrophomina phaseoli (Taub.) Ashby; canker, ashystemrot.
  Florida.
Meloidogyne sp.; rootknot nematode. (The host C. juncea
  very tolerant and C. spectabilis apparently immune used in
  crop rotation to "clean up" rootknot.) C. America.
Microsphaeria diffusae Cke. & Pk.; powdery mildew. W.
  Indies, S. America.
Mycena citricolor (Berk. & Curt.) Sacc. infects any inoculated
  crotalaria on leaf or green stem. Costa Rica.
Nectria sp.; stem disease. Brazil.

Oidium erysiphoides Fr. f. crotalariae Cif. & Frag.; powdery
  mildew. Subtropical USA, C. America, Venezuela, Peru.
Parodiella perisporioides Speg. = P. grammodes (Kuntze)
  Cke.; severe foliage disease (with minute black bodies
  peppered on leaves). C. America, S. America.
Phakopsora crotalariae (Diet.) Arth. = Uredo crotalariae
  Diet. (?) or U. crotalariae-vitellinae Rang.; rust. S.
  America.
Phyllosticta crotalariae Speg.; numerous brown spots. Para-
  guay.
Phymatotrichum omnivorum (Shear) Dug.; dryland rootrot.
  Texas.
Phytophthora sp.; plant baserot of C. spectabilis. Peru.
Pleospora herbarum (Pers.) Rabh.; leaf spot on old plants.
  Neotropics.
Pratylenchus sp.; nematode in rootzone. Peru.
Pyrenochaeta sp.; leafspot. Subtropical USA.
Rhizoctonia spp. (? solani, ? microsclerotia); stem canker,
  webrot. Subtropical USA, Peru, W. Indies, C. America.
Sclerotium rolfsii (Curzi) Sacc. = Pellicularia rolfsii (Sacc.)
  West; soil-line rot. Subtropical USA, W. Indies, Argen-
  tina, Cuba.
Septoria crotalariae Viég.; stem and leaf spots. Brazil.
Uredo (see Phakospora)
Uromyces decoratus Syd.; black rust. Brazil, Costa Rica.
Viruses: Citrus Infectious Variegation, by inoculation in
  California. Aphid-borne common Cowpea Mosaic, in W.
  Indies where it has been called Systemic Mottle. Crotalaria
  Phyllody, in Florida, Puerto Rico, Costa Rica, El Salvador
  (in the last two countries, reported as Little Leaf). Cro-
  talaria Mottle Mosaic in Puerto Rico, C. America. Tobacco
  Streak (Brazilian) by inoculation and by natural infection in
  Brazil.

CROTON spp. (Note: This is not the same as the small bush, the
  garden "croton" that is Codiaeum, which is given attention under
  that generic name in this volume.) About 30 species of rather
  rapid-growing, small- to medium-sized wild trees sometimes
  planted for ornament, living fence, temporary agricultural shade;
  in some cases the aromatic leaves and fruits are used by Indians
  as flavoring for foods and drinks; "copalchí" Sp. Po., "wild
  croton tree" En., numerous uncommonly used Indian names.
  (Diseases listed here are from a few reports, personal observations, and without special note of the species of the host.)
  Arthuria columbiana (Kern. & Whet.) Cumm.; leafrust. Co-
    lombia.
  Asterina spp.; black moldy leafspots. Widespread.
  Baeodromus dominicana. Thir. & Kern; rust. W. Indies.
  Bubakia spp., a number of species; tree rusts. W. Indies,
    S. America.
  Cassytha filiformis L.; pungent smelling dodder. Puerto Rico.
  Catacauma serra-negrae Viég.; scabby leaflesion. Brazil.

Cephaleuros virescens Kunze; algal leafspot. Brazil.
Cercospora crotonophila Speg., C. tiglii, P. Henn., C.
 manaoensis P. Henn., C. rubida Muller & Chupp, leaf-
 spots. W. Indies, C. America, S. America.
Colletotrichum (? gloeosporioides); anthracnose. General.
Cuscuta americana L.; common dodder. Puerto Rico.
Elsinoë .venezuelensis Bitanc.; spot anthracnose, scab. Vene-
 zuela, nearby countries.
Eudarluca australis Speg.; hyperparasite on rusts. Argentina.
Irenopsis crotonis (Stev. & Tehon) Stev.; black mildew. W.
 Indies.
Meliola spp.: M. bicornis Wint. var. constipata Speg. (= M.
 constipata Speg.), M. longispora Stev., M. malacotricola
 Speg.; all black mildews. Scattering but fairly general.
Phakopsora crotonicola Kern, Thur. & Whet. = Bubakia
 argentinensis (Speg.) Jacks. & Holw.; rust. S. America.
Phoma paraguariensis Speg.; black stem and leaf lesions. C.
 America, S. America.
Phorodendron sp.; an upright, woody parasitizing bush. C.
 America.
Phthirusa pyrifolia Eich.; parasitizing pendant woody shrub
 (with prehensile leaftips). C. America.
Phyllachora globispora Speg.; black spot, tarspot (on some
 host species not shiny and in some more infection of stems).
 Costa Rica, S. America.
Phyllosticta portoricensis Young; spotting, sometimes diffuse,
 on leaf and stem. Puerto Rico.
Puccinia crotonicola Speg.; rust. S. America.
Puiggarina crotonis Speg. = Phyllachora crotonis (Speg.) Sacc.;
 leaf disease, spotting. S. America.
Septobasidium castaneum Burt. var. draconianum Viég.; felt.
 Brazil.
Stemphyliomma crotonis Toro; black mildew. W. Indies.

CUCUMIS MELO L. An annual herb originating long ago in semi-
 dry Eastern tropics now adapted to dryer but not desert areas of
 tropical America, plants characterized by vines that bear sweet
 musky-flavored often netted fruits shipped and eaten locally;
 "melón rugosa," "melón" Sp., "melão de cheiro" Po., "cantaloupe" En., and C. SATIVUS L.; lushly growing annual vine with
 smooth fruits used in fresh condition or as pickles; "pepino" Sp.,
 "maxixe" Po., "cucumber" En. (On the whole both species are
 susceptible to the same diseases except as noted.)
  Acremonium sp.; yellow leafspot. Trinidad.
  Alternaria alternata (Fr.) Keiss = A. tenuis Nees; black edge-
   of-leaf, spot-blackening on fruits of cantaloupe. W. Indies,
   C. America. A. cucumerinum (Ell. & Ev.) Elliott = A.
   brassicae (Berk.) Sacc. f. nigrescens Pegl.; rare dryseason
   black leafspot. (Spores large.) El Salvador, Guatemala,
   Peru, Mexico, W. Indies. (Note: On the whole, Alternarias
   are not abundant in the tropics where I have worked; however, the Cucurbitaceae seem to be commonly affected under
   certain not well understood conditions.)

Aphelenchus arvense Bast.; nematode severe on cucumber. Brazil.

Botryodiplodia sp. = Diplodia sp.; stem canker. Occasional in C. America, S. America.

Cassytha americanum L.; green dodder on cucumber. Puerto Rico.

Ceratocystis paradoxa Mor.; stem rot. W. Indies.

Cercospora citrullina Cke. = C. cucurbitae Ell. & Ev.; large leafspot. Fairly general.

Choanephora cucurbitarum (Berk. & Rav.) Thaxt.; blossom blight. Scattered in Subtropical USA, W. Indies.

Cicinnobolus cesatii dBy. (a hyperparasite of Erysiphe on cantaloupe). Argentina.

Cladosporium herbarum (Pers.) Lk.; dryseason infection on old leaves. C. America, W. Indies.

Colletotrichum lagenarium (Pass.) Ell. & Halst. = C. orbi-
culare (Berk. & Curt.) Arx; rarely leafspot, severe fruit spot and rot under warm-moist conditions. Rather general. (Note: It is worth noting that this Colletotrichum is especially adapted to cucurbits but without the broad host range of C. gloeosporioides; this lends itself to the control measure of using non-cucurbits in a well carried out croprotation regime--and I have seen this profitably followed in Panama.)

Corynespora cassiicola (Berk. & Curt.) Wei; leafspot, petiole lesion. Florida, Honduras, Costa Rica.

Curvularia lunata (Wakk.) Boed.; irregular leaf injuries. C. America.

Cuscuta americana L.; yellow dodder. Puerto Rico, C. America.

Diplodia sp. (see Botryodiplodia)

Erwinia carotovora (L. R. Jones) Holl.; market softrot of fruit. Collar rot in field of rare occurrence. E. tracheiphila (E. F. Sm.) Holl.; bacterial wilt. Scattered in mountain gardens of Guatemala, Honduras, Costa Rica.

Erysiphe cichoracearum DC. = Oidium (see below); mid dry-season powdery mildew. General.

Fusarium moniliforme Sheld.; secondary foliage dryrot. Scattered in C. America in warmest areas. F. oxysporum Schl. f. sp. melonis Sny. & Hans.; wilt. In warm, friable, somewhat acid soils, mostly on cantaloupe. F. solani (Mart.) Sacc.; fruit rot, secondary invasion in rotted stems. Reported from a few places in W. Indies, C. America, S. America.

Helicotylenchus nannus Stein.; nematode found near cantaloupe roots. Brazil.

Isaria sp. (conidial stage of Cordyceps); insect parasite, spores from leaf. S. America.

Meliola sp.; black mildew on cantaloupe. Costa Rica.

Meloidogyne sp.; rootknot. Scattered in W. Indies, Subtropical USA, C. America, S. America.

Mucor sp.; market fungus fruit rot. W. Indies.

Mycena citricolor (Berk. & Curt.) Sacc.; on cucumber, leaf-
    spot under diseased coffee tree. Puerto Rico.
Mycosphaerella citrullina (C. O. Sm.) Gros.; gummystem
    and fruitrot. Florida, Puerto Rico, C. America, Jamaica,
    Cuba, Virgin Islands.
Oidium ambrosiae Thuem. (conidial state of Erysiphe);
    powdery mildew, often common on upper leaf surface.
    Widespread.
Penicillium sp.; on dying leaves, accompanies some fruit
    softrots in markets. C. America.
Phymatotrichum omnivorum (Shear) Dug.; dryland rootrot.
    Texas.
Phytophthora spp. (several listed, that require specialists to
    separate species. In my own isolations I have recovered
    many variable forms and I am uncertain of some reports);
    fruitrots, damping-off. Puerto Rico, El Salvador, Guate-
    mala, Panama, Argentina, Chile, Cuba, Costa Rica.
Pseudomonas lachrymans (E. F. Sm. & Bryan) Car.; angular
    leafspot. Puerto Rico, W. Indies, Pacific coast of C.
    America, S. America.
Pseudoperonospora cubensis (Berk. & Curt.) Rostow; downy
    mildew, leaf collapse, fungus seen mostly on under surface
    of leaves. General, except in wettest areas.
Puccinia cucumeris P. Henn.; rust. Brazil.
Pythium spp.; damping-off. W. Indies, C. America, S.
    America.
Rhizoctonia solani Kuehn = Thanatephorous cucumeris (Frank)
    Donk; seedling damping-off and a fruit soil-rot. Subtropi-
    cal USA, W. Indies, C. America, Mexico, S. America.
    R. microsclerotia Matz = Pellicularia microsclerotia (Pat.)
    Rog.; web blight, only on cucumber. El Salvador, Puerto
    Rico, Costa Rica.
Sclerotinia sclerotiorum dBy.; = Whetzeliana sclerotiorum
    (dBY.) Corf & Dumont; field and market fruitrot, plant
    base disease. W. Indies, Subtropical USA, Uruguay.
Sclerotium rolfsii (Curzi) Sacc. (basidiospore stage is Pelli-
    cularia rolfsii (Sacc.) West, seen in Florida and Puerto
    Rico); soil-line infection. Panama, Florida.
Septoria cucurbitacearum Sacc.; cool season leafspot. Bolivia,
    Florida, Guatemala.
Sphaerotheca sp.; leaf mildew. Uruguay, Chile, California.
"Spiral group nematode"; ? root attack. Chile.
Stemphylium cucurbitacearum Osner; leafspot. On cucumber
    in Trinidad, S. America.
Trichoderma sp.; large root and stem infection. Puerto
    Rico, Florida, Panama.
Viruses: Common Cucumber Mosaic Virus (CMV); in sub-
    tropical USA, Puerto Rico, Mexico, many parts of C.
    America and S. America. CMV celery/commelina strain,
    more virulent than common CMV, coexistent with diseased
    Commelina, in Florida, C. America. Cantaloupe Mosaic
    in C. America, Puerto Rico. Watermelon Mosaic Virus in

locations scattered in W. Indies, C. America. Banana
Mosaic. C. America. Unidentified aphid-borne cucumber
leaf harshness and fruit stunt. El Salvador, Costa Rica.

CUCURBITA spp. The subjects of this section, are natives of
Western Hemisphere warm lands and are very important foods
for millions both in metropolitan areas and in corners of
aboriginal occupation. Not only for the nutritious fruits and
seeds consumed, but as well for the "fat" roots, lush shoots,
and tendrils. Folk of tropical America since before recorded
times have selected these vegetables for local, sometimes very
isolated, growing. They have been given literally hundreds of
Indian and other common names. Two species are especially
used: C. MAXIMA Duch. produces on harsh vines the hard-
shelled, better-keeping edible fruits for storage; "zapallo, "
"cidracayote, " "anyama" Sp. , "abóbora, " "abóbora comum"
Po. , "hard squash, " "winter squash" En. and C. PEPO L. ,
compact to short vines rather softer shelled various sized nutritious fruits; "calabazo, " "ayote, " "pipián" Sp. , "ayoté, "
"abóbra menina" Po. , "vegetable marrow, " "pumpkin" En.
(Diseases generally attack both species.)
  Acremonium spp. , hyperparasites on mildews. Scattered.
  Alternaria spp. , large and small spores. Blackened disease
      lesions on leaf and peduncle, some fruit spot. Especial
      note on this in Subtropical USA, Puerto Rico, El Salvador,
      Guatemala, Brazil, but is probably Neotropical.
  Ascochyta cucumis (Ell. & Ev.) Elliott (perfect state is
      Mycosphaerella citrullina (C. O. Sm.) Gross.); gummy
      stem, black fruitrot. Subtropical USA, scattered in C.
      America, Brazil.
  Botryodiplodia theobromae Pat. (= Diplodia sp.); stem-end-
      rot of fruit, peduncle rot. Subtropical USA, Puerto Rico,
      Brazil.
  Botrytis cinerea Pers. ; fruit moldyrot in storage. Subtropical
      USA, Peru, Chile, Ecuador.
  Ceratocystis paradoxa (Dade) Mor. ; on old stems, branch
      wilt, mild disease. Subtropical USA, in El Salvador found
      on plants near second growth jungle.
  Cercospora cayaponiae Stev. & Solh. (can expect it on culti-
      vated cucurbits, thus far reported only from Willd.); leaf-
      spot. Brazil. C. citrullina Cke. = C. cucurbitae Ell. &
      Ev. ; small leafspots. Subtropical USA (found infrequently),
      W. Indies, C. America, Brazil.
  Choanephora cucurbitarum Thaxt. ; blossom blight, blossomed
      rot of young fruits. Brazil, Puerto Rico.
  Cicinnobolus cesatii dBy. ; (hyperparasite of Oidium). El
      Salvador.
  Colletotrichum lagenarium (Pass.) Ell. & Halst. (also re-
      ported as C. orbiculare (Berk. & Curt.) Arx; leaf and
      fruit spot, streak spot of stem. Widespread.
  Diplodia (see Botryodiplodia)
  Erwinia carotovora (L. R. Jones) Holl. ; market fruitrot fol-
      lows harvest injury. General.

Erysiphe (see Oidium)

Fusarium oxysporum Schl. f. sp. cucumerinum Owen; wilt.
  Brazil, Argentina. F. solani (Mart.) Sacc. f. sp. cucurbitae Sny. & Hans.; rootrot, fruit wetrot. Chile, Brazil,
  Argentina, C. America. (F. roseum-type secured in
  numerous isolations in El Salvador, Puerto Rico.)

Macrophomina phaseoli (Taub.) Ashby; a stem disease. Brazil.

Meloidogyne spp.; rootknot. General and often severe in old
  land repeatedly replanted. (In Brazil identified as M.
  javanica Chit.)

Mycena citricolor (Berk. & Curt.) Sacc.; leafspot under
  diseased coffee and artificial infections by inoculation.
  Costa Rica, Puerto Rico.

Mycosphaerella (see Ascochyta)

Nectria peponicola Speg.; stem disease, on pumpkin and
  squash shells. Argentina.

Oidium ambrosiae Th., O. cichoracearum DC., O. erysiphoides
  Fr., and some others, all names given the widely occurring
  conidial state of the ascogenous fungus Erysiphe cichoracearum DC. (The latter, under some conditions, never develops); powdery or white mildew, ash disease of living
  leaves. Neotropics.

Phyllosticta cucurbitacearum Sacc. (? = P. orbicularis Ell.
  & Ev.); leafspots. Brazil.

Phymatotrichum omnivorum (Shear) Dug.; dryland rootrot.
  Texas, (uncertain report from Mexico).

Phytophthora capsici Leonian; wilt and white fruitrot of hard
  squash in Chile, Venezuela, and Argentina. P. citrophthora Leonian wilt. Argentina.

Pseudoperonospora cubensis Rostow; downy mildew. General.
  Often most severe on introduced cucurbitas, not usually extreme in damage on native varieties.

Pythium aphanidermatum (Eds.) Fitz.; blossomend rot and
  cottony leak. Mexico, Peru, Brazil. P. irregulare Buis.,
  P. myriotylum Drechs. (the most common), P. vexans
  dBy.; all fruitrots. Brazil, C. America.

Rhagadolobium cucurbitacearum (Rehm) Th. & Syd. (? =
  Asterotextis cucurbiterum (Rehm) Arx); stromatic leafspot.
  W. Indies, S. America.

Rhizopus nigricans Ehr.; market fruitrot. Chile, Brazil.

Sclerotinia sclerotiorum (Lib.) dBy. = Whetzeliana sclerotiorum (dBy.) Korf & Dumont; market fruitrot. Brazil,
  Subtropical USA.

Sclerotium rolfsii (Curzi) Sacc. but its perfect state not reported; occasional on rotted fruit in markets. Brazil,
  Guatemala, Puerto Rico, Panama.

Septoria cucurbitacearum Sacc.; (on hard squash) fruit lesion,
  leafspot. Bolivia.

Sphaerotheca castagnei Lév. (? = S. humuli Burr.); mildew
  on squash leaves in Chile.

Stemphylium ilicis Teng. also reported S. cucurbitacearum
  Osner; stem and leaf spot on squash. Argentina.

Uromyces hellerianus Arth.; a severe rust on a wild cucurbit
in El Salvador. Deserves study preparatory to its possi-
ble attack of closely related cucurbits.
Viruses (not extensively studied): Common Cucumber Mosaic
Virus (CMV) in Subtropical USA. CMV, virulent Celery/
Commelina strain in Florida, Puerto Rico, C. America,
S. America often in scattered locations at edges of fields.
Typical Squash Mosaic Virus in Venezuela, El Salvador,
Panama (from symptoms). Watermelon Mosaic Virus in
Subtropical USA, Venezuela.
Xanthomonas cucurbitae (Bryan) Dows.; bacterial spot. Sub-
tropical USA. I have seen (but not isolated bacteria from)
bacterial leafspots on squashes in Cuba, Puerto Rico, El
Salvador, Panama.

CYATHEA ARBOREA (L.) J. E. Sm. A tree fern used in orna-
mental plantings. There are a few wild species but this has
been the important planted one I was asked to look at for dis-
ease; "elecho arboreo" Sp., "feto de arvore" Po., "tree fern"
En.
Botrytis sp.; fungus on decadent leaftips. Puerto Rico.
Cephaleuros (? minimis); very small leafspot and apparent
deep infection at base of leaflets. Puerto Rico. C.
virescens Kunze; common algal leafspot. Puerto Rico,
Cuba, Guatemala.
Cladosporium herbarum Lk.; decline in old leaves. C.
America.
Cuscuta americana L.; dodder. Puerto Rico (on a young plant
in wet mountains).
Decline due to poor soil, too much sun, too close to lime-
shedding walls, lack of moisture, very little soil humus.
W. Indies.
Diaporthe sp.; associated with leaf decline. W. Indies.
Diplodia sp.; midrib disease. W. Indies. (Isolations when
cultured produced no spores.)
Fusarium (roseum-type); infection of injured leaf. W. Indies,
C. America.
Griggsia cyathea Stev. & Dalb.; irregular black leafspot.
Puerto Rico.
Helminthosporium sp.; black leafspot. El Salvador, Guate-
mala, Costa Rica, Puerto Rico.
Nectria filicina (Cke. & Harkn.) Sacc.; leafstalk lesions.
Guatemala.
Pestalotia sp.; spot on old leaves. Semiparasitic. General.
Rosellinia bunodes (Berk. & Br.) Sacc.; rootrot. Guatemala,
Costa Rica, Puerto Rico.
Stagonospora sp.; on old leaves, semiparasitic. California.
Strigula complanata Fée; milk-white spots. Puerto Rico.

CYCAS REVOLUTA Thunb. A tree with palm-like leaves used in
funeral and other religious rites; "palma de sagú," "cica" Sp.,
"fern palm," "sago palm" En., "palma da sagu," "palma da
fúnebre" Po.

Alternaria alternata Keiss = A. tenuis Nees; secondary black
    smudge on decadent leaf. Rare occurrence in Costa Rica.
Anabaena cycadeae Reinke; the common root alga (benign en-
    dophyte). Puerto Rico, Panama, Costa Rica.
Azotobacter sp. ; associated with root problems. Florida,
    occasional in C. America.
(? Capnodium spp.); sootymolds. C. America, W. Indies.
Cassytha filiformis L. ; green dodder. Puerto Rico (by inocu-
    lation).
Colletotrichum sp. (under special conditions ascogenous stage
    Glomerella cingulata (Ston.) Sp. & Schr. is produced);
    anthracnose, blight. Subtropical USA, Puerto Rico, Virgin
    Islands, Panama.
Corticium/Pellicularia (spores and mycelium found); scanty
    growth on leaf. El Salvador.
Cuscuta americana L. ; gold thread dodder. Puerto Rico (on
    inoculation).
Geotrichum coccophilum Speg. ; disease of insects on leaves.
    Brazil.
Hendersonia togniniana Pol. ; on leaf. S. America.
Hormodendron cladosporioides Sacc. ; on leaf, weak parasite.
    C. America.
Nostoc sp. ; algal infection of roots. Subtropical USA.
Penicillium sp. ; leaftip invasion on weak plant. El Salvador.
Pestalotia cycadis Allesch. and P. palmarum Cke. ; leaf
    blotches, blighting. Subtropical USA, W. Indies, C.
    America.
Phoma bresadolae Sacc. ; black blight. Subtropical USA.
Rhabdospora sp. ; leaf decline. Florida.
Sirococcus cycadis Speg. ; on wilted dry leaf. Argentina.
Strigula complanata Fée; milk-white spot. Puerto Rico.
Trichoderma (? viride), isolated from dying pot plant. Costa
    Rica.

CYCLANTHERA PEDATA Schr. An annual food plant of Indians
    from Mexico to northern Peru. Fruits are the part eaten, both
    raw and cooked; "caihua, " "caiba, " "pepino crespa" Sp. ,
    "achoccha" Andean Ind. Practically nothing is known of diseases
    on this crop. Some personal observations as follows:
    Colletotrichum sp. ; fruit spot. Ecuador.
    Diplodia sp. ; stem disease. Specimens from Mexico.
    Erwinia carotovora (L. R. Jones) Holl. ; market softrot of
        fruit. Guatemala.
    Oidium sp. ; white mildew. C. America.
    Rhizoctonia (? solani); damping-off. Mountain garden in
        Panama.
    Viruses: Both common Cucumber Mosaic Virus (CMV), and
        CMV virulent strain Celery/Commelina in Costa Rica.

CYDONIA OBLONGA Mill. Introduced from temperate zone Europe,
    this exotic fruit tree grows under only very special ecological
    conditions in the tropical Americas; "membrillo" Sp. , "marme-
    leiro" Po. , and "quince" En.

Agrobacterium tumefaciens (E. F. Sm.) Conn; crowngall.
Colombia, Chile, Argentina.

Ascochyta sp.; an uncommon leafspot. S. America.

Botryosphaeria dothidea (Moug.) Ces. & de Not. (for B. quercuum see Physalospora); stem canker. Subtropical USA.

Capnodium sp.; sootymold. Occasional.

Cercospora cydoniae Ell. & Ev. = C. mali Ell. & Ev. (C.
tomenticola Sacc. ? = Cladosporium sp.); leafspot.
Brazil, Venezuela, Chile.

Cladosporium sp. (similar to C. herbarum and not a Cercospora); on decadent leaves. S. America.

Colletotrichum gloeosporioides Penz. (also reported as: Gloeosporium fructigenum Berk. or G. rufomaculans Theum.)
is conidial state of the ascogenous Glomerella cingulata
(Stonem.) Sp. & Schr.; bitter rot of fruit, leafspot. S.
America.

Coniothecium sp.; semiparasite on leaf. S. America.

Corticium salmonicolor Berk. & Br.; pink disease of branch.
Brazil.

Cytospora microspora (Cda.) Rab. = Valsa leucostoma Pers.;
bark canker, blight. Chile.

Diplocarpon maculatum (Atk.) Jorst. = Fabraea maculata
(Lév.) Atk. (the conidial state is Entomosporium maculatum
Lév.); leaf eyespot, blight, fruit blackspot. Argentina,
Chile, Brazil, Uruguay.

Erwinia amylovora (Burr.) Winsl. et al.; fireblight. Subtropical USA, Colombia.

Exoascus bullatus (Berk.) Sadeb. = Taphrina bullata (Berk.)
Tul.; leaf blister. Chile.

Gloeosporium perennans Zell. & Childs conidial state of the
ascogenous Neofabraea perennans (Zell. & Childs) Kienh.;
perennial canker. Cold locations in S. America.

Glonium microsporum Sacc.; small round black leafspots. S.
America.

Helotium uliginosum Fr.; branch canker. Brazil.

Hendersonia sp.; weak parasite on leaves. S. America.

Lichens and mosses, accumulations in some areas. Colombia.

Meliola sp.; black mildew. S. America.

Meloidogyne sp.; rootknot. Texas, Chile.

Monilinia fructigena (Aderh. & Ruhl.) Honey (the conidial
state is Monilia fructigena Pers.), and Monilinia laxa
(Aderh. & Ruhl.) Honey (its conidial state is Monilia laxa
(Ehr.) Sacc. & Vogl.) are two brownrots of fruit and twig,
and flower blights. Brazil, Chile, and nearby countries.
(Sometimes listed as Sclerotinia disease.)

Mycosphaerella sp. (considered the ascogenous stage of a
Cercospora); fruitspot and leafspot. S. America.

Myxosporium mali Bres.; a branch canker. Peru.

Pellicularia koleroga Cke.; web and thread blight on leaves
and branch tips. Subtropical USA, S. America in scattered locations.

Penicillium expansum Lk.; bluemold storage rot. General.

Pestalotia sp.; dieback complex.   Specimens from Chile.
Phyllosticta cydoniicola P. Henn.; circular brown leafspot.
    Brazil, Argentina, Chile.
Physalospora obtusa (Schw.) Cke. = P. cydoniae Arn. (coni-
    dial state is Botryosphaeria quercuum (Schw.) Sacc.); fruit
    blackrot, twig and stem canker, leafspot.   Subtropical
    USA, Uruguay, other parts of S. America.
Rosellinia sp.; root disease.   Colombia.
Sclerotinia (see Monilinia)
Septogloeum cydoniae (Mont.) Pegl.; angular leafspot.   Argen-
    tina.
Sphaceloma sp.; spot anthracnose.   Brazil.
Sphaeropsis cydoniae Cke. & Ell. = S. cydoniae Arn. (be-
    lieved by some to be Physalospora obtusa); blackrot and
    stem canker.   Peru.
Stigmatea mespili Sor.; small swollen leafspot.   S. America.
Valsa (see Cytospora)

CYMBOPOGON CITRATUS (DC.) Stap. and C. NARDUS (L.) Rend.
    Lemongrass and citronella grass respectively; both yield fragrant
    oils and are natives of Old World tropics but are small steady
    sources of income in some parts of the Western tropics; "grama
    oriental" Sp., "citronella grass," and "lemon grass" En.   (All
    diseases listed seem to be on both host species.)
    Alternaria sp.; infrequent irregular injury on leaf and stem.
        Guatemala.
    Curvularia sp.; common old-leaf disease.   General.
    Helminthosporium sacchari (de Haan) Butl.; eyespot.   Florida,
        El Salvador, Guatemala.
    Himantia stellifera Johnst.; webblight on leaf and sheath.
        Puerto Rico, Costa Rica, Florida, Guatemala.
    Melanconium sacchari Mass.; stalk and leafsheath infection.
        W. Indies, C. America.   (This appears to have been a
        problem years ago when rind disease from this fungus was
        serious on sugarcane, but both diseases are now not easily
        found on their hosts.)
    Myriogenospora paspali Atk. = M. atramentosa (Berk. &
        Curt.) Diehl; tangletop.   Florida.
    Papularia vinosa (Berk. & Curt.) Mason; infection of dead
        parts of leaf.   Puerto Rico.
    Pleospora doidgeae Petr.; leafspot.   C. America.
    Rhizoctonia sp. (threads without small sclerotia); an aerial
        web blight.   Puerto Rico, Guatemala.
    Spittle and chlorosis from effects of spittle bug (insect seen
        inside spittle deposits).   Guatemala.
    Virus: the common Sugarcane Mosaic Virus clearly identified
        in El Salvador, Guatemala and Brazil.

CYNARA SCOLYMUS L.   An introduction for food production from
    the Near East-Europe; "artichoke" En., "alcachofra" Po.,
    "alcaucil," "alcachofa" Sp.
    Ascochyta cynarae Maf.; yellow-gray leafspots.   Chile, Peru.

Botrytis cinerea Pers.; mold. California.
Capnodium sp.; sootymold. Scattered in S. America.
Cercospora grandissima Rangel and C. obscura Heald & Wolf;
    leafspots. Brazil (first organism), California, Texas
    (second organism).
Diplodia sp.; stem canker. Chile, Brazil, Argentina.
Erysiphe cichoracearum DC. = Oidium sp.; white mildew.
    Chile, Argentina, Brazil.
(Fusarium spores, scraped from dead leaf with redrot sent
    from Peru to me in Panama.)
Penicillium italicum Wehm.; mold in markets. Chile (and
    other countries of S. America?).
Phyllosticta cynarae Sacc.; spreading grayspot. Argentina,
    Chile.
Red rot of leaf, undetermined cause (material limited and in
    poor condition). Peru.
(Rhizoctonia mycelium on dead leaf, see Fusarium note.)
Septocylindrium cynarae Speg.; a heavy mildew-like fungus
    on leaves. Argentina.
Tylenchorhynchus sp.; nematode from root zone, ? parasitic.
    Peru.

CYNODON DACTYLON Pers.   An important grass for pasture,
    lawn, and erosion-control.   In spite of common names it is not
    of Antillean origin but is from Eurasia; it has wide tropical
    adaptability but does not withstand long periods of intense cold;
    "grama de bermuda, " "zacate de gallina" Sp., "gramado de
    bermuda, " "capím de bermuda" Po., "bermuda grass" En.
    Algal greenscum, smothering of lawn. Puerto Rico.
    Asterina (see Dimerosporium)
    Cercospora seminalis Ell. & Ev.; spot on leaf and glume.
        Guatemala, Texas.
    Cochliobolus carbonum Nelson; blight. Mexico.
    Coniosporium (see Papularia)
    Corticium sasakii (Shirai) Matsu (= Rhizoctonia grisea (J. A.
        Stev.) Matz); leafbanding. Subtropical USA, W. Indies, El
        Salvador, Costa Rica.
    Darluca filum (Biv.) Cast.; spores of hyperparasite on Puc-
        cinia. Puerto Rico and no doubt elsewhere.
    Dimerosporium erysiphoides Ell. & Ev. = Asterina sp., often
        scanty black growth, leaf moldyness. S. America, W.
        Indies.
    Erysiphe graminis DC.; powdery mildew. California.
    Helminthosporium cynodontis Marig.; leaf spot-and-blight.
        Subtropical USA, W. Indies, C. America. H. giganteum
        Heald & Wolf (= Drechslera gigantea Ito); zonate eyespot.
        Subtropical USA, W. Indies, C. America. H. stenospilum
        Drechs. (= Cochliobolus stenospilum Mats. & Yama;
        purple halo leafspot. Florida, Puerto Rico.
    Leafspotting, yellowish angular spots. Undetermined. Proba-
        bly potash starvation, possible virus. Florida, Central
        America.

Leptostromella cynodontis Sacc.; on leaves.   S. America.
Meloidogyne sp.; rootknot.   Subtropical USA.
Papularia arundinis (Fr.) Cda. (= Coniosporium rhizophilum
   (Pers.) Sacc.); stolon disease.   S. America, Subtropical
   USA.
Phyllachora cynodontis (Sacc.) Niessl and P. graminis (Pers.)
   Fckl. var. cynodonticola Speg.; large tarspots.   S.
   America, Puerto Rico, Cuba, Mexico, Bermuda, Subtropi-
   cal USA.
Phyllosticta cynodonicola P. Henn.; leafspot.   Brazil.
Physarum cinereum Pers.; slime mold.   Florida, W. Indies.
Puccinia cynodontis (Lacr.) Desm.; brown rust.   General.
   P. graminis Pers.; stem rust.   Subtropical USA.
Pyricularia grisea Sacc. is conidial state, the perfect or
   ascogenous state is a Ceratosphaeria; gray spot and rot at
   node. Guatemala, W. Indies.
(? Pythium sp.); in lawn diseased spots of grass.   Panama.
Rhizoctonia sp. (not solani); brownpatch in lawns.   Subtropical
   USA, C. America.
Sclerotinia sp.; called dollarspot of lawn in Florida.   (This
   disease I have also questionably attributed to Pythium sp.
   in Panama.)
Sclerotium portoricense Stev.; leaf and stem disease.   Puerto
   Rico. S. rolfsii (Curzi) Sacc. = basidial state is Pelli-
   cularia rolfsii (Sacc.) West; ground line rot.   Florida.
Slime molds, several (see Physarum for one that is deter-
   mined). W. Indies, C. America.
Sorosporium syntherisma (Pk.) Farl.; headsmut.   Texas,
   California, Mexico, Guatemala.
Ustilago cynodontis P. Henn.; headsmut.   Subtropical USA.
   U. paraguariensis Speg.; culmsmut.   Paraguay, Brazil,
   Uruguay.
Virus: ? Sugarcane ratoon stunt virus, typical stunt and dis-
   colored vasculars in bermuda grass at edge of diseased
   plot of sugarcane. El Salvador, Costa Rica.

CYPHOMANDRA BETACEA (Cav.) Sendt.   The woody but small
   tree-like (3-4 meter) South American producer of much relished
   tomato-like fruit sold in markets in many parts of Latin
   America, harvested both from back yard plantings and from the
   wild; "palo de tomate" Sp., "macieira de tomate" Po., "tree
   tomato" En. (has not been given much intensive disease study).
   Aecidium tafiense Lindq.; leaf rust.   Argentina.
   Cephaleuros virescens Kunze; algal leafspot, malformed young
      leaves.
   Cercospora sp.; leafspot.   Costa Rica.
   Corynebacterium michiganense (E. F. Sm.) Jens.; bacterial
      canker.   California.
   Cronartium uleanum Syd.; leafrust.   Brazil.
   Diplodia sp. = ? Botryodiplodia theobromae Pat.; part of the
      dieback complex.   Costa Rica.
   Meloidogyne sp.; mild rootknot.   Costa Rica.

Nonparasitic epiphytic fungi, bacteria, algae, lichens.   Costa
   Rica, El Salvador, Honduras.
Pythium sp. (uncertain type of fructification); damping-off.
   Costa Rica.
Rhizoctonia solani Kuehn = Thanatephorous cucumeris (Frank)
   Donk; rootrot, sclerotia on base of stalk.   Costa Rica.
Rosellinia bunodes (Berk. & Br.) Sacc.; black rootrot.   Costa
   Rica.
Septoria lycopersici Speg.; leafspot when inoculated.   Peru.
Strigula spp.; lichen spots, several types.   C. America.
Viruses: Common Cucumber Mosaic Virus (CMV) in C.
   America.   Potato Virus Y in Brazil.   Sunflower Virus in
   Uruguay.   Sunflower Virus in Argentina.   Tobacco Streak,
   Brazilian type in Brazil.

DACTYLIS GLOMERATA L.   Introduced from Europe, where it is a
   successful pasture and orchard grass; "pasto ovillo" Sp., "pas-
   ture grass," "orchard grass" En.
Calviceps purpurea (Fr.) Tul.; ergot on seed (which is a poison
   to livestock).   Chile.
Cochliobolus carbonum Nelson; footrot.   California.
Footrot (not the above, undetermined complex of several fungi).
   Puerto Rico.
Fusarium spp.; plants declined and haulms dead.   Chile and
   other countries in S. America.
Meloidogyne sp.; root disease.   Florida, W. Indies.
Nematodes at roots, more than one kind.   Chile.
Phyllachora bromi Fckl. and P. graminis (Pers.) Fckl.; tar-
   spot.   S. America, C. America.
Puccinia graminis Pers.; rust.   Chile, Subtropical USA.
Scolecotrichum graminis Fckl.; leaf disease.   Brazil.
Ustilago striiformis (West.) Niessl; smut.   S. America.

DAHLIA spp.   Origin of these tuberous rooted ornamentals is
   Mexico and Northern Central America where some grow to small
   tree-like flowering plants in moist mountains, but there appear
   to be some dozen species of which a common one cited in the
   American tropics is D. VARIABILIS (Willd.) Desf.; "dalia" Sp.
   Po., "dahlia" En.
Aecidium dahliae Syd.; leafrust.   Mexico.   A. dahliae-maxonii
   Cumm.; leafrust.   Guatemala.
Alternaria sp. (medium-size spore); leafspot (not common,
   sometimes growers considered it insect damage).   Guate-
   mala and Costa Rica mountain areas.
Aphelenchoides parietinus Stein.; root nematode.   Brazil.
Armillaria mellea Vahl = Armilariella mellea (Fr.) Karst.;
   rootrot.   California.
Aspergillus spp.; on old flowers and stored roots.   C.
   America, S. America.
Botrytis cinerea Pers.; moldy bud, flower, and tuber mold.
   Brazil, Argentina, C. America.
Cercospora grandissima Rang.; leafspot.   Subtropical USA,
   W. Indies, C. America, S. America.

Choanephora americana Moell.; blossom blight.   Florida,
    Puerto Rico.
Cladosporium sp.; leaf decline on old plants, spores re-
    covered from dried flowers.   General.
Coleosporium dahliae Arth.; leaf rust.   Mexico.
Cuscuta americana L.; dodder from laboratory inoculation.
    Puerto Rico.
Entyloma calendulae (Oud.) dBy. f. dahliae Viég. and E.
    dahliae Syd.; leafsmuts, brownspots.   Brazil, Argentina,
    Venezuela, Colombia, El Salvador.
Erysiphe (see Oidium)
Fusarium oxysporum Schl. (? f. sp. dahliae); systemic wilt.
    W. Indies.   F. solani (Mart.) Sacc.; base and stem dry-
    rot.   W. Indies.
Helicotylenchus nannus Stein.; root nematode.   Brazil.
Meloidogyne sp.; rootknot.   Brazil, Subtropical USA.
Oidium sp. sometimes identified as O. erysiphoides Fr.
    (ascigerous stage not often seen but is Erysiphe cichora-
    cearum DC.); mildew.   General (except not found in rain-
    forest areas).   Also E. polygoni DC. has been identified
    in dahlia as a mildew in the W. Indies.
Phyllosticta dahliaecola Brum.; leafspot.   Brazil.
Phymatotrichum omnivorus (Shear) Dug.; dryland rootrot.
    Texas.
Pseudomonas solanacearum E. F. Sm.; bacterial wilt.   Sub-
    tropical USA, Costa Rica in low mountains, Panama, Cuba,
    Brazil.
Sclerotinia sclerotiorum (Lib.) dBy. = Whetzeliana sclerotiorum
    (Lib.) Korf & Dumont; stemrot, blossom blight.   Argentina,
    Central America in high mountain locations.
Sclerotium rolfsii (Curzi) Sacc.; rot at ground line.   Sub-
    tropical USA, C. America, S. America.
Tylenchus davainii Bast.; a nematode found at roots.   Brazil.
Viruses: Dahlia Mosaic (DMV), mottling and stunt, occur-
    rence is widespread.   Sunflower Virus in the Argentine.
    Common Tobacco Streak Virus (TSV) in Argentina.   TSV,
    Brazilian strain in Brazil.   Tomato Spotted Wilt Virus
    (TSWV) in Argentina.

DALBERGIA spp.   Forest trees that furnish fine cabinet materials
such as the so-called "rosewood."
    Catacauma dalbergiicola P. Henn. Th. & Syd.; stromatic
        leafspot.   S. America.
    Cephaleuros virescens Kunze; algal leafspot.   Puerto Rico.
    Ceuthocarpon dalbergiae Rehm; on foliage.   S. America.
    Dothidea pulchella Speg. = Munkiella pulchella Speg.; leaf-
        spot.   S. America, C. America.
    Fomes spp.; woodrot of both logs and cut lumber.   Wide-
        spread.
    Helminthosporium lavrense Viég.; lesions on foliage.   S.
        America.
    Meliola bicornis Wint.; black mildew.   W. Indies, C.
        America.

Mycosphaerella devia Petr. & Cif.; leafspot.   Widespread.
Pestalotia albomaculans P. Henn.; leafspot, twig lesion.
    General.
Phoradendron spp.; coarse woody parasitic bushes on host
    branches.   El Salvador, Costa Rica, Guatemala.
Phthirusa spp.; vinelike woody parasites on host branches.
    C. America, W. Indies.
Phyllachora dalbergiae Niessl, P. perforans (Rehm) Sacc. &
    Syd.; both cause tarspots and shothole.   Brazil, C.
    America.
Phyllosticta dalbergiae Syd., P. nivea Syd.; small leafspots.
    Brazil.
Polyporus sp.; logwood rot.   C. America.
Poria sp.; stump rot.   Guatemala, Honduras, Puerto Rico.
Puiggarina myiocoproides Speg. = Phyllachora sp.; leafspot.
    S. America.
Sphaerophragmium dalbergiae Diet.; rust.   Cuba.
Strigula complanata Fée; white lichenspot.   Neotropical.
Uredo dalbergiae P. Henn.; leafrust in Brazil.   U. nidulans
    Syd.; powdery leafrust in Bolivia.
Xylaria sp.; tree base attack.   Costa Rica.

DATURA spp.   Grown for ornament and medicine.   Indians use
    certain species in religious services, and in some areas certain
    species are employed for crop-land rotation purposes; "figueira
    do inferno" Po., "chamísco," "reine-de-la-noche" Sp., "sweet
    queen-of-night," "trumpet flower" En.
    Aecidium daturae Kern., Thur. & Whet.; rust.   Widespread.
    Alternaria crassa (Sacc.) Rands (distinct from A. solani that
        is sometimes incorrectly reported); leafspot, rare pod
        disease.   W. Indies, Costa Rica, El Salvador, S. America.
        A. daturae Fautr. in Brazil may be a race of this fungus.
    Cercospora daturicola (Speg.) Ray; leafspot.   General.
    Diplodia sp.; stem disease.   C. America, Florida.
    Erysiphe (see Oidium)
    Meloidogyne sp.; rootknot nematode.   Costa Rica.
    Oidium erysiphoides Fr. the usually seen conidial state of
        Erysiphe cichoracearum DC.; mild mildew, seasonal oc-
        currence.   Neotropics.
    Phomopsis venenosa (Sacc.) Trav. & Spess.; stem lesions.
        S. America.
    Phyllosticta julia Speg.; leafspot.   Argentina, Paraguay,
        Brazil.
    Phymatotrichum omnivorum (Shear) Dug.; dryland rootrot.
        Texas.
    Pleosphaerulina argentinensis Speg.; leaf lesions.   Argentina.
    Pseudomonas solanacearum E. F. Sm.; bacterial wilt.   Hon-
        duras.
    Rhabdospora venenosa Speg.; on stems that are killed.
        Argentina.
    Rhizoctonia sp.; unusual scanty mycelium on healthy leaves,
        occasional invaded areas on old leaves (of Datura suaveolens
        Humb. & Bonp.).   Costa Rica.

Sclerotium rolfsii Sacc.; soil-line rot. Costa Rica.

Septoria daturae Speg. = Rhabdospora daturae (Speg.) Kuntze;
    leafspots in S. America. S. lycopersici Speg. producing
    small but profuse spots, serves as a holdover source of
    disease for nearby tomato fields. Puerto Rico, El Salvador,
    Costa Rica, Panama, and reported in Texas.

Thielaviopsis basicola (Berk. & Br.) Ferr.; black rootrot.
    Subtropical USA, Costa Rica.

Viruses: Common Tobacco Mosaic Virus (TBM) is widespread.
    Cymbidium Mosaic Virus in Florida. A virus mottle said
    to be Colombian in origin reported in Bolivia, Colombia,
    Ecuador.

DAUCUS CAROTA L. var. SATIVA DC. This widely grown vegetable root crop is of Old World origin; "zanahoria" Sp., "canoura"
Po., "carrot" En.

Alternaria alternata (Fr.) Keiss. = A. tenuis Nees; leaf
    disease complex, the fungus semi-parasitic. Scattered.
    A. dauci (Kuehn) Groves & Sko.; leafspot, root lesion. C.
    America. A. carotae (Ell. & Lan.) J. A. Stev. & Welm.;
    leafspot. C. America, Ecuador. (Note: Instances were
    seen in which Alternaria problems on carrots were much
    more severe in the tropics than in the temperate zone. So
    many times these fungi are rare in the tropics. I felt I
    depended too much on temperate zone findings and wished
    I could see, from and in the tropics where I have been, a
    complete reworking of both the mycology and pathology of
    the Alternarias on carrot; more search for perfect stages,
    and attention given to comparisons with Stemphylium,
    Helminthsporium, and Curvularia. I found on carrot leaves
    in Panama spores of a possible Helminthosporium, and on
    carrot in dry parts of Honduras and Costa Rica spores of
    Curvularia mixed with spores of Alternaria of both large
    and small dimensions.)

Aphelenchus avenae Bast.; root nematode. Brazil.

Botrytis cinerea Pers.; graymold market rot. Occasional.

Cercospora carotae (Pass.) Solh.; leafspot. Subtropical USA,
    cooler areas of W. Indies, C. America, Brazil and Argen-
    tina.

Cuscuta spp.; dodders. C. America. (The slower growing
    dodder, Cassytha, seldom if ever reaches a carrot field
    before it is harvested.)

Diaporthe arctii (Lasch.) Nits.; stem disease on seed plants.
    Guatemala mountain area.

Ditylenchus dipsaci Filip.; nematode. Brazil.

Erwinia carotovora (L. R. Jones) Holl.; softrot. Wherever
    carrot roots are shipped or marketed.

Fusarium spp.; storage rot complex. Mexico, Puerto Rico.

Helicotylenchus nannus Stein.; a nematode reported in Brazil.

Heterodera rostochiensis Wor.; nematode (so-called "Golden
    nematode" of potato). Peru.

Meloidogyne spp.; causing disfiguring rootknot troubles. M.

exigua Goeldi determined in Brazil and Chile.　M. incog-
nita Chitw. in Subtropical USA, Texas, Mexico.
Monhysteria stagnalis Stein.; nematode.　Brazil.
"Nematodes" unidentified; root injury.　Nicaragua, Chile, Peru,
　　Panama, Ecuador.
Phoma daucina Speg.; on foliage of wild carrot.　Argentina.
Phyllosticta sp.; leaf disease.　Peru.
Phymatotrichum omnivorum (Shear) Dug.; rootrot.　Louisiana,
　　Texas and Arizona in USA.
Rhizoctonia microsclerotia Matz; web-blight of leaves.　Puerto
　　Rico, Costa Rica.　R. solani Kuehn mycelial stage of
　　Thanatephorous cucumeris (Frank) Donk; damping-off, root-
　　canker, shoulder strangle, scurf, basalrot.　C. America.
Rhizopus nigricans Ehr.; black storagerot, wet-soil rot.
　　Mexico, Brazil, Uruguay, El Salvador.
Sclerotinia sclerotiorum (Lib.) dBy. = Whetzeliana sclerotiorum
　　(Lib.) Korf & Dumont; rainyseason fieldwilt, store wetrot.
　　Argentina, Mexico, Brazil, Chile, occasional in C. America.
Sclerotium rolfsii (Curzi) Sacc. = Pellicularia rolfsii (Sacc.)
　　West; collapse, scattered in W. Indies, Honduras, Chile,
　　in Uruguay at times severe.
Septoria apii (Briosi & Cav.) Chester; leafspot.　Brazil, Ar-
　　gentina.
Streptomyces scabies (Thaxt.) Waks. & Henr.; root scab.
　　Argentina.
Thielaviopsis basicola (Berk. & Br.) Ferr.; black root.　Sub-
　　tropical USA, El Salvador by inoculation.
Tylenchorhynchus sp.; nematode.　Peru.
Tylenchus davianii Bast.; nematode.　Brazil.
Viruses: Common Cucumber Mosaic Virus (CMV) in Panama.
　　CMV virulent celery/commelina strain in Florida, Panama.
　　Sunflower virus, Argentina type in Argentina.

DELONIX REGIA (Bojer) Raf., = POINCIANA REGIA Bojer.　A gift
　　from Africa, a great and handsome tree having abundant finely
　　pinnate leaves that round up a broad, flat domed top shape, set
　　off brilliantly with fiery red flowers during the long flowering
　　time; it is important to nurserymen; "flamboyán" Sp., "flam-
　　boyant" Po., "royal poinciana" En.
　　Agrobacterium tumefaciens (E. F. Sm.) Conn.; crowngall in
　　　　nursery plants.　Florida.
　　Armillaria mellea Vahl = Armilariella mellea (Fr.) Karst.;
　　　　rootrot.　Panama.
　　Botryosphaeria dothidea (Moug.) Ces. & de Not.; small-limb
　　　　canker, dieback.　Subtropical USA, W. Indies.
　　Cassytha filiformis L.; green dodder (on young tree branch by
　　　　inoculation).　Puerto Rico.
　　Cephaleuros virescens Kunze; algal leafspot.　Costa Rica,
　　　　Panama, Brazil, W. Indies.
　　Cercospora guanicensis Young; leafspot.　Puerto Rico.
　　Ciliciopodium caespitosum Speg.; on a branch.　Paraguay.
　　Clitocybe tabescens Bres. = Armilariella tabescens (Scop.)

Sing.; a rootrot. Florida. (This fungus reported from
work in Florida to attack 213 species of plants in 137
genera. It can be seen in numerous items in this book.
A wide host range is also known for Armillaria mellea to
which it is remarkably similar.)

Colletotrichum also called Gloeosporium (see Glomerella)

Cuscuta spp.; yellow dodders. Found here and there in
poorly maintained nurseries.

Fomes spp.; trunk and stump rots accompanied by conks.
Panama, Colombia, Florida.

Ganoderma lucidum Karst.; trunk base attack of old trees.
Panama, C. America.

Glomerella cingulata (Stonem.) Sp. & Sch. (conidial state is
Colletotrichum sp.); pod drop and leaf disease, sometimes
forming of ascocarps along pod sutures. C. America, W.
Indies, Panama.

Marasmiellus icterinus Sing.; branch infection. Florida.

Meliola sp.; black mildew. Panama.

Pestalotia sp.; leaf and petiole spot. Widespread.

Phoradendron sp.; bushy parasite. Panama.

Phthirusa parvifolia Eich.; semi-pendant bush parasite.
Jamaica, Cuba, El Salvador.

Phymatotrichum omnivorum (Shear) Dug.; rootrot. Texas.

Phytophthora sp.; root disease in farm nursery. El Salvador.

Psittacanthus calyculatus (DC.) Don. and P. americanus Mart.;
often large bush parasites causing "wood rose" at attach-
ment to host. Guatemala, El Salvador.

Strigula complanata Fée; common lichen spot on leaf and
petioles. W. Indies, C. America.

DESMODIUM spp. There are 25 or more species encountered in
Latin America, important as cover crops, as browse for livestock, and in soil building programs. Species characters vary
from low growing and small to large, some are used in farming
as annuals others in perennial stands, some are grazed growing
compatibly with pasture grasses; "desmodio," "amor seco" Sp.,
"desmodio," "mermelado de cavallo," "foragem verde" Po.,
"tropical clover," "desmodium" En. (Unless noted especially on
one species, all parasites listed have been reported from numerous Desmodiums.)

Aecidium citrinum (Syd.) Gae., A. desmodii P. Henn. (may
now be old synonyms of such as Uromyces, Puccinia,
Uredo, Phakopsora); rusts. General.

Alternaria sp.; uncommon spots on basal leaves. On several
Desmodiums. El Salvador, Costa Rica.

Botrytis sp.; seedling disease. Honduras.

Cassytha filiformis L.; slender green-dodder. Attacks any
Desmodium. Puerto Rico.

Cephaleuros virescens Kunze; algal scurf spot (especially on
Desmodium tortuosum DC.). Puerto Rico, Costa Rica.

Cercospora desmodiicola Ell. & Kell. var. leiocarpa Gonz.
Frag. & Cif.; leafspot. W. Indies. C. meibomiae Chupp;

leafspots on Desmodium incanum DC. Widespread. C. melaleuca Ell. & Ev.; stemspots on Desmodium tortuosum DC. in Florida.

Chaetoseptoria wellmanii J. A. Stev.; leafspots large and sometimes cause severe defoliation. On several Desmodiums. C. America.

Cladosporium infuscans Thuem.; leaf disease and collapse. Subtropical USA, C. America.

Cuscuta americana L.; common dodder. On any Desmodium if parasite is present. W. Indies. C. haughtii Yunc.; dodder. Ecuador, Gallapagos Islands.

Fusarium oxysporum Schl.; systemic wilt. Trinidad. F. (roseum-type); part of dieback complex. El Salvador. F. solani (Mart.) Sacc. (Nectria haematococca Berk. & Br.); from diseased plant base. Trinidad, Puerto Rico.

Irenina meibomiae Stev.; black mildew (on D. frutescens Sch.). S. America.

Isariopsis caespitosa Petr. & Cif. (apparently not I. griseola although I could not distinguish microscopically between the former from Desmodium and the latter from Phaseolus); angular leafspot. El Salvador, Dominican Republic.

Meliola spp. seemingly several: M. bicornis Wint., f. heterotricha Speg., M. constipata Speg., M. desmodii Kar. & Rou., M. trinidadensis Stev. & Tehon, maybe others; all black mildews. Some may be only on certain Desmodiums but this has to be studied further. American tropics.

Meloidogyne sp.; rootknot nematode. Florida, Jamaica.

Microsphaera diffusa Cke. & Pk. = Oidium sp., powdery mildew. Neotropics.

Mycena citricolor (Berk. & Curt.) Sacc.; leafspot (near diseased coffee trees). On D. affine Sch., D. canum Schi. & The., D. triflorum DC. All in Costa Rica. No doubt any host species is susceptible.

Oidium (see Microsphaeria)

Parodiella perisporioides (Berk. & Curt.) Speg. (? = P. paraguayensis Speg.); leaf spot. On any Desmodium. Neotropics.

Phakopsora meibomiae Arth.; rust (on D. incanum DC. and D. tortuosum DC.). Widespread.

Phomopsis sp. (possibly Diaporthe); stem and root disease. Dominican Republic.

Phyllachora stevensii Syd.; tarspot. On a few Desmodiums. Panama, W. Indies, C. America. P. diocleae P. Henn.; tarspot. S. America.

Phyllosticta desmodiiphila Speg.; leafspot. On several Desmodiums. W. Indies, S. America. P. meibomiae Seav.; leafspot (on D. rhombifolium DC.). Florida.

Phymatotrichum omnivorum (Shear) Dug.; dryland rootrot. Texas.

Puccinia sp.; rust (on D. scorpiurus Desv.). S. America.

Ramularia desmodii Cke.; mealy leafspot. Subtropical USA.

Sclerotium rolfsii (Curzi) Sacc. = the spore bearing Pellicularia rolfsii (Sacc.) West. ; soil-line rot. Subtropical
  USA, S. America.
Stagonospora desmodii Ell. & Ev. ; stem disease (on D. tortuosum). Florida, Puerto Rico.
Synchytrium citrinum Gaeum. (= Woroninella citrina Syd.);
  leafgall. Scattered in S. America.
Thecophora deformans Dur. & Mont. ; somewhat uncommon
  seedsmut. Cooler parts of Neotropics.
Uredo spp. : U. desmodiicola Speg. = Uromyces hedysari-
  paniculati (Schw.) Farl. , U. microtheles Speg. , U. sub-
  hyalina Speg. All rusts, apparently S. American.
Uromyces spp. (in this rust genus the first species is described from Guatemala, others from S. America): U.
  antiguanus Cumm. (on D. orbicularis Sch.), U. castaneus
  Syd. (on D. incanum DC.), U. desmodii Cke. (on Desmodium sp.), the common U. hedysari-paniculata (Schw.)
  Farl. on any Desmodium and generally distributed, U.
  huallagensis P. Henn. (on Desmodium sp.), U. orbicularis
  Diet. (on D. adscendous DC.), D. malmei Sch. , U. cenuis-
  tipes Diet & Holw. (on D. uncinatum DC.).
Viruses on various Desmodiums: Common Bean Yellow Mosaic
  ? = Bean Dwarf Mosaic of Brazil in El Salvador. Desmo-
  dium mottle mosaic, unidentified, in C. America. Cowpea
  Aphid-borne Mosaic, in W. Indies. Argentine Sunflower
  Virus in Argentina. Severe stunt, evidently virus, but not
  identified, in Costa Rica. Witches' Brooming, evidently
  virus but not identified, in Costa Rica.

DIANTHUS CARYOPHYLLUS L. This, the only species considered,
  was brought into the American tropics from Europe and North
  America, and has been for a long time an important source of
  income by large florists and for a multitude of small growers;
  these popular cut flowers are found, in season, in all major and
  many minor markets of Latin America; "cuevo" Po. , "clavel"
  Sp. , "carnation" En.
    Alternaria dianthi Stev. & Hall; rare blight, stemcanker, leaf-
      spot. Subtropical USA, W. Indies in cool season, C.
      America in mountain-slope gardens, S. America. A. dian-
      thicola Neerg. , in Jamaica.
    Amphichaeta diffusa Batista; on stems and leaves. S. America.
    Ascochyta dianthi (Alb. & Schw.) Berk. ; small leafspots.
      Brazil, Argentina.
    Botrytis cinerea Pers. ; graymold of buds and market flowers,
      more common in rainyseason. C. America, Argentina,
      Colombia, Chile.
    Cercospora dianthae Muller & Chupp; typical leafspot. Brazil,
      El Salvador.
    Colletotrichum dematium (Fr.) Grove = C. dianthi Weste. ;
      anthracnosis. Nicaragua, Brazil, Guatemala, Chile.
    Cuscuta sp. ; dodder. El Salvador.
    Diplodina dianthi Batista; infectious stem-and-leaf disease.
      Brazil.

(? Erwinia carotovora?); a slow-acting white bacterium in cut stem ends of flowers.   El Salvador.

Fusarium spp., a number of them aiding in stemrot complex from which isolations indicated F. roseum Lk., F. semitectum Berk. & Rav., F. equiseti (Cda.) Sacc., F. scirpitype are all involved.   Seem occasional and scattered.   F. oxysporum Schl. f. sp. dianthi (Drill. & Del.) Sny. & Hans.; wilt.   Florida, Costa Rica (were in potted plants held at a warm location).   Argentina.

Heterosporum echinulatum (Berk.) Cke.; annular ringspot of leaf, shoot, flower.   Colombia, Chile, Mexico, Argentina, Peru.

Meloidogyne sp.; mild rootknot.   In warm soils of Florida, Chile, Guatemala, Venezuela, but probably in other countries where effects are masked by low temperatures of garden soils.

Oidium sp. = Erysiphe sp.; mildew.   Guatemala, Florida, Mexico.

Phoma caryophylla Cke.; stem lesion.   S. America in cool areas.

Phyllosticta sp.; leafspot.   Brazil, Argentina.

Phytophthora citrophthora (Sm. & Sm.) Leon. and P. parasitica Dast.; footrot.   Argentina.

Pseudomonas sp.; bacterial plant wilt.   Mexico, Guatemala.

Pythium debaryanum Hesse; cutting disease.   Guatemala.

Rhizoctonia solani Kuehn (no sporestage found); cutting disease. Puerto Rico.

Rosellinia bunodes (Berk. & Br.) Sacc.; black rootrot.   Costa Rica, Guatemala in both countries in newly cleared jungle land.

Sclerotium rolfsii (Curzi) Sacc. = sporuliferous Pellicularia rolfsii (Sacc.) West; soil-line rot.   Florida and Subtropical USA, Uruguay, Argentina.

Septoria dianthi Desm.; a common yellow leafspot.   Chile, Guatemala, Venezuela, Mexico, Nicaragua, Uruguay.   (Possibly S. dianthophila Speg. is synonym of S. dianthi.)

Uromyces caryophyllinus (Schrank) Wint. is the name most often used in the American tropics for this rust, but it is believed U. dianthi (Pers.) Niessl, may be the most correct according international rules; the common rust.   Chile, Subtropical USA, Mexico, Bolivia, Argentina, Venezuela, Peru, Brazil, Jamaica.   Probably in every country where the crop is grown.

Virus: an undoubted mosaic mottling, sometimes severe with old leaves and young leaves showing streak symptoms.   Not studied as to source, properties or other identity.   El Salvador, Guatemala.

DIEFFENBACHIA PICTA Sch. and D. SEGUINE Sch.   Large-leaved natives among Western tropical ornamentals, grown also for so-called luck; "rábano cimarrón," "buen-suerte," "mata puerco" Sp., "aninga do paru," "rabanete venenosa" Po., "dumb-cane"

En. (The diseases listed below appear to attack both host species of this favored perennial.)

Cephaleuros virescens Kunze; algal leafspot.   Brazil.

Colletotrichum sp. conidial stage of Glomerella cingulata (Stonem.) Sp. & Sch. = G. cynta St. ; anthracnose canker. General but seldom severe.

Erwinia carotovora (L. R. Jones) Holl. and E. diffenbachiae McF. ; softrot, stem and leaf collapse.   Florida, Puerto Rico.

Mancha acuosa or water spot, from lack of shade.

Meliola dieffenbachiae Stev. = Irenina aracearum Stev. ; black mildew spot.   British Guiana, Puerto Rico, Virgin Islands, Cuba, Santo Domingo, Ecuador, Panama, Costa Rica, Guatemala.

Phyllosticta colocasiae Hoehn. ; leafspot with center kernels. Not common but seen in W. Indies, C. America, S. America.

Phytopthora palmivora Butl. ; stemrot.   California, Puerto Rico.

Pythium splendens Braun; rootrot.   Florida.

Virus: Mosaic said to be Dasheen mosaic in Florida.

Xanthomonas dieffenbachiae (McCul. & Pirone) Dows. ; bacterial leafspot.   Florida, El Salvador, Costa Rica.

DIGITARIA DECUMBENS Stent.   This excellent pasture grass is a relative of some poor Western Hemisphere grasses, was obtained in the Orient and, according to tropical livestock men in the Americas, has shown special adaptability as to growing on unpromising soils, in hot conditions, and is relished by grazing animals; "pasto pangola" Sp. Po., "pangola grass" En.

Ceratosphaeria sp. = conidial stage is Pyricularia grisea Sacc. ; culm infection, leafspot, blight.   C. America, Brazil.

Cercospora sp. ; leafspot.   Puerto Rico.

Cochliobolus carbonum Nelson; footrot.   Mexico.

Curvularia spp. ; leaf and stem disease.   Puerto Rico.

Fusarium sp. ; spores secured from apparently secondary decay of leaves.   Puerto Rico.

Helminthosporium sp. ; leafspot.   Puerto Rico.

Lethal browning, possibly physiological or toxins from fungi. British Guiana.

Mycosphaerella maydis (Pass.) Lindau; large lesions on leaves. W. Indies, C. America, northern S. America, Surinam.

Puccinia oahuensis Ell. & Ev. = Uredo syntherismae Speg. ; rust.   Honduras, Puerto Rico, Surinam, Ecuador, Brazil. (Some clones of host appear resistant.)

Virus, Pangola Grass Stunt studied by Dirven and Van Hoof in Surinam is not only there but in Puerto Rico, British Guiana (? Florida).   It is transmitted by a sucking insect, Sogata.

Volutella sp. ; a leaf disease.   Puerto Rico.

DIOSCOREA ALATA L.   A vine grown for its much-valued nutritious,

tropical tuberous root.   Some species of this plant are being in-
creasingly exploited for medicinal purposes; "igname," "nyame,"
"ñame" Sp., "ignahme," inhame," "caracultivado" Po., "ala,"
"chinese yam," "true yam" En.
Aecidium dioscoreae Lindq. and A. puracense Petr. ; leaf-
   rusts. Argentina.
Alternaria sp. ; leafspot. Venezuela.
Aspergillus sp. ; flowermold. Panama.
Belondira porta Thorne; nematode. Puerto Rico.
Botrytis sp. ; storage rootrot. C. America.
Cassytha filiformis L. ; green dodder. Puerto Rico.
Catacauma glaziovii (P. Henn.) Tr. & Syd. = Phyllachora
   graziovii P. Henn. ; black rough stromatic leafspots. Brazil.
Cephaleuros virescens Kunze; algal leafspot. Brazil.
Cercospora spp. ; all produce leafspots. C. brasiliensis Av. -
   Sac. reported in Brazil, Dominican Republic, Venezuela,
   C. carbonacea Miles, C. contraria Syd. in Brazil, C. dio-
   scoreae Ell. & Mart., C. pachyderma Syd. reported in
   much of W. Indies, C. America, some in S. America.
Colletotrichum gloeosporioides Penz. (seems probably there is
   a race especially adapted to Dioscorea); anthracnose.   Neo-
   tropics. (Note: Twice I secured ascocarps of Glomerella,
   apparent perfect stage of Colletotrichum, from old vines in
   Costa Rica.)
Curvularia sp. ; from lesions on vine. Guatemala.
Cuscuta americana L. ; yellow dodder. Puerto Rico.
Diaporthe phaseolorum (Cke. & Ell.) Sacc. var. batatatis
   (Harter & Field) Wehm. ; dryrot of stem and root.   El
   Salvador.
Didymaria fulva Ell. & Ev. ; leaf disease. Guatemala.
Didymella dioscoreae (Berk. & Curt.) Sacc. stem lesion.   Sub-
   tropical USA.
Diplodia = Botryodiplodia; vine dieback, tuber rot. Panama,
   Puerto Rico, St. Thomas, Brazil, C. America. (I found
   an asco-state of the parasite, only on old vines in W.
   Indies.)
Fusarium oxysporum Schl. (I suggest here f. sp. dioscoreae);
   wilt, discolor of tuber vasculars. Puerto Rico, Cuba,
   Guatemala, Panama. F. solani (Mart.) Sacc. ; secondary
   invasion in old stem and leaf lesions. Guatemala, Vene-
   zuela.
Goplana dioscoreae (Berk. & Br.) Cumm. (? = Aecidium);
   rust. Argentina. G. ecuadorica Syd. ; rust. Ecuador.
Helminthosporium sp. ; round leafspot. W. Indies. (When a
   spot is secondarily enlarged in Puerto Rico it is by such
   fungi as Fusarium, Colletotrichum, Nigrospora, Pestalotia,
   and the spot spreads and becomes very irregular in out-
   line.)
Himantia corticalus Viégas; sterile mycelium on foliage.
   Brazil.
Meliola sp. ; black mildew. Puerto Rico.
Meloidogyne exigua Goeldi and M. incognita Chitw. ; root
   nematodes. The first in Brazil, second in Guatemala.

Mycena citricolor (Berk. & Curt.) Sacc.; leafspot.   Puerto
   Rico.
Mycosphaerella sp.; ascocarps in a leafspot, possibly an old
   lesion from Cercospora.   Nicaragua.
Penicillium spp.; market rootrots.   S. America, Panama,
   Nicaragua.
Phyllachora ulei Wint.; circular shiny black spots.   Brazil,
   Trinidad, Colombia, Mexico, Costa Rica, Guyana, Panama,
   Dominican Republic, Puerto Rico, Venezuela.   (For an-
   other Phyllachora report see Catacauma.)
Phyllosticta dioscoreae Cke., P. dioscoreicola Brun., P.
   dioscoreae-daemonae P. Henn.; red-brown leafspots.
   Brazil, W. Indies.
Puccinia valida Arth.; a brown rust.   Mexico, C. America.
Ramularia dioscoreae Ell. & Ev.; leaf and stem spot.   C.
   America.
Rhizoctonia (soil inhabiting); rootrot, leaf on infested soil.
   W. Indies, C. America.   Rhizoctonia (aerial); web blighting
   in wet weather, target spotting in dryer condition.   El
   Salvador, Panama, Costa Rica, Mexico, Guatemala, Ecua-
   dor, Venezuela.   (Apparently, this is not Himantia, which
   see above.)
Rosellinia bunodes (Curzi) Sacc. = Pellicularia rolfsii (Sacc.)
   West; soil-line rot.   Scattered in many countries.
Scutellonema blaberum Andr.; root nematode in W. Indies.
   S. dioscoreae Lorde.; root nematode in Brazil.
Septoria versicolor Pat.; a typical white leafspot.   Ecuador.
Sphenospora pallida (Wint.) Diet.; yellow rust.   Brazil, Vene-
   zuela, Panama, Costa Rica, Trinidad.
Uredo dioscoriicola Kern, Cif. & Thu.; orange rust.   Puerto
   Rico, Cuba, Brazil.   The U. pallatangae Jorst. is a simi-
   lar rust in Ecuador.
Viruses: Dioscorea mottling mosaic, otherwise unidentified,
   in W. Indies, C. America, S. America.   Brazilian tobacco
   streak by inoculation in Brazil.
Volutella ciliata (Alb. & Schw.) Fr.; on leaves.   S. America.
Xiphinema americanum Cobb; large nematode in rhizosphere.
   Brazil.

DIOSPYROS KAKI L. f.   A subtropical medium-sized fruit tree
   from the Orient where it has been grown since ancient times,
   and is now introduced to cooler, somewhat dry areas of the
   Neotropics, where it is slowly becoming well known and more
   used; "caquí," "kaki" Sp., Po., "japanese persimmon" En.
   Aecidium sp.; leafrust.   Brazil.
   Agrobacterium tumefaciens (E. F. Sm.) Conn; crowngall.
      Subtropical USA.
   Botryosphaeria dothidea Ces. & de Not.; branch canker.   Sub-
      tropical USA.
   Capnodium sp.; sootymold.   Widespread.
   Cephaleuros virescens Kunze; algal leafspot.   Brazil.
   Cephalosporium diospyri Crand.; wilt.   Subtropical USA.
   Cercospora kaki Ell. & Ev.; leafspot.   Widespread.

Colletotrichum gloeosporioides Penz.; anthracnose. Probably wherever grown.

Dieback, severe (cause undetermined); isolations unsuccessful. Costa Rica.

Elsinoë diospyri Bitanc. & Jenk.; spot scab and anthrax. S. America.

Fomes sp.; tree heartrot. W. Indies.

Leptothyrium pomi (Mont. & Fr.) Sacc.; fruit flyspeck. W. Indies, C. America.

Melasmia falcata Syd.; black stromatic leafspot. Brazil.

Oidium sp.; white mildew on new growth. Panama.

Oospora sp.; yeasty fruitrot. General.

Pellicularia koleroga Cke.; branch pinkdisease, leafrot. Subtropical USA, C. America, S. America.

Phaeosaccardinula diospyricola P. Henn.; black scab on leaves. Brazil.

Phyllosticta biformis Heald & Wolf; leafspot. Subtropical USA, Chile.

Rhizopus nigricans Ehr.; fruit rot. General.

Rosellinia sp.; rootrot. Colombia, Venezuela.

Strigula spp. (several types); lichen spots on leaves. C. Rica.

DIPTERYX ODORATA Willd. A medium-sized leguminous tree, native to the northeastern part of South America, is planted in semi-wild clumps or portected in woodlands and harvested as a forest product. Bears pods with a single seed which after processing is the fragrant "tonka bean" of commerce, used in perfumery and in flavoring of tobacco and foods.

Botryodiplodia theobromae Pat. = Diplodia sp.; possibly from dieback of small branches (typical spores from pod scraping). Guyana.

Cephaleuros virescens Kunze; algal leafspot. Honduras.

Colletotrichum gloeosporioides Penz. (asco-state unreported); anthracnose spots on leaves and pods. Honduras.

Fusarium sp.; from dieback twig. Honduras, Costa Rica.

Loranthus sp. (young plant found before flowering); bushy Phanerogamic parasite. Panama.

Marasmius sp.; cord blight. S. America.

Pellicularia koleroga Cke.; web-thread causing foliage rot. Costa Rica, Guyana.

Penicillium sp.; isolation from old pod in museum, source unknown.

Pestalotia sp.; leaf and stem lesion. C. America.

Phoma-like leafspot. Honduras.

Phyllosticta sp.; leafspot. S. America.

Rhizoctonia solani Kuehn = Thanatephorous cucumeris (Frank) Donk; stem disease. Guyana, Venezuela.

Strigula complanata Fée; white lichen-spot. C. America.

DRACAENA see CORDYLINE

DRIMYS WINTERI Forst. Tree grown for its spicy character, the

harvested aromatic bark is used in medicine and sold in markets where it is called "canelo" Sp. Po., and "winters bark" En.

Acremonium araucanum Speg.; on senescent leaves. Argentina.

Actinothyrium drimydis Speg.; rounded large leafspots. Chile.

Antennaria scoriadea Berk.; moldy leaf disease. Chile.

Asterinella drimydis (Lév.) Speg.; spots of black mildew. S. America.

Caliciopsis clavata (Lév.) Fitzp.; branch canker. C. America, S. America.

Calonectria inconspicua Wint.; on diseased stem. S. America.

Corynelia tropica (Auers. & Rab.) Starb.; leafspots with sunken stroma. Chile.

Depazea drimydis Berk.; foliage disease. Chile.

Dimerosporium tropicale Speg.; black mildew. Argentina.

Dothidea drimydis Lév. = Crotone drimydis (Lév.) Th. & Syd. = Munkiella drimydis (Lév.) Speg.; leafspot and blast. S. America, Costa Rica.

Dothiorella winterii Speg.; branch lesion. Argentina.

Fusarium sp.; isolation from dieback. Panama.

Helminthosporium orbiculare Lév.; leaf blast. S. America.

Hymenochaete tabacina (Sow.) Lév.; canker. C. America, S. America.

Hysterium foliicolum Fr.; erumpent leafspot. S. America.

Lembosia drimydis Lév. = Leveillella drimydis (Lév.) Tr. & Syd. = Asterina compacta Lév.; radiating leafspot disease. Chile.

Leptothyrium drimycola Speg.; small speck leafspots. S. America.

Meliola compacta (Lév.) Speg., M. corallina (Mont.) Sacc., M. crustacea Speg. = Irenina crustacea (Speg.) Stev.; all are black mildews. Chile, Costa Rica, Panama.

Mycosphaerella drimydis (Berk.) Sacc.; large rounded leafspot. Brazil, Chile. (I found apparently typical sporocarps of Mycosphaerella on large rounded spots without surface spores, on leaves collected for me in Panama.)

Parasterinella drimydis (Lév.) Speg.; leafspot. Argentina.

Pestalotia valdiviana Speg.; leafspot, twig disease. S. America.

Phyllosticta drimydis Speg. and P. winteri Speg.; leafspots. S. America.

Pleonectria vagans Speg.; on diseased old bark. Chile.

Poria ferruginosa (Schrad.) Fr.; on dead branch. Chile.

Septoria drimydis Mont. and S. winteri Speg.; leafspots. S. America.

Sphaeronaema clavatum Lév.; twig disease. Chile.

Stuartella drimydis Rehm; leaf blight. Chile.

DOLICHOS LABLAB L. Introduced from Old World tropics, an edible bean borne on vines that often act as good cover of fragile soils, also used as ornamental flowering climber; "frijol caballero," "frijol lablab" Sp., "feijao fradinho," "feijao lablab" Po., "lablab," "hyacinth bean" En.

Alternaria alternata (Fr.) Keiss. = A. tenuis Nees; leaf-edge
  blackening of ragged leaf. A few occurrences in dry sub-
  tropical USA, El Salvador, Puerto Rico.
Cercospora spp. all cause severe leafspots: 1. C. canescens
  Ell. & Mart. in Subtropical USA, W. Indies, C. America,
  Venezuela. 2. C. cruenta Sacc. (? = C. canescens), in
  C. America, Brazil. 3. C. dolichi Ell. & Ev. in Trini-
  dad. 4. C. wildemanii Syd. in Trinidad.
Chaetoseptoria wellmanii J. A. Stev.; circular leafspot.
  Mexico, C. America.
Colletotrichum sp. (not lindemuthianum or gloeosporioides);
  stem lesions, triangular leafspot, El Salvador.
Erysiphe polygoni DC.; white powdery mildew. W. Indies,
  C. America, Brazil. Probably general.
Leveillulla taurica Arn.; a mildew. Nicaragua.
Meloidogyne sp.; rootknot. Subtropical USA, Costa Rica.
Mycena citricolor (Berk. & Curt.) Sacc. (by inoculation);
  large leafspot. El Salvador.
Parodiella perisporioides Speg.; black mildew. On specimens
  in letters sent from Ecuador, Subtropical USA, Costa
  Rica.
Pellicularia (slender fastgrowing mycelium and no sclerotia,
  very small basidia and basidiospores); leaf disease.
  Panama, Puerto Rico.
Phakopsora vignae Arth. = P. pachyrhiza Syd. (see Physopella);
  rust. Barbados, Puerto Rico.
Phymatotrichum omnivorum (Shear) Dug.; rootrot. Texas.
Physalospora sp. = Botryosphaeria quercuum (Schw.) Sacc.
  (from moist chamber studies); dieback. El Salvador, Costa
  Rica.
Physopella concors (Bres.) Arth. (? Phakopsora pachyrhizae
  Syd.); brown leafrust. Puerto Rico, Brazil, Guatemala.
Puccinia dolichi Arth.; leafrust. Cuba.
Rhizoctonia (? solani); zonal leafspot, targetspot stem lesion.
  Costa Rica. Rhizoctonia of aerial character see Pelli-
  cularia.
Septoria sp.; leafspot. Brazil.
Uromyces phaseoli (Pers.) Wint. f. typica Arth.; typical rust-
  spots with yellow halos. Fairly general.
Viruses: Undetermined mild mottle mosaic in C. America.
  Cowpea Mosaic induced by inoculation in Trinidad. Bra-
  zilian Tobacco Streak induced by inoculation in Brazil.

DURIO ZIBETHINUS L. The "durian" is a large fruit tree introduced from the Orient; there are only a few in Latin America.
  (As seen on specimen trees, in Honduras and Costa Rica, a
  small number of diseases are as follows:)
    Cephaleuros virescens Kunze; algal leafspot.
    Cladosporium (? herbarum); spores found on old misshapen
      leaf.
    Corticium salmonicolor Berk. & Br.; pink disease, near in-
      fected coffee trees.

Fruit decays, fungus and yeast.
Leaf speck (Leptothyrium-like).
Penicillium sp. on fallen unripened fruits.
Poria sp. on fallen branch.
Strigula complanata Fée; lichen leafspot.

ECHINOCHLOA COLONUM Lk. An Old World introduction, pro-
duces grain even when growing under poor conditions, some use
it as a forage grass; "arroceillo," "grama pintada," "paja
americana" Sp. Po., "jungle rice," "barnyard grass" En.
   Alternaria alternata (Fr.) Keiss. = A. tenuis Nees; blackening
      of old foliage and seed heads. Puerto Rico.
   Cercospora echinochloae J. J. Davis; leafspot. Florida,
      Guatemala, Venezuela.
   Claviceps balansioides Moell.; seed disease (poisonous for
      consumption by man or beast). Brazil.
   Colletotrichum graminicola (Ces.) G. W. Wils.; leaf disease.
      In Brazil (where it is reported on Echinochloa crusgalli
      Beauv.).
   Curvularia geniculata (Tracy & Earle) Boed.; stem disease.
      Costa Rica.
   Fusarium sp.; wilt. Puerto Rico.
   Helminthosporium sp.; leafblast. El Salvador, Guatemala,
      Panama.
   Meloidogyne sp.; rootknot. Subtropical USA, Costa Rica.
   Puccinia flaccida Berk. & Br.; rust on Echinochloa crusgalli
      in Brazil. P. recondita Rob.; rust. Brazil.
   Pyricularia grisea Sacc. (conidial stage); seedling disease.
      W. Indies, Subtropical USA.
   Rhizoctonia microsclerotia Matz; web blight. Puerto Rico,
      Costa Rica, Subtropical USA.
   Ustilago globigena Speg.; smut. Puerto Rico, Venezuela.
   Viruses: Rice Hoja Blanca Virus in Texas, Florida, Vene-
      zuela, Cuba, Mexico, C. America, Colombia, Surinam.
      Sugarcane Mosaic Virus in Subtropical USA, W. Indies,
      Surinam.

ELAEIS GUINEENSIS Jacq. Not too long ago this palm that produces heavy burdens of fruits which have an oil in the epicarp
and an oil in the seed, was introduced to the Western tropics
from Africa and has been a welcome addition to agriculture in
the Neotropics; "palma africana," "palma de aciete" Sp.,
"dendezeiro" Po., "african oil palm" En.
   Achorella attaleae Stev.; leafspot. Panama.
   Algae (mixed species); smothering growth in seedbed of
      seedlings. Costa Rica.
   Bacterium (unidentified); part of fruit fall complex. Colombia,
      Honduras.
   Cephaleuros virescens Kunze; algal leafspot, mild appearance,
      mostly in moist regions. Honduras, Nicaragua, Brazil.
      (In detailed work I could not find this alga on oil palm in
      Puerto Rico although I found abundant lichen spotting.)

Ceratocystis sp. ; darkened leaf petiole tissues, leaf collapse.
Colombia, Costa Rica.
Cercospora elaedis Stey. ; leafspot.   Surinam, Venezuela.
Chlorosis and bronzing of leaf (in large part potash deficiency).
W. Indies, C. America, S. America.
Crown disease (undetermined cause); death of center, growth
curvature.   Venezuela.
Curvularia sp. ; stem lesion, petiole spot.   Colombia, Hon-
duras, Costa Rica.
Diplodia (Botryodiplodia) sp. ; leaf lesion, fruit drop from
diseased rachis.   Puerto Rico.
Fomes noxius Corner; root and base rot.   Nicaragua, Hon-
duras.   (Other Basidiomycete heart rots occur but unde-
termined.)
Fusarium oxysporum Schl. f. sp. elaeis Toovey; wilt.   Suri-
nam, Nicaragua.   Fusarium (undetermined species); root
decay, leaf collapse, fruit peduncle disease.   C. America,
Colombia.
Ganoderma pseudoferrum Overeem & St. ; base and trunk
decay.   Costa Rica, Surinam, Puerto Rico, Nicaragua.
Graphiola phoenicis (Moug.) Poit. ; false smut spot of leaf.
C. America, W. Indies, Colombia.
Helminthosporium sp. ; brown spot of leaflets.   Colombia, W.
Indies, C. America.
Little leaf (some kind of minor element deficiency).   Colombia.
Marasmius sp. (no sporocarps found); black-thread blight.
Nicaragua, Costa Rica.
Meliola elaeis Stev. ; black mildew spots.   General.
Pestalotia palmarum Cke. ; sunken lesion of petiole, leaflet
spot, ragged leaf edge.   Surinam, Guyana, Puerto Rico,
Haiti, Cuba, Jamaica, Nicaragua, Venezuela, Nicaragua,
Ecuador.
Phoma sp. ; in leaf lesion complex.   Puerto Rico.
Phycopeltis sp. ; algal scaly plate.   Puerto Rico.
Phyllosticta argyrea Speg. ; leafspot and speckle.   Brazil,
Colombia, Venezuela.
Phytophthora palmivora Butl. ; budrot.   Nicaragua, Honduras,
Guatemala, Colombia, Panama.
Rhadiaphelenchus cocophilus Cobb; redring (nematode disease),
bud dwarf, tree death.   Venezuela, Brazil.
Rhizoctonia sp. ; brown mycelial growth on seedling base.
Puerto Rico.
Rosellinia sp. ; root injury.   Costa Rica.
Schizophyllum commune Fr. ; senescent-frond disease.   Puerto
Rico, Nicaragua.
Sphaerognomia elaeicola Batista; leaf disease.   Brazil.
Strigula complanata Fée; white lichen spotting.   Practically
general.   Not seen in dryest areas.
Tapering stem (undetermined physiological cause).   Scattered
in Honduras, Nicaragua, Guatemala.
Ustulina (? zonata); woodrot.   Nicaragua.
Virus; Palm Virus causes death of bud.   Venezuela.

ERAGROSTIS spp.   Tenacious, soil building, wiry grasses, one
bends over acting as a special erosion control plant and is
called "lovegrass" En., "hierba amor" Sp.
   Balansia epichloe (Weese) Diehl; black ring, sterility.   Sub-
    tropical USA.
   Cladosporium spongiosum Berk. & Curt.; foliage disease.   S.
    America.
   Claviceps purpurea (Fr.) Tul.; ergot.   Chile.
   Curvularia (? geniculata); culm and leaf disease.   El Salvador.
   Fusarium sp. in isolation work studying Curvularia disease.
    El Salvador.
   Helminthosporium giganteum Heald & Wolf; eyespot.   Sub-
    tropical USA, El Salvador.
   Phyllachora eragrostidis Chard.; tarspot.   W. Indies, C.
    America (probably in S. America).
   Tilletia eragrostidis Clint. & Ricker; covered smut.   Scattered
    in Subtropical USA, C. America, S. America.   T. eremo-
    phila Speg.; smut.   Argentina.
   Uromyces eragrostidis Tracy; rust.   Mexico, Subtropical USA,
    Puerto Rico, Mexico, Bolivia, Ecuador, Nicaragua.

EREMOCHLOA OPHIUROIDES Hack.   A grass introduction from
Oriental tropics, an increasingly popular perennial lawngrass in
the West Indies and Central America; "centipede grass" En.,
"grama de ciempiés" Sp., "erva centopeia" Po.
   Alternaria alternata (Fr.) Keiss. = A. tenuis Auct.; on very
    old leaf.   Puerto Rico, Costa Rica.
   Bacterial isolates consistently secured from yellowed plants,
    no proof of pathogenicity.   Puerto Rico.
   Colletotrichum graminicola (Ces.) Wils.; anthracnose of mild
    effect.   Widespread.
   Curvularia geniculata (Tracy & Earle) Boed. and C. lunata
    (Waak. & Boed.) Boed.; leaf and culm attack.   Subtropical
    USA, W. Indies.
   Helminthosporium gramineum Rabb.; leaf spot-blast.   Puerto
    Rico, Costa Rica.
   Nigrospora sp.; culm disease.   Puerto Rico.
   Rhizoctonia solani Kuehn = Thanatephorous cucumeris (Frank)
    Donk (basidial stage found in rainy seasons); large spread-
    ing spots in lawn.   Puerto Rico, Costa Rica, Guatemala.
   Sclerotinia sp.; delimited lawn-spot.   Florida.
   Spirogyra sp. (and other algae); wet-ground disease, choking.
    Puerto Rico.
   Strigula complanata Fée; lichen-spot under trees with the
    disease.   Puerto Rico.
   Torula sp.; black stem.   Puerto Rico.

ERIOBOTRYA JAPONICA (Thunb.) Lindl.   An introduction from
Eastern tropics, much grown in semi-dry areas, a popular,
more or less medium-sized fruit tree, often a heavy producer;
"níspero del japón" Sp., "ameixa do japão" Po., "loquat" En.
   Alternaria sp.; an uncommon leafspot.   Cuba, Puerto Rico,
    Venezuela, Bahamas.

Aposporiopsis saccardiana Mariani; a foliage disease.   S.
    America.
Armillaria mellea Vahl = Armilariella mellea (Fr.) Karst.;
    rootrot.   California, Guatemala.
Botryosphaeria dothidea (Moug.) Ces. & de Not.; stem canker.
    California, Chile.
Capnodium spp. (and other related genera) sootymolds.   Neo-
    tropical.
Cephaleuros virescens Kunze; algal leafspot.   Often in moist
    areas.
Cercospora circumscissa Sacc.; leafspot.   Costa Rica, Brazil
    (probably all Neotropics).
Clitocybe tabescens Bres. = Armilariella tabescens (Scop.)
    Sing.; rootrot.   Florida, Mississippi in USA.
Colletotrichum gloeosporioides Penz. (= Glomerella cingulata
    (Stonem.) Sp. & Schr.); anthracnose, flower blight, wither-
    tip.   Florida, Texas, California, W. Indies, C. America.
    (See Gloeosporium.)
Corticium salmonicolor Berk. & Br.; pink branch-disease.
    Scattered in the Neotropics.
Diplodia sp. (= Botryosphaeria quercuum (Schw.) Sacc.); die-
    back.   Subtropical USA, C. America.
Entomosporium maculatum Lév. the conidial form of Diplocar-
    pon maculatum (Atk.) Jorst.; the fruit and leafblotch.   Sub-
    tropical USA, C. America, Brazil, Uruguay.
Epiphytic microorganisms (not determined as to genera or
    species and apparently on, but non-parasitic: bacteria,
    fungi, microliverworts, algae, nematodes) on three- and
    four-year old leaves and twigs.   Panama.
Erwinia amylovora (Burr.) Winsl. et al; fireblight.   Subtropi-
    cal USA, Guatemala, suggestive symptoms seen in Vene-
    zuela.
Fomes lamoensis (Murr.) Sacc. & Trott. that is slow in
    spreading in the field; brownroot and honeycomb woodrot.
    Brazil, Guianas.
Fusicladium eriobotryae (Cav.) Sacc.; leaf, stem and fruit
    scab disease.   Argentina, Uruguay, Subtropical USA, C.
    America.
(Gloeosporium eriobotryae Speg. apparently synonymous with
    Colletotrichum gloeosporioides, but as seen in C. America
    the ashy leaf and fruit spot seems to be disease from a
    special race.   It is also reported from at least Argentina,
    Uruguay, Brazil, and Colombia.)
Glonium parvulum (Gerard) Cke.; trunk disease.   S. America.
Hendersonia eriobotryae Speg.; on leaves.   Argentina.
Leptosphaeria puttemansii Maubl.; speck-like spots on leaf
    and fruit.   Brazil, C. America (probably widespread).
Melanomma sordidissimum Speg., on branch wood.   Argentina.
Meliola sp.; black mold.   El Salvador, Panama.
Meloidogyne sp.; rootknot nematode.   Florida.
Microthyrium minutissimum Thuem.; leaf infections.   Brazil.
Mycena citricolor (Berk. & Curt.) Sacc.; at times a serious
    defoliating leafspot.   Guatemala, El Salvador, Venezuela.

Nectria vulgaris Speg. = Creonectria ochroleuca (Schw.) Seav.;
   dieback. Costa Rica, Argentina.
Pellicularia koleroga Cke.; thread- and web-blight. Costa
   Rica.
Pestalotia spp. (most commonly reported as: P. funerea
   Desm. and P. longi-aristata Maubl.); leafspots, stem
   lesions. Subtropical USA, Puerto Rico, Costa Rica, Argen-
   tina, Brazil.
Phaeoseptoria eryobotryae Rang.; on living leaves. Brazil.
Phyllosticta eriobotryae Thuem.; grayish brown leafspot.
   Subtropical USA, C. America, Argentina, Brazil, Vene-
   zuela. P. fusiformis Nic. & Agg.; leafspot. Argentina.
   P. uleana Syd.; large circular leafspot. Brazil.
Phymatotrichum omnivorum (Shear) Dug.; said to be a dryland
   rootrot. Texas, believed to have been seen in Mexico but
   not proved.
Physalospora gregaria Sacc. = Botryosphaeria quercuum
   (Schw.) Syd.; on dead twigs. Costa Rica, El Salvador.
Phytophthora cactorum (Leb. & Cohn) Schr.; collar-rot.
   California.
Pleomelogramma argentinense Speg.; branch and trunk wood-
   rot. Argentina.
Rhizoctonia (solani? no basidial stage, mycelium in culture or
   in nature did not exhibit sclerotial structures); cutting
   disease. Costa Rica.
Rosellinia bunodes (Berk. & Br.) Sacc.; black root-rot. Costa
   Rica.
Septobasidium saccardinum (Rangel) Mar.; not parasitic on
   host but mycelial mats are found covering scale insects.
   Guatemala, Venezuela, Panama, Brazil, Argentina.
Septoria eriobotryae Maf.; leafspot. S. America.
Spheropsis eriobotryae Speg. = Haplosporella eriobotryae
   (Speg.) Petr.; a fruit disease, and attacks leaf. C.
   America, S. America.
Strigula complanata Fée; the common white lichen-spot. W.
   Indies, C. America, S. America. (Ecology is not under-
   stood but this lichen on this host requires at times wet con-
   ditions with, apparently, dryseason between rainy periods.)
Virus: Leafmottle and twig streaking similar to a mosaic.
   Not studied. Panama.
Xanthomonas pruni (E. F. Sm.) Dows.; bacterial spot. Brazil.

ERIOCHLOA POLYSTACHYA HBK. A forage grass, common in the
West Indies (it was collected by the von Humboldt-Bonpland expe-
dition), and it is well adapted to the American tropics; "carib
grass" En., "malojilla" Sp. (This grass has not been given
much attention by pathologists.)
   Balansia (see Ephelis)
   Darluca filum (Biv.-Bern.) Sacc. = the perfect state is Eudar-
      luca australis Speg.; non-parasitic on host, but is hyper-
      parasite of Puccinia in the W. Indies.
   Ephelis japonica P. Henn. (conidial stage of a Balansia); head
      disease. Puerto Rico.

Helminthosporium sp. ; large leafspot.   Florida, W. Indies.
Nigrospora oryzae (Berk. & Br.) Petch; mostly on dead leaf
    tissues.   Neotropics.
Phyllachora eriochloae Speg. ; shiny leaf-tarspots.   W. Indies,
    S. America.
Puccinia substriata Ell. & Barth. = Uredo eriochloae Speg.
    (alternate occurance on solanums and at times on tobacco);
    rust.   Venezuela, W. Indies, probably widespread.   (P.
    substriata is also on Eriochloa punctata Ham. and of com-
    mon occurrence causing a mild rust disease on this host
    species.   If a rust race should develop by mutation that
    made it more severe in its affect on tobacco, the Eriochloa
    grasses would need special study as alternate hosts.)
Sphaerodothis luquillensis Chard. ; black stromatic leafspot.
    Puerto Rico.

ERVUM see LENS

ERYTHRINA GLAUCA Willd.   A good-sized tree used in somewhat
    wet areas to absorb excessive soil moisture during rains and for
    shade in the dryseason, important agricultural shade; "poro, "
    "bucare de agua, " "bucayo" Sp. , "suina" Po. , "swamp immor-
    telle" En.   E. POEPPIGIANA Cook is a larger tree that is an
    excellent source of leaf mulch and withstands two to three
    prunings a year to regulate shade; "poro-poro, " "bucare gigante"
    Sp. , "mulunger" Po. , "immortelle" En.   From both species tall
    stakes that strike root are extensively used for living fences.
    (Over 40 species of the genus are known, some are flowering
    bushes and vines, one tree species is protected and planted by
    Indians for its large edible seeds, flowers of some are eaten
    both cooked and fresh, and decoctions from some species are
    regularly employed by jungle peoples to stun fish.   The two
    species named above are similar in their disease susceptibilities.
    A number of branch, trunk and woodrot fungi reported on
    ERYTHRINA CRISTA-GALLI L. by Spegazzini in Argentina give
    indication of such probabilities on the immortelles in other parts
    of the tropics.)
    Armillaria mellea Vahl = Armilariella mellea (Fr.) Karst. ;
        rootrot.   Guatemala, Honduras.
    Botryodiplodia theobromae Pat. = Diplodia sp. ; dieback of
        twigs.   Neotropics.
    Calostilbe striispora (Ell. & Ev.) Seav. ; bark disease (spread
        by workmen removing spines on trunk).   Venezuela, Suri-
        nam, Costa Rica.
    Cephalosporium lecanii Zimm. ; not parasite of host, attacks
        the insect Pseudococcus.   Puerto Rico.
    Cercospora erythrinae-tomentosae Hansf. ; large leafspot.
        Brazil, Venezuela, C. America.   C. pittieri Syd. ; leaf-
        spot and defoliation.   Venezuela.
    Ceriomyces spongia Speg. (on Erythrina crista-galli L.);
        trunk rot.   Argentina.
    Cladosporium herbarum Lk. ; on senescent leaf.   El Salvador.

Colletotrichum sp.; anthracnose of leaf and pod.   El Salvador.
Diaporthe corallodendri Speg. (on E. crista-galli): a branch-
    wood rot.   Argentina.
Dicheirinia binata (Berk.) Arth.; a leafrust. Grenada, Trini-
    dad, Puerto Rico, Cuba, Costa Rica, Guatemala, Panama,
    Colombia.
Elsinoë sp.; spot with ascocarps on leaf and twig.   C.
    America.
Fomes lamaoensis (Murr.) Sacc. & Trott. = F. noxius
    Corner; brown rot disease, honeycomb woodrot (slow field
    spread). W. Indies, C. America, S. America. F. lig-
    nosus (Kl.) Bres.; root disease, friable rot of wood (fast
    field spread). W. Indies, C. America, S. America.
Fusarium sp.; part of dieback complex. Costa Rica, Guate-
    mala.
Ganoderma lucidum Karst. and G. pseudoferreum Ov. & St.;
    rootrot and trunk base attack. C. America.
Hypocrea corticioides Speg., H. ibicuyense Speg., H. platense
    Speg. (all on E. crista-galli, causing branch and log rots
    and all in Argentina).
Hysterium atro-purpureum Speg. (on E. crista-galli); log rot.
    Argentina.
Irpex atro-purpureum Speg. = Xylodon atropurpurense (Speg.)
    Kuntze (on E. cristo-galli); log rot. Argentina.
Loranthus robustissimum Trel.; large parasitic bush on un-
    pruned branches of bucare. Costa Rica, El Salvador.
Marasmius sp.; horse-hair blight on jungle trees of all
    species. Costa Rica, Panama.
Meliola bicornis Wint.; black moldspot of leaf. W. Indies,
    C. America, Brazil, Colombia, Panama. M. crenatissima
    Syd. and M. erythrinae Syd.; thin moldspots. Panama,
    Puerto Rico.
Meloidogyne sp.; rootknot nematode on growing post roots
    without notable damage to post. Costa Rica.
Mycena citricolor (Berk. & Curt.) Sacc.; leafspot on young
    trees in same area with diseased coffee and cinchona.
    Costa Rica.
Mycosphaerella erythrinae Koord. = M. erythrinae Stev.;
    large and irregular leafspots. Panama, Cuba, Trinidad,
    C. America.
Nectria cinnabarina Tode; stem canker. C. America. N.
    vulgaris Speg. (on E. cristo-galli); dieback. Argentina.
Nummularia (probably two species), on dead and dying fencing.
    Costa Rica, Puerto Rico, Honduras.
Oryctanthus botryostachys Eich. and O. ruficaulis Eich.;
    small parasitic bushes on tree branches. Surinam.
    Orycanthus sp., in Costa Rica.
Pellicularia koleroga Cke.; thread-web blight. Costa Rica,
    Puerto Rico.
Phorodendron gracilispicum Trel.; bush parasite on tree
    branch. Costa Rica, Colombia, Venezuela.
Phthirusa spp.; vine-like bushy parasites on tree branches

(parasite specimens unobtainable for species identity without
felling tree hosts).   Seen in Costa Rican jungles and in old
second growth bush areas.

Phyllosticta australis Speg.; gray-white leafspot.   Brazil,
    Argentina.   P. erythrinicola Young; small white leafspot.
    Puerto Rico.

Ravenelia platensis Speg. (on E. crista-galli); rust.   Argen-
    tina.

Rosellinia bunodes (Berk. & Br.) Sacc. and R. pepo Pat.;
    black and white rootrots respectively.   W. Indies, Colombia,
    Venezuela (probably in all countries where these agricultural
    shade trees are grown).

Sclerotium eurotioides Speg. = S. argentinum Sacc. & Speg.
    (on E. cristo-galli); trunk rot.   Argentina.

Septobasidium spp.; felt growth accompanying insects.   Guade-
    loupe.

Sphaerostilbe repens Berk. & Br.; wetstain.   W. Indies.

Stereum atro-zonatus Speg. (on E. crista-galli); rot of tree
    trunk.   Argentina.

Strigula complanata Fée; algal leafspot.   General.

Struthanthus dichotrianthus Eich. and S. syringifolius Mart.;
    both weeping woody vine-like parasites on branches of host.
    Surinam.

Uredo erythrinae P. Henn.; rust.   S. America.

Uromyces erythrinae Lagh.; leafrust.   Ecuador, C. America.

Virus.   A mottle and malforming virus studied in Venezuela.
    A typical leaf mottle and shortening of internodes from a
    virus.   C. America.

Weakspot of leaf.   C. America, W. Indies.

Witches' broom.   Probably fungus causation.   Kills large
    branches.   Venezuela, Colombia, Trinidad.   (The cacao
    witches' broom organism Marasmius perniciosus was iso-
    lated once from a dead and dying small branch of Erythrina
    glauca in Trinidad.)

ERYTHROXYLON COCA Lam.   Native to the Andes, a woody bush
    over three meters high with neatly formed shiny leaves and red
    fruits.   There are several species of the genus, but the only
    one of them commercially important, here given, is a producer
    of habit-forming chewable leaves that contain cocaine and other
    alkaloids; "coca" Sp. Po., "cocaine" En., and many Indian
    names.   (Note: There are very isolated places in the Andes
    where the only source of revenue for people is the meticulously
    harvested and dried leaves of coca.   Since it is frowned upon
    because it is a source of cocaine, it requires a combination of
    courage and academic interest for it to be given any scientific
    attention at all as to growing problems or diseases.   The re-
    ports from which the list below is compiled are all of practi-
    cally accidental observations, some hitherto unpublished.   No
    matter how much the Indians need the money from it, studies
    on its troubles have no official sanctions anywhere.)

    Armillaria mellea Vahl = Armilariella mellea (Fr.) Karst.;
        rootrot in lowland. plants.   Guatemala, Panama.

Bubakia erythroxylonis (Graz.) Cumm.; rust. W. Indies, S.
   America.
Capnodium spp.; fumago, sootymold. Costa Rica, Honduras,
   Peru, Venezuela.
Cephaleuros virescens Kunze; algal leafspot. (In moist areas
   and protected mountain coves.) Brazil, Peru, Costa Rica,
   Puerto Rico.
Cercosporella cocae Speg.; leaf infections. Argentina.
Chlorosis from weak growth due to combined effects from
   undetermined root fungi and minor element deficiencies.
   Ecuador.
Colletotrichum gloeosporioides Penz. = C. cocae Speg.; an-
   thracnosis and dieback induced after excess leaf harvest.
   Neotropical occurrence.
Corticium salmonicolor Berk. & Br.; pink disease on speci-
   men bushes. Costa Rica.
Cylindrosporium sp.; isolated from transluscent leafspots that
   I once diagnosed as possibly of bacterial cause. Costa
   Rica.
Diplodia sp. = Botryodiplodia theobromae Pat.; tip dieback.
   Peru, Costa Rica. (Conidial strain of parasite, no perfect
   stage ever produced in moist chambers.)
Fomes spp.; brown rootrot. S. America.
Hypocrella sp.; on leaves. Brazil.
Meliola erythroxylifoliae Bat. & Sil.; black mildew. Brazil,
   Peru, Costa Rica.
Mycena citricolor (Berk. & Curt.) Sacc.; leafspot (in moist-
   land shaded plantings). Peru, Costa Rica.
Nectria erythroxyfoliae Viégas; on stems. Brazil.
Phoradendron spp.; P. martianum Trel., P. pulleanum Kra.,
   and P. surinamense Pul.; all bushes parasitizing the host
   and all noted in Surinam.
Phyllachora usteriana Speg.; shiny leafspot. Brazil.
Phyllosticta erythroxyli Graz.; leafspot. Bolivia, Peru, Brazil,
   Argentina.
Phytophthora sp. (disease it caused not given). Argentina.
Porella sp.; micro-liverwort, leafblemish. On specimen
   bush in Costa Rica.
Protomyces cocae Speg., leaf blister. Argentina.
Puccinia erythroxyli Viégas; rust. Brazil.
Ravenelula boliviensis Speg.; a leaf rust. Bolivia.
Rhizoctonia sp. (probably solani); strangle of seedling tree.
   Peru. Rhizoctonia sp. (aerial type); mycelial strands on
   bark. Costa Rica.
Rosellinia bunodes (Berk. & Br.) Sacc.; a black rootrot.
   Puerto Rico, Costa Rica, Peru, Ecuador.
Sphaerella erythroxyli Speg.; on leaves. Argentina.
Strigula complanata Fée; white lichen leafspot. Peru, Costa
   Rica, Puerto Rico.
Uredo erythoxylonis Graz. (? synonym of Puccinia erythroxyli)
   = Bubakia erythroxylonis (Graz.) Cumm.; leafrust. Cuba,
   Puerto Rico, Peru, Brazil, Bolivia, Colombia, Costa Rica.

Virus: Witches' Broom. Peru, Bolivia.
Wood and root decay from undetermined parasitic Basidiomycete causing white dryrot. Peru.

EUCALYPTUS GLOBULUS Labill. This is a highly successful introduction from Australia to all countries of the American
tropics. There are over 500 species of this tree genus, most
of which are lofty growers, and the one in the Neotropics most
studied agronomically and pathologically is E. globulus, of which
many fine selected varieties are known; it is given much attention in Brazil, where large plantations grown in coffee-exhausted
areas produce needed soil cover and erosion control, reforestation, supplies of kitchen wood and charcoal, and lumber; "ecua-
lypto" Sp. , "eucalipto" Po. , "eucalyptus" En.

Agrobacterium tumefaciens (E. F. Sm.) Conn; crowngall,
  warty stem. Colombia, Argentina, California.
Alternaria sp. (sometimes reported as Macrosporium); large
  leafspot. Peru, rare reports from Costa Rica and Guate-
  mala.
Aphelenchoides parietinus Stein. ; rhizosphere nematode.
  Brazil.
Aphelenchus avenae Bastian; root nematode. Brazil.
Arachnomyces flavidus Speg. ; on fallen leaves. Argentina.
Armillaria mellea Vahl = Armilariella mellea (Fr.) Karst. ;
  shoestring rootrot. California, S. America.
Axonchium amplicolle Cobb; a root nematode. Brazil.
Bagnisiopsis eucalypti Dearn. & Barth. ; twig disease. Cali-
  fornia.
Botryosphaeria dothidea (Moug.) Sacc. & Trott. ; branch canker,
  twig blight. Subtropical USA, W. Indies, C. America. In
  Argentina reported also as a root and collar rot.
Botryotrichum villosum Speg. ; on dead bark. S. America.
Botrytis cinerea Pers. ; twig disease, and foliage blight on
  new flush. California, Florida, Argentina, Chile, Peru.
  B. fusca Sacc. is reported in California.
Byssotheciella eucalyptina Syd. ; on foliage. S. America.
Cephaleuros virescens Kunze; algal leafspot. Costa Rica,
  Guatemala, Brazil.
Cephalothecium sp. ; saprophytic. ? Peru.
Ceratocystis pilifera (Fr.) Mor. ; sapwood stain. Subtropical
  USA, W. Indies, Mexico, Brazil.
Cercospora epicoccoides Cke. & Mass. ; a common leafspot.
  Colombia, Peru, Argentina. The different C. eucalyptii
  Cke. & Mass. is a large angular leafspot. Brazil.
Ceuthospora molleriana Petr. ; on withered leaves. Argentina.
Chelisporium hysteroides Speg. ; on twigs and stakes. S.
  America.
Chlorosis, indication of root problems: poor drainage, fungus
  root disease, lack of iron, excess salt or lime, etc.
  California, Guatemala, Colombia.
Cladoderris platensis Speg. = Merulius sordidus Berk. &
  Curt. ; branch and trunk disease. Argentina.

Cladosporium sp. = Heterosproium sp.; often found on deca-
    dent leaves. Neotropics.
Claudopus argentinensis Speg.; log rot. Argentina.
Clavaria intricatissima Speg. = Ramaria intricatissima (Speg.)
    Corner; capsule decay. S. America.
Clitocybe tabescens Bres. = Armilariella tabescens (Scop.)
    Sing.; rootrot. Subtropical USA.
Colletotrichum gloeosporioides Penz. (= C. eucalypti Bitanc.,
    Gloeosporium capsularum Cke. & Hark., and G. rhipidium
    Speg.); anthracnose of leaf, capsule, and branch. Neo-
    tropics. (See Glomerella below.)
Coniothecium eucalypti Thuem.; branch disease. Ecuador. C.
    platense Speg.; stake rot. Argentina.
Coniothyrium eucalyptii Gonz. Frag., and C. leprosum Fairm.;
    both are branch diseases. Colombia, W. Indies.
Coremium glaucum Lk.; leaf disease. California.
Corticium spp.: C. calceum Fr., C. epiphyllum Pers., and
    C. leve Pers. (Note: I am uncertain about the naming
    of these three Corticiums, but I do not question the find-
    ings. Presence of such fungi (on jungle growths, as I
    know), are sometimes parasitic, sometimes semiparasitic,
    and sometimes wholly epiphytic organisms and they spread
    for a short while to numerous cultivated trees and finally
    disappear.) Described as parasites on foliage, found in
    scattered locations. Subtropical USA, C. America, S.
    America. C. salmonicolor Berk. & Br.; pink disease and
    leafrot. Scattered in Neotropics.
Cryptosporium ceuthosporioides Cke. & Harkn. and C. eucalypti
    Cke. & Harkn.; branch disease. California.
Cylindrocladium scoparium Morg. f. brasiliensis Bat. & Cif.;
    seedling damping-off in Brazil, occasional trunkrot in
    Argentina.
Cylindrosporium sp.; a foliage disease. Peru.
Cytospora australiae Speg. and C. eucalyptina Speg.; on dead
    and dying branches. Argentina.
Damping-off in seedbeds (complex of many organisms) con-
    sidered by many as the most important single problem in
    Eucalyptus culture. Brazil, Argentina, Colombia.
Dendrophoma australasica Speg.; on fallen leaves. Argentina.
Diaporthe cubensis Bruner; gummy canker. Cuba. D. euca-
    lypti Harkn., and D. medusaea Nits.; diseases of foliage.
    Subtropical USA.
Dicandranidion argentinensis Speg.; fungus on bark. Argentina.
Didymospheria circinnanas Harkn.; leaf attack, while D.
    epidermis Fr. causes a branch disease. Both in California.
Diplodia australiae Speg. = Metadiplodia acerina (Lév.) Zamb.
    (a bark disease), D. eucalypti Cke. & Harkn. (on wood of
    branches), D. microspora Sacc. (leaf attack), D. nova-
    hollandiae Speg. (causes dieback), D. tenuis Cke. & Harkn.
    (branch attack). All in dryer parts, but not deserts, of
    the Neotropics.
Ditylenchus dipsaci Filip.; nematode from rootzone. Brazil.

Dothiorella acervulata Speg.; originally found on bark of dead branch. Argentina.

Endothia havanensis Bruner; branch and twig canker. Cuba, Surinam.

Eutypa heteracantha Sacc.; log and stump decay. S. America.

Eutypella ludibunda Sacc.; bark disease. S. America.

Fairmaniella leprosa (Fairm.) Petr. & Syd.; erumpent canker. S. America.

Fomes applantus Gill. and F. robustus Karst.; heartrots. Subtropical USA, El Salvador. (There are other species of Fomes reported from other than the Eucalyptus of this list.)

Fracchiaea heterogenea Sacc. = F. subcongregata (Berk. & Curt.) Ell.; small fruiting bodies on dead stems. S. America, Guatemala, Mexico.

Fusarium lateritium Nees (the conidial state of Gibberella lateritium (Nees) Sny. & Hans.); seedling and twig infection. Argentina. The F. oxysporum Schl. f. sp. aurantiacum (Lk.) Wr. (sometimes reported as f. sp. eucalypti) is a seedling wilt and blight. Subtropical USA, W. Indies, Chile. F. solani (Mart.) Sacc.; is an inflorescence rot. El Salvador.

Ganoderma sessile Murr.; delimited rot of base, caries. Argentina.

Gerronema (see Omphalia)

Gleosporus rhypidium Speg. = Dictyopanus pussilus Sing.; mushroom rootrot. Argentina.

Glomerella eucalyptides Av.-Sac. (? asco-stage of Colletotrichum spp. which see above); on dieback. S. America.

Gloniopsis argentinensis Speg.; on lumber and cut stakes. Argentina.

Harknessia eucalypti Cke. & Harkn.; leaf attack. Colombia. H. uromycoides Speg. = Melanconium uromycoides Speg.; leaf and twig blight. California, S. America.

Helminthosporium eucalypti Speg.; wood infection. Argentina.

Helotium immarginatum Karst. and H. pallescens Fr.; on dead wood. S. America.

Hendersonia eucalypti Cke. & Harkn.; branch attack and leafspot. Subtropical USA, Peru, C. America.

Heterosporium (old name for Cladosporium).

Hypocrea consimilis Ell.; twig disease. California.

Hypoxylon subeffusum Speg.; on dead branch. S. America.

Kalmusia argentinensis Speg. and K. eucalyptina Speg.; on cut stakes in Argentina and as weak parasites on dying stems. C. America, S. America.

Laestadia eucalypti Speg.; on fallen leaves and bark. S. America.

Lecanidion australe Speg. = Patellaria atrata (Hedw.) Fr.; bark fungus on old branches. Argentina, Brazil.

Lepiota hiatuloides Speg.; on rotted log. Argentina.

Loranthaceae (Phanerogamic bush-like parasites) seen scattered in tops of a number of these trees in C. America,

Brazil and Colombia. No identifications as to genera or
  species of these parasites.
Macrophoma molleriana (Th.) Berl. & Vogl. ; leaf disease.
  California.
Marasmius dichromophus Speg. = M. nigripes (Schw.) Sing.,
  stump rot. Argentina.
Melanconium globosum Cke. & Harkn. ; branch disease. Wide-
  spread.
Meloidogyne sp. ; rootknot nematode. Chile.
Microthyrium eucalypticola Speg. ; mold on leaves. Argentina
  and probably widespread.
Monochaetia desmazierii Sacc. ; leaf disease. California.
Mycena citricolor (Berk. & Curt.) Sacc. ; leaf infection in
  laboratory. Puerto Rico. M. polygrammoides Speg. ; log,
  leaf, and stem decay. Argentina.
Mycosphaerella molleriana (Th.) Lind. var. megalospora Cam. ;
  leafspot. California, Brazil, Florida.
Nectria eucalypti (Cke. & Harkn.) Berk. & Vogl. ; branch
  attack. California.
Nummularia sp. ; under dead bark of post. Puerto Rico.
Oidium sp. ; powdery mildew and leaf crinkle. Peru, Brazil,
  Argentina.
Omphalia bruehi Speg. = Gerronema struckertii (Speg.) Sing. ;
  on fruits. Argentina.
Orbilia rubella (Pers.) Karst. ; on fallen branches. S.
  America.
Pestalotia inquinans Cke. & Harkn. , P. monochaeta Desm.
  and P. truncata Lév. , are leaf diseases in California,
  Peru, and Brazil, while P. molleriana Th. is both a leaf
  disease and causes kill of seedling tops in new plantations
  in Uruguay.
Pezizella carneo-rosea Sacc. and P. oenotherae (Cke. & Ell.)
  Sacc. ; leaf and twig disease. California.
Phaeocreopsis hypoxyloides (Speg.) Sacc. & Syd. ; on atrophied
  branches in cool and dry areas. S. America.
Phoma eucalyptidea Th. ; a large leafspot. Brazil.
Phragmodothidea eucalypti Dearn. & Barth. ; on bark. Cali-
  fornia.
Phyllachora eucalypti (Speg.) Petr. ; leafspot. Widespread.
Phyllosticta eucalypti Ell. & Ev. = P. extensa Sacc. & Syd. ;
  leafspots. California.
Phymatotrichum omnivorum (Shear) Dug. ; rootrot. Texas,
  Mexico.
Physalospora latitans Sacc. and P. suberumpens Ell. & Ev. ;
  both on dieback. California, Florida, Brazil.
Phytophthora cinnamomi Rands (rootrot in California), P.
  parasitica Dast. (branch and stem lesion in Argentina and
  Brazil).
Pleonectria megalospora Speg. ; a bark disease. Brazil,
  Guatemala.
Polyporus spp. all cause brown heart and butt rot diseases,
  some reported: 1. P. caseicarnis Speg. in S. America.

2. P. gilvus (Schw.) Fr. in Subtropical USA. 3. P. hirsutus Fr. in Subtropical USA, and in C. America. 4. P.
pinsitus Fr. in Uruguay. 5. P. schweinitzii Fr. in Argentina, Mexico, Subtropical USA, and W. Indies. 6. P. sul-
phureus Bull. in Brazil, California and Argentina. 7. P.
versicolor L. in Subtropical USA.

Polystictis celottianus Sacc. & Manc.; woodrot. S. America.

Poria cocos (Schw.) Wolf; a root infection, "tuckahoe" development. Florida. P. vesispora (Pers.) Rom.; on dead
wood. Florida.

Puccinia psidii Wint.; rust. Reported from Brazil, probably
in nearby countries.

Pythium ultimum Trow.; seedling damping-off. Guatemala,
Argentina, Brazil (where it is often associated with Fusarium sp.).

Ramaria intricatissima (Speg.) Corner; canker. Argentina.

Rhizoctonia solani Kuehn = Thanatephorous cucumeris (Frank);
damping-off. Neotropics.

Rosellinia spp.; rootrots. C. America, S. America.

Schizophyllum commune Fr.; on dead branches. El Salvador,
Guatemala.

Schizoxylon lichenoides Speg.; on living diseased branch.
Argentina.

Scleroderma tuberoideum Speg. = S. radicans Lloyd; rot of
stump and base of live tree. S. America, C. America.

Scleroderris eucalypti (Cke. & Harkn.) Sacc.; on branches.
California.

Sclerotinia fuckeliana (dBy.) Fckl.; inflorescence-and-fruitrot.
Chile, Guatemala.

Sclerotiopsis australasica Speg. = Pilidium australasium
(Speg.) Petr.; rot of decaying leaves. Argentina.

Sclerotium citrinellum Speg.; decay of leaves. Argentina.

Septobasidium curtisii (Berk. & Desm.) Boed. & Stein.; felt
fungus. Subtropical USA, Colombia (twigs with "felt" sent
to me with the notation they came from several species of
Eucalyptus in Colombia and Uruguay).

Septonema eucalipticola Speg.; leaf disease of nursery in
Uruguay. S. minutum Ber. & Rou.; branch attack in
Colombia. S. multiplex Berk. & Curt.; causing lesion on
trunk in California.

Septoria ceuthosporioides (Cke.) Sacc. and S. mortolensis
Penz. & Sacc.; leafspot and defoliation. California, Peru.

Septosporium scyphophorum Cke. & Harkn.; on leaf. California.

Sphaeronema eucalypti Cke. & Harkn.; branch disease. California.

Stagonosporium platense Speg.; on cut stakes without bark.
Argentina.

Stereum albomarginatum Schw. (in C. America and S.
America), and S. hirsutum Willd. (in California and Uruguay); both are active parasites of branch and trunk on
poorly growing trees, in some cases causing death.

Trametes pulchra Speg.; woodrot.   Brazil and Argentina at
    least, in S. America.   I saw Trametes sp. in W. Indies
    and C. America.
Tylenchus davianii Bastian; rhizosphere nematode.   Brazil.
Uredo sp. (possibly Puccinia psidii see above); an old rust
    report.   Brazil.
Urophlyctis sp.; blight of lush shoots.   Uruguay.
Valsa eucalypti Cke. & Harkn.; dieback.   California.
Verticillium alboatrum Reinke & Berth.; seedling wilt.   Peru.
Virus: Eucalyptus Chorosis Virus in Argentina.
Xylaria apiculata Cke.; on dead wood.   Widespread.
(Note: As the reforestation programs of Latin America pro-
    gress, Eucalyptus trees will be increasingly grown, under-
    stood, and much studied.   Of those parasites listed there
    are about 15 recognized as fairly dangerous in this tree
    at this time.   Many of the diseases I have given are minor
    at present, and a number have not been given because of
    lack of space.   However, as new areas are planted, old
    and hitherto benign native forest and jungle parasites will
    find Eucalyptus.   Certain of the now known mild parasites
    of the tree crop will become virulent by mutation and
    natural selection.   It can be promised that as the years
    pass there will be more serious disease attack on Eucalyptus.

EUCHARIS GRANDIFLORA Planch.   An Andean bulbous plant, grace-
    ful with large flowers, much enjoyed as a pot plant and in
    gardens; "lirio amazonica" Sp. Po., "amazon lily" En.
    Aecidium delicatum Arth. = Puccinia roseanae Arth; brown
        leafrust.   General.
    Botrytis cinerea Pers.; occasional graymold and blight.   W.
        Indies, C. America, Florida.
    Cassytha filiformis L. and Cuscuta americana L.; both dod-
        ders.   Puerto Rico.
    Leaf scald, physiological.   Puerto Rico.
    Penicillium sp.; bulb rot and on old leaf edge.   Puerto Rico.
    Red elongated spot on leaf and flower stalk (not fungus or
        bacterial).   El Salvador.
    Rhizoctonia (? solani), brown mycelial weft over plant base
        but causing no infection.   Puerto Rico.
    Rootrot, undetermined cause.   El Salvador.
    Stagonospora curtisii (Berk.) Sacc.; red rounded leafspot.
        Subtropical USA, El Salvador, Guatemala.
    Uredo eucharidis P. Henn.; brown rust (possibly Puccinia see
        above).   Ecuador, Peru.
    Virus.   Eucharis Mosaic Virus in W. Indies, C. America.

EUCHLAENA MEXICANA Schrad.   Native of Mexico-Central
    America, a branching ancient grain plant over three meters
    high, is a close relative of maize, and apparently an integral
    part of the genetic make-up of maize, for which reason diseases
    on this ancient grain plant are of special concern.   It produces
    hard grains that when pounded are cooked by Indians, the stalks
    and leaves used as animal forage; "teosinté" Sp. Po. En.

Angiopsora (see Physopella)

Claviceps tripsaci Stev. & Hall; ergot.   Scattered in C.
America.

Cochliobolus heterostrophus Drechs.; leafblotch and seedling
blight.   Mexico.

Helminthosporium turcicum Pass.; blight.   Widespread.

Physoderma maydis Miyabe; brownspot of culm and spike.
Subtropical USA, Mexico, W. Indies, C. America (occurrence on this host acts as a holdover source to go onto
maize).

Physopella zeae Cumm. & Ram. (= Angiopsora zeae Mains,
synonym is Puccinia pallescens Arth.); the highland rust
(covered leaf rust).   Guatemala, El Salvador, Trinidad,
Honduras, Jamaica, Costa Rica.

Puccinia polysora Underw.; leafrust (small pustule).   Found
in warm areas.   P. sorghi Schw.; the more common leafrust (larger, elongated pustule).   Neotropics in warm areas.

Sclerospora sorghi West. & Upp.; downy mildew, multiple
shoots.   Texas, Mexico.

Ustilago euchlaenae Arcang. (in Uruguay), U. kellermanii
Clint. (in Guatemala), and U. zeae (Beck.) Ung. = U.
maydis (DC.) Cda. (in Argentina); all three are smuts and
the occurrence of the latter acts as a holdover source for
this parasite to go onto maize.

Viruses: Cucumber Mosaic Virus of the virulent Celery-Com-
melina strain in Florida.   Maize Dwarf or Stunt Mosaic
Virus with Mycoplasma-like bodies in Venezuela, Mexico,
Guatemala.   (Both probably much more widespread.)

Xanthomonas stewartii (E. F. Sm.) Dows.; bacterial wilt.
Mexico, C. America.

EUGENIA spp.   This pantropical genus has about 1000 species but
I have had to deal with only three.   First, E. AROMATICA
Baill. (synonyms are E. CARYOPHYLLATA Thunb., CARYO-
PHYLLUS AROMATICUS L.), introduced experimentally from the
Orient, its sun-dried flower buds are the renown cloves of commerce; "clavo" Sp., "cravo da índia," "guabijú" Po., "clove"
En.   Second, E. JAMBOS L. (synonyms are JAMBOSA VULGARIS
DC., CARYOPHYLLUS JAMBOS Stokes), from the Eastern
tropics, is now well naturalized and spread in the Western
tropics, it grows easily from seeds and stakes and is used as
windbreak, aid in reforestation, charcoal wood, fruit; "manzana-
rosa," "jambos" Sp., "pomarrosa," "maciera da rosa" Po.,
"rose-apple" En.   Third, E. UNIFLORA L., from Brazil and
now in much of tropical America, the bush-like fruit tree;
"arrayán," "pitánga," "guinda" Sp., "batinga," "pitangueira"
Po., "surinam cherry" En.   (Except where noted the diseases
are all on three host species.)

Actinothecium callicola Speg.; leafspot.   Brazil.

Antennaria scoriadea Berk.; black mold of foliage.   Chile.

Asteridium eugeniae Mont.; black leafspot.   Puerto Rico.

Asterina spp. (all leafspots, sometimes of star-like

conformation): A. colliculosa Speg. in Puerto Rico, A.
fawcetti Ryan in Mona island of Puerto Rico, A. myrciae
Ryan in Puerto Rico, A. pelliculosa Ryan in Chile, A.
sylvatica Speg. in Brazil.
Asterinella cylindrothecia (Speg.) Th. and A. puiggarii
(Speg.) Th.; leafspots. W. Indies.
Botryosphaeria dothidea (Moug.) Ces. & de Not.; branch
canker. W. Indies, Venezuela, Panama.
Calonectria rubropunctata Rehm; scab of stem. S. America.
Capnodium (? citri); sootymold. Cuba, Puerto Rico, Panama,
El Salvador, Costa Rica.
Catacauma semi-lunata Chard.; leafscab. W. Indies.
Cephaleuros virescens Kunze; algal leafspot. Probably in al-
most any country, reports at least from Brazil, Florida,
Puerto Rico, Cuba, Trinidad.
Cercospora eugeniae (Rang.) Chupp; leafspot. Cuba (reported
to me), Costa Rica (my microscopic examination of leaves
in my yard showed no spores on large and characteristic
spots). Brazil.
Clitocybe tabescens Bres. = Armilariella tabescens (Scop.)
Sing.; rootrot in subtropical USA.
Clypeosphaeria chilensis Speg.; on dead branches. Chile.
Colletotrichum gloeosporioides Penz. (= C. eugeniae Rang.)
conidial stage, also seen the asco-state Glomerella cingu-
lata (Stonem.) Sp. & Schr.; anthracnoses of leaf and twig,
branch lesion, flower injury, fruit disease. Neotropics.
Coniothyrium trigonicolum Rang.; leaf disease. Brazil, and
other nearby countries.
Cyphella paraguayensis Speg.; on dying branch wood. Para-
guay, Argentina, Brazil.
Dictyochorina portoricensis Chard. (on wild Eugenia); leaf-
spot. Puerto Rico.
Didymosphaeria eugeniicola Speg.; twig and branch decay.
Chile.
Dothidea granulosa Kl.; on foliage. Chile.
Elsinoë chilensis Bitanc. and E. pitangae Bitanc. & Jenk.
(asco-states of Sphacelomas); spot anthracnose-scab. S.
America.
Erinella calospora Pat. & Gaill. (on jambos); leaf scabby-
spot. S. America.
Gemma, pink and oblong when young becoming flat after in-
fecting leaf in rainforest (on manzana-rosa). Puerto Rico,
Costa Rica.
Helminthosporium asterinoides Sacc. & Syd. (on wild Eugenia);
leaf disease on under surface. Brazil.
Horse hair blight (wiry black cords but unidentified). Costa
Rica.
Hypoxylon subeffusum Speg., H. marginatum (Schw.) Berk.,
and H. truncatum (Schw.) Miller; all branch and trunk wood
diseases. Paraguay and other parts of S. America, in
Costa Rica secondary fungi on manzana-rosa.
Lasmenia balansae Speg.; stroma on leaf. Paraguay, Brazil.

Limacinia scoriadea (Berk.) Keiss.; a sootymold. Chile.
Loranthaceae, many mistletoe-like parasites. No collections
 made for determinations. Scattering occurrence.
Marasmius sarmentosus Berk. = Crinipellis sarmentosus
 (Berk.) Sing.; on dead wood and leaves. Puerto Rico,
 Cuba, S. America, El Salvador.
Meliola spp., are black mildews and some named: M.
 brasiliensis Speg. in Brazil, M. helleri Earle in Puerto
 Rico, M. laxa Gaill., in S. America and M. valdiviensis
 Speg. in Chile.
Melophia spp. (= Linochora spp.), all leafspots, examples:
 M. arechavaletai Speg., report in Uruguay, M. eugeniae
 Ferd. & Winge, in W. Indies, with M. macrospora Speg.,
 and M. nitens Speg., both in Paraguay.
Merulius chilensis Speg.; woodrot on wild Eugenia. Chile.
Microcyclus labens Sacc. & Syd. = Polystomella granulosa
 (Kl.) Th. & Syd.; granular stroma on leaf. Chile.
Micropeltis albo-marginata Speg.; speck spots on underside
 of leaf. S. America.
Mycena citricolor (Berk. & Curt.) Sacc.; leafspot sometimes
 but not always near diseased coffee. Costa Rica, Panama.
Mycosphaerella eugeniae Rehm (on pitanga); large leafspot.
 Brazil.
Myiocopron valdivianum Speg. = Peltella valdivianum (Speg.)
 Stev.; leaf disease. Chile.
Napicladium fumago Speg. and N. myrtacearum Speg.; both
 sootymolds, fumaginas. First in Chile, second in Para-
 guay.
Negeriella chilensis P. Henn.; leaf disease. Chile.
Oidium sp.; white mildew. Cuba, Panama. (Not readily seen
 on glaucus foliage. Occurred as scanty growth.)
Pestalotia sp.; delimited spots on leaf and petiole. Brazil,
 Costa Rica. Probably general.
Phaeophleospora eugeniae Rang. (on pitanga); leaf infection.
 Brazil.
Phaeoseptoria eugeniae Viég. (on pitanga); leafspot. Brazil.
Phyllachora spp. (all black stromatic somewhat shiny leaf-
 spots): P. angustispora Speg., in Argentina, P. biareolata
 Speg., in Brazil, Colombia, Paraguay, W. Indies, P.
 eugeniae Chard., in Puerto Rico, P. mullerii Chard., in
 Brazil, P. phylloplaca Th. & Syd. in Brazil, P. vimulosa
 Speg., in Costa Rica, and P. whetzelia Chard., in Puerto
 Rico.
Phyllosticta eugeniae Young; brown leafspot. Puerto Rico.
 P. icarahyensis Rang. (on pitanga); leafspot. Brazil. P.
 myrticola Speg.; leafspot. Brazil.
Polystomella (see Microcyclus)
Puccinia spp., several rusts, examples are: P. barbacensis
 Rangel, P. cambucae Putt., P. cumula Arth. & Cumm.,
 P. eugeniae Rangel, P. grumixamal Rangel, P. jambulana
 Rangel, but the most common, widespread and attacking all
 cultivated Eugenias, is P. psidii Wint., the "rose-apple
 race."

Rosellinia bunodes (Berk. & Br.) Sacc.; rootrot. W. Indies,
    C. America. (One single collection of typical R. pepo
    Pat. secured in Costa Rica.)
Septobasidium atratum Pat. (on jambos); feltgrowth. Guade-
    loupe.
Septoria eugeniarum Speg. = S. eugenicola Speg.; leafspot.
    S. America. (Sometimes these two fungus species are re-
    ported as distinct.)
Seynesia chilensis Speg.; moldspot on leaf. Chile.
Sphaerodothis balansae (Tassi) Hoehn.; stroma on underside of
    leaf. Paraguay.
Strigula complanata Fée; milky lichenspot. Neotropics.
Trichosphaeria paraensis Syd.; branch disease. Brazil.
Trichothyrium chilensis Speg.; hyperparasite on black moldy-
    spot. Chile.
Uncinula australis Speg.; leaf mildew on pitanga. Paraguay.
Uredo spp. apparently synonyms of Puccinia psidii.
Vivianella chilensis Speg.; on dead branches. Chile.
Zernia clypeata Petr. in Brazil.

EUPHORBIA PULCHERRIMA Willd. A popular tropical woody and
  herbaceous shrub with showy flowers, its origin is Central
  America and Mexico but it is now wide spread; "pascua," "pascua de jardín" Sp., "poinsetía añao" Po., "poinsettia," "garden
  pride" En.
    Armillaria mellea Vahl = Armilariella mellea (Fr.) Karst.;
      rootrot. California.
    Botryodiplodia theobromae Pat. = Diplodia theobromae (Pat.)
      Now.; dieback, cancer of stem, infection after cutting
      flowers. Neotropical.
    Botryosphaeria dothidea (Moug.) Ces. & de Not.; stem disease.
      Texas, C. America.
    Cassytha filiformis L.; green dodder infection in laboratory.
      Puerto Rico.
    Cercospora euphorbiae Pat. = C. euphorbiae-pubescentis
      Unam.; leafspot. C. America, S. America. C. pulcher-
      rima Tharp; leafspot. Texas, Brazil. C. rubida Muller
      & Chupp; leafspot. Brazil.
    Clitocybe tabescens Bres. = Armilariella tabescens (Scop.)
      Sing.; rootrot. Subtropical USA.
    Colletotrichum gloeosporioides Penz. (asco state is Glomerella
      cingulata (Stonem.) Sp. & Sch.); anthracnose diseases.
      Widespread.
    Corynebacterium poinsettiae Starr & Pirone; bacterial canker
      and leafspot. S. America scattered occurrence.
    Cuscuta americana L.; yellow dodder infection in laboratory.
      Puerto Rico.
    Diplodia (see Botryodiplodia)
    Fusarium oxysporum Schl. (? f. sp. pulcherrimae); typical
      wilt. Puerto Rico. F. solani (Mart.) Sacc.; rootrot.
      Puerto Rico.
    Meloidogyne sp.; rootknot nematode. Puerto Rico.

Oidium sp. (= Sphaerotheca euphorbiae Salm.); powdery
   mildew. El Salvador, Guatemala. (Immature perithecia
   found in mountain garden in El Salvador.)
Phymatotrichum omnivorum (Shear) Dug.; rootrot. Dry parts
   of Subtropical USA.
Puccinia euphorbiae P. Henn.; leafrust. Mexico. P. intumes-
   cens Holw.; swelled rust sori on leaves. Mexico, W.
   Indies.
Pythium spp. (unidentified species); rootrots, cutting failure.
   Costa Rica.
Rhizoctonia solani Kuehn = basidial state is Thanatephorous
   cucumeris (Frank) Donk; damping-off of seedlings and cut-
   tings. Puerto Rico, U.S. Virgin Islands, Florida.
Sclerotium rolfsii (Curzi) Sacc. = perfect state is Pellicularia
   rolfsii (Sacc.) West; soil-line rot. Puerto Rico (in nursery).
Sphaceloma poinsettiae Jenk. & Ruehle; spot anthracnose,
   scab. Florida, Puerto Rico, St. Croix, El Salvador,
   Nicaragua.
Thielaviopsis basicola (Berk. & Br.) Ferr.; a black rootrot.
   Puerto Rico.
Uromyces euphorbiae Cke. & Pk.; rust. Chile, Puerto Rico,
   Louisiana, Florida, Argentina, Bolivia, Peru, Ecuador
   (probably general).
Virus, an undetermined, and not studied, mosaic in El Salva-
   dor, Guatemala.
Witches' broom, cause unknown. C. America.

EUTERPE EDULIS Mart., and E. OLERACEA Mart. Often success-
   fully grown in the wild where certain other crops fail, these
   palms produce delicious edible and tender hearts; "palmito" Sp.,
   "palmeira," "acae" Po., "palm heart tree" En.
Cephaleuros virescens Penz.; algal leafspot. Brazil, Honduras.
Diplodia euterpes Syd. and D. palmicola Syd.; leafrib diseases
   and sometimes stemcanker. S. America.
Dothidina palmicola (Speg.) Th. & Syd.; leaf infection. Puerto
   Rico, S. America.
Eutypa euterpes Syd.; leaf disease. S. America.
Ganoderma sp.; base rot of old tree. Puerto Rico.
Geococcus-like accumulation of algal masses in leaf angles.
   W. Indies.
Glomerella cingulata (Stonem.) Sp. & Sch. the conidial state
   is a Colletotrichum; anthracnosis. Scattered.
Graphiola phoenicis (Moug.) Poit.; scabby raised pustules on
   leaf. Neotropics.
Hyaloria pilacre Moel.; wood rot. S. America.
Meliola furcata Lév.; black mildew. W. Indies, S. America.
Nectria euterpes Moel.; leaf disease. S. America.
Oidium sp.; superficial white mildew. S. America, C.
   America.
Pestalotia palmarum Cke.; leaflet, midrib, and seed spot.
   Puerto Rico, S. America, C. America.
Pilacrella delectans Moel.; woody decay. S. America.

Rosellinia euterpes Rehm; rootrot. S. America. R. bunodes
(Berk. & Br.) Sacc.; rootrot. C. America.
Strigula spp. (various colors and characters); lichen leafspots.
Neotropics.
Trentepohlia sp.; epiphytic algal tufts. Panama.
Tryblidiella rufula (Spreng.) Sacc. = Rhytidhysterium rufulum
(Spreng.) Petr.; disease of midrib and dying leaflet. W.
Indies, S. America.
Typhula trailli Berk. & Cke.; leaf disease. S. America.
Xylaria palmicola Wint.; root disease. S. America.

FAGOPYRUM ESCULENTUM Moench. This grain crop introduced
from Europe is actually more or less an exotic in Latin
America; its importance is still questionable, and it has had
little attention for diseases; "trigo negro" Sp., "trigo sarraceno,"
"trigo maurisco" Po., "buckwheat" En.
Acrothecium lunatum Wakk.; small leafspots. S. America.
Ascochyta sp.; leafspot. S. America.
Cercospora fagopyri Chupp & Muller; large leafspot. Vene-
zuela, Brazil, Peru.
Macrophomina phaseoli (Taub.) Ashby = Sclerotium bataticola
Taub.; root and stem charcoalrot. Brazil.
Meloidogyne sp.; nematode rootknot. Brazil.
Nigrospora sacchari (Speg.) Mason; weak parasite causing
infections. Trinidad. N. sphaerica (Sacc.) Mason in
Andes plantings.
Oidium sp. (probably Erysiphe polygoni DC.); powdery mildew.
Peru.
Pellicularia (see Rhizoctonia)
Phyllosticta sp.; leafspot. Brazil.
Ramularia rufomaculans Pk.; leafspot and mildew-like surface
growth. S. America.
Rhizoctonia solani Kuehn = Thanatephorous cucumeris (Frank)
Donk (also reported as Pellicularia sp.); stemrot, and
seedling damping-off. Peru, Brazil.
Sclerotium bataticola (see Macrophomina)

FAGUS see NOTHOFAGUS

FEIJOA SELLOWIANA Berg. A large leaved shrub-like fruit tree,
native to South America and used considerably there but not
much studied as to diseases; "goibeira serrana" Po., "guyaba
cimarrón" Sp., "feijoa" En.
Botryosphaeria dothidea (Moug.) de Not.; stem canker from
inoculation in California.
Botrytis cinerea Pers.; a fruitrot. California.
Catacauma feijoae Th. & Syd.; stromatic spot on leaf. Brazil.
Causalis sp.; leaf disease. Brazil.
Colletotrichum gloeosporioides Penz. = its asco-state is
Glomerella cingulata (Stonem.) Sp. & Sch.; fruitrot. Texas,
California, S. America.
Helminthosporium feijoae Cif.; leafblight. Brazil.

Pellicularia koleroga Cke. ; branch and leaf rot.   Florida,
    Costa Rica.
Penicillium expansum Lk. ; fruitrot.   California.
Phanerococcus feijoae Th. & Syd. ; leaf disease.   Brazil.
Phymatotrichum omnivorum (Shear) Dug. ; dryland rootrot.
    Texas.
Rhynchomeliola pulchella Speg. ; a rare black mildew.
    Paraguay.
Schizothyrium hypodermoides Rehm; black mildew causing
    mesophyll invasion.   Brazil.
Sclerotium erysiphoides Speg. ; leafspot and blight.   Brazil.
Sphaceloma psidii Bitanc. & Jenk. ; spot anthracnose, scab.
    Florida.

FESTUCA spp.   Various pasture grasses, some used in lawns.   The
    diseases listed may be often widespread, but not reported from
    all countries, and where serious in Central America can so
    weaken the Festucas that other grasses crowd them out; "capim"
    Po., "hierba" Sp., "grass" En.
    Calothyrium antarcticum (Speg.) Stev. ; leafspots.   S. America,
        C. America.
    Chaetodermium chlorochaetum Speg. ; leaf-and-culm disease.
        C. America, S. America.
    Claviceps purpurea (Fr.) Tul. ; ergot.   Bolivia, Chile, Argen-
        tina, C. America.
    Dictyothyrium perpusillum Speg. ; leaf-and-culm disease.   S.
        America.
    Helminthosporium dictyoides Drechs. ; leafblast.   Brazil.   H.
        sativum Pam., King & Bakke; plant rot and collapse.
        Argentina, C. America.
    Leptopeltina antarctica Speg. = Stomiopeltis antarctica (Speg.)
        Arx; a black leaf fungus.   Argentina.
    Linospora magellanica Speg. ; leafspot.   Brazil, Chile, Argen-
        tina.
    Lophodermium arundinaceum Chev. and L. oxyascum Speg. ;
        leaf disease.   Chile, and nearby countries.
    Micromastia tripnospora Speg. ; leafrot fungus.   S. America.
    Morenoina australis (Speg.) Th. ; leaf streak-spot.   S. America.
    Nematodes (unidentified) from rhizosphere in Chile.
    Niptera fuegiana Speg. ; scabspot on leaf.   S. American colder
        areas.
    Ophiobolus graminis Sacc. = Gaeumannomyces graminis
        (Sacc.) Arx & Olive; typical takeall disease.   S. America,
        Costa Rica.
    Pleospora forsteri Speg. ; ragged leaves.   Widespread.
    Pratylenchus sp. ; nematode found in root zone.   Chile.
    Pseudomonas coronafaciens (Elliott) Stev. f. atropurpurea
        (Reddy & Godk.) Stapp; chocolate-spot.   Argentina.
    Puccinia clematidis (DC.) Lagh. = P. recondita Rob. (on F.
        octoflora Walt.) in Chile, P. coronata Cda. is widespread,
        P. festucae Plow. in Chile, ? other parts of S. America,
        P. graminis Pers. (on F. arenaria Lam.) in Chile, P.

mellea Diet. & Neg. in Chile, P. piperi Rick (on F. australis Nees) in S. America, C. America, Subtropical USA,
(all above Puccinias are rusts).

Rhizoctonia (aerial type possibly microsclerotia but no sclerotia found); web-blight. W. Indies, C. America.

Rhynchosporium secalis (Oud.) J. J. Davis; leafblotch. Argentina.

Sclerotium clavus DC. and S. rolfsii (Curzi) Sacc.; sclerotium
rots. Scattered, occurrence limited to warm and moist
soils.

Tilletia fusca Ell. & Ev. var. patagonica Hirsch.; smut.
Patagonia in S. America.

Tylenchorhynchus sp.; nematode secured in rootzone. Chile.

Uromyces cuspidatus Wint.; rust. General and probably on
most Festucas. U. fuegianus Speg.; leafrust (on F. purpurascens Hook f.). Chilean Islands.

Ustilago festuca-tenellae P. Henn. = U. mulfordiana Ell. &
Ev.; flower head smut (on F. tenella). Mexico. U.
sphaerocarpa Syd.; seed smut. Mexico.

FICUS spp. [but see F. CARICA, following entry]. These species
are such as: the strangling fig, F. AUREA Nutt.; the large-
leaved one multiplied by nurserymen in the tropics to be grown
for ornament often in tubs in temperate zone homes and con-
servatories, F. ELASTICA Roxb.; a graceful tree used in roadside and park plantings, F. NITIDA Thunb.; and a handsome
small-leaved climber on tree trunks, rocks, and masonry walls,
F. PUMILA L. Unless otherwise stated the disease noted occurs
on wild species of Ficus.

Agrobacterium tumefaciens (E. F. Sm. & Towns.) Conn;
crowngall on F. aurea and F. elastica in Brazil.

Anisochora topographica (Speg.) Th. & Syd.; irregular stromatic leafspots on F. pumila in Brazil.

Auricularia sp.; on old wood of wild fig. Costa Rica.

Botryodiplodia theobromae Pat. also reported as Diplodia
theobromae (Pat.) Now. = Physalospora rhodina Berk. &
Curt. (Botryosphaeria quercuum (Schw.) Sacc.) the asco-
stage; dieback. Neotropical distribution.

Capnodium spp. (I have collected on wild figs numerous examples of these "fumagos" or sootymolds, and on attempt-
ing to make authoritative diagnoses as to genera and species was often unable to come to clear cut decisions.
These sootymolds can be found more or less generally.)

Cassytha filiformis L.; olive-green dodder. W. Indies.

Catacauma brittonia Chard. on F. pumila in Cuba, and C.
portoricensis Chard. = Phyllachora hoyosensis Petr., on
F. nitida and wild species in W. Indies. Both scab-like
leafspots.

Cephaleuros virescens Penz.; algal spot-scab. W. Indies,
C. America, S. America in moist areas.

Ceratocystis fimbriata Ell. & Halst.; woodstain and branch
death. Puerto Rico.

Cercospora bolleana Speg. = Mycosphaerella bolleana Higgins;
  rusty colored leafspots. Brazil, Venezuela, El Salvador
  (probably general). C. urostigmatis P. Henn.; large leafspots. Brazil, Venezuela, C. America.
Clitocybe tabescens (Scop.) Bres. = Armilariella tabescens
  (Scop.) Sing.; rootrot on F. nitida and F. elastica in Subtropical USA.
Colletotrichum gloeosporioides Penz. = ascospore state is
  Glomerella cingulata (Stonem.) Sp. & Schr. on many wild
  species, seen and isolated from cultivated F. nitida, F.
  aurea, F. elastica. Neotropics.
Corticium salmonicolor Berk. & Br.; pink disease reported
  from only F. elastica, F. nitida, F. pumila. W. Indies,
  C. America.
Cuscuta americana L.; the fast growing sturdy yellow dodder
  on F. elastica and F. pumila in Costa Rica. Cuscuta spp.;
  slender dodders on F. pumila in C. America and W.
  Indies.
Eutypa erumpens Mass.; trunk lesions sometimes fatal on F.
  nitida in W. Indies, Trinidad.
Fomes lamaoensis Murr. (honeycomb woodrot) and F. lignosis
  Klo. (friable woodrot) on F. nitida in C. America.
Fumago oosperma Speg.; sootymold (there are numerous un-
  determined sootymolds forming black pellicle-like epiphytic
  growths) occurs wherever sucking insects are prevalent.
Fusarium solani (Mart.) Sacc.; isolated from branch rot on
  F. nitida in Puerto Rico.
Heterodera fici Kir.; rootknot on F. elastica. Florida.
Hypoxylon borinquensis Hoehn. and H. stygium (Lév.) Sacc.
  both are stem lesions. W. Indies.
Langsdorffia hypogaea Mart.; a remarkable root parasite on
  several Ficus species. (Note: This Phanerogamic parasite and other species of it are native to jungle growths
  and develop underground root-stem structures engorged
  with stored nutrients, which peasants and indigenes dig and
  dry, using them for candles often called "siejos.") Reports are from Mexico to southern Brazil.
Loranthacea of various species and genera are seen on Ficus
  spp. These spectacular parasites on wild weed trees have
  been given little intensive study. They accumulate on tree
  branches along with the epiphytic ferns, bromeliads, mosses, tilandsias, orchids and the like.
Meliola eriophora Speg.; black mildew on wild trees and
  probably on cultivated. Neotropics.
Meloidogyne sp.; rootknot on F. elastica. Subtropical USA.
Melophia costaricensis Speg.; leaf disease. Costa Rica.
Microthyrium microspermum Speg.; black mildew. Paraguay.
Munkiella topografica Speg. = Anisochora topografica (Speg.)
  Th. & Syd.; on foliage. Paraguay.
Mycena citricolor (Berk. & Curt.) Sacc.; leafspot. Any spe-
  cies of Ficus is susceptible, with leaf spots producing in-
  fection bodies which is not true of some suscepts. C.
  America, W. Indies.

Mycosphaerella pittierii Syd.; large leafspot. S. America.
Nectria cinnabarina (Tode) Fr.; branch dieback on F. nitida.
    Brazil.
Nummularia bulliardii Tul. and N. discreta Tul.; branch
    canker of F. nitida in W. Indies.
Oidium sp.; mild appearing whitish mildew on wild Ficus,
    scattered occurrence.
Ophiodothella fici Bess. and O. floridana Chard. (which may
    be the same parasite); black leafspot on F. aurea, and
    F. nitida, and wild species. Subtropical USA, Cuba, El
    Salvador, Puerto Rico.
Phoma sp.; on foliage parts. Scattered.
Phylactaena ficuum P. Henn.; yellow leafspots. Subtropical
    USA, El Salvador, Brazil.
Phyllachora spp. these all cause delimited black spots on
    leaves, some spots said to be dull others shiny but these
    distinctions do not always hold over a period of aging.
    (It is of note that Ophiodothella, see above, is considered
    by some Phyllachora.) A few species: P. aspideoides
    Sacc. & Berl., P. cayennensis Th. & Syd., P. effigurata
    Syd., P. ficicola Allesch. & P. Henn., and P. vinosa
    Speg. = Puiggarina vinosa Speg. Scattered in occurrence,
    from Subtropical USA to southern Brazil.
Phyllosphere epiphytic microorganisms, specially noted on
    leaves several years old. One finds bacterial colonies,
    hyphal growths on surfaces, algae, lichens, minute liver-
    worts. In moist areas.
Phyllosticta spp. on various wild and cultivated figs in W.
    Indies, Subtropical USA, S. America, probably in C.
    America. A few reported: P. doliariaecola Bat. & Silv.
    in Brazil, P. physopella Dearn. & Bar. in W. Indies, P.
    roberti Boy. & Jac. in W. Indies, Bahamas and Subtropical
    USA. In some cases spots are numerous and small, in
    some, as for example the last named, spots are large and
    target-board type.
Physalospora atractina Syd.; small leafspots. Brazil. P.
    hoyae Hoehn.; leafspot, Puerto Rico.
Poria sp.; stump decay, and on old trunk of living tree of F.
    elastica invaded near dead stump. Puerto Rico.
Psittacanthus calyculatus (DC.) Don.; a woody and large, red
    flowered bush (mistletoe-like) on F. glabrata HBK., and
    the primary parasite, then hyperparasitized by the smaller
    again mistletoe-like parasite with greenish flowers, the
    Phoradendron piperoides Trel. El Salvador.
Pythium spp.; cutting failure of F. elastica.
Ramularia sp.; leafspot on F. aurea, F. elastica and wild
    species in W. Indies. (? = Mycosphaerella?)
Rhizoctonia (aerial type); leafblight of F. pumila and F.
    nitida in Florida, Louisiana, Puerto Rico.
Rosellinia spp.; rootrots of wild and planted Ficus trees,
    scattered in general.
Schizophyllum commune Fr.; trunk lesions and branch canker
    of F. nitida. Puerto Rico.

Sclerotium rolfsii (Curzi) Sacc. = Pellicularia rolfsii (Sacc.)
  West the sclerotial state, causing soil rot, basidial state
  in form of non-parasitic surface growth on underside of
  leaves of F. pumila in Florida.
Septobasidium spp.; felt growths over insect colonies. On
  wild figs. Scattered.
Stomatochroon lagerheimii Palm; algal systemic leaf infections
  on wild figs. C. America.
Strigula complanata Fée; white lichen spotting on F. nitida,
  F. elastica, F. aurea, F. pumila, and wild types. Neo-
  tropical.
Struthanthus marginatus Bl.; mistletoe-like parasite. Brazil.
Trentepohlia sp.; hairy brownish algal epiphytes on wild figs.
  C. America.
Uredo ficicola Speg. = Physopella fici (Cast.) Arth. also
  given as Uredo fici Cast. var. guaropiensis Speg. =
  Physopella ficina Arth.; common rust. Neotropical distri-
  bution.
Venturia sp.; leafspot. Mexico.

FICUS CARICA L. [see also FICUS spp., preceding entry]. Known
  in Middle East tropics long before ancient Egyptian days, it was
  considered the first fruit tree ever planted by prehistoric man.
  Known in literature and held in religious veneration, it was
  brought to and grows in the Western tropics where it produces
  much; "higuero" Sp., "figueira" Po., "fig" En.
Agrobacterium tumefaciens (E. F. Sm. & Towns.) Conn;
  crown gall. Subtropical USA, Honduras.
Alternaria alternata (Fr.) Keiss. = A. tenuis Nees; market
  fruit-blemish, senescent-leaf attack. Subtropical USA,
  Honduras, and probably other countries. A. fici Farn.;
  leafspot. Subtropical USA, Guatemala, Peru, Brazil.
Aphelenchus avenae Bast.; root nematode. Brazil.
Armillaria mellea Vahl = Armilariella mellea (Fr.) Karst.;
  rootrot. Subtropical USA.
Ascochyta caricae Rab.; leafspot. C. America, Educador.
Basidiomycete (isolated from four dying cuttings, white hyphae
  of cultured organism produced no sporocarps), Puerto Rico.
Botryosphaeria ficus (Cke.) Sacc.; trunk canker. Ecuador.
  B. dothidea (Moug.) Ces. & de Not.; dieback, branch
  canker. Honduras, Subtropical USA.
Botrytis cinerea Pers.; fruitrot, stem canker. Chile, Sub-
  tropical USA.
Capnodium sp.; sootymolds. C. America, S. America, W.
  Indies.
Cercospora bolleana (Th.) Speg. (= Mycosphaerella bolleana
  Higg.); leafspot. Ecuador, Mexico, Dominican Republic,
  Subtropical USA, S. America. Probably C. elasticae
  Zimm. and C. fici Heald & Wolf are synonymous with C.
  bolleana.
Cerotelium fici (Butl.) Arth.; the common figrust. Neotropics.
  (Note: The binomial of common figrust is also given as

Physopella fici Arth. and Uredo fici Cast; it is possible
this rust's designation is not absolutely settled. Rust on
species of Ficus are reported with such names as:
Puccinia sp., Physopella ficina, Kuehneola fici, Uredo
ficina, and U. ficicola; perhaps the whole figrust problem
deserves renewed study using the most modern methods.)
Cladosporium herbarum Pers.; fruitspot, leafedge disease.
  W. Indies, Subtropical USA, C. America.
Clitocybe tabescens (Scop.) Bres. = Armilariella tabescens
  (Scop.) Sing.; rootrot. Florida.
Colletotrichum gloeosporioides Penz. = C. elasticola Tass.
  (It is suggested, since it is such a well adapted anthracnose,
  it is convenient to make a trinomial as follows: C. gloeo-
  sporioides Penz. f. sp. elasticae Tass. that would set it
  apart from many other anthracnoses.) The asco-state is
  Glomerella cingulata (Stonem.) Sp. & Schr.; foliage an-
  thracnose, cutting failure, fruitrot. Specially studied in
  Ecuador, Brazil, Costa Rica, and Florida.
Collybia velutipes (Curt.) Quél.; tree basalrot. Chile.
Corticium salmonicolor Berk. & Br.; branch pink-disease
  and leafrot. Scattered in C. America, Subtropical USA.
Dematophora (see Rosellinia)
Diptherephora communis deMan; rootzone nematode. Brazil.
Diplodia sp. (Botryodiplodia sp.); fruitrot. Guatemala.
Ditylenchus dipsacii Filip.; rootzone nematode. Brazil.
Eutypella fici Ell. & Ev.; leaf disease, dieback. C. America,
  Subtropical USA.
Fumago vagans Fr.; sootymold. Subtropical USA, Ecuador.
Fusarium graminearum Schwabe also given as 1. F. roseum
  Lk. conidial form of Gibberella zeae (Schw.) Petch; fruit-
  rot, twigblight. Subtropical USA, Costa Rica, Argentina,
  Brazil. 2. F. lateritium Nees conidial form of G. baccata
  (Wallr.) Sacc.; twig disease. Subtropical USA, Costa Rica.
  3. F. moniliforme Sheld. (my study indicated it was F.
  moniliforme Sheld. var. subglutinans Wr. & Reink. but I
  saw no perfect asco-state); internal fruitrot. California,
  Costa Rica (fruitrot seen in dry western part).
Glomerella (see Colletotrichum)
Helicotylenchus nannus Stein.; nematode found near roots.
  Brazil.
Macrophoma fici Alm. & Cam.; stem canker, fruit dryrot.
  Subtropical USA, Brazil, Ecuador.
Meloidogyne spp. (M. exigua Goeldi, M. incognita Chitw. and
  others); rootknot nematodes. Subtropical USA, Brazil, C.
  America.
Monhysteria stagnalis Bastian; a rhizosphere nematode.
  Brazil.
Myriangium argentinum (Speg.) Sacc. & Syd. = Phymatosphae-
  ria argentina Speg.; fungus parasite of scale insect. Argen-
  tina.
Oospora sp.; fruit sour-rot, yeastyness. Subtropical USA,
  probably Neotropics.

Ormathodium fici Tims & Olive; brown leafspot.   Louisiana.
Paratylenchus hamatus Thorne & All. ; rhizosphere nematode.
    California.
Pellicularia koleroga Cke. ; thread-web blight.   Subtropical
    USA, C. America, W. Indies (scattered in moist areas),
    reported with some uncertainty from Trinidad.
Phomopsis cinerescens (Sacc.) Trav. ; stem canker.   Sub-
    tropical USA, Brazil.
Phyllachora brittonia (Chard.) Seav. (= Catacauma brittoniana
    Chard.); spots of raised black stroma on lower leaf surface.
    Brazil, Cuba, Puerto Rico.
Phyllosticta sycophila Th. ; large brown leafspot.   Brazil,
    Chile.
Phyophthora carica (Hara) Hori, P. citrophthora (Sm. & Sm.)
    Leon. , and P. palmivora Butl. ; all cause rootrot, leaf-
    spot and fruitrot.   Ecuador, Brazil, C. America.
Physalospora abdita Stev. = P. rhodina Cke. (Botryosphaeria
    quercuum (Schw.) Sacc.); dieback.   Subtropical USA, W.
    Indies, Costa Rica.
Pratylenchus sp. ; nematode in rootzone.   Brazil.
Psittacanthus cuneifolius Bl. ; woody bush as a parasite on fig
    tree branch.   Argentina.
Rhizoctonia (aerial type); leaf web-disease.   Costa Rica, W.
    Indies.   R. solani Kuehn = Thanatephorous cucumeris
    (Frank) Donk; cutting disease, collar rot.   Widespread.
Rhizopus nigricans Ehr. ; fruitrot.   S. America, C. America,
    Subtropical USA.
Rosellinia bunodes (Berk. & Br.) Sacc. (= Dematophora);
    rootrot.   Mexico, El Salvador, Brazil.
Saccharomyces sp. (see also Oospora above); yeasty fruitrot.
    Neotropics.
Schizophyllum commune Fr. ; woodrot of branches.   California,
    C. America.
Sclerotinia sclerotiorum (Lib.) dBy. = Whetzeliana sclerotio-
    rum (Lib.) Korf & Dumont; stem canker, flush blast,
    branch disease.   Subtropical USA, Chile, C. America, W.
    Indies.
Sclerotium rolfsii (Curzi) Sacc. (no basidial stage of parasite
    seen on host); cutting rot at soil line.   Florida.
Septobasidium pseudo-pedicellatum Burt; felt, occasional
    canker.   Subtropical USA, W. Indies.
Tubercularia vulgaris Tode = Nectria cinnabarina Fr. ; canker,
    twig disease, coral rot.   Subtropical USA, Costa Rica.
Trichothecium roseum (Pers.) Lk. ; fruitrot.   Subtropical USA.
Virus: Fig Mosaic Virus causing both typical mottle and leaf
    malformation in Argentina, Brazil, Uruguay, Subtropical
    USA.
Xiphinema index Thorne & All. ; rhizosphere nematode.   Cali-
    fornia.

FLACOURTIA spp.   Small trees--some species are spiny bushes--
    used as hedges and for production of fruits; called "ciruella" Sp. ,
    and "governors plum" En.

Apodanthes flacourtiae Karst. ; systemic in root and stem, a
   Phanerogamic parasite (first found in last century, occur-
   rence now is rare).   Venezuela.
Caudella oligotrichia Syd. ; leaf spot.   S. America.
Leaf crinkle (believed to be minor element deficiency).   Hon-
   duras.
Limacinia sp. ; sootymold.   Costa Rica.
Meliola xylosmicola Garces; black mildew.   Colombia.
Oidium sp. ; white mildew.   Costa Rica.
Phyllosticta sp. ; small leafspot.   C. America.
Root collapse (isolations revealed non-sporulating mycelium
   of an unknown Basidiomycete).   Costa Rica.

FORTUNELLA MARGARITA Swin.   A small tree with oval-shaped
   "small orange fruits" called "cumquat" or "kumquat" Sp. Po.
En.   Susceptible to many of the diseases of the orange, but for
reasons of economics not given major attention.
   Alternaria citri Ell. & Pierce; fruit blackrot, leafspot.   Sub-
      tropical USA, Puerto Rico, Guatemala.
   Botryodiplodia theobromae Pat. = B. quercuum (Schw.) Sacc.
      also called Diplodia = Physalospora rhodina Cke. ; dieback,
      gumming of branch, fruit peduncle rot.   Florida, C.
      America.
   Cephaleuros virescens Kunze; algal leafspot.   Most often in
      moist warm areas.
   Colletotrichum gleosporioides Penz. = asco-state is Glomerella
      cingulata (Stonem.) Sp. & Schr. which I recovered in moist
      chamber on diseased twigs after long incubation.   El Salva-
      dor, Subtropical USA, Widespread.
   Cuscuta campestris Yunck. ; dodder.   Florida.
   Diaporthe citri Wolf; fruit stemend rot.   California.
   Elsinoë australis Bitanc. & Jenk. ; spot anthracnose.   Brazil.
   Fusarium spp. isolations from secondary dieback and leaf
      disease in Florida, Costa Rica and El Salvador.
   Leptothyrium pomi (Mont. & Fr.) Sacc. ; fruit flyspeck.   C.
      America.
   Mycosphaerella sp. ; large leafspot.   Florida.
   Phoma socia Wolf; twig disease.   Subtropical USA, Puerto
      Rico.
   Phyllosticta citricola Hori; leafspot.   Subtropical USA, Vene-
      zuela.
   Phytophthora sp. (Undetermined), from diseased rootlets of
      potted seedlings.   Florida.
   Secondary dieback in aged tree, apparently foliage outgrowing
      roots.   A long list of saprophytic and semi-parasitic fungi,
      recovered in moist chamber studies.   El Salvador, Florida,
      Costa Rica.
   Tylenchulus semipenetrans Cobb; root nematode.   Florida.
   Virus: Citrus Tristeza Virus (unsatisfactory stock for budding)
      in Brazil.
   Xanthomonas citri (Hasse) Dows. ; bacterial canker.   S.
      America.

FRAGARIA CHILOENSIS Duch. var. ANANASSA Bail. Prehistoric
indigenes grew a kind of strawberry in an island, Chiloe, off
the coast of Chile, from which seeds were taken to Europe and
North America and the present much used garden varieties in
the world are crosses of this with a few lesser species. Commercial types are mostly good in cool tropics, plants are low
perennial herbs producing the fragrant berry; common names
are "fresa," "frutilla" Sp., "amendoim," "morangueiro" Po.,
"strawberry" En.

    Aphelenchoides fragariae Chr.; nematode dwarf and crinkle.
      Subtropical USA, W. Indies. A. parietinus Stein.;
      rhizosphere nematode. Brazil.
    Armillaria mellea Vahl = Armilariella mellea (Fr.) Karst.;
      rootrot. California.
    Ascochyta fragariae Lib.; leafspot. S. America.
    (Bacteria causing human diarrhea from contaminations by
      pickers when harvesting, eliminated by requiring careful
      supervision and cleanliness. Scattered.)
    Belanolaimus longicaudatus Rau; nematode root disease.
      Florida.
    Blackroot, various causes involved including semi-parasitic
      microorganisms, poor soils, old age, lack of shade.
      Mountain gardens in Guatemala, El Salvador, and probably
      other countries.
    Botryodiplodia theobromae Pat. (Diplodia); crownrot, root
      disease, runner canker. Widespread.
    Botrytis cinerea Pers.; leaf graymold, fruit brownrot, fruit
      mummy. Probably general in occurrence, severe at
      times in Colombia, Brazil, Honduras, Guatemala.
    Capnodium spp.; sootymolds. Puerto Rico, Cuba, C. America,
      S. America.
    Cercospora vexans Massal.; large leafspot. Venezuela, Sub-
      tropical USA. Specimens of typical leafspot collected in
      Puerto Rico but no spores found.
    Cladosporium herbarum Lk.; on fading leaves, in lesions on
      petioles, withered fruit. Scattered in various cool dry
      areas in C. America.
    Colletotrichum gloeosporioides Penz. = ascospore state is
      Glomerella cingulata (Stonem.) Sp. & Schr. (this latter
      not recovered by me), the conidial state is sometimes re-
      ported as Gloeosporium fragariae Mont. and C. fragariae
      Brooks; fruitrot, girdle, stolon disease. Reported es-
      pecially from Brazil and Florida, but is widespread.
    Dendrophoma obscurans (Ell. & Ev.) Ander. it has been called
      Phoma sp.; angular leafspot, stolon injury, plant blight.
      W. Indies, Subtropical USA, Brazil, Uruguay, Argentina.
    Diplocarpon earliana (Ell. & Ev.) Wolf; scorch of leaf. Sub-
      tropical USA, Brazil, W. Indies, C. America, probably
      other countries.
    Diplodia (see Botryodiplodia)
    Discohainesia (see Pezizella)
    Ditylenchus dipsaci Filip.; root nematode. Brazil, California.

Frommea mexicana Mains; leaf rust. Mexico.
Fusarium solani (Mart.) Sacc. and F. roseum Lk.; involved
    in root disease complex. Chile, Venezuela, isolates from
    Panama.
Helicotylenchus nannus Stein.; rhizosphere inhabiting nema-
    tode. Brazil.
Meloidogyne sp.; rootknot. Colombia, Florida.
Mollisia earliana (Ell. & Ev.) Sacc.; leafscorch. Argentina.
Monhystera stagnalis Bastian; nematode in rhizosphere. Brazil.
Mycena citricolor (Berk. & Curt.) Sacc.; leafspot. Costa
    Rica.
Mycosphaerella fragariae (Tul.) Lind. (the conidial state, often
    most commonly seen is Ramularia tulasnei Sacc.); the most
    common red leafspot, blackseed, smallpox. Neotropics.
Nematodes of so-called "spiral group"; in rhizosphere. Chile.
Neotylenchus abulbosus Stein.; nematode yellows. California.
Oidium fragariae Hartz is often reported as Oidium sp. and
    is probably a Sphaerotheca.
Pezizella oenotherae (Cke. & Ell.) Sacc. = Discohainesia
    oenotherae (Cke. & Ell.) Nannf.; leaf and stolon rot, a
    fruit rot. Subtropical USA, Brazil, C. America.
Phoma (see Dendrophoma)
Phyllosticta fragaricola Desm. & Rob.; leaf disease. Sub-
    tropical USA.
Phytophthora cactorum (Leb. & Cohn) Schr.; bud, flower and
    fruit disease. Brazil. P. fragariae Hickm.; red stele.
    Puerto Rico, Panama. P. parasitica Dast.; collapse.
    Argentina.
Procephalobus mycophilus Stein.; root and leaf nematode.
    Brazil, Mexico, Florida.
Pseudhalenchus minutus Tarj.; root nematode. Florida.
Pythium ultimum Trow.; blackroot. Subtropical USA, Costa
    Rica, El Salvador.
Ramularia (see Mycosphaerella)
Rhizoctonia solani Kuehn = Thanatephorous cucumeris (Frank)
    Donk; rootrot, center rot, hard fruit. Subtropical USA,
    Brazil, C. America, W. Indies, Peru. An unusual
    Rhizoctonia with abundant thin aerial webs of light color
    causing leaf blight. Mountain garden in El Salvador.
Rhizopus nigricans Ehr.; common fruit-softrot, leak, gray-
    mold. Often seen in markets. Widespread.
Sclerotinia sclerotiorum (Lib.) dBy. = Whetzeliana sclero-
    tiorum (Lib.) Korf & Dumont; crownrot. Puerto Rico,
    Florida.
Sclerotium rolfsii (Curzi) Sacc. (basidial state not seen);
    moldy rot. Brazil, Subtropical USA.
Sphaerotheca (? humuli) = Oidium sp.; white mildew. In
    cool areas of Guatemala, Argentina, Colombia, Panama,
    probably more or less general.
Tylenchorhynchus sp.; nematode recovered from rhizosphere.
    Chile.
Verticillium alboatrum Reinke & Berth.; wilt. Brazil.

Viruses: Strawberry Mottle in Subtropical USA, Guatemala.
Strawberry Stunt in W. Indies, C. America, S. America.
Tomato Spotted Wilt killing plants in Brazil. Strawberry
Witches' broom in W. Indies, C. America, S. America.
Xanthomonas fragariae Ken. & King.; bacterial angular leaf-
spot. Florida.
Xiphinema sp.; nematode from rhizosphere. Chile.

FURCRAEA GIGANTEA Vent. The relatively gigantic plant of the
tropical American agaves, grown for its tall stiff sword-like
leaves with hard bast fiber in them, the fibers on extraction and
processing are much used and exported; "pita grande," "piteira"
Po., "cabuya," "henequén" Sp., "mauritius-hemp," "west indies
hemp" En.
Alternaria sp. (sometimes incorrectly reported as Macro-
sporium); an uncommon mild leafspot. Colombia, Panama,
W. Indies.
Amphisphaeria fourcroya P. Henn.; black moldyspots. S.
America.
Bacterial (sometimes given as Phytomonas or Xanthomonas)
species, often systemic, have been isolated from fiber, in
C. America and Colombia.
Botryodiplodia theobromae Pat. (Diplodia); leaf common black-
rot, usually large lesions. Puerto Rico, Panama, Guate-
mala, S. America, El Salvador.
? Capnodium sp.; a thick layer of sootymold. Colombia,
Puerto Rico.
Cephalothecium sp.; on leaves. Colombia.
Ceratocystic sp.; stem vascular and fiber stain, plant kill.
Colombia.
Cercospora fourcroyae Obr. & Botero; leafspot. Brazil,
Colombia.
Cladochytrium sp.; on leaf. Colombia.
Colletotrichum agave Cav. more correctly C. gloeosporioides
Penz. = Glomerella cingulata (Stonem.) Sp. & Schr.; stalk
lesion, anthracnose. Brazil, Colombia, C. America.
Coniothyrium concentricans Desm.; concentric zoned leafspot.
Colombia.
Corticium salmonicolor Berk. & Br.; leaf disease, felpo.
Costa Rica, El Salvador, Colombia.
Didymosphaeria pachytheca Sacc. & Syd.; on stem and leaves.
S. America.
Dothydella paryii (Farl.) Th. & Syd.; leafspot (semi-radiate).
Neotropics.
Echidnodella fourcroyae Ryan; leaf disease. Puerto Rico.
Fusarium oxysporum Schl. (? f. sp. furcraea ?); wilt.
Colombia. Fusarium of roseum-type isolated from
diseased leaf and dry fiber. El Salvador.
Gemma-producing fungus (a Basidiomycete); leaf lesion. In
rainforest area in Guatemala.
Geococcus-like green algal coating on bottom leaves in El
Salvador and Puerto Rico.

Leptosphaeria sp.; small discrete but sunken leafspots.
Colombia, Honduras.
Lichen, a thin gray epiphytic pellicle. C. America.
Liverworts, microtypes growing on leaves in moist areas
in C. America.
Macrophoma sp.; black leafspot. Colombia.
Metasphaeria sp.; leaf fungus. Colombia.
Mirandia furcroyae Toro; mold growth on leaf. S. America.
Nectria suffulta Berk. & Curt. on wild Furcraea; stem
disease. Cuba, Bermuda, Puerto Rico, Haiti, Mexico.
Nigrospora oryzae Petch; on senescent leaves. C. America.
Perisporia sp.; black mildew. Colombia, C. America.
Phoma concentrica Desm. (? same as Coniothyrium concen-
tricans Desm. ?); is a concentric zoned leafspot. Colombia.
P. furcroyae Thu.; black lesion on stem. Puerto Rico,
Brazil.
Pleomassaria furcroyea Speg.; black mildew. Colombia,
Brazil, Argentina.
Pleospora herbarum Rab.; circular leafspot. S. America,
C. America.
Rabenhorstia furcroyae Pass.; leafspot. S. America.
Rosellinia bunodes (Berk. & Br.) Sacc.; rootrot. El Salvador.
Thielaviopsis sp.; rootrot. Colombia.
Trichocladium olivaceum Mass.; leaf decay. El Salvador.
Trichothecium roseum Lk.; leafrot. C. America, Colombia.
? Virus. Mosaic-like mild mottle in El Salvador and
Colombia.

GARCINIA MANGOSTANA L. In the twentieth century this handsome Oriental tree of medium stature has spread slowly to most
of our countries, and those who open the velvet brown crusty
fruit say of the fragrant flesh within that it is the world's most
delicious; and give it such common names as "mangostína" Sp.,
"mangostão" Po., "mangosteen" En.
Botrytis sp.; fruitrot. Honduras, W. Indies, ? S. America.
Capnodium spp.; sootymolds. Widespread.
Cephaleuros virescens Kunze; algal leafspot. W. Indies, C.
America.
Colletotrichum gloeosporioides Penz. = Glomerella cingulata
(Stonem.) Sp. & Schr.; anthracnose. General.
Knyaria garciniae Frag. & Cif. = Tubercularia garciniae P.
Henn.; canker. Dominican Republic.
Mycena citricolor (Berk. & Curt.) Sacc.; leafspot. Costa
Rica.
Nectria sp.; branch canker. C. America.
Pellicularia koleroga Cke.; thread-web blight. Argentina,
Puerto Rico.
Pestalotia espaillati Cif. & Frag.; sunken leafspot. Dominican
Republic.
Phyllosticta sp.; leafspot. C. America, W. Indies.
Polyporus sp.; on old tree a basal disease. Honduras.
Strigula complanata Fée; the common white lichen spot. C.
America, W. Indies.

Struthanthus spp. ; woody bush-like parasites.   Costa Rica.
Trentepohlia sp. ; algal hair on old branch ends.   Puerto Rico.

GARDENIA JASMINOIDES Ellis = G. FLORIDA L.   The fragrant
flowered shrub originally from China popular in much of the
Western tropics in cool locations, important economically to
growers as a cut flower; "jasmín, " "jasmín del cabo" Sp. ,
"gardenia" Po. , "gardenia, " "cape jasmine" En.
  Botryodiplodia theobromae Pat. (Diplodia) = Botryosphaeria
    quercuum (Schw.) Sacc. in its asco-state, is also given
    as Physalospora rhodina Cke. ; dieback, flower-cut disease.
    C. America, Puerto Rico.
  Botrytis cinerea Pers. ; bud drop, flower disease (not physio-
    logical).   W. Indies.
  (? Capnodium citri Desm. and other genera and species);
    sootymolds.   Neotropics.
  Cassytha filiformis L. ; green dodder infection by inoculation.
    Puerto Rico.
  Cephaleuros virescens Kunze; scab spots.   Widespread.
  Cladosporium herbarum Lk. ; disease of old leaves and old
    buds.   W. Indies, S. America.
  Corticium salmonicolor Berk. & Br. ; top disease.   Costa Rica.
  Cuscuta americana L. ; yellow dodder, natural infection.
    Puerto Rico.
  Diplodia (see Botryodiplodia)
  Meliola amphitricha Fr. ; black mildew.   General.
  Meloidogyne incognita Chit. ; rootknot.   Subtropical USA, W.
    Indies.
  Mycena citricolor (Berk. & Curt.) Sacc. ; leafspot.   Puerto
    Rico.
  Oidium sp. ; white mildew.   Seen in many countries, it may
    be Erysiphe polygoni DC, which is reported from Texas in
    the USA.
  Rhizoctonia sp. ; bush base disease.   Puerto Rico.
  Strigula complanata Fée; lichen spot.   General.
  Xanthomonas maculifolium-gardinae (Ark. & Barrett) Elliott;
    bacterial leafspot.   California.

GENIPA AMERICANA L.   Native of the Western tropics, a large
tree with russet fruits, a valued dietary addition much marketed
in many countries and although it is important it has not been
much studied for diseases; "jagua, " "genip" Sp. , "genipap" Po. ,
"genipa" En.
  Alternaria sp. = Macrosporium sp. ; leafspot.   W. Indies.
    (I have watched for this leafspot but never encountered it,
    possibly because of local conditions.)
  Amazonia tehoni Toro; black mildew.   Puerto Rico.
  Asterina genipae Ryan; superficial black mildew spots.   W.
    Indies.
  Cephaleuros virescens Kunze; algal scab.   Brazil, Costa Rica,
    Honduras, (possibly much more widespread).
  Gonatobotrys blihiae Frag. & Cif. f. sp. genipae Frag. & Cif. ;
    on leaf.   Dominican Republic.

Meliola asterinoides Wint.; a common black mildew.   General.
Phyllachora genipae Stev. & Dal.; tarspot.   W. Indies, C.
    America.
Polystomella pulcherrima Speg.; leaf disease.   Dominican
    Republic.
Sphaceloma genipae Bitanc., S. lago-santensis Bitanc. &
    Jenk., and S. morinda Bitanc. & Jenk.; spot anthracnoses.
    Brazil.

GERANIUM   see   PELARGONIUM

GERBERA JAMESONII Bolus.   Long ago brought from Africa, now
    a well-adapted plant for garden beds and cut flowers in many
    parts of the American tropics; "margarida africano," "gerbera"
    Po., "gerbera," "margarita africana" Sp., "african or transvaal
    daisy," "gerbera" En.
    Botrytis cinerea Pers.; graymold flower blight.   C. America.
    Cercospora gerberae Chupp & Viégas; common leafspot.   W.
        Indies, S. America.
    Colletotrichum gloeosporioides Penz. = Glomerella cingulata
        (Stonem.) Sp. & Schr.; stemrot, leaf collapse.   C.
        America.
    Erysiphe cichoracearum DC. = Oidium sp.; powdery mildew.
        Neotropics.
    Meloidogyne sp.; rootknot.   Subtropical USA.
    Phyllosticta gerbericola Bat.; leafspotting.   Brazil.
    Phytophthora spp.; rootrot.   Scattered.
    Rhizoctonia sp.; rootrot.   C. America.
    Sclerotinia sclerotiorum (Lib.) dBy. = Whetzeliana sclero-
        tiorum (Lib.) Korf & Dumont; plantrot.   S. America, C.
        America.
    Septoria gerberae Syd.; blotch spot.   Surinam, C. America.

GLADIOLUS HORTULANS Bailey.   This complex cultigen from
    crossings of many varieties is an economically important bulbous
    plant giving spikes of brilliant flowers for both internal and
    international trade; "gladiolo" Sp. Po., "espadaña" Sp., "gladio-
    lus" En.
    Alternaria alternata (Fr.) Keiss = A. tenuis Nees; scanty
        occurrence of leafspot, moldy blackening.   W. Indies, Sub-
        tropical USA, Guatemala.   Alternaria sp. (large spored,
        different from the above); leafspot found one year, 1945,
        did not recur.   El Salvador.
    Botrytis cinerea Pers.; flower and plant mold.   El Salvador.
        B. gladiolorum Timm. (perhaps this and the above are
        synonyms); mold of leaf, flower, corm.   C. America, W.
        Indies, Subtropical USA.
    Cladosporium herbarum Lk.; weak plant leaf disease.   Scat-
        tered.
    Curvularia lunata (Wakk.) Boed.; brownspot on leaves and
        flowers.   Florida, El Salvador, Puerto Rico.
    Cuscuta spp.; yellow dodders.   C. America.

Fusarium oxysporum Schl. f. sp. gladioli (Mass.) Sny. &
 Hans.; bulb dryrot, systemic wilt. Several areas in C.
 America, subtropical USA. F. solani (Mart.) Sacc.; rot
 of sideroots and collar. Colombia, Peru.
Meloidogyne sp.; rootknot. Florida, W. Indies, C. America
 (probably S. America also).
Penicillium gladioli McCull. & Thom; corm rot. General.
 Penicillium sp.; a secondary rot following Fusarium. W.
 Indies, C. America.
Phyllosticta gladioli Ell. & Ev. and P. gladioloides Bat.;
 leaf disease. Latter in Brazil, former in C. America.
Pseudomonas marginata (McCull.) Stapp; scab, neck-rot, bac-
 terial leafspot. Widespread.
Rhizoctonia spp. (both aerial and terrestrial types); leaf
 disease, often causes rainyseason wetrot of bulb. Puerto
 Rico, Guatemala, Panama, El Salvador, Guatemala.
"Rust" or "roya"; in reality not true rust, caused by several
 semiparasitic fungi in poorly grown plants. El Salvador,
 Puerto Rico.
Sclerotium rolfsii (Curzi) Sacc. = Pellicularia is the basidial
 form but I have never recovered it on gladiolus; soil blight.
 Scattered in Subtropical USA, W. Indies occasionally.
Septoria gladioli Pass.; leafspot, corm hardrot. Neotropical.
Stemphylium sp.; red leafspot. Florida, Peru, Brazil.
Stromatina gladioli (Drayt.) Whet. = Sclerotium gladioli Mass.;
 corm dryrot, leaf and stalk rots. General.
Verticillium sp.; wilt. Chile.
Viruses: Bean Yellow Mosaic Virus in Puerto Rico. Cucum-
 ber Mosaic Virus var. Celery/Commelina in W. Indies,
 Florida, probably scattered generally. Gladiolus mottle
 and stunt, undetermined but symptoms not the two just
 named, in W. Indies, S. America.

GLIRICIDIA SEPIUM Steud. (synonyms: G. SEPIUM HBk. and
 LONCHOCARPUS SEPIUM DC). A native tree readily multiplied
 with seed and stakes, it is used importantly in shade agriculture,
 and its curly-grained wood is valued in fine cabinet work, furniture, ox wagon-beds; "palo de sombra," "gliri" Sp. Po.
 Botryodiplodia theobromae Pat. (Diplodia) = Botryosphaeria
  quercuum (Schw.) Sacc.; common dieback. Neotropical.
 Ceratocystis fimbriata Ell. & Halst.; woodstain. Colombia.
  (Found in Costa Rica near cacao trees with same symptoms.)
 Cercospora atropurpurascens Chupp; leafspot. Brazil, Venezuela. C. gliricidiae Syd.; leafspot. Puerto Rico, Trinidad, C. America. C. gliricidiasis Gonz. Frag. & Cif.;
  large leafspots, defoliation. W. Indies.
 Colletotrichum sp. (similar to gloeosporioides but spores
  large and no asco-stage found in El Salvador studies);
  foliage and fruit anthracnose. Widespread.
 Corticium salmonicolor Berk. & Br.; leafrot, branch bark
  infections. El Salvador.

Cyphella villosa Pers. ; small fungus cup associated with dieback. C. America, S. America.

Fomes spp. ; rootrots. Costa Rica.

Fusarium decemcellulare Brick, (ascospore state is Calonectria rigidiuscula (Berk. & Br.) Sacc.); stem canker, twig-
tip disease. El Salvador.

Ganoderma sp. ; rot of old tree base. El Salvador.

Gomosis, complex of semiparasites (isolations revealed a
Rhizoctonia, a Phoma, and a Pestalotia on a gummy branch
in Costa Rica). Widespread.

Isariopsis sp. ; leafspot. El Salvador.

Lenzites palisoti Fr. ; woodrot. S. America.

Loranthus spp. ; large mistletoe-like parasites. El Salvador.

Meliola sp. ; discrete black mildew spots. C. America.

Nectria sp. ; branch cankers. General.

Pellicularia koleroga Cke. ; thread-web blight. Puerto Rico,
Panama.

Phomopsis gliricidiae Syd. = P. citri Fawc. (= Diaporthe
medusaea Nits.); dieback. Widespread.

Phthirusa angulata Krause, P. theobromae Eich., Phthirusa
sp. ; all more-or-less recumbent Phanerogamic parasites.
Surinam and C. America.

Poria sp. ; on dead branches. C. America.

Psittacanthus calyculatus Don. ; mistletoe-like parasite (produces "wood rose" where branch is attacked). El Salvador.

Rhizoctonia (probably solani); cutting or stake bottom-disease
inhibiting rooting. El Salvador.

Rosellinia bunodes (Berk. & Br.) Sacc. and R. pepo Pat. ;
rootrots. W. Indies, C. America.

Trametes meyenii Klo. ; woodrot on felled trees. S. America.

GLYCINE MAX (L.) Merr. (= SOJA MAX Piper and PHASEOLUS
MAX L.). A native of China-Japan that is well adapted to the
Americas; its seeds or beans more and more used as protein
food for man and fed to farm animals in the American tropics;
"soya," "frijol soya" Sp., "soja," "feijão soja" Po., "soybean"
En.

Alternaria (relatively large spore); rare leafspot. Puerto
Rico. A. alternata (Fr.) Keiss. = A. tenuis Nees; infection in one field after wind damage. El Salvador.

Aphelenchoides parietinus Stein. ; nematode found in rhizosphere. Brazil.

Aspergillus sp. ; seedmold. El Salvador.

Ascochyta phaseolorum Sacc. = Phoma exigua Desm., and
A. pisi Lib. ; leafspots. Brazil, Venezuela.

Botryodiplodia pallida Ell. & Ev. (see Physalospora).

Cassytha filiformis L. ; green dodder. Occasional in W.
Indies.

Cephalosporium gregatum All. & Chamb. ; brown stemrot.
Mexico.

Cercospora spp. ; C. glycine Cke. ; white spot in Venezuela,
C. kakuchii Matsu & Tomo. ; purple spot in Central

America and Venezuela, C. sojina Hara = C. daizu Miura,
the asco-state is Mycosphaerella phaseolicola (Desm.)
Sacc. ; frogeye leafspot is noted specially in Venezuela in
Subtropical USA and in Central America but is probably
widespread, and C. stevensii Young; leafspot in Brazil.
Chaetoseptoria wellmanii J. A. Stev. ; large leafspot and
severe defoliation (crop is highly susceptible).   Costa Rica,
El Salvador, Panama, Venezuela.
Colletotrichum gloeosporioides Penz. (the conidial state culture
I could not use to successfully cause disease in coffee or
cucumber or orange but was successful on soybean) the asco-
state is Glomerella cingulata (Stonem.) Sp. & Schr. ; leaf
and petiole disease, perfect state on dead stems.   Picked
up in Florida, Costa Rica, Venezuela, Panama.   C. linde-
muthianum (Sacc. & Magn.) Bri. & Cav. ; elongate spots
on leaf, twig, pod.   Brazil.
Corynespora casiicola Wei; leaf targetspot, bean bumspot.
Subtropical USA, Nicaragua, probably scattering in general.
Cuscuta americana L. ; yellow dodder.   W. Indies.
Diaporthe sojae Leh. = D. phaseolorum (Cke. & Ell.) Sacc. ;
canker of stem and pod.   Neotropics.
Ditylenchus dipsaci Filip. ; stem disease from nematode.
Brazil.
Erysiphe polygoni DC. = Oidium; the powdery white
mildew.   Latter is common where crop is grown, the
perfect state in a few instances, reported as collected in
cooler parts of Puerto Rico, Florida, Venezuela.
Fusarium oxysporum Schl. f. sp. tracheiphilum (E. F. Sm.)
Sny. & Hans. ; wilt in certain friable somewhat acid, warm
soils.   Subtropical USA, S. America (I have not seen it in
C. America).   F. semitectum Berk. & Rav. = F. roseum
Lk. ; basal rot.   Trinidad.   Fusarium spp. ; dry bean injuries, damping-off, branch diseases, leafspot enlarge-
ments, pod-end disease.   Scattered.
Helicotylenchus nannus Stein. ; nematode in rhizosphere.
Brazil.
Helminthosporium vignicola (Kaw.) Olive; leaf target spot,
seed injury in pod.   Widespread.
Leafspots (undetermined as no spores recovered from spot
studies.   Various types: gray with yellow halo, black
speckle, small translucent spot becoming brown, "slitting"
between main veins, dark leaf edge, blotch) in Central
America.
Macrophomina phaseoli (Maubl.) Ashby; gray stem.   Subtropical USA, Puerto Rico, Brazil.
Meloidogyne spp. ; rootknot.   Scattered, probably in all
countries, where the soybean is produced in large quantity.
(Note: A list of some of the species of Meloidogyne reported as attacking soybean in Brazil is of interest: M.
arenaria Chitw. , M. exigua Goeldi, M. hapla Chitw. , M.
incognita Lorde. , M. inornata Lorde. , and M. javanica
Chitw.   The pathologist-nematologist's challenge is to

determine if there are external symptom differences in the
effects on the host, and to find if there are any differences
needed for control measures; this is both a practical and
an academic problem.)

Monhysteria stagnalis Stein. ; a rootzone nematode, found in
Brazil.

Mycena citricolor (Berk. & Curt.) Sacc.; leafspot by inoculation.   Costa Rica.

Myrothecium roridum Tode; enlarged leafspot following Colletotrichum.   Subtropical USA, C. America.

Nematospora coryli Pegl.; seed yeastspot.   Subtropical USA,
Guatemala, El Salvador.

Oidium sp. (see Erysiphe)

Penicillium spp. ; seed moldyness.   General.

Peronospora manshurica (Naum.) Syd. ; downy mildew.   El
Salvador, Louisiana in USA, Colombia, Brazil.

Phakopsora sojae Saw. = P. pachyrhize Syd. ; brown rust.   W.
Indies.

Phoma sp. ; blackstem canker.   Subtropical USA, W. Indies.
(See Ascochyta.)

Phyllosticta sp. ; leafspot.   C. America.

Phymatotrichum omnivorum (Shear) Dug. ; dryland rootrot.
Texas.

Physalospora rhodina Cke. = Botryosphaeria quercuum (Schw.)
Sacc., with conidial state as Botryodiplodia theobromae Pat.
or B. pallida Ell. & Ev. ; on dead and dying stems.   W.
Indies, C. America, probably S. America.

Pratylenchus pratensis Filip. and P. steineri Lorde. , Zam. ,
and Boock; root disease nematodes.   Brazil.

Pseudomonas glycinea Coer.; bacterial blight-wilt.   Subtropical
USA, Venezuela.   P. solanacearum E. F. Sm. ; bacterial
wilt.   Puerto Rico, C. America.   P. tabaci (Wolf & Foster)
Stev. ; bacterial leafspot.   Argentina.

Pythium sp. ; damping-off.   C. America.

Rhizobium leguminosarum Bald. & Fred; desired root nodules.
Widespread.   (If not present must be introduced.)

Rhizoctonia (? microsclerotia, at least aerial forms); two
kinds of leafspots, web blights.   Scattered in C. America
also reported from Bolivia and Subtropical USA.   R. solani
Kuehn. = Thanatephorous cucumeris (Frank) Donk; root and
collar rot.   General.

Sclerotinia sclerotiorum (Lib.) dBy. = Whetzeliana sclerotiorum
(Lib.) Korf & Dumont; plant collapse.   El Salvador.

Sclerotium rolfsii (Curzi) Sacc. (basidial Pellicularia not
found); soil-line rot, wilt.   Subtropical USA, Puerto Rico,
Venezuela.

Viruses: Soybean Mosaic that causes severe chlorosis,
mottle, crinkling of leaf, pod marking in many countries.
Soybean Stunt that causes stunt, mottle, rugosity, witches-
brooming in C. America, and Tobacco Ringspot that results
in mottle and bud blight in Subtropical USA, Brazil.

Xanthomonas phaseoli (E. F. Sm.) Dows. var. sojense

(Hedges) Starr. & Burk. ; pustular spot.   Argentina, Nicaragua, Subtropical USA, probably more widespread.
Xiphinema americanum Cobb and X̲. campinense Lorde. ; root
ectonematodes.   Brazil.
(Note: I have not included here about 50 semi-parasites and
saprobes that have been seen, but, up to the present, considered unworthy of a disease list.)

GOSSYPIUM HIRSUTUM L.   Cultivated cotton.   (Note: The crop is
the oldest source of natural fiber in tropical America--used by
ancient Mexicans, Caribs, and Incas.   The botanical name G̲.
hirsutum is used here, but with its white fiber, and a few
others, some with brown fibers and mixtures of them, are all
much produced.   In perpetually warm habitats, cottons are perennials varying from shrubs to small trees.   Lints in capsules, or
bolls, are colored browns through tans to white.   Western Hemisphere tropical species I have compared for possible disease differences are G. BARBADENSE L. , G. PERUVIANUM Cav. , and
G. MEXICANUM Tod.)   Common names for the plant are "algo-
donero" Sp. , "algodoeira" Po. , "cotton" En.   Diseases listed below are not restricted to either cultivated or wild-growing plants.
Agrobacterium tumefaciens (Riker et al.) Conn; crowngall.
Peru.
Alternaria alternata Keiss. = A. tenuis Nees; black spot,
capsule discoloration.   A mild effect, scattered reports.
A. macrospora Zimm. = A. longipedicellata (Reich.) Snow. ;
seedling leaf blast, large spot, capsule spot.   Scattered
reports.
Amoebosporous sp. (old report of severe attacks, associated
with a Chytrid); dry toprot.   W. Indies.
Ascochyta gossypii Ell. & Ev. ; canker, wet-season blight,
ashy spot.   Colombia, Subtropical USA, Mexico, Ecuador,
Guatemala.   A. phaseolorum Sacc. = Phoma exigua Desm. ;
ashy leafspot.   Brazil.
Ashbya gossypii (Ashby & Now.) Guill. ; internal boll stain.
W. Indies, Trinidad, Guatemala, S. America.
Aspergillus spp. (all cause rot of capsules), A. flavus Lk. ,
in dryer parts of Subtropical USA, A. niger v. Tiegh. , in
Colombia, Peru, Costa Rica, Brazil, and A. wentii Weh. ,
in Argentina where it also causes seedling injury.
Bacillus gossypina Sted. ; a capsule gomosis, barba, spot.
Colombia, Subtropical USA, W. Indies, Brazil, Peru,
Dominican Republic.
Belonolaimus longicaudatus Rau; root-rot nematode.   Peru.
Botryodiplodia theobromae Pat. and B̲. gossypina Cke. (see
Physalospora)
Botryosphaeria dothidea (Moug.) Ces. & de Not. ; stem and
boll disease.   Subtropical USA, on wild cotton tree G̲.
mexicanum in El Salvador.
Brachysporium sp. ; seedling blight.   Subtropical USA.
(? Capnodium sp.); sootymold (some times apparently following
insect attacks, but in other cases seem supported by a

natural exudate from underside of healthy cotton leaf). C.
America.

Cassytha filiformis L.; a green dodder, artificial infection.
Puerto Rico.

Cercospora gossypina Cke. synonym is C. althaeina Sacc.,
the ascospore state is Mycosphaerella gossypina (Atk.)
Earle; rounded zonal leafspot and capsule attack. Mexico,
Subtropical USA, W. Indies, Colombia, Ecuador, Venezuela, Brazil.

Cercosporella (see Mycosphaerella)

Chaetomium brasiliense Bat. & Pont.; old leaf and stem
attack. Brazil.

Choanephora conjuncta Couch; flower disease. Subtropical
USA. C. cucurbitarum (Berk. & Rav.) Thaxt.; both a
flower and a capsule disease. Subtropical USA.

Cladosporium herbarum Pers.; serious capsule infection on
weakened tissue following other diseases. Neotropics.

Colletotrichum gossypii South. asco-state Glomerella gossypii
Edg. are synonyms of C. gloeosporioides Penz. with its
asco-state of G. cingulata (Stonem.) Sp. & Sehr.; common
anthracnose. Neotropics.

Criconemoides sp.; nematode found in rhizosphere. Peru.

Curvularia lunata (Wakk.) Boed.; capsule disease. Panama,
El Salvador, Puerto Rico.

Cuscuta sp.; a fast growing slender yellow dodder. El
Salvador.

Cytospora (see Valsa)

Diplodia (see Botryodiplodia)

Doassansia gossypii Lagerh.; small leaf-smut sori. Brazil,
Ecuador.

Epicoccum sp.; on leaf. Brazil.

Fomes lamaoensis Murr. (? = Trametes sp.); pivotal-root
rot. Peru, C. America.

Fumago vagans Fr. (also see Capnodium); a thick and scaly
sootymold. W. Indies, Nicaragua, Guatemala.

1. Fusarium decemcellulare Brick (ascospore state is a
Calonectria, not often collected or reported); dieback. W.
Indies, C. America. 2. F. moniliforme Sheld. (ascospore
state is a Gibberella); capsulerot, blast of small plants,
dryrot of main root, isolated from inside seed in Brazil.
W. Indies, C. America, S. America at least in Peru,
Colombia and Brazil. 3. F. oxysporum Schl. f. sp.
vasinfectum (Atk.) Sny. & Hans (no asco-state); typical
wilt, enters roots in warm slightly acid friable soils in
Mexico, C. America, W. Indies, Subtropical USA, Argentina, Colombia, Ecuador, Bolivia, Peru. (There are
varieties resistant to this Fusarium, which, however may
become susceptible if rootknot is also in the field.)
4. F. roseum Lk. = F. concolor Reink. (has associates
such as F. avenaceum (Cda.) Sacc., F. equisiti (Cda.)
Sacc., F. scirpi Lam. & Fautr., and the F. semitectum
Berk. & Rav.); wet-season canker of stem, capsule redrot,

seedling damping-off. Neotropics. (Note: F. roseum under certain conditions develops an ascospore state, but it has not been determined what all those conditions are; I have seen the asco-carps only twice, on soil and trash at dead plant bases in poorly drained parts of two fields in El Salvador.) 5. F. solani (Mart.) Sacc. with its asco-state of Nectria haematococca Berk. & Br.; stem and seedling disease in moist seasons, root disease in acid soil. W. Indies, Trinidad, Nicaragua, C. America, Argentina, Panama.

Geotrichum sp.; wetseason secondary infection of capsules. C. America.

Glomerella (see Colletotrichum)

Helicotylenchus sp. in Bolivia and H. nannus Stein. in Brazil, both are rhizosphere inhabiting nematodes.

Hormodendrum cladosporioides (Fres.) Sacc.; part of capsule rot complex. Peru, Brazil.

Hymenochaete sp.; sporocarps on dying trunk of wild tree-like cotton G. mexicanum in El Salvador.

Internal capsule ("boll") disease follows bug puncturing. Several microorganisms found: undetermined bacteria. Nematospora, Nigrospora, Phoma-like, Pestalotia, Fusarium.

Kuehneola (see Phakopsora)

Lenzites sp.; branch disease on two-year old cotton trees in mountains. Panama.

Lightning damage (field areas of killed and dying plants in circular pattern about an evident center where it was struck, if planting is young, plants at edges of area may recover). C. America.

Macrophomina phaseoli (Taub.) Ashby = mycelial state is Rhizoctonia bataticola Butl.; stemblight. Scattered in Subtropical USA and Nicaragua.

Meloidogyne sp.; rootknot (adds to virulence of attack when combined with root-inhabiting fungi such as Fusarium, Thielaviopsis, and Verticillium). Panama, Peru, Brazil, Guatemala, probably scattered in general.

Metaphelenchus rhopalocerous Stein.; nematode in rhisosphere. Brazil.

Monilia sitophila (Mont.) Sacc. = Neurospora sitophila Shear & Dodge; capsule disease. Brazil. (Something similar, but not fully identified, in El Salvador and Nicaragua.)

Mycena citricolor (Berk. & Curt.) Sacc.; leafspot (close to diseased coffee tree and by inoculation). Costa Rica.

Mycosphaerella areola Ehrl. & Wolf (conidial state is Septocylindrium areola Berk. & Curt, also given as Ramularia areola Atk. or R. gossypii (Speg.) Cif., but first named as Cercosporella gossypina Speg.); the typical areolate brown leafspot. Scattered in W. Indies, Subtropical USA, S. America, C. America.

Nectria cinnabarina Tode; stem canker. Subtropical USA, C. America.

Nematospora coryli Pegl. and N. gossypii Ashby & Now.; both
are capsule diseases. W. Indies, Brazil, probably widespread.

Neocosmopora vasinfecta E. F. Sm. (Some have confused
this, and equated it, with "Fusarium vasinfectum," however, I find the two are very different. Neocosmopora is
air-borne and rain-splashed, is a stem and branch killing,
partly systemic, fungus that produces red perithecia on the
surface of dead stems while the Fusarium apparently produces no perithecia or asco-state of any sort.) Neocosmopora has been reported from both the West Indies and
South America.

Nigrospora sphaerica (Sacc.) Mason; black capsule disease.
Neotropical but scattered.

Oidium tenellum (Berk. & Curt.) Lind. = Rhinotrichum tenellum Berk. & Curt.; Olpitrichum macrosporum Lind. = R.
macrosporum (Farl.) Sumst.; both secondary diseases of
capsules. Subtropical USA.

Oryctanthus sp.; mistletoe-like branch parasite, on unidentified
large wild-growing tree cotton possibly G. barbadense with
brown lint. El Salvador.

Ovulariopsis gossypii Wakef. is the conidial state of Erysiphe
malachrae Seaver, powdery mildew, may be the most common cotton disease in the tropics. Neotropics.

Penicillium sp.; damping-off, capsule infection. Scattered in
general.

Peronospora gossypina Av.-Sac.; downy mildew of fruits.
Brazil.

Pestalotia spp.; leaf, stem, capsule spots. Neotropics.

Phakopsora gossypii (Arth.) Hirat.; common rust. Neotropical. (Note: In the Neotropics, cotton rust was first
found in Ecuador and from there it spread to the West
Indies, next step subtropical USA, and then north. Thus
far it has not been devastating as it is only severe on lower
leaves that are largely past vigorous activity. Without re-
gard to synonymy, but for literature reference purposes,
some names given in the Western tropics for cotton rust
are: Aecidium desmium Berk. & Curt., A. gossypii Ell.
& Ev., Cerotelium desmium (Berk. & Br.) Arth., C.
gossypii Arth., Kuehneola gossypii Arth., Phakopsora
desmium (Berk. & Br.) Cumm., P. gossypii Dale, Phragmidium sp., Puccinia cacabata Arth. & Holw., P. schedon-
nardi Kel. & Sw., P. stakmanii Presl., Uredo gossypii
Lagh. It is believed at this time that Phakopsora desmium
(Berk. & Br.) Cumm., Puccinia cacabata Arth. & Holw.,
and P. stakmanii Presl. may be acceptable binomials in
addition to Phakopsora gossypii (Arth.) Hirat.)

Phlyctaena gossypii Sacc.; stem canker. Subtropical USA,
Puerto Rico.

Phoma gossypii Sacc.; seedling blight, stem disease. Subtropical USA, hillside plantings in C. America.

Phyllosticta gossypina Ell. & Mart.; leafspot with stippling.

Subtropical USA, Mexico, Puerto Rico, Panama, Brazil,
Bolivia. P. malkoffii Bub.; leafspot with stippling. Sub-
tropical USA, W. Indies, Brazil.
Phymatotrichum omnivorum (Shear) Dug.; dryland rootrot.
Mexico, Texas.
Physalospora rhodina Cke. = Botryosphaeria quercuum
(Schw.) Sacc. the ascospore state (asci borne in partly
buried perithecia on dead arm tissue) of such conidial
forms as Botryodiplodia theobromae Pat. and B. gossipina
Cke. also given as Diplodia; black capsule and dieback, at
times associated with pinkrot. Widespread.
Phytophthora spp. attacking capsules: P. cactorum (Leb. &
Conn) Schroet., in W. Indies and El Salvador, P. para-
sitica Dast. in W. Indies, and P. nicotianae de Haan var.
parasitica (Dast.) Waterh., in Subtropical USA, Trinidad,
Puerto Rico, S. America.
Pleospora herbarum (Pers.) Rab. and P. nigricantia Atk.;
both are weak parasites of leaves on poorly grown plants.
Scattered.
Polyporous sp.; diseased old pivotal root of tree cotton. El
Salvador. Costa Rica.
Pratylenchus steineri Lorde., Zam., & Boock; nematode in
rhizosphere. S. America.
Pythium spp.; cause damping-off. Neotropics.
Rhizoctonia crocorum (Pers.) DC. is mycelial state of the
basidium bearing fungus, Helicobasidium purpureum (Tul.)
Pat.; the black or purple rot of main pivotal root. W.
Indies, Trinidad. R. solani Kuehn = Thanatephorous
cucumeris (Frank) Donk; damping-off, soreshin, basal rot
of young maturing plants, rot of lateral roots. Neotropical.
Rhizoctonia of aerial types, possibly they belong to micro-
sclerotia but sometimes characters do not fulfill descrip-
tions of it; web-blight, long gossamer streamers, leafrot.
C. America.
Rhizopus nigricans Ehr. = R. stolonifer (Ehr.) Lind.; asso-
ciated with black and pink capsule rot. Subtropical USA,
Colombia, C. America, Brazil, Peru, Argentina.
Rosellinia bunodes (Berk. & Br.) Sacc.; rot of large lateral
roots, wilt. El Salvador, Costa Rica.
Rotylenchus reniformis Lindf. & Oli.; root nematode. Sub-
tropical USA, Peru.
Schizophyllum alneum (L.) Schr. = S. commune Fr.; root and
stem disease of old plants. Subtropical USA, C. America,
S. America.
Sclerotium rolfsii (Curzi) Sacc. (basidial form is Pellicularia
rolfsii (Sacc.) West); base rot and wilt of young plants.
Peru, Venezuela, Colombia, Guatemala, Subtropical USA.
Septocylindrium (see Mycosphaerella)
Septoria gossypina Cke.; stem disease. Subtropical USA, C.
America (may be general).
Shedding of capsules or bolls, sometimes called a special
disease. Said to be due to minor element deficiencies of

calcium, potash, zinc, and others, even an excess of manganese and salt may cause shedding, but both root and stem diseases of many sorts also result in early capsule drop.

Spermophthora gossypii Ashby & Now. ; internal capsule decay. Jamaica.

Stilbum nanum Mass. f. gossypina Av. -Sac. ; capsule rot. Brazil.

Thielaviopsis basicola (Berk. & Br.) Ferr. ; blackrot of pivotal (or central) root.   Especially severe if root is also attacked by rootknot nematode.   Subtropical USA, Peru.

Trichoderma viride Pers. ; seed mold, capsule disease. Scattered in most of the tropics.

Trichothecium roseum Lk. ; secondary capsule rot.   Subtropical USA.

Tylenchorhynchus dubius Thorne and T. claytoni Stein. ; root nematodes.   Subtropical USA, Peru.

Tylenchus davianii Bast. ; nematode in rhizosphere.   Brazil.

Uredo (see note following Phakopsora)

Valsa gossypii Cke. imperfect state of Cytospora which causes root rot.   Scattered.

Verticillium alboatrum Reinke & Berth. (and the V. dahliae Kleb. reported from Brazil); vascular attack causing wilt, xylem element wilt.   Scattered but in many countries.

Viruses: Abutilon Mosaic Virus in Brazil.   Cotton Anthocyanosis in Brazil.   Cotton Leaf Curl and Crumple in Puerto Rico, Guatemala, El Salvador.   The rare Cotton Veinal Mosaic in Brazil.   Cotton Leaf Mottle, scattered.   Cotton Yellow Dwarf from Sida rhombifolia with severe yellow mottling in Venezuela, El Salvador.   Euphorbia Mosaic, which is a severe mottle in Brazil.   Sunflower Virus (Argentine), in Argentina.   Tobacco Streak (Brazilian) in Brazil.

Xanthomonas malvacearum (E. F. Sm.) Dows. ; angular leafspot, blackarm, stem canker, capsule spot.   (Believed one of the two most universally encountered cotton diseases.) Neotropical.

Xiphinema americanum Cobb; nematode in rhizosphere.   Peru.

GUILIELMA GASIPES HBK. (sometimes given as BACTRIS UTILIS L.).   A tropical American warm-country palm, slender trunked with and without thorns, bearing yearly several heavy clusters of nutritious fruit; its wood is dense, used for staves, arrows and lances, and for needles.   (The formidable trunk thorns seen so much have resulted in neglect of disease and horticulture study of this good food plant but thornless varieties are being selected); "pejibaya, " "pejibeye" Sp. Po. , "pixbae, " "pupunha" Indian, "peach palm" En.

Auerswaldia palmicola Speg. ; A. guilielmae P. Henn. ; scab-spots on leaves.   Widespread.

Botryodiplodia sp. (Diplodia); leaf disease following harvest injury.   Honduras.

Budrot. Cause undetermined. Honduras, Costa Rica, Ecuador.

Capnodium sp. ; sootymold. General.

Cephaleuros sp. (not virescens); small green algal leafspots. Costa Rica.

Chaetomium sp. ; decay of leaf lamina in moist chamber. Puerto Rico.

Colletotrichum gloeosporioides Penz. = Glomerella cingulata (Stonem.) Sp. & Schr. ; anthracnose of leaf, peduncle and fruit. General.

Discomycete (unidentified) on rotting fallen fruit peduncle. Honduras.

Leafscab, undetermined, large and gray color. C. America.

Lembosia sp. ; black leaf-blemish. Costa Rica.

Meliola sp. ; black mildew spotting. C. America.

Pestalotia sp. ; leaf lamina spot, sunken midrib lesion. Honduras.

Phaeochora guilielmae P. Henn. ; dark swollen leaf spots. S. America.

Phyllosticta sp. ; shiny leaf spot. C. America, W. Indies.

Root rot (unidentified and not common) seen in Honduras and Costa Rica.

Strigula complanata Fée; white lichen. General.

Xylaria sp. ; disease of young leaf midrib. Ecuador.

HELIANTHUS ANNUS L. A coarse almost wild annual crop originally from ancient Mexico, a large-seeded selection whose grains are processed and eaten like nut meats, the attractive edible oil is extracted for cooking and commerce, and the grain is used in livestock food; "girasol, " "mirasol" Sp., "girassol" Po., "sunflower" En. (The pathology of this crop has been given only minor attention. Small amounts of H. TUBEROSUS L. are grown, and are probably susceptible to most if not all diseases listed below.)

Aecidium spp. ; rusts (imperfect stages) reported from various countries. (See also Coleosporium and Puccinia.)

Agrobacterium tumefaciens (Riker et al.) Conn; crowngall. A report from Brazil.

Albugo tragopogonis Pers. ; white-rust. Brazil, Argentina, Peru, Uruguay.

Bacterial stem and leaf disease, unidentified but found severe in small plots in Costa Rica.

Botrytis cinerea Pers. ; flower-head gray mold. Rainy season disease. Many countries.

Cassytha filiformis L. ; green dodder. Florida.

Cephalothecium (see Trichothecium)

Cercospora helianthi Ell. & Ev. ; common leafspot. Neotropics. C. helianthicola Chupp & Viégas; red-brown leafspot. Brazil and probably nearby countries.

Coleosporium helianthi (Schw.) Arth. ; one of the rusts. Subtropical USA, El Salvador and probably in many areas.

Cuscuta spp. ; yellow dodders. Widespread.

Erysiphe cichoracearum DC; asco-state of Oidium, see below.
Fusarium equiseti (Cda.) Sacc.; root and collar disease.
　　Uruguay, Chile, Peru. F. oxysporum Schl.; a non-sys-
　　temic rootrot. Uruguay, Argentina. F. semitectum Berk.
　　& Rav.; occasional stem canker. S. America, Mexico.
　　F. solani (Mart.) Sacc.; common stem canker. Argentina,
　　W. Indies, Guatemala, Uruguay, Nicaragua.
Lodging and weak neck, in part lack of nutrition, in part from
　　infection by semi-parasitic organisms. Widespread.
Macrophomina phaseoli (Taub.) Ashby; ashy stem disease.
　　Nicaragua, Guatemala.
Meloidogyne sp.; rootknot. General.
Oidium ambrosiae Th. (the conidial condition of Erysiphe);
　　usually found as white powdery mildew. Neotropics.
Penicillium spp.; seed molds. General.
Phyllachora ambrosiae (Berk. & Curt.) Sacc.; shiny-black
　　leaf spot. Brazil.
Phymatotrichum omnivorum (Shear) Dug.; dryland wilt. Texas,
　　Mexico.
Plasmopara halstedii (Farl.) Berk. & deT.; downy mildew
　　(cool area moist season disease). Chile, Texas.
Pseudomonas solanacearum E. F. Sm.; bacterial wilt. Puerto
　　Rico, Panama.
Puccinia helianthi Schw.; common rust. Mexico to Argentina.
Pythium debaryanum Hesse and P. ultimum Trow; damping-
　　off. Scattered.
Ramularia helianthi Ell. & Ev.; leaf disease. Subtropical
　　USA (dry parts).
Rhabdospora helianthicola (Cke. & Harkn.) Sacc.; a stem
　　disease. California.
Rhizoctonia solani Kuehn = Thanatephorous cucumeris (Frank)
　　Donk; damping-off. Scattered in General.
Rosellinia bunodes (Berk. & Br.) Sacc.; rootrot only on H.
　　tuberosus in S. America.
Sclerotinia minor Jagger; crownrot. Chile. S. sclerotiorum
　　(Lib.) dBy. = S. libertiana Fckl. = Whetzeliana sclero-
　　tiorum (Lib.) Korf & Dumont; crownrot, stalk rot. Uru-
　　guay, Chile, Peru, Brazil, Argentina.
Sclerotium rolfsii (Curzi) Sacc. (no basidial state reported);
　　ground-line rot of young plants. Peru.
Sphaeria dematinna Pers.; an anthracnose. S. America.
Trichothecium roseum Lk. = Cephalothecium roseum Cda.;
　　spores of hyperparasite. Chile.
Tylenchorhynchus sp.; nematode found in rhizosphere. Chile.
Viruses: Sunflower Virus causing mottle, dwarfing, sterile
　　seedhead in Argentina. Mycoplasma infection listed as
　　Aster Yellows that causes severe chlorosis, dwarfing, mal-
　　formation and sterility in California. This host is suscepti-
　　ble to many viruses on inoculation in the laboratory.

HELICONIA BIHAI L. The 30-40 species of this plant are tropical
　American, have exotic blooms carried on tough slender stems

along which are broad handsome leaves that are harvested and plaited into mats, made into beds, walls of houses and sides of rooms, and used as thatch; a salable wax is patiently scraped from leaf undersides and this is used to waterproof, preserve leather, burnish metal jewelry, tools, weapons, pottery and woodcarvings, and it serves as an anti-rust treatment in wet-jungle communities; "bihai" and many other names used by Indians.

Algae (undetermined, green and plentiful in several perpetually moist situations); superficial leaf and stem discoloration. Puerto Rico, Costa Rica, Panama, Ecuador, Brazil.

Cephaleuros sp.; small algal leafspots. Ecuador.

Colletotrichum gloeosporioides Penz. conidial state of Glomerella cingulata (Stonem.) Sp. & Schr. (both present); infections along main veins. Puerto Rico, El Salvador. Probably widespread.

Fusarium oxysporum Schl. f. sp. cubense (E. F. Sm.) Sny. & Hans. on several wild Heliconias; systemic wilt. C. America, Colombia. F. solani (Mart.) Sacc.; isolated from diseased root. Puerto Rico.

Glomerella (see Colletotrichum)

Guignardia heliconiae Gonz. Frag. & Cif.; leaf blotch. W. Indies, Costa Rica.

Helminthosporium torulosum (Syd.) Ashby; leafspot, petiole lesion. C. America, S. America.

Meliola spp., all produce black mildew on leaves: M. heliconiae Stev. in Panama, El Salvador, Puerto Rico, M. macrospora Baker & Dale in W. Indies, M. musae (Kunze) Mont. in Panama. These are probably much more widely distributed.

Metaspheria heliconiae Gonz. Frag. & Cif.; a discrete leaf-spot. W. Indies, Panama, C. America, Puerto Rico.

(Mycena citricolor: Heliconia spp. appear immune from infection, and these plants grown in rows act as means to halt its spread between experimental coffee areas.)

Pratylenchus musicola Filip.; nematode in root zone. Costa Rica.

Pseudomonas solanacearum E. F. Sm.; bacterial wilt. C. America, Panama.

Puccinia heliconiae (Diet.) Arth.; most common rust. General. (It should be noted that the hyperparasite Eudarluca australis Speg. attacks this Puccinia.)

Pyrenobotrys heliconiae (P. Henn.) Th. & Syd.; leafspots (stromatic with prominent perithecia). Reported in Brazil, probably widespread.

Sooty molds, perhaps several, in many locations.

Strigula complanata Fée; milky white lichen-spot. Neotropics.

? Virus, mottling and leaf malformation, probably Commelina strain of Cucumber Mosaic virus.

HEVEA BRASILIENSIS Muell.-Arg. A large South American tree, in the wilds 35-40 meters tall but maintained in plantations

18-20 meters, grown in thousands of hectares in at least 12
Neotropical countries and in recent years given increased attention in Latin America.   Its unique harvest is the milky latex
dripping into cups fastened at the end of shallow slanting cuts
which are renewed about every two days downward on the tree
trunk and on healing form a band of scar-tissue bark called the
tapping panel which is susceptible to special diseases.   (From
the Orient and Occident are reported a total of over 350 fungus
parasites, about 15 loranthaceae, at least four parasitic algae,
and several nematodes.   The list given below, of only about 115
diseases, is strictly from the Western Hemisphere and excludes
some semi-parasites and saprophytes and gives no latex or stored
rubber troubles.)   Common names: "hule, " "jebe" Sp. , "serin-
gueira" Po. , "rubber" En.

Alternaria alternata (Fr.) Keiss. = A. tenuis Nees; rare
   border disease of stunted leaves on young trees.   Costa
   Rica.   Alternaria sp. (large spored); leafspot.   Guatemala,
   Mexico.
Apophaeria (see Melanopsammopsis)
Aschersonia spp. ; parasites on rubber insects.   Scattered.
Ascocyta heveae Petch; leaf-edge disease.   Costa Rica,
   Brazil, Peru.
Aspergillus spp. ; on seed collected for planting.   General.
Botryodiplodia elasticae Petch; leafspot.   W. Indies, Amazon
   Valley of S. America.   B. theobromae Pat. which is the
   conidial and parasitic form of the perfect state fungus
   Botryosphaeria quercuum (Schw.) Sacc. ; budding failure in
   nursery, fruit spot and rot, fruit mummy, moldy panel,
   dieback following sunscald, seedling top disease in poorly
   cared-for seedbed.   Neotropics.   Botryodiplodia sp. (undescribed but apparently a different form if not a different
   species); an especially severe tapping panel disease, rot
   following injury of seedlings or budded stock in nurseries.
   Honduras, Costa Rica, Panama, Peru, and probably other
   countries.   (Note that these parasites are also reported
   as Diplodia.)
Brown bast of panel from trunk wounds and excessive tapping.
   Scattered.
Capnodium brasiliense Putt. and C. lanosum Cke. ; sootymolds.
   Probably general, reports from W. Indies, C. America, S.
   America (Amazon valley).
Catacauma huberi (P. Henn.) Th. & Syd. = Phyllachora huberi
   P. Henn. ; black crust and tarspot on leaf.   C. America,
   S. America (Amazon valley).
Cephaleuros virescens Kunz; algal leafspot.   Probably in any
   country that grows this rubber.
Ceratocystis fimbriata Elliott & Halst. (probably at least a
   variety should be recognized and could be designated,
   heveae); sapstain, panel-rot disease.   C. America, Mexico,
   Brazil.   (Note: Studied by J. B. Carpenter in Costa Rica
   (Plant Disease Reporter 38: 334-7. 1954), this fungus on
   rubber had specially large perithecia with pluriannulate necks.)

Cercospora heveae Vinc.; leafspot disease.   C. America, W. Indies, S. America.

Chlorosis (various causes among them dry condition, shallow soil, minor element deficiencies) in C. America, probably in many countries but of scattered occurrence.

Choanephora cucurbitarum (Berk. & Rav.) Thaxt.; flower disease.   Scattered, troublesome to rubber breeders.

Cladosporium herbarum (Pers.) Lk. (there may be a specially virulent race); in fruitrot complex, leaf edge disease, young plant wilt.   S. America.

Coccospora aurantiaca Wallr. both conidia and sclerotia; on weak seedlings.   S. America.

Colletotrichum gloeosporioides Penz. the conidial state of Glomerella cingulata (Stonem.) Sp. & Sch.; common anthracnose, fruit spot and rot, leafspot (in Brazil one type causes shot hole, in Costa Rica another type differing from others infects common anthracnose leafspots causing abnormal spot spread); twig disease.   Neotropical.

Conyothyrium sp.; a gray leafspot.   S. America (Amazon valley).

Corticium salmonicolor Berk. & Br. = conidial state is Necator decretus Mass.; branch disease.   C. America, S. America.

Cryptosporium sp.; swollen twig.   S. America.

Crystallocystidium (see Stereum)

Curvularia sp.; dieback after sunscorch.   C. America.

Dendrographium sp.; mold of panel.   S. America.

Diaporthe heveae Petch; branch canker.   Honduras.

Didymella sp.; leafspot.   Mexico, Costa Rica, Guatemala.

Dimerosporium heveae Charles; dark leafmold.   S. America.

Diplodia (see Botryodiplodia)

Dothidella (see Melanopsammopsis)

Elsinoë heveae Bitanc. & Jenk.; spot anthracnose.   Brazil.

Eutypa ludibunda Sacc.; panel disease.   S. America (Amazon valley).

Fomes spp.:   F. hornodermus Mart., a rootrot in S. America.   F. inflexibilis (Berk.) Cke. a rootrot in S. America.   F. lamaoensis Sacc. & Trott. = F. noxius Corner, a brown rootrot, widespread.   F. lignosus (Kl.) Bres. = Polyporus lignosus Kl., the white rootrot, stump, and log decay, occasional in S. America and C. America.   F. marmoratus (Berk. & Curt.) Cke. a rootrot in S. America.   F. semitostus Berk., a rootrot in S. America.

Fusarium roseum Lk.; panel mold.   Mexico, Costa Rica, Guatemala.   F. solani (Mart.) Sacc.; panel rot.   Mexico.

Fusicladium (see Melanopsammopsis)

Ganoderma amazonense Weir and G. australis Weir, both cause rootrot in Amazon valley of S. America.   G. lucidum Karst.; rootrot.   S. America.   G. pseudoferreum Ov. & Steinm.; red rootrot.   Costa Rica, Brazil.

Gloeosporium alborubrum Petch; dieback, blight of flushleaf, shot hole leafspot.   W. Indies, S. America (Amazon valley),

Costa Rica. G. heveae Petch (apparently much different
from the above); gray leafspot, rimspot of tapping groove.
S. America, Costa Rica. (Note: Both G. alborubrum and
G. heveae by some have been made synonymous with
Colletotrichum gloeosporioides, but at least in Costa Rica
they seemed to me to be different from each other and
deserved separation from C. gloeosporioides.)

Guignardia heveae Syd. = G. heveae Gonz. Frag. & Cif. =
    G. hevicola Cif. ; leaf disease. Dominican Republic, W.
    Indies.

Helicobasidium compactum Boed. ; a felt sheath on root over
    sap sucking coccids, when abundant may cause collar canker
    of seedling. Guatemala, Brazil, Mexico, Costa Rica.

Helminthosporium heveae Petch; birds-eye spot. Florida,
    Mexico, C. America, Brazil.

Hirsutella verticillioides Charles and Hypocrella sp. ; these
    confuse as they are not real plant disease fungi but are
    parasites of insects on rubber stems. Scattered reports.

Hymenochaete noxia Berk. (very similar to Stereum); woodrot.
    S. America.

Hypoxylon sp. ; panel disease. S. America (Amazon valley).

Kretzschmaria coenopus (Fr.) Sacc. ; rot of fallen branches.
    S. America, W. Indies, Costa Rica.

Leptosphaeria sp. ; a leafspot. S. America.

Loranthaceae (numerous unidentified genera and species); all
    are mistletoe-like parasites. Peru, Brazil. Many on wild
    trees in Amazon valley.

Marasmius equirinus Muell. ; shiny black thread, horsehair-
    blight of foliage. W. Indies, Costa Rica, Guatemala, S.
    America. M. pulcher Petch; white colored thread horse-
    hair-blight of foliage. C. America, S. America (Amazon
    valley). M. sarmentosus Berk. ; brown colored cord, foliage
    threadblight. S. America.

Megalonectria pseudotrichia (Schw.) Speg. ; branch canker. S.
    America.

Melanopsammopsis ulei (P. Henn.) Stahel, may be a very ac-
    ceptable binomial and is based on the asco-state (synonyms
    are: Aposphaeria ulei P. Henn. is name of pycnidial state,
    Dothidella ulei P. Henn., a name given to the perithecial
    state was for a long time the most acceptable name while
    Fusicladium macrosporium Kuy is a name given to the
    conidial state, now Microcyclus ulei (P. Henn.) Muller &
    Arx is the most recent and possibly best name for this para-
    site, Passalora heveae Mass. is another name of the conidial
    state, and Scolicotrichum heveae Vinc. is another name of
    conidial state); very serious rubber leafspot, its classic com-
    mon name in USA, English, and European usage is "South
    American Leaf Disease. " Neotropical, where not too dry.

Meliola amphitrichia Fr. ; black mildew. W. Indies, C.
    America, subtropical USA. M. hevea Vinc. ; blackmildew.
    S. America.

Monascus ruber v. Tiegh. ; on ripe fruit. C. America.

Mycena citricolor (Berk. & Curt.) Sacc. = Omphalia flavida
Maubl. & Rangel; gemma producing leafspot. Peru, in
Costa Rica by natural infection and by inoculation.
Nectria diversispora Petch; branch and stem canker. Brazil.
Nematodes (unidentified) feeding on parasitic fungi such as
Rhizoctonia near host roots. Costa Rica.
Nidularia australis Tul. ; developed on dead seedling. Costa
Rica.
Nummularia anthracnodes (Fr.) Cke. and N. cincta Ferd. &
Win. ; wound diseases. C. America, Haiti, Dominican
Republic, S. America (Amazon valley).
Oidium heveae Stein. ; white mildew. Brazil. (Seen in São
Paulo state in 1958, by 1959 had been well eradicated.)
Omphalia flavida (see Mycena)
Ophiobolus heveae P. Henn. ; on leaf, gray-disease. Subtropical
USA, W. Indies, S. America (Amazon valley).
Parodiella melioloides Wint. ; black mold on leaves. C.
America, S. America.
Parodiopsis perae Arn. ; leaf patch. Widespread.
Pellicularia filamentosa (Pat.) Rog. (aerial type "Rhizoctonia");
target leafspot. Bolivia, Brazil, Costa Rica, Honduras,
Peru, Ecuador, Guatemala. P. koleroga Cke. ; thread web-
blight, leafrot. Scattered.
Penicillium spp. ; fruit rot on seeds and reduced germination.
General.
Peniophora (see Stereum)
Periconia heveae J. A. Stev. & Imle = P. manihoticola Viégas;
a seedling stem-disease and leafblight. Mexico, C. America,
Brazil.
Pestalotia palmarum Cke. ; seedling collar-disease, also sec-
ondary in various leafspots. Neotropics.
Phoma sp. ; branch and twig canker. Brazil.
Phomopsis heveae Boed. ; seedling disease, twig dieback, re-
ported in Brazil as a trunk rot. Florida, W. Indies,
Brazil.
Phoradendron piperoides Trel. ; mistletoe-like parasite. Suri-
nam (see note above about Loranthaceae).
Phycopeltis sp. ; on leaf twig. Costa Rica.
Phyllachora (see Catacauma)
Phyllosticta heveae Zimm. ; reddish-brown leafspot. S. Ameri-
ca, C. America, Florida, Trinidad, Mexico.
Phytopthora palmivora Butl. ; abnormal leaf fall, branch canker,
fruit and twig rot, panel black-stripe rot. Neotropics. A
few Phytophthora spp. (not the above); seedling diseases,
budding problems, panel disease. Scattered.
Polyporus zonalis Berk. and other unidentified species; root
wound-rot, pocket rot of trunk. S. America (Amazon
valley).
Porella sp. ; leafy liverwort causes seedling smothering. In
wet areas of Costa Rica and Nicaragua.
Poria spp. all are wood rots: P. albocincta Cke. & Mass., P.
borbonica Pat. , P. floridana Sacc. & Trott. in W. Indies,

P. graphica Bres. in C. America, W. Indies, P. hypo-
brunnea Petch causes red rootrot in W. Indies, C. America,
S. America, and P. vincta Cke. causes red-strand-rot in
S. America (Amazon valley).
Pythium sp.; patch canker on low part of panel. Guatemala.
Rhizoctonia solani Kuehn (no basidial state seen or reported);
nursery seedling diseases. C. America.
Rosellinia bunodes (Berk. & Br.) Sacc.; black rootrot. Scat-
tered in many countries, R. pepo Pat.; white rootrot. Less
common but seen in Costa Rica, probably occurrence is
scattered. R. puiggarii Speg.; rootrot. S. America.
Schizophyllum umbrinum Berk.; on old branches a bark disease.
Haiti, Dominican Republic.
Scolecotrichum (see Melanopsammopsis)
Seedling unknown disease, individual plants die, cause often
may be partly minor element deficiency, sometimes weakly
semi-parasitic microorganisms are involved such as bac-
teria, fungi, nematodes. C. America.
Septobasidium spp. all are symbiotic with scale insects and
all reported from S. America: S. fumigatum Burt, S.
heveae Couch, S. rhabarbarinum (Mont.) Bres., S. strato-
sum Couch.
Septoria sp.; spots of living leaves. Colombia.
Sphaerella heveae Petch; seen as leafedge infection. Costa
Rica.
Sphaeropsis heveae Gonz. Frag. & Cif.; branch dieback.
Dominican Republic.
Sphaerostilbe repens Berk. & Br.; stinking rot, red rhizo-
morph. W. Indies, Guatemala, S. America (Amazon
valley).
Stereum papyrinum Mont. = Peniophora papyrina Mont.; wood-
rot. S. America. S. umbrinum Fr. = Crystallocystidium
portentosum (Berk. & Curt.) Rick; woodrot. S. America.
Stilbum cinnabarinum Mont.; panel canker. S. America
(Amazon valley).
Strigula complanata Fée; white lichen leafspot. General.
Sunscald, leads to various infections from weak parasites,
often follows after weeks of overcast weather, on nursery
plantings and on older trees. Costa Rica, Guatemala
(probably elsewhere).
Thyridaria tarda Banc.; on stems. S. America.
Torrubiella rubra Pat. & Lagerh.; has caused confusion, an
insect parasite (see Hirsutella note). Dominican Republic,
Haiti.
Trametes spp. all woodrots of widespread distribution: T.
corrugata (Pers.) Bres., T. cubensis (Mont.) Sacc., and
T. hydnoides (Sw.) Fr., T. serpens Fr.
Ustulina vulgaris Tul. and U. zonata Sacc.; both cause root
disease and panel infection. In Costa Rica in addition
these cause charcoal tree-butt rots. C. America, S.
America.
Valsa sp.; in dieback complex. Costa Rica.

"Wintering" (abnormal leaf fall early in season), physiological
upset from weather, poor nutrition, so-called "wet feet,"
and some budded stocks are more susceptible than others.
Often results in weak parasites like Cladosporium attacking
leaves that are slow to drop. Scattered occurrence.

Xylaria hystrix Berk. and X. thwaitesii Cke.; dieback, seedling basal canker, panel rot, root disease. Neotropical
(the first named reported mostly from Brazil).

Zygosporium paraensis Vinc.; leaf shot hole and leafspot. W.
Indies, Brazil, Peru.

HIBISCUS ABELMOSCHUS see HIBISCUS ESCULENTUS

HIBISCUS CANNABINUS L. An annual crop, a fairly recent fiber
producing introduction from Eastern tropics; "cañamo" Sp.,
"kenaf," "roselle hemp" En. H. SABDARIFFA L. known first
in the Old World, now used in the New, grown for its red swollen
and edible fruit-like flower calyx; "roselle," "jamaica sorrel,"
"tropical cranberry" En. (Unless otherwise noted diseases listed
are reported from both hosts.)

Alternaria sp.; small leafspot, black stemlesion. Guatemala,
Subtropical USA, Mexico. (Rather rare.)

Botryodiplodia theobromae Pat. = Botryosphaeria quercuum
(Schw.) Sacc. (on tropical cranberry); canker and dieback.
S. America.

Botryosphaeria hibisci Sacc. and B. dothidea (Moug.) Ces. &
deNot. canker, fiber-stem rot. Subtropical USA.

Botrytis cinerea Pers.; leaf, petal and calyx moldy disease.
Peru, Subtropical USA, Guatemala.

Cephaleuros virescens Kunze; scurf. W. Indies, Florida.

Cercospora abelmoschi Ell. & Ev. (on jamaica sorrel); calyx
spot and leafspot. S. America. C. hibisci Tracy &
Earle; leafspots. Subtropical USA, W. Indies, C. America.
C. hibiscina Ell. & Ev. (apparently quite different from
other Cercosporas and occurs on kenaf); diffuse infection on
underside of leaf, severe defoliation. Brazil, Mexico,
Puerto Rico. C. malayensis Stev. & Solh. on jamaica
sorrel; leafspot. S. America.

Chlorosis from various causes (not viruses) mostly are from
minor element deficiencies, shallow soils, etc. on kenaf,
especial stunting, bud kill, leaf yellowing and bronze.
Cuba.

Choanephora infundibulifera (Curr.) Sacc.; flower blight. Subtropical USA, in wet season on flowers in El Salvador.

Clitocybe tabescens Bres. = Armilariella tabescens (Scop.)
Sing.; rootrot. Florida.

Colletotrichum gloeosporioides Penz. (= C. hibisci Polac.);
the common anthracnosis, stem canker. Widespread.

Erysiphe sp. (perfect state of Oidium) on kenaf; powdery
mildew. Guatemala highlands.

Fusarium culmorum (W. G. Sm.) Sacc. on kenaf; stem infection disease. Argentina. F. moniliforme Sheld. (asco-stage

not reported) on tropical cranberry, seedpod disease. Argentina, Trinidad, W. Indies. F. solani (Mart.) Sacc. (with Nectria state found on diseased plant bases in Nicaragua); plant collapse, root disease, cortex rot in kenaf. El Salvador, Nicaragua, Trinidad, W. Indies (lowland plantings).

Helminthosporium sp. on kenaf; a large leafspot. Peru.

Irenopsis coronata Speg. on tropical cranberry; black mildew. W. Indies.

Kuehneola malvicola (Speg.) Arth.; rust. Subtropical USA, W. Indies.

Leveillula taurica Arn.; mildew. Cuba, Florida, C. America.

Lightning strike; circular area of plants killed in field, some secondary fungus infections of plants at edge of blasted area where plants were injured although not killed but made susceptible to certain weak parasites such as Phytopthora sp., Pythium sp. seen in Cuba, Nicaragua.

Meloidogyne spp. (both M. incognita Kof. & Wh. and M. exigua Goeldi given as species). First in W. Indies and C. America, latter in Brazil.

Microsphaera euphorbiae Pk.; powdery mildew. Subtropical USA, Puerto Rico, S. America.

Oidium sp.; common white powdery mildew, and see Erysiphe or possibly Microsphaera as perithecial stages. Neotropics and somewhat seasonal.

Pellicularia filamentosa (Pat.) Rog. (not typical of the common soil inhabiting mycelial Rhizoctonia as this special Pellicularia bears abundant basidia spores and is an aerial form); large zonal leafspot. Cuba, Puerto Rico, El Salvador, Guatemala, Bolivia. (Note: This organism appears identical in morphology and field existence with the target leafspot organism on Hevea rubber, but cross inoculations have not been attempted; there is as yet not any understood morphological character that objectively describes the aerial character of the P. filamentosa here noted.)

Phyllosticta sp. on kenaf; leafspot. Peru.

Phymatotrichum omnivorum (Shear) Dug. on tropical cranberry; rootrot. Texas, New Mexico.

Physalospora fusca Stev. and P. obtusa (Schw.) Cke. = Botryosphaeria quercuum (Schw.) Sacc. (both perfect stages of Diplodia or Botryodiplodia spp.); found as branch disease and stem canker. Florida.

Phytophthora parasitica Dast.; collar and rootrot. Cuba, Texas, Puerto Rico, Florida, El Salvador.

Pseudomonas solanacearum E. F. Sm. on tropical cranberry; bacterial wilt. Subtropical USA, W. Indies.

Pythium sp. (both typical mycelial invasion and gemma-like bodies found); root disease. Cuba, El Salvador.

Rhizoctonia solani Kuehn = Thanatephorous cucumeris (Frank) Donk (this is terrestrial in habit); damping-off. Widespread.

Rosellinia bunodes (Berk. & Br.) Sacc., the black rootrot and R. pepo Pat., the white rootrot in El Salvador from field inoculation studies on the two hosts.

Sclerotinia sclerotiorum (Lib.) dBy. = Whetzeliana sclerotiorum (Lib.) Korf & Dumont, on tropical cranberry; a
  stem disease and a calyx rot in the market.   El Salvador.
  On both hosts: S. rolfsii (Curzi) Sacc. ; seedling collapse,
  collar rot.   Cuba, Florida.
Sclerotium bataticola Taub. = Macrophomina phaseoli (Taub.) Ashby
  on tropical cranberry; ashy stemblight.   Cuba, El Salvador.
Trichoderma lignorum Harz. on tropical cranberry; a severe
  rootrot. Cuba.
Viruses: Brazilian Tobacco Streak in Brazil.   Cotton Anthocyanosis in Brazil is from the weed Sida, a brilliantly
  chlorotic leaf spotting and at times severe stunt, may be
  widespread but seen specially in El Salvador, Florida,
  Cuba, Puerto Rico, Venezuela, Virgin Islands, Brazil, and
  Guatemala.   A distinct virus mottle with mild leaf malformation, typical mild mosaic at times.   W. Indies, C.
  America, probably in some parts of S. America.

HIBISCUS ESCULENTUS L. (= ABELMOSCHUS ESCULENTUS
  Moench [note: I regret including the synonym here and have
  done so solely because occasional references use the, to me,
  incorrect generic name Abelmoschus, which confuses the hosts
  "gumbo" and "gambrette"; the species name moschatus is of distinct meaning and designates the character of the fragrant seeds
  of "gambrette" which is the H. abelmoschus].   From the Old
  World tropics, an annual plant whose pods are a valued vegetable; "chimbombo, " "quimbombó" Sp. , "quiabo, " "quiabeiro"
  Po. , "gumbo, " "okra" En. , and H. ABELMOSCHUS L. (=
  ABELMOSCHUS MOSCHATUS Moench); also from the Old World
  tropics, produces highly valued seed with specially musky odor,
  in the Western tropics, grown mostly in the island of Martinique;
  "abelmosco" Po. Sp. , "musk mallow" En. , "ambrette" French.
  Diseases on one host seem to occur on the other, except in a
  few instances which are indicated in the list below.
  Acrostalagmus albus Pruess on gumbo; confusing fungus not
    disease of plant but a parasite of sucking insect.   Puerto
    Rico.
  Alternaria alternata Keiss. = A. tenuis Nees; injury reported
    by others but not seen by me, of old leaf and black streak
    of pod on both hosts.   W. Indies, including plants on island
    of St. Thomas.
  Ascochyta abelmoschi Harter on gumbo; rather common leaf-
    spot. Neotropical.   A. phaseolorum Sacc. = Phoma exigua
    Desm. , leafspot.   Brazil.   (These two Ascohytas may well
  . be the same fungus.)
  Aspergillus spp. ; flower injury and pod injury on gumbo in C.
    America, probably widespread.
  Botryodiplodia theobromae Pat. = asco-state is Botryosphaeria
    quercuum (Schw.) Sacc. ; podrot, stem canker.   Neotropics.
  Botrytis cinerea Pers. ; stem and pod mold.   Neotropics, re-
    ported only from gumbo, the B. hortensis Ell. & Ev. , is a
    somewhat rare spreading leafspot.   Jamaica (on gumbo but
    may be on gambrette).

Cercospora abelmoschi Ell. & Ev. = C. brachypoda Speg.
  and C. hibisci Tracy & Earle (apparently synonyms) on
  gumbo, cause typical leafspot. Virgin Islands, in all W.
  Indies islands, Trinidad, Brazil, Santo Domingo, Haiti.
  C. hibiscina Ell. & Ev. reported on gumbo, believed also
  on gambrette but no proof; attacks lower leaf surface, effuse
  growth. Brazil, Mexico, Puerto Rico, Venezuela, Haiti,
  Virgin Islands. C. malayensis Stev. & Solh. on both hosts;
  leafspot. W. Indies, Jamaica, Venezuela, Colombia, Subtropical USA.
Choanephora conjuncta Couch on gumbo; decay of fallen flowers.
  Subtropical USA, C. America. C. cucurbitarum (Berk. &
  Rav.) Thaxt. also on gumbo, on flower petals sometimes
  before falling off plant. Puerto Rico, Florida.
Cladosporium herbarum Lk. on gumbo; leaf edge infections,
  and on moldy pods in markets of C. America, probably
  widespread.
Colletotrichum gloeosporioides Penz. the conidial stage of
  Glomerella cingulata (Stonem.) Sp. & Sch.; anthracnose.
  (some workers give the conidial form the name C. hibisci
  Póll.). In the West Indies anthracnose isolations, from
  the weed, Sida sp., and from gumbo diseased branch tissues, were most severe in causing gumbo anthracnose, but
  the coffee Colletotrichum was mild to nonpathogenic on Sida.
  It is notable that Ciferri considered it a physiologically different strain, and the race causing branch, twig and bud-tip
  anthracnose of both the Sida and the H. esculentus could
  well be given and I suggest the designation C. gloeosporioides
  Penz. f. sp. sidae. Anthracnose is Neotropical in occurrence.
Diaporthe phaseolorum Sacc. f. sp. sojae (Lehm.) Wehm., on
  gumbo; pod disease. Subtropical USA.
Erysiphe cichoracearum DC. on gumbo; mildew. Brazil, and
  in irrigated gardens on dry slopes in Guatemala and El
  Salvador.
Fusariums all from gumbo as follows: F. moniliforme Sheld.
  fruiting on seed in ripe pods in Trinidad. F. oxysporum
  Schl. footrot in Trinidad and Florida. F. semitectum Berk.
  & Rav. spores on surface of seed in Trinidad. F. vasin-
  fectum Atk. (= F. oxysporum Schl. f. sp. vasinfectum
  (Atk.) Sny. & Hans.); causing wilt in Brazil and El Salvador.
Fusoma hibisci Gonz. Frag. on gumbo; a stem disease.
  Dominican Republic.
Helminthosporium lanceolatum Poole; leaf disease on gumbo.
  Dominican Republic.
Meliola hibisci (Spreng.) Gonz. Frag. & Cif.; without doubt on
  gumbo, probably on both, the black mildew. Neotropical
  but most often in moist areas.
Meloidogyne sp. on both hosts; rootknot. Scattered.
Oidium abelmoschi Thuem.; mildew. Neotropical
Ophiobolus consimilis Ell. & Ev. on gumbo; from old dead
  stem. Subtropical USA.

Paratylenchus sp. on gumbo; nematode in rhizosphere (injurious?). Subtropical USA.

Phoma okra Cke. on gumbo; black stem lesion.   Subtropical USA.

Phyllosticta hibiscina Ell. & Ev. on gumbo and probably gambrette; leafspot. Subtropical USA, W. Indies, El Salvador, Brazil.

Podtip shrivel and blackening.   Cause probably physiological, found on gumbo in mountain garden in El Salvador.

Pseudomonas solanacearum E. F. Sm. (both species of host susceptible); bacterial wilt.   El Salvador, W. Indies.

Puccinia malvacearum Bert. ; rust of gumbo.   Scattered in S. America.

Pythium ultimum Trow; damping-off.   W. Indies.

Rhizoctonia solani Kuehn = Thanatephorous cucumeris (Frank) Donk; damping-off, sancocho.   W. Indies, C. America, Subtropical USA.

Sclerotium rolfsii (Curzi) Sacc. (basidial stage not seen or reported) on gumbo but gambrette is also said to be susceptible.   Subtropical USA, W. Indies, C. America.

Trichotecium subfulvum Ell. & Ev. on gumbo; a rare leafspot.   Jamaica.

Verticillium alboatrum Reinke & Berth. on gumbo; wilt.   Subtropical USA.

Viruses of gumbo (no virus reports available on gambrette): Abutilon mosaic virus in Brazil, Brazilian Tobacco Streak in Brazil, Euphorbia Mosaic in Brazil, Sida yellow variegation virus in W. Indies, and Tobacco Leaf Curl in Puerto Rico.

HIBISCUS ROSASINENSIS L.   Origin said to be China, popular in Western tropics; woody, sometimes tree-like bush with glamorous flowers that when cut long withstand wilting without water; "clavelón," "amapola" Sp., "hibisco" Po., "hibiscus," "china rose" En.   (Note: Plant pathologists occasionally encounter other arborescent plants of the genus, such as the flowering H. SYRIACUS L. and the fiber bearing H. TILIACEUS L.   These seem to have many of the same diseases as those that affect the common ornamental hibiscus.)

Agrobacterium tumefaciens (E. F. Sm. & Town.) Conn; crown-gall.   Rare occurrences in Subtropical USA, Panama, Cuba.

Algae (unidentified) excessive growth inhibited rooting of cuttings.   Costa Rica.

Alternaria macrospora Zimm. ; rare leaf disease.   El Salvador.

Armillaria mellea Vahl. = Armilariella mellea (Fr.) Karst. ; mushroom rootrot.   California, Panama.

Botryodiplodia spp. or Diplodia, conidial stages of Botryosphaeria.

Botryosphaeria hibisci (Schw.) Sacc. ; branch canker.   Subtropical USA.

Botrytis sp. ; flower bud-tip and flower-petal rot.   Puerto Rico.

Bud chlorosis disease, ending in bud death.   Probably physiological.   Cuba.

Capnodium spp. (e.g. C. brasiliense Putt., C. citri Berk. &
Desm. and C. coffeae Pat.); sootymolds in plantation areas
where special crops are grown.   The reports are indicative
of some confusion of pathologists as to the true mycological
discrepancies between the species of these fungi that are
subsisting on superficial excretions from plant lice and such
insects.   Occurrence of the black "fumagoes" can be listed
as Neotropical.

Capnodium hibisci Auct. a conidial form, is a very specific
kind of Capnodium organism (see Microxyphiella).

Cassytha filiformis L.; green dodder, love-vine.   W. Indies.

Cephaleuros virescens Kunze; algal leafspot, scurf.   Noted
especially in Brazil and W. Indies but seen scattered in
Ecuador, Colombia, and C. America.

Cercospora abelmoschi Ell. & Ev. = C. hibisci Tracy & Earle
and C. hibiscina Ell. & Ev. ; both cause leafspots and diffuse growths in Subtropical USA, W. Indies.   (In most
cases leafspots not numerous, but one spot can destroy a
leaf.) Probably present in S. America.

Choanephora conjuncta Couch; disease on flower petals.   Brazil,
W. Indies, C. America, Subtropical USA. C. infundibuli-
fera (Curr.) Sacc. is both leaf disease and flower blight.
Subtropical USA.

Clitocybe tabescens Bres. = Armilariella tabescens (Scop.)
Sing. ; mushroom rootrot.   Subtropical USA.

Colletotrichum gloeosporioides Penz. is the conidial form of
Glomerella cingulata (Stonem.) Sp. & Schr. (latter found on
retained dieback branches); anthracnose and dieback.   Neotropics.

Corticium salmonicolor Berk. & Br. ; branch pinkdisease.
Costa Rica.

Cuscuta americana L. ; common yellowdodder, spaghetti vine.
W. Indies, C. America.

Gemmiferous fungus (not Mycena), tawny to pink, oblong and
turgid infection bodies; leafspot.   In a rainforest area garden
next to jungle in Costa Rica.

Meliola triumfettae Stev. ; black mildew.   W. Indies.

Meloidogyne sp. ; rootknot.   General but scattered and not
severe enough to be of concern.

Microxyphiella hibiscifoliae Bat. & Nasc., the asco-form of
the conidial fungus Capnodium hibisci Auct. (tufts of black
hairs about base of mainvein, on underside of leaf); leaf
disease.   Puerto Rico, El Salvador, Cuba, Brazil.

Mycena citricolor (Berk. & Curt.) Sacc. ; luminescent leafspot.
Puerto Rico (natural infection).

Nectria cinnabarina Tode; branch canker, dieback.   Subtropical
USA, Brazil, C. America, W. Indies.

Pellicularia sp. (aerial type, not seen before); epiphytic on
leaf, stem, becomes parasite only on old leaf tissue.   Costa
Rica. P. koleroga Cke. ; thread-web blight.   Subtropical
USA, W. Indies, C. America.

Pestalotia sp.; part of dieback complex.   Costa Rica, Puerto
     Rico.
Phthirusa sp.; phanerogamic parasite.   El Salvador.   (P.
     theobromae Eichl. is reported on a tree-like hibiscus
     probably H. siriacus, in Surinam.)
Physalospora fusca Stev., (P. obtusa Cke., and P. rhodina
     Cke. = Botryosphaeria quercuum (Schw.) Sacc.) the perfect
     stage of Botryodiplodia spp.; on dieback branches.   C.
     America, W. Indies, Florida.
Pseudomonas solanacearum E. F. Sm.; bacterial wilt.
     Florida.   The very different P. syringae V. Hall; leafspot.
     Scattered in Subtropical USA.
Puccinia malvacearum Bert.; rust, defoliation.   El Salvador.
Pythium/Phytophthora; cutting rot and root failure of cuttings.
     Puerto Rico.
Rhizoctonia sp. (possibly same as the Pellicularia with aerial
     character that I studied in Costa Rica); foliage disease.
     El Salvador.   R. solani Kuehn = Thanatephorous cucumeris
     (Frank) Donk, distinctly terrestrial type; cutting disease,
     nursery bed collar rot.   Puerto Rico.
Rosellinia bunodes (Berk. & Br.) Sacc.; black rootrot, and R.
     pepo Pat.; white rootrot.   Both in C. America, probably
     elsewhere.
Sclerotinia sclerotiorum (Lib.) dBy. = Whetzeliana sclero-
     tiorum (Lib.) Korf & Dumont; flowerbud rot and drop.
     Costa Rica.
Sphaerostilbe repens Berk. & Br.; root disease, insects asso-
     ciated with injury.   W. Indies.
Strigula complanata Fée; lichen white-spot.   Neotropics.
Struthanthus sp.; woody vine-like parasite (acts like a mistle-
     toe).   El Salvador.
Trentepohlia sp.; brownish hair-like tufts of alga on old
     branches.   Costa Rica, Guatemala.
Trichoderma sp.; in dieback complex after pruning.   Costa
     Rica.
Tryblidiella sp. = Rhytidhysterium rufulum (Spreng.) Petr.;
     in dieback complex.   Puerto Rico, Costa Rica.
Xylaria sp.; root disease.   Cuba, Puerto Rico.

HIBISCUS SABDARIFFA L.   see   HIBISCUS CANNABINUS L.

HIBISCUS SYRIACUS   see   HIBISCUS ROSASINENSIS (note)

HIBISCUS TILIACEUS   see   HIBISCUS ROSASINENSIS (note)

HIEROCHLOE ANTARCTICA R. Br., H. REDOLENS Roem. & Schul.,
     and H. UTRICULATA Kunth.   Grasses grown in cool-dry ex-
     posures and harvested by Indian mat and basket weavers to them
     the much valued pliable, wear resistant, fragrant grasses;
     "hierba del cesta," "ratonera" Sp., "erva de cesta" Po., "basket
     grass" En. and many Indian names.   (The designation Southern
     Andes for occurrences includes findings in mountains of at least

Chile, Argentina, and Brazil, possibly also part of Paraguay.
Part of the disease list was for problems from a missionary.
It seems from fragmentary information that most diseases given
are on all the species named.)

Dictyothrium perpusillium Speg.; leaf disease.   S. America
    (Southern Andes).
Dilophospora chilensis Speg.; caulm rot.   S. America, es-
    pecially Chile.
Entosordaria fuegiana Speg.; on living plant.   S. America
    (Southern Andes).
Helotium herbarum Fr.; leafspot.   S. America, especially
    Chile.
Hendersonia hierochloae (Speg.) Sacc. & Trott.; small leaf-
    spot.   Chile.
Leptosphaeria fuegiana Speg.; leaf and caulm disease.   S.
    America (Southern Andes).
Lophodermium oxyascum Speg.; leaf collapse.   Chile.
Napicladium valdivianum Speg.; leaf scab-spot.   S. America.
Pleospora chilensis Speg.; leaf and stem spot.   S. America
    (Southern Andes).
Puccinia coronata Cda.; rust.   S. America (Chile).
Sclerotium clavus DC.; sclerotial rot.   S. America (Chile).
Sphaerulina fuegiana Speg.; leafspot.   S. America (Southern
    Andes).
Stagonospora utriculata Allesch.; leaf lamella collapse.   S.
    America especially Chile.
Volutella bombycina Speg.; stem rot.   S. America (Andes).

HIPPEASTRUM see AMARYLLIS

HIPPOCRATEA OBTUSIFOLIA Roxb. and H. OVATA Lam.   The
    water extracts from these are used by knowledgeable peasantry
    as insecticides against body and head lice on man.   Plants grown
    near peasant houses and protected in wild growth are tropical
    American, vine like and may be somewhat ornamental shrubs;
    "matapiojo," "bejuco prieto" Sp., "louse vine" En., "matapiolho"
    Po.
    Acrocladium andinum Petr.; on foliage.   S. America.
    Aecidium hippocrateae Diet.; rust.   Scattered.
    Agrobacterium tumefaciens (E. F. Sm. & Towns.) Conn;
        crowngall.   Florida.
    Asterina sp.; a black spot.   Widespread.
    Asterinella sp.; a leafspot.   Widespread.
    Botryorhiza hippocrateae Whet. & Olive; gall forming rust.
        Puerto Rico, Cuba, C. America, Trinidad.
    Meliola spp. cause black mildews.   M. amphitrichia Fr. in
        W. Indies and C. America.   M. guaranitica Speg. in S.
        America.   M. torulipes Cif. in Neotropics.
    Microthyriella uleana Syd.; leafspot.   S. America.
    Microthyrium hippocrateae (Ryan) Toro; black mildew and
        superficial spot.   Puerto Rico.
    Pestalotia (? funerea); leaf blotch and stem lesion.   Puerto
        Rico.

HOLCUS see SORGHUM

HORDEUM VULGARE L.   This grain crop brought from Europe to
   Western tropics is produced in limited cool and somewhat dry
   areas, not in warm, wet parts; it is called "cebada" Sp.,
   "cevada" Po., "barley" En.
   Alternaria (large spored); leaf disease in Brazil, on seed in
      Colombia.
   Aspergillus spp.; eight or ten species on stored grain.   Noted
      especially in Colombia, but they are widespread.
   Cladosporium gramineum Pers. and C. herbarum Lk.; both
      aid in severe seed moldyness, the second species also
      causes leaf injury in Brazil, both Neotropical in occur-
      rence.
   Claviceps purpurea (Fr.) Tul.; ergot.   Ecuador, Argentina,
      Chile, Colombia.   (Possible in any tropical barley crop
      and because of its poisonous effect should be watched for
      most carefully.)
   Cochliobolus carbonum Nelson; leafspot and rot.   Mexico.
   Colletotrichum graminicola G. W. Wils.; conidial state of
      the asco-state Glomerella tucumanensis (Speg.) Mull. (=
      Physalospora tucumanensis Speg.); anthracnose, seedrot.
      Subtropical USA, Guatemala, S. America.
   Epicoccum neglectum Desm. and E. nigrum Lk.; plant injury
      in a complex of associates.   Brazil, Colombia, and other
      countries in S. America.
   Erysiphe graminis DC.; powdery mildew.   Nearly wherever
      barley is grown.   E. graminis DC. f. sp. hordei Mar.;
      powdery mildew.   Guatemala, Ecuador, Peru, Argentina.
      (Conidial state is Oidium monilioides Lk.)
   Fusarium graminearum Schwabe is the conidial form of
      Gibberella saubinetii Sacc. = G. zeae Sacc.; head rot,
      scab (dangerous for use as animal food).   Scattered in S.
      America, C. America.   F. roseum Lk. and F. culmorum
      (W. G. Sm.) Sacc.; plant-base and stem rot.   Colombia,
      El Salvador, probably in several countries.
   Gibberella (see Fusarium graminearum)
   Glomerella (see Colletotrichum)
   Helminthosporium gramineum (Rab.) Eriks.; leaf stripe.
      Widespread, the common H. sativum Pam. Ki. & Bakke;
      blotch, scald.   S. America.   H. teres Sacc., its asco-
      state is Pyrenophora teres Drechs.; leaf blight.   S.
      America.
   Heterosporium avenae Oud.; leafmold.   Brazil.
   Hormodendrum sp.; moldy grain.   Colombia.
   Marssonina graminicola (Ell. & Ev.) Sacc. and M. secalis
      Oud.; both cause leafscald.   Argentina.
   Micrococcus tritici Prill.; grain shrivelling.   Argentina, Brazil.
   Mucor sp.; from musty grain.   Colombia.
   Oidium (see Erysiphe)
   Ophiobolus graminis Sacc. and O. cariceti Sacc.; footrots.
      Mexico, S. America (Andean).

Penicillium spp.; from stored grain. Colombia.
1. Puccinia hordei Otth = P. anamola Rostr.; probably most
   common leafrust. S. America, Mexico, C. America. 2.
   P. hordeicola Lindq.; rust (on Hordeum lechleri Sch.) in
   Argentina. 3. P. glumarum Eriks. & E. Henn.; glume and
   stemrust. General. 4. P. graminis (Pers.) f. sp. secalis
   Eriks. & E. Henn.; stemrust. Chile, Peru. 5. P. graminis
   pers. f. sp. tritici Eriks. & E. Henn.; leafrust. Argentina,
   Colombia. Found in Argentina on Hordeum jubatum L. var.
   pampeanum Haum. 6. P. montanensis Ell. & Ev.; rust
   (on Hordeum lechleri) in Argentina. 7. P. simplex Eriks.
   & E. Henn.; leafrust (on H. distichon L.) in Chile. 8. P.
   tornata Arth. & Holw.; leafrust (on both H. andinum Trin.
   and H. muticum Presl.) in Bolivia. (Note: The Puccinia
   rusts have more chance to mutate in Mexico and Central
   and South America than in the north temperate zone. I
   made no attempt to indicate for Latin America the rust race
   numbers, and for reasons of space a few Puccinia species
   and some subspecies of at present lesser concern are not
   included in this list of eight rusts. Another point: much
   work in progress in breeding cereals for the tropics has
   indicated that some of the small grains will be very much
   changed to adapt them for production in lower and warmer
   areas, in which case genetic studies on hosts and rusts
   will be more than ever important.)
Pyrenophora (see Helminthosporium teres)
Pythium spp.; damping-off, seedling blights. Especially in
   low lands, wet and warm soils. Noted especially in experimental planting in shallow soil in El Salvador. Reported
   from Guatemala, Colombia, Peru.
Ramularia hordei McAlp.; a brown leafspot. Brazil, Peru.
Rhynchosporium secalis (Oud.) Davis; scald, leafblight. Widespread in C. America occurs at cold altitudes and in Colombia and Argentina.
Sclerotium rolfsii (Curzi) Sacc. (the asco-state is Pellicularia
   rolfsii (Sacc.) West); plant collapse. Florida. Sclerotium sp.; causes plantrot. Colombia.
Septoria avenae Frank; leafspot, and S. nodorum Berk.; leafspot and glume disease. Both on Hordeum distichon L.
   Brazil. S. halophila Speg.; leafblotch. Argentina. Septoria sp. Guatemala.
Stemphylium sp.; spores recovered from stored grain. Co-
   lombia.
Ustilago spp. (all cause smuts): U. bullata Berk. Argentina.
   U. hordei (Pers.) Lagerh. the covered smut occurrence.
   Neotropical. U. jensenii Rostr. Chile. U. nigra Tapke
   Colombia. U. nuda (Jens.) Rostr., the loose smut occur-
   rence. Widespread. U. spegazzinii Hirsch., the flag
   smut in S. America. U. tritici Rostr. Peru.
Viruses: Barley Stripe Mosaic in Colombia, Enanismo or
   stunt in Ecuador, Colombia, the chlorotic Rice Hoja Blanca
   Virus in Colombia, Brazil, and Venezuela, and a leaf mottle
   (unidentified) in Guatemala.

Xanthomonas transluscens L. R. Jones, Johns., & Reddy var.
cerealis Dows.; bacterial stripe. Argentina, Colombia.

HUMULUS LUPULUS L. The attractive European vine increasingly
grown in a few tropical countries producing for brewers the
processed fruiting spikes; "hops" En., "lúpulo" Po. Sp. (Its
diseases have had only minor attention thus far in the American
tropics.)
  Capnodium sp.; sootymold, fumago. Colombia, Brazil, Argen-
    tina.
  Cercospora cantuariensia Salmon & Wr.; large circular leaf-
    spots. Colombia.
  Fusarium sp.; footrot. Colombia.
  Meloidogyne sp.; nematode rootknot. California.
  Mycosphaerella erysiphina (Berk. & Br.) Kirch.; leafspot.
    California, in Guatemala typical leafspots but without spores
    for positive identity.
  Phyllosticta humuli Sacc. & Speg.; leafspot. Guatemala.
  Pseudoperonospora humuli (Miy. & Tak) G. W. Wils.; downy
    mildew. Colombia, Argentina, Brazil, Guatemala.
  Sphaerotheca humuli (DC.) Burr.; powdery mildew. Brazil,
    Argentina.
  Verticillium alboatrum Reinke & Berth.; wilt. Argentina.
  Virus: mosaic type in Colombia.

HURA CREPTANS L. A native, handsome warm country tree planted
often as an ornamental specimen, drops explosive hard-rough
fruits; "javillo," "tronado" Sp., "acacu" Po., "sandbox tree" En.
  Cercospora hurae Stev. and C. huricola Chupp; leafspots.
    Widespread.
  Colletotrichum gloeosporioides Penz. is the conidial state and
    no asco-form found. (Sometimes reported as C. curvisetum
    Stev.); common anthracnose leafspot. Neotropical.
  Cuscuta americana L.; yellow brittle dodder. Puerto Rico.
  Dieback, cause undetermined. Saprophytic bacteria and fungi
    isolated from twigs and branches. Costa Rica.
  Fomes lamoensis Murr.; trunk rot of cut or fallen trees. W.
    Indies, C. America.
  Helminthosporium capense Thuem. and H. hurae P. Henn.;
    leafspotting that may become confluent. S. America (es-
    pecially noted in Brazil).
  Irenopsis hurae (Syd.) Stev. = Meliola hurae Syd.; black mil-
    dew. W. Indies, Panama.
  Leptosphaeria hurae Pat.; leafspot, often small lesions. C.
    America.
  Meliola (see Irenopsis)
  Parodiopsis hurae (Arn.) Baker & Dale and P. perae Arn.;
    black mildew spots. C. America, Colombia, W. Indies,
    Trinidad.
  Poria sp.; on fallen branch. Puerto Rico.
  Valsaria sp.; branch canker. S. America.

HYDRANGEA HORTENSIA DC. = H. MACROPHYLLA Ser. Impor-
tant for ornamental plantings and cut flowers, perennial shrub
with large flower heads; "hortensia" Sp. Po., "hydrangia" En.
  Armillaria mellea Vahl = Armilariella mellea (Fr.) Karst.;
    rootrot. California.
  Botrytis (? cinerea); mold on fading flowers. C. America
    in mountains.
  Cephaleuros virescens Kunze; algal leafspot. Brazil.
  Cercospora hydrangeae Ell. & Ev.; common leafspot. Sub-
    tropical USA, W. Indies, C. America, Brazil, Argentina.
    (Seldom severe).
  Cuscuta sp. (a slender type); dodder. Guatemala.
  Erysiphe (see Oidium)
  Glomerella cingulata (Stonem.) Sp. & Sch. = its conidial state
    is Colletotrichum gloeosporioides Penz.; anthracnose.
    Widespread.
  Meloidogyne sp.; rootknot. Puerto Rico.
  Oidium hortensia Jorst. = Erysiphe polygoni DC. which is the
    asco-state (reported from Chile); white powdery mildew.
    More or less Neotropical noted especially in Colombia,
    Chile in S. America and seen in Costa Rica, El Salvador
    and Guatemala in C. America.
  Rootrot, unusually friable gray tissues, cause undetermined.
    Costa Rica, Guatemala.
  Sclerotium rolfsii (Curzi) Sacc. = Pellicularia rolfsii (Sacc.)
    West; rot at ground level. Florida.

HYMENOCALLIS CARIBAEA Herb. Native to the Caribbean region,
a popularly grown ornamental bulbous plant, with showy flowers;
"lirio" Sp., "açucena" Po., "spiderlily" En.
  Asterinella hippeastri Ryan; black mildew spots. Puerto Rico.
  Cercospora pancratii Ell. & Ev. = asco-state is Mycosphae-
    rella aggregata Earle; leafspot (often red). Subtropical
    USA, W. Indies, C. America.
  Curvularia trifolii (Kauf.) Boed.; leafspot, leaf edge infection.
    Puerto Rico.
  Glomerella sp. (not G. cingulata) small sporocarps = the
    conidial fungus is Colletotrichum sp. reported as Gloeo-
    sporium hemerocallidis Ell. & Ev.; anthracnose lesions.
    Subtropical USA, W. Indies, (probably widespread).
  Mycosphaerella (see Cercospora)
  Phyllosticta (see Stagonospora)
  Puccinia habranthii Diet. & Neg., P. reichei Diet. & Neg.,
    and P. roseanae Arth.; all rusts. S. America.
  Spores recovered of apparently weakly parasitic fungus species
    of Aspergillus, Botrytis, Curvularia, small Helminthospo-
    rium and a Nigrospora, subsisting on fallen flowers. El
    Salvador.
  Stagonospora curtisii (Berk.) Sacc.; redspot. Widespread.
    (This disease is also reported as caused by Phyllosticta
    hymenocallidis Seav. but apparently it is Stagonospora.)

Virus symptoms of mosaic type common and general, unidentified.

HYPARRHENIA RUFA Stapf. A tough, important pasture and ground-cover grass that grows under adverse conditions.
   Botryodiplodia theobromae Pat.; isolate of it infected a leaf
      when inoculated in the laboratory but I obtained no proof
      of disease from it in the field. El Salvador.
   Cerebella andropogonis Ces.; black head. S. America, C.
      America.
   Claviceps sp.; ergot (small dimensions). C. America, S.
      America.
   Curvularia sp.; leaf and stem spot. El Salvador.
   Fusarium sp.; isolated from dying clump of plants, but culture
      was not infective. El Salvador.
   Helminthosporium sp.; leafspot. C. America, S. America.
   Inflorescence disease (not black head or ergot), undetermined
      cause; sterility. El Salvador.
   Leafspot symptoms suggested Phyllachora but no confirmation.
   Puccinia eritraensis Paz. and P. levis (Sacc. & Bizz.) Magn.;
      leafrusts. C. America, S. America.
   Rhizoctonia (slender chlamydospores); basal rot. El Salvador.
   Virus symptom of stunt and browned vasculars at stem nodes.
      Seen adjacent to sugarcane suffering from ratoon stunt.
      Costa Rica.

ILEX PARAGUARIENSIS St. Hil. A famous South American broad-leaved evergreen that is grown wild and cultivated, is a shrub to a small tree whose processed caffeine-containing leaves are produced mostly in Paraguay, Uruguay, southern Brazil, and Argentina. (Note: All diseases indicated in this list were reported from S. America, mostly from the Argentine collections of Spegazzini. [When he started to work on such an important, much grown crop, in ancient stands and in long wild-gathered semi-jungle cultures, but at the same time never touched by any mycologist-pathologist, it must have been to the pioneer specialist in these things, like a new world in itself, like the realization to him of a delightful dream of discovery of natural splendor hitherto reached by no man.]) Crop is increasingly popular sold internally and exported; "yerba maté," "maté" Sp. Po., "paraguay tea" En., "yerva do palos" Po., and, depending on quality, "caa gazu," "caa míri," "caa cuy" Po.
   Alternaria (a new combination?) (see Macrosporium)
   Acanthonitschkea argentinensis Speg.; branch and stem lesion.
   Apiospora yerbae Speg.; on dying branches.
   Asterina mate Speg. = Clypeolella mate (Speg.) Theiss.; a
      black mildew.
   Blitridium mate Speg.; on foliage branch dieback.
   Botryodiplodia (see Diplodia)
   Cercospora ilicicola Maubl., C. mate Speg. = C. mate (Speg.)
      Mar., C. yerbae Speg.; leafspots.
   Cercosporina mate Speg. = Cercospora ruste Speg., unusual
      large leafspot. S. America.

Coccomyces yerbae Speg.; on decaying leaves.
Colletotrichum yerbae Speg. = C. gloeosporioides Penz. but
  possibly it may be a special race; anthracnose and leaf
  disease. S. American countries, Costa Rica.
Coniothyrium mate Speg., C. maticola Speg., and C. yerbae
  Speg.; twig and log decays. S. America, C. America.
Corticium incarnatum (Pers.) Fr.; stem disease.
Cryptodiaporthe mate (Speg.) Wehm.; on leaf and stem.
Cryptosphaerella mate Speg.; on decorticated rotting branches.
Cylindrocladium scoparium Morg.; leaf disease.
Cylindrosporium sp.; stem canker. Paraguay.
Cyphella albo-violascens (Alb. & Schw.) Karst.; woodrot.
Cytosporella yerbae Speg.; dieback and twig rot.
Cytosporina yerbae Speg.; on leaf.
Diaporthe mate Speg. = Cryptodiaporthe mate Speg. and D.
  yerbae Speg. = D. eres Nits.; dieback diseases and on
  rotten twigs.
Diatrypella missionum Speg.; rot of dead branch in dieback
  complex.
Dictyosporium yerbae Speg.; woodrot.
Didymosphaeria yerbae Speg.; cause of branch canker.
Dimerosporium decipiens (de N.) Sacc. var. yerbae Speg.;
  black mildew on log.
Diplodia yerbae Speg. = Nemadiplodia anomala (Mont.) Zam-
  bet., probably should be Botryodiplodia yerbae a new
  combination; severe dieback.
Endoxyla yerbae Speg.; a stem or branch disease.
Epicoccus sp.; on leaf.
Eutypa ludibunda Sacc.; wood infection of old trees.
Fusarium sp.; secondary twig infection.
Gibberella saubinettii (Mont.) f. sp. mate Speg. as given in
  Viégas classic list in the section taken from Spegazzini's
  work on maté (seems it is more probably G. cyanogena
  (Desm.) Sacc., and the conidial form is probably Fusarium
  sulphureum Schl. with probable f. sp. mate); twig and
  branch disease and on trunk.
Glonium microsporium Sacc. var. minor Speg.; on trunk and
  branches.
Harpographium yerbae Speg., agaric on rotting branch.
Helminthosporium yerbae Speg. = Corynespora yerbae (Speg.)
  Ell.; branch disease.
Hendersonia mate Speg.; on dry twigs (? after harvest).
Hypholoma appendiculatum Bull.; an agaric on dead wood.
Hypoxylon rubigenosum (Pers.) Fr. var. microcarpum Speg.;
  bark disease.
Irenina ingaeicola (Speg.) Stev. = Meliola ingaeicola Speg.;
  leaf black mildew.
Leptosphaeria paraguariensis Maubl., and L. yerbae Speg.;
  killing branches.
Macroplodiella maticola Speg.; on dying branches.
Macrosporium yerbae Speg. should be corrected to Alternaria
  yerbae (Speg.) Joly; leaf and branch disease.

Marasmius-like sporocarp from diseased root.
Massariella yerbae Speg.; on dead branch wood.
Megalonectria yerbae Speg.; branch canker and dieback.
Melanomma mate Speg.; on tree trunk.
Melanopsamma yerbae Speg. = Abaphospora yerbae (Speg.)
    Kirsch.; log rot.
Meliola yerbae Speg.; black mildew.
Mycosphaerella ilicicola Maubl.; large leafspot.
Myiocopron yerbae Speg.; black mildew growth on dead and
    dying twigs.
Nectria sphaeriicola Speg.; hyperparasite of fungi on logs.
Odontia crustosa Pers. var. tropicales Speg.; woodrot, hyd-
    num-like sporocarp.
Ozonium atro-umbrinum Speg.; on bark.   Paraguay.
Paracapnodium pulchellum Speg.; leaf sootymold.
Peckia mate Speg.; on old and spent leaves.
Pellicularia koleroga Cke.; thread-web blight.   (Note: Occur-
    rence of this fungus in wild South American tree growth,
    of which maté is a part, indicates to me the fungus must
    be a truly pantropical parasite.   Reading about its past
    history in Africa and the Orient, and seeing it there in
    full bloom, has made me think that at one time it was
    possibly brought to the Western tropics from the Old
    World.   I believe it is originally from both tropics, East-
    ern and American.)
Pestalotia paraguariensis Maubl.; leaf and twig lesion.
Phaeobotryosphaeria yerbae Speg. = Botryosphaeria disrupta
    (Berk. & Curt.) Arx; on dying and dead branches.
Phaeomarsonia yerbae Speg.; leaf disease.
Phoma yerbae Speg.; on twigs.
Phyllosticta mate Speg. = ? P. yerbae Speg.; spots on dying
    and diseased leaves, defoliation.
Phytophthora sp.; stem/leaf disease.
Psathyrella disseminata Pers.; mushroom on dead wood.
Pythium spp.; young plant collapse.
Root infections from both Rhizoctonia sp. and Rosellinia sp.
    C. America, S. America.
Solenia villosa Fr. var. eximia Sacc. and S. villosa Fr. var.
    subochracea Speg.; branch decay, if as old trunk disease
    could produce boletus-like sporocarp.
Spermatoloncha maticola Speg.; hyperparasite of black mildew
    on foliage.
Sphaerella iliciola Maubl.; stem and leaf disease.
Sphaeromyces maticola Speg.; branch disease.
Sphaerulina yerbae Speg.; on leaf.
Stagonospora yerbae Speg.; twig disease causing dieback.
Stilbopeziza yerbae Speg.; on dying branches.
Stilbum sp.; branch canker.
Strickeria mate Speg. = Teichospora mate (Speg.) Sacc.; tree
    trunk disease.
Stysanus yerbae Speg.; on stems.
Target leafspot (observed on processed leaf), undetermined
    cause.   Brazil.

Thyridium yerbae Speg.; on decorticated branch and trunk
    wood.
Tryblidium guaranticium Speg. = Tryblidiella spegazzinii
    Rehm, and T. guaraniticum Speg. var. major Speg.; die-
    back. (Both names may now be considered as Rhytidhy-
    sterium discolor Speg.)
Valsa yerbae Speg.; branch disease.
Valsaria clavatiasca Speg. and V. insitiva Ces. & deN.;
    dieback and on dead branches.
Venturia missionum Speg.; dieback.
Weakspot (cause unknown) observed on extracted dry-leaf ma-
    terial. Peru.
Winterella yerbae Speg.; on branch wood.
Zignoella yerbae Speg.; on logs (? hyperparasite of alga?).

IMPERATA BRASILIENSIS Trin. and I. CONTRACTA Hitch. Coarse
    strong grasses, soil retainers in eroding pastures and along road-
    ways, used in thatch for ranchos and bedding for men and ani-
    mals; "bálago," "paja grueso" Sp., "sapé" Po., "satintail,"
    "imperial grass" En.
    Aschersonia spp.; fungus parasites on insects. Widespread.
    Dictyocharisa arundinellae Chard.; leaf disease. Colombia.
    Dictyochlorella sp.; leaf and caulm spot. S. America.
    Mainsia imperialis (Speg.) Lindq.; leafrust. Argentina.
    Metasphaeria phyllachoracearum Petr.; leafspot. S. America.
    Myriogenospora paspali Atk.; leaf disease. Brazil.
    Phaeopeltasphaeria panamensis Stev.; leafspot. C. America,
        Panama.
    Phyllachora graminis (Pers.) Fckl. in Brazil and P. oxyspora
        Starb. in West Indies; both tarspots on leaf.
    Puccinia infuscans Arth. & Holw. in Guatemala and P.
        kaernbachia (P. Henn.) Arth. = P. posadensis Sacc. &
        Trott.; in W. Indies, Brazil, both are leafrusts.
    Xanthomonas axonoperis Starr & Garces; bacterial gomosis.
        Colombia.

INDIGOFERA ANIL L. and I. ENDECAPHYLLA Jacq. The genus
    is pantropical and for millenia the famous internationally market-
    able indigo blue dye was secured from species of it. Chemical
    substitute now almost completely replaces the natural product
    except for very special purposes, and some is still in use by
    certain peasantry and artisans. The leguminous nature and its
    field characters make some species valuable in ground cover be-
    tween plantation tree crops and for other agronomic practices;
    "aníl" Sp. Po., "indigo" En.
    Aspergillus sp.; on stored seedpods. El Salvador.
    Botryodiplodia theobromae Pat. (= Diplodia sp.); dieback of
        old plants. Neotropics.
    Capnodium spp.; sootymolds. Under partial tree shade. C.
        America.
    Cassytha filiformis L.; green dodder. Puerto Rico.
    Chaetoseptoria wellmanii J. A. Stev.; round leafspot. C.
        America.

Colletotrichum (? gloeosporioides); leaf injury following insect
  feeding. Costa Rica.
Corticium salmonicolor Berk. & Br. ; on foliage. Costa Rica,
  Puerto Rico.
Cuscuta americana L. ; yellow dodder. Puerto Rico.
Darluca sp. = Eudarluca sp. ; hyperparasite on rust. El
  Salvador.
Fusarium spp. ; secondary infections. Widespread.
Mycena citricolor (Berk. & Curt.) Sacc. ; leafspot obviously
  spread from overhanging diseased coffee tree. Puerto
  Rico.
Oidium sp. ; white mildew in dryseason. Costa Rica.
Parodiella perisporioides (Berk. & Curt.) Speg. Colombia.
  P. perisporioides (Berk. & Curt.) Speg. f. sp. microspora
  Th. & Syd. ; leaf infections. Colombia, C. America.
Puccinia sp. ; a yellow rust. Brazil.
Pythium spp. ; damping-off. Neotropical.
Ravenelia indigoferae Tran. ; a brown leafrust. Neotropics.
  R. laevis Diet. & Hol. ; brown leafrust. Mexico. R.
  schroeteriana P. Henn. ; brown leafrust. Argentina.
Rhizoctonia (aerial type but no aerial sclerotia); foliage disease.
  Costa Rica. R. solani Kuehn. = its basidial state is typical
  Thanatephorous cucumeris (Frank) Donk; footrot, damping-
  off. W. Indies, C. America.
Sclerotium coffeicola Stahel; leaf targetspot. Surinam.
Uromyces indigoferae Diet. & Hol. ; rust. Subtropical USA,
  Mexico, C. America, Venezuela.

INGA spp. Important shade plantation crops and used in reforestation, with many names, but some species have specifically ap-
plied vernacular names that are given in parentheses; there are
over 50 species in the Western tropics and about six have at-
tracted special pathological attention: I. EDULIS Mart. ("pepe-
tón, " "guama, " "guamo") used for agricultural shade and edible
seeds. I. LAURINA Willd. ("cujinicuíl, " "spanish ash, " "guama")
used in agricultural shade and lumber. I. PATERNO Harms.
("guamá, " "fruta del mico") shade in plantations, pulp of pod
eaten as fruit. I. PRUESSII Harms. "quijiniquíl, " "cujín, "
"spanish ash") used for plantation shade, wood for lumber, pods
have edible seeds. I. SPURIA Humb. & Bonpl. ("guama, "
"inca tree, " "pepeto, " "humboldt's tree, " "west indian ash")
valued as agricultural shade, produces lumber, its twisted pods
contain much valued and gathered edible seeds. I. VERA Willd.
("guama, " "guabá, " "inca tree, " "pepeto") used in coffee and
cacao shade and for wood. Diseases given below on the whole
probably attack all species employed in shade agriculture; a few
special occurrences are noted.
  Actinothyrium ingae Bat. ; on leaf lamina. Brazil.
  Antennaria spp. ; sooty molds. Scattered.
  Arthrobotryum ingae P. Henn. ; stem infection. S. America.
  Asteridium elegantissimum Rehm; black mildew (Asterina-like).
    S. America.

Auricularia polytricha (Mont.) Sacc. = Hirneola polytricha Fr. ;
   on fallen trunk, woodrot.   S. America, C. America.
Belonopsis ingae Seav. ; leafspot.   Neotropics.
Bitzea (see Chaconia)
Botryodiplodia spp. also reported as Diplodia; dieback, pod
   rots, leaf diseases.   Neotropics.
Calothyrium ingae Ryan; leaf disease.   C. America.
Catacauma (see Phyllachora)
Cephaleuros virescens Kunze; common algal leafspot.   General.
Ceratospermopsis xylopiae Bat. ; fumago.   Brazil.
Cercospora ingae Obr. -Bot. ; leafspot.   Colombia (described
   as "typical Cercospora spots" as seen in C. America and
   W. Indies, but no spores found on surfaces).
Chaconia ingae (Syd.) Cumm. (synonyms: Bitzea ingae (Syd.)
   Mains, Maravalia ingae Syd. , Ravenelia ingae (P. Henn.)
   Arth. , R. whetzelii Arth. , Uredo ingae P. Henn. , Uromyces
   ingaeiphilus Speg. , U. porcensis Mayor, U. pulverulentis
   Speg.); the most common leaf and stem rust, may cause
   malformations.   Neotropical.
Colletotrichum sp. = Glomerella cingulata (Stonem.) Sp. &
   Schr. ; anthracnose.   Widespread.
Corticium salmonicolor Berk. & Br. ; pink disease.   C.
   America, S. America.
Creonectria tucumanensis (Speg.) Chard. = Nectria tucumanen-
   sis Speg. ; canker.   Argentina.
Cuscuta sp. ; hyperparasite of Loranthus sp. on I. vera.   Costa
   Rica.
Diatractium ingae (Allesch.) Syd. = Scolecodothiopsis ingae
   Stev. ; large dark leafspots.   W. Indies, Panama, S.
   America, C. America.
Dicheririna superba Jacks. & Holw. ; leafrust.   C. America,
   S. America.
Dimerosporium ingae P. Henn. ; black mildew on wild species
   of Inga.   S. America.
Diplodia (see Botryodiplodia)
Elsinoë (see Sphaceloma)
Eriomycopsis tenuis Syd. ; on foliage.   S. America.
Fusarium spp. ; stem and twig cankers, dieback complex.   C.
   America, Colombia, W. Indies.
Gleocapsa sp. ; alga in chronically "wet branch cleft. "
   Panama.
Gloniella ingae Rehm; leafspot.   S. America.
Guignardia ingae Stev. ; leafspot and stemspot.   W. Indies, C.
   America.
Helicobasidium compactum Boed. ; woodrot.   S. America.
Henningsomyces nigrescens (Rehm) Th. & Syd. ; woodrot.   S.
   America.
Hirneola (see Auricularia)
Hydnum sp. ; on old roots of fallen tree.   Costa Rica.
Irenina ingaecola (Speg.) Stev. ; black mildew.   Neotropics.
Irenopsis toruloides Stev. ; black mildew.   W. Indies.
Linospora guaranitica Speg. ; leafspot.   Brazil.

Loranthus sp. (from the ground resembled L. robustissimum);
   large leaved mistletoe-like parasites on I. spuria, I. vera,
   I. edulis, and I. sp.   Costa Rica.
Marasmius equirinus Mul.; black horse-hair blight and M.
   pulcher Petch; white horse-hair blight.   In Panama on I.
   paterno.
Maravalia (see Chaconia)
Melasmia ingae Stev.; leaf disease.   Puerto Rico.
Meliola spp.   All cause black mildew diseases: M. chagras
   Stev. in Panama, M. harioti Speg. in S. America, M.
   ingaecola Speg. = Irenina ingaecola (Speg.) Stev. in S.
   America, M. toruloidea Stev. = Irenopsis toruloidea Stev.
   in W. Indies.   On shade above coffee, at times it induces
   unwanted defoliation in the dryseason.
Microstroma ingaicola Lamk.; common witches' broom on I.
   vera.   Neotropical.
Microxyphiella sp. (Capnodium-like black hyphae in tufts at
   base of main leaflet vein on I. edulis).   Costa Rica.
   (Note: The material I secured was late in the last year I
   was in Costa Rica and it had asco-carps that were immature; they compared somewhat favorably with the Microxyphiella on Hibiscus rosa-sinensis, but I did not make exact
   measurements.)
Mycena citricolor (Berk. & Curt.) Sacc. (= Stilbella flavida
   P. Henn. or Omphalia flavida Maubl. & Rang.); the spec-
   tacular and luminescent leafspot organism.   Scattered in
   Peru, Costa Rica, Puerto Rico.
Mycosphaerella maculiformis (Pers.) Schr. on fallen leaves
   (perfect state of a Cercospora ?); Subtropical USA, W.
   Indies, probably General.
Nectria sp. (and also see Creonectria); branch canker.
   Colombia.
Nematospora sp.; pod disease.   El Salvador, Guatemala.
Oidium sp.; powdery mildew, ash disease.   Colombia, Guate-
   mala.
Omphalia (see Mycena)
Oryctanthus cordifolius Eichl.; woody stemmed mistletoe-like
   parasite.   Guayanas, C. America.
Parodiopsis spp.   All cause black mildews but on the whole
   rather inconspicuous in effect, and almost on all Inga
   trees; moreover, these specific names may be synonymous:
   P. ingarum Arn., P. melioloides Maubl., P. rubra Baker
   & Dale, and P. stevensii Arn.   Widespread.
Pellicularia sp. (aerial type, no microsclerotia); target leaf-
   spot.   Peru, C. America.   P. koleroga Cke.; leafrot.
   Costa Rica, Guatemala.
Perisporina manaoensis P. Henn., P. roupalae Viégas and
   P. truncatum Arn.; all black mildews.   First and second
   reported from Brazil, third from W. Indies.
Pestalotia spp.; petiole, stem and leaf lesions.   Neotropics.
Phaeosaccardinula tenuis Chard.; sootymold.   W. Indies.
Phoradendron robustissimum Eich.; large and heavy mistletoe-

like bushy parasites, some hyperparasitized (for example
one was attacked by Oryctanthus cordifolius Urb. which was
in turn attacked by a black mildew and an anthracnose and
another P. robustissimum was attacked by O. cordifolius
that in turn had on it diseased stems with ascocarps of
Glomerella, some leaves were diseased by Mycena leafspot,
and a sootymold was over all).   Costa Rica.
Phyllachora amphibola Syd. = Catacauma ingae Chard.; shiny-
    black leafspot.   C. America, W. Indies.
Phyllosphere microorganisms.   From a moderately wet loca-
    tion in a narrow valley of Costa Rica branches of I. edulis
    with specially old but unspotted leaves were taken twice for
    microscopic examination, and on them were seen: non
    parasitic bacteria, simple celled surface algae, small lichen
    colonies, a minute leafy liverwort, slender non-parasitizing
    hyphal threads of three types (hyaline, amber color, dark
    brown), a few elongate fungus spores, at an old petiole
    attachment to a stem I recovered sluggish moving nematodes
    and a number of very small fungus structures that were
    sclerotia.
Phyllosticta molluginia Ell. & Hol.; leafspot.   Argentina.
Pilostyles ingae Hook.; root and trunk attack by small
    Phanerogamic parasite.   Colombia.
Pod diseases in the form of spots and tip rots from various
    unidentified organisms.   Neotropical.
Polystomella repanda Speg.; leaf disease.   Brazil.
Poria sp., on old dead exposed root of I. vera.   Costa Rica.
Psittacanthus calyculatus Don. on I. laurina; a tall woody
    mistletoe-like parasite.   (One such parasite was hyper-
    parasitized by a Phoradendron in turn super-hyperparasitized
    by a Struthanthus, on it another Psittacanthus was secured
    with a hyperparasitism by a smaller Psittacanthus and by a
    Cuscuta.)   El Salvador.   P. clusiifolius Eichl. and P.
    cucullaris Bl.; woody mistletoe-like parasites.   Surinam.
Ravenelia (see Chaconia)
Rootfailures, in some cases they follow results of digging holes
    near old shade trees; also it is apparent that root tissue
    collapse often follows heavy shade pruning of a tree.
Rosellinia bunodes (Berk. & Br.) Sacc. the black rootrot and
    R. pepo Pat. the white rootrot.   More or less General,
    somewhat scattered, the former most common in C.
    America.   R. cuprea Rick; rootrot.   S. America.
Schizophyllum sp. near ends of exposed dead host roots in
    jungle clearing.   Guatemala.
Scolecodothiopsis (see Diatractium)
Scolecopeltis ingae Toro; black leafspot.   Puerto Rico.
Septobasidium castaneum Burt. var. draconium Viégas and
    S. saccardinum (Rang.) March.; stem-bark diseases.   S.
    America.
Septoideum stevensii Arn.; leaf disease.   W. Indies.
Septoria ingae Viégas; a leafspot.   Brazil.
Sphaceloma sp.; spot anthracnose (the perfect state would be
    Elsinoë).   Brazil.

Spirogyra sp.; algae in chronic "wet branch-cleft" of I. spuria.
Puerto Rico.
Stereum ochroleucum Fr.; woodrot. S. America.
Stilbella (see Mycena)
Stomatochroon sp.; alga, systemic in leaf. Costa Rica.
Struthanthus marginatus Blu. and S. orbicularis Blu.; mistle-
toe-like recumbent woody parasites. El Salvador.
Thallochaete ingae Th.; a leaf disease. S. America.
Trabutia sp.; leafspot of crusty nature. El Salvador.
Uredo mogy-mirim Viégas; rust. Brazil. (Also see Chaconia.)
Uromyces (see Chaconia)
Verticillium candidulum Sacc. var. ingae Speg.; wilt of young
trees. S. America.
Virus: A mottling and chlorosis in C. America and Venezuela
and a witches' broom in Venezuela.
Xenochora ingae (Stev.) Petr.; stem and leaf disease. S.
America.
Xylaria spp.; associated with dieback, woodrot, rootrot.
Widespread.
Zukalia fusispora Pat.; on foliage. S. America.

IPOMOEA BATATAS (L.) Lam. An important food plant characterized by vine foliage with fat tubers below ground, native to
tropical America but grown worldwide in subtropical and tropical
lands, somewhat neglected as a subject for intensive agriculture
in Latin America as it is still much plagued with diseases; it
is perennial and potentially a highly productive root crop; "batata
cenaurea," "camote," "chaco" Sp., "batata doce," "batatas,"
"batata-dolce" Po., "batatá," "mabí," "cumar," "apichu" Ind.,
"sweet potato" En.
Acrocystis (see Streptomyces)
Aecidium (see Coleosporium)
Albugo ipomoea-panduratae (Schw.) Swing.; common leafblister,
white rust. In cool areas of Neotropics. A. convolvulacearum Speg.; white rust. Neotropics. A. minor Speg. =
Cystopus convolvulacearum Ott. & Speg.; white rust. Do-
minican Republic, Venezuela, Brazil.
Alternaria alternata (Fr.) Keiss. (small spores); a few in
leafspot complex. Occasional. A. solani (Ell. & Mart.)
Sor.; footrot. Mexico.
Aphelenchoides parietinus Bast.; root nematode. Brazil.
Aspergillus niger v. Tiegh.; storage blackmold rot. C.
America, W. Indies.
Botryodiplodia = Diplodia tubericola Taub. the conidial state
of the ascomycete Botryosphaeria quercuum (Schw.) Sacc.;
wetseason blackrot. Neotropical.
Cassytha filiformis L.; green dodder. Puerto Rico, Florida.
Catenularia megalospora Speg.; stem disease. S. America.
Ceratocystis fimbriata (Ell. & Halst.) Elliott (= Sphaeronema
fimbriatum (Ell. & Halst.) Sacc.); stem and root blackrot.
Neotropical. (This organism from sweet potato seems dif-
ferent in pathogenicity from the two races on coffee and
rubber.)

Cercospora bataticola Cif. & Brun.; a large leafspot. Brazil,
  Dominican Republic, Cuba, W. Indies. C. cordobensis
  Speg.; leafspot. Brazil. C. ipomoeae Wint.; leafspot.
  Venezuela, Puerto Rico. C. stuckertiana Syd.; leafspot.
  Argentina. C. viridula Ell. & Ev.; leafspot. Trinidad.
Chlorosis with stunt, and bud death, from unknown cause.
  Cuba, Virgin Islands.
Choanephora conjuncta Couch and C. cucurbitarum Thaxt.;
  flower blast and leafmold. Scattered in W. Indies, C.
  America, Subtropical USA.
Coleosporium ipomoeae (Schw.) Burr. (synonyms: Uredo
  ipomoeae Schw., Aecidium dominicanum Gonz. Frag. &
  Cif.); common rust. Neotropical.
Colletotrichum gloeosporioides Penz. (asco-state is Glomerella
  cingulata (Stonem.) Sp. & Schr.); anthracnose, sometimes
  benign invasion. Neotropics.
Cuscuta americana L.; common yellow dodder. C. America,
  W. Indies.
Diaporthe batatis Hart. & Field; dryrot of stem and root.
  Subtropical USA.
Dothiorella convolvuli Gonz. Frag. & Cif.; stem and vine
  canker. Dominican Republic.
Elsinoë batatas Viégas & Jenk. (conidial state is Sphaceloma
  batatas Saw.); stem and leaf scab-spot. Brazil, Mexico,
  Puerto Rico.
Erwinia carotovora (L. R. Jones) Holl.; market root-softrot.
  Widespread.
Erysiphe polygoni DC. (conidial state, usually found, is
  Oidium erysiphodes Fr.); the powdery white mildew.
  Mostly mild and in dryer areas, is of widespread distri-
  bution.
Fusarium oxysporum Schl.; "shallow rot" of outer layer on
  tuber. El Salvador, Puerto Rico, Costa Rica, Subtropical
  USA. (It was found in infection tests of cultures in El
  Salvador that there was no systemic invasion.) However,
  F. oxysporum Schl. f. sp. batatas (Wr.) Sny. & Hans.;
  severe wilt, typical systemic stem and root-vascular
  disease. El Salvador, Costa Rica, W. Indies, Florida
  (Isolation in El Salvador readily invaded stem vasculars).
  F. roseum Lk.; woody storage rot. Subtropical USA,
  Mexico, Costa Rica.
Guignardia convolvuli Gonz. Frag. & Cif. (conidial form is
  Macrophomina convolvuli Gonz. Frag. & Cif.); stem
  disease, charcoal storage-rot. Probably scattered generally
  but special reports from Dominican Republic, Brazil.
Helicotylenchus nannus Stein.; rootzone nematode. Brazil.
Himantia corticalis Viégas (non-sporulating deeply colored
  rhizomorphs); root disease. Brazil.
Macrophomina phaseoli (Taub.) Ashby = Sclerotium bataticola
  Taub. (compare with Guignardia?); charcoal-rot. Scat-
  tered, general distribution.
Meliola clavulata Wint. and M. malacotricha Speg.; both black

mildews. Neotropics. M. clavulata Wint. f. sp. batatae
Stev. is an apparently special race found in Dominican
Republic. M. ipomoeae Earle. Puerto Rico and Virgin
Islands.

Meloidogyne spp. ; rootknot. Jamaica, Brazil.

Monhysteria stagnalis Bastian; root nematode. Brazil.

Monilochaetes infuscans Ell. & Halst. ; common scurf.
Widespread.

Mycena citricolor (Berk. & Curt.) Sacc. ; leafspot by inoculation. Puerto Rico.

Nectria ipomoeae Halst. ; on surface of killed stems. West
Indies.

Oidium (see Erysiphe)

Penicillium spp. ; bluemolds in storage. Widespread.

Pestalotia batatae Ell. & Ev. ; tuber pustules in markets.
W. Indies, C. America.

Phyllosticta spp. (all cause purple haloed brown leafspots);
P. batatas (Th.) Cke. in Venezuela and Brazil. P.
bataticola Ell. & Mart. in Brazil. P. stuckeri Speg.
Widespread.

Plenodomus destruens Harter; footrot. Subtropical USA, W.
Indies.

Pseudomonas solanacearum E. F. Sm. ; bacterial wilt.
Mexico.

Puccinia spp. (leafrusts): P. insignis Holw. Mexico. P.
macrocephala Speg. S. America generally. P. puta
Jacks. & Holw. (= Aecidium distinguendum Syd. and/or
P. distinguenda Jacks. & Holw.). Brazil, Colombia,
Ecuador, Peru, Venezuela.

Pyrenophora terrestris (Hans.) Gor., Walk. & Lars. ; pink-
rootrot. Subtropical USA, Puerto Rico.

Pythium spp. often P. ultimum Trow; market leaf-rot.
Scattered.

Rhizoctonia solani Kuehn. (this mycelial state causes the
diseased conditions of stem canker, sprout and tuber rot).
Subtropical USA, W. Indies, C. America, Mexico, and
probably in S. America. (The perfect or basidiospore state
when found was determined as Thanatephorous cucumeris
(Frank) Donk and its sporulating structures were at the
time on superficial hymenium on three-year-old plants in
El Salvador.

Rhizopus nigricans Ehr. ; storage black softrot. Widespread.
R. stolonifer (Ehr.) Lind. ; storage rot. C. America in
cooler mountain localities.

Rosellinia bunodes (Berk. & Br.) Sacc. and R. pepo Pat. ;
rootrots. Not common, often in newly cleared jungle soil
or at the edge of coffee or cacao plantings. W. Indies,
C. America.

Schizophyllum commune Fr. ; root dryrot. Puerto Rico, Costa
Rica.

Sclerotium rolfsii (Curzi) Sacc. = Pellicularia rolfsii (Sacc.)
West; neckrot, collapse. Florida, Argentina, Dominican
Republic, Cuba.

Sclerotium (undetermined species); rootrot.   Puerto Rico,
    Dominican Republic.
Sphaceloma (see Elsinoë)
Sphaeronema = Ceratocystis.
Streptomyces ipomoea (Person & Mart.) Waks. & Henr. (=
    Acrocystis batatas Ell. & Halst. = Actinomyces ipomoea
    Pass. & Mart.); pox, tuber pitting, scurfy spot.   Wide-
    spread.
Trichodorus sp.; rootzone inhabiting nematode.   Subtropical
    USA.
Tylenchorhynchus sp.; rootzone inhabiting nematode.   Sub-
    tropical USA.
Tylenchus davianii Bastian; root disease nematode.   Brazil.
Uredo (see Coleosporium)
Viruses: viruses are common causing typical mosaic, stunt,
    rosette, corkyness and necrotic spot symptoms.   In my
    inoculations the virulent Cucumber Mosaic Virus (the
    Celery-Commelina strain) caused severe mottle and stunt.
    Viruses may be a potent reason sweet-potatoes are such
    poor producers in some areas.
Weakspot on leaves, cause unknown.   Puerto Rico.
Xiphinema americanum Cobb; root feeding nematode.   Brazil.

JACARANDA ACUTIFOLIA Humb. & Bonpl. one of some 50 species
    (Brazil-Argentina trees).   Imposing and famous ornamental tree
    noted for its fern-like foliage and abundant blue flowers; "jaca-
    randá" Po. Sp., "jacaranda" En.
    Aecidium spp. (leafrusts in S. America): A. circinatum
        Wint., A. jacarandae P. Henn., and A. puttemansianum
        P. Henn.
    Armillaria mellea Vahl = Armilariella mellea (Fr.) Karst.;
        rootrot.   Subtropical USA.
    Cephaleuros virescens Kunze; algal leafspot.   Brazil, C.
        America (seen in El Salvador, Guatemala, Honduras causing
        early defoliation in dry seasons).
    Cladosporium herbarum Lk.; recovered from pods.   Argentina.
    Clitocybe tabescens Bres. = Armilariella tabescens (Scop.)
        Sing.; rootrot.   Puerto Rico.
    Fomes lamaoensis Murr.; on old trunk, stump rot.   Scattered.
    Hydnum sp.; on dead wood.   Honduras.
    Irene sp.; dark mildew spots.   Honduras.
    Leaf fall (unusual and unseasonable from cause undetermined).
        Puerto Rico, Panama.
    Meliola amphitricha Fr.; black mildew.   Panama.
    Phyllachora jacarandae Bat.; leaf disease.   Brazil.
    Root diseases, various causes but organisms undetermined.
        Scattered.
    Rusts (see Aecidium)
    Strigula complanata Fée; white lichen spot.   Puerto Rico.

JAMBOSA   see   EUGENIA

JANIPHA   see   MANIHOT

JASMINUM GRANDIFLORA L. (a vine), and J. PUBESCENS Willd.
(a bush). Both introduced, much grown ornamentals and although there are several species these two seem most popular;
"jasmim" Po., "jazmín" Sp., "jessamine" "jasmin" En.
Botryodiplodia theobromae Pat. is the conidial state also
    called Diplodia, (it has an asco-state Botryosphaeria quer-
    cuum (Schw.) Sacc. sometimes reported as Physalospora
    rhodina Berk. & Curt.); dieback, branch disease. Puerto
    Rico, Virgin Islands, Panama, Honduras.
Calonectria spp. (asco-carps on canker-diseased stems of J.
    pubescens). Puerto Rico, Costa Rica.
Cassytha filiformis L. on both bush and vine host; green
    dodder. W. Indies, Florida.
Cephaleuros virescens Kunze; algal leafspot. Widespread (in
    moist areas).
Cercospora jasminicola Chupp & Muller; leafspot, blast.
    Honduras, Guatemala, Brazil, Puerto Rico, Venezuela.
Choanephora infundibulifera (Curr.) Sacc.; blossom blight.
    Neotropics.
Cladosporium sp.; leaf-tip disease and leaf-edge rot. Puerto
    Rico.
Clitocybe tabescens Bres. = Armilariella tabescens (Scop.)
    Sing.; rootrot. Subtropical USA.
Colletotrichum gloeosporioides Penz.; the common anthrac-
    nose on leaf and twig. (The asco-state, Glomerella cingu-
    lata (Stonem.) Sp. & Schr., is occasional in wet seasons.)
    Widespread.
Cuscuta americana L.; coarse yellow dodder. Puerto Rico.
Diplodia (see Botryodiplodia)
Elsinoë jasmini Bitanc. & Jenk.; scab. Brazil.
Fusarium oxysporum Schl.; root disease. Argentina. Fusarium
    (several unidentified species); stem cankers. C. America.
Ganoderma sp. (sporocarps severely eroded); on very old stump
    of J. grandiflora. Puerto Rico.
Haplosporella jasmini Ell. &. Ev.; twig disease. Subtropical
    USA.
Marasmius ramealis Bull.; root disease. Subtropical USA (dry
    areas).
Meliola amphitricha Fr.; black mildew. Neotropics (not severe).
Meloidogyne sp.; rootknot. Subtropical USA, Puerto Rico.
Phoma domestica Sacc.; on foliage. S. America.
Phomopsis sp.; stem gall. Subtropical USA.
Phyllosticta jasmini Sacc. and P. jasminensis Bat. & Vit.;
    leaf spotting. Brazil.
Rosellinia bunodes (Berk. & Br.) Sacc. and R. pepo Pat.;
    rootrots. C. America, W. Indies.
Sclerotium rolfsii (Curzi) Sacc. the sclerotial state of Pelli-
    cularia rolfsii (Sacc.) West; basal rot. Florida.
Strigula compalanta Fée; white lichen spot. Common and wide-
    spread.
Trichothyrium oleaceae Gonz. Frag. & Cif.; small leafspots.
    W. Indies.

Uromyces comedens Syd.; leafrust.   Neotropical (mild usually).
Virus: distinct symptoms of typical mosaic mottle and stunt.
   Puerto Rico.

JUGLANS REGIA L.   Introduced from Eurasia, it is the famed nut
   tree somewhat ill adapted to warmer-wet tropics but grows in
   selected parts of the mountains of Mexico, South and Central
   America (J. JAMAICENSIS DC., a lesser tree, is said to come
   from the West Indian mountains); "nogal" Sp., "nogueira" Po.,
   "walnut" En.
   Armillaria mellea Vahl = Armilariella mellea (Fr.) Karst.;
      rootrot. Chile, California.
   Botryodiplodia sp.; dieback.   Occasional.
   Botryosphaeria ribis Gross. & Dug. = B. dothidea (Moug.)
      Ces. & de Not.; canker.   Brazil.
   Cacopaurus epacris Allen & Jens. and E. pestis Thorne;
      rhizosphere inhabiting nematodes.   California.
   Cephaleuros virescens Kunze; algal leafspot on J. jamaicensis
      DC.   Costa Rica.
   Ceratocystis sp.; woodstain disease.   Argentina.
   Cercospora forsteriana Chupp & Viéges; leafspot.   Brazil.
      (Some "typical Cercospora spotted leaves" were collected
      from J. jamaicensis tree in Costa Rica but on microscopic
      examination no spores found.)
   Cladosporium sp.; causing somewhat diffuse leafspot on J.
      jamaicensis.   Costa Rica.
   Colletotrichum sp.; anthracnose spots.   Guatemala.
   Corticium salmonicolor Berk. & Br.; branch rot on J.
      jamaicensis.   Costa Rica, Guatemala.
   Criconemoides sp.; rootzone nematode.   Chile.
   Cryptovalsa nitschkei Fckl.; on injured branch.   Chile.
   Dothiorella gregaria Sacc.; dieback.   Subtropical USA, Chile.
   Exposporina fawcetti E. E. Wils. = pycnidial state is Hender-
      sonula toruloidea Natt.; branch canker.   S. America, Cali-
      fornia.
   Gnomonia juglandis (DC.) Trev. = Marssonina juglandis
      Magn.; anthracnose leafblotch.   Chile, Argentina.
   Hendersonia juglandina Speg.; twig disease.   Chile.
   Microstoma juglandis (Bereng.) Sacc.; white mold.   General.
   Mycena citricolor (Berk. & Curt.) Sacc.; leafspot on J.
      jamaicensis.   Guatemala.
   Oidium sp.; powdery white mildew.   Scattered.
   Phoma juglandicola Speg.; on foliage.   Chile.
   Phytophthora sp.; inkrot of root.   S. America.
   Pleospora multimaculans Heald & Wolf; leaf disease.   Texas,
      Argentina.
   Pratylenchus sp.; rootzone nematode.   Chile.
   Rosellinia necatrix Prill.; rootrot.   Chile.
   Stereum umbrinum Berk. & Curt.; trunk disease.   S. America.
   Strigula complanata Fée; lichen leafspot on J. jamaicensis.
      Costa Rica.
   Trichodorus sp. and Tylenchorhynchus sp.; both rootzone
      nematodes.   Chile.

Valsa juglandicola Speg.; branch rot. Chile.
Xanthomonas juglandis (Pierce) Dows.; blight, bacterial leaf
    disease. Chile, Argentina, Mexico.
Xiphinema sp.; rootzone nema. Chile.

KENTROPHYLLUM see CARTHAMUS

LACTUCA SATIVA L. Although one of the best known and long
.cultivated salad plants in the Old World, it is only recently becoming important in the American tropics as losses in crops
have been great from adverse growing conditions and diseases;
"lechuga" Sp., "alface" Po., "lettuce" En.
    Albugo sp.; white rust. Subtropical USA, Guatemala.
    Alternaria sonchi Davis; large leafspot in a few locations in
        El Salvador, Guatemala, Subtropical USA, W. Indies.
    Aphelenchoides parietinus Stein.; root nematode. Brazil.
    Botrytis cinerea Pers.; common graymold rot. Neotropics.
    Bremia lactucae Regel = B. ganglioniformis (Casp.) Shaw;
        common downy mildew. Widespread (in cool gardens).
    Cercospora longissima Sacc. (synonyms: C. lactucae J. A.
        Stev. and C. lactucae P. Henn.); leafspot, blast. Neo-
        tropics.
    Criconemoides sp.; nematode collected near roots. Peru.
    Cuscuta sp.; dodder. El Salvador.
    Dicaeoma hieraciatum (Jacks.) Arth. on wild species of
        host. (See Puccinia.)
    Erwinia carotovora (L. R. Jones) Holl.; market softrot.
        Common and widespread.
    Erysiphe cichoracearum DC. = Oidium sp.; typical powdery
        mildew. Subtropical USA, Ecuador, Chile, Peru, C.
        America, Venezuela.
    Fusarium solani (Mart.) Sacc.; rootrot, damping-off. Ecuador,
        Panama, Guatemala.
    Helicotylenchus nannus Stein.; nematode from around roots.
        Brazil.
    Heterodera rostochiensis Wr.; root infecting nematode. Peru.
    Leafspot (undetermined cause) medium sized, rounded, red
        corona, not fungus. Costa Rica.
    Marssonina panattoniana (Berl.) Magn.; leaf ringspot, severe
        anthracnose. Florida, Ecuador, Chile, Jamaica.
    Meloidogyne exigua Goeldi and other species; rootknot. Wide-
        spread.
    Monhysteria stagnalis Stein.; nematode in rootzone. Brazil.
    Mycena citricolor (Berk. & Curt.) Sacc.; diseased by inocula-
        tion (leafspot spreads laterally in leaf). Costa Rica.
    Phymatotrichum omnivorum (Shear) Dug.; rootrot. Arizona,
        Texas, Mexico.
    Physarum cinereum Pers.; acting as an epiphyte. Puerto Rico.
    Phytophthora sp. (mycologically difficult at times to be certain
        in mountains of Costa Rica, whether parasite was Phytoph-
        thora or Pythium); irregular bottom rot. Costa Rica, Ecua-
        dor, Guatemala, Panama.

Pleospora herbarum (Pers.) Rabh.; leafspot.   Brazil, Florida.
Pratylenchus sp.; nematode found in rhizosphere.   Peru.
Pseudomonas marginalis (N. A. Brown) Stev.; leafblight.
    Argentina.
Puccinia arthurella Trott.; rust.   Trinidad, Venezuela.   P.
    hieracii (Schum.) Mart.; rust.   California.   (Another rust,
    Dicaeoma, was reported many years ago on wild species
    of Lactuca in W. Indies and it is possible all of these may
    be the same organism.)
Pythium sp.; damping-off (also see Phytophthora).   Neotropics.
Rhizoctonia solani Kuehn is the parasitic mycelial state of
    Thanatephorous cucumeris (Frank) Donk; damping-off and
    bottom-rot.   Neotropical.
Sclerotinia sclerotiorum (Lib.) dBy. = Whetzeliana sclerotiorum
    (Lib.) Korf & Dumont; watery softrot.   Probably in all
    countries, but special reports from at least Ecuador, Argen-
    tina, Chile, Peru, Subtropical USA, and Dominican Repub-
    lic. (The slightly different S. minor Jagger is reported in Chile
    in addition to S. sclerotiorum.)
Sclerotium rolfsii (Curzi) Sacc., its spore bearing state is
    Pellicularia rolfsii (Sacc.) West; soil rot, plant collapse.
    Texas, Florida.
Septoria lactucae Pass.; myriads of small red spots, blast.
    In several countries (often most severe in medium cool lo-
    calities).
Stemphylium botryosum Wallr.; leafspot.   Argentina.
Viruses: The Commelina-Celery Strain of Cucumber Mosaic
    Virus in Costa Rica.   Lettuce Head deformation in Uruguay.
    Typical lettuce Mosaic Virus in Argentina, Brazil, Jamaica,
    Puerto Rico, Venezuela.   Tobacco streak (Brazilian) in
    Brazil.   Tomato Spotted Wilt in Brazil.

LAGERSTROEMIA INDICA L.   The profusely blooming tall bush to
    multi-stemmed short tree with such names as "mirto riza,"
    "astromelia," "astromero," "júpiter" Sp. Po., "crape myrtle"
    En., and L. SPECIOSA (L.) Pers. = L. REGINAE Roxb., a
    large, often tall tree, noted for its long flowering season, wood
    very desirable where available and known, the tree famed for its
    abundant blossoms; "flor de la reina," "reina de las flores,"
    Sp. Po., "queen crapemyrtle," "queen of the flowers" En.
    (Both species are natives of the Orient, diseases, with a few ex-
    amples as noted, attack both species.)
        Botryodiplodia theobroma Pat., which is the conidial state of
            Botryosphaeria quercuum (Schw.) Sacc., also given as
            Physalospora rhodina Berk. & Curt.; branch-rot after
            cutting flowers on crape myrtle, dieback of both species
            (note that some disease lists give such synonymous names

as Diplodia and Lasiodiplodia). Puerto Rico and wide-spread.

Botryosphaeria dothidea (Moug.) Ces. & deNot.; branch disease (only on L. indica). Subtropical USA?, W. Indies.

Calonectria rigidiuscula (Berk. & Br.) Sacc. The typical gall only on L. speciosa is exactly like the cacao green-point gall. In moist chambers when galls finally began disintegration Fusarium spores were recovered. These were long curved macrospores believed to be F. decem-cellulare Brick. Puerto Rico.

Capnodium spp.; sootymolds. Widespread on both hosts.

Cephaleuros virescens Kunze; algal leafspot called rust (re-ports only on L. speciosa). W. Indies, C. America.

Cercospora lythracearum Heald & Wolf, the perfect state is Mycosphaerella lythracearum Wolf; leafspot and blotch mostly on L. indica, but on L. speciosa this leafspot is larger, black and becomes scald-like. W. Indies, C. America, Subtropical USA.

Clitocybe tabescens Bres. = Armilariella tabescens (Scop.) Sing.; mushroom root rot on L. indica. Subtropical USA.

Diplodia (see Botryodiplodia)

Erysiphe lagerstroemiae West = Oidium sp.; white mildew of flower branches only on L. indica. Often severe at start of rains. Widespread.

Fusarium (see Calonectria)

Glomerella cingulata (Stonem.) Sp. & Schr. (ascocarps fre-quent in rainy seasons on certain type of dieback of both hosts, isolations produced typical cultures of Colletotri-chum); dieback. General.

Leptothyrium sp.; leafspeck only on L. speciosa. Puerto Rico.

Loranthus sp.; parasitizing bush on L. speciosa. Puerto Rico.

Mycena citricolor (Berk. & Curt.) Sacc.; leafspot on L. indica in field near diseased coffee, on L. speciosa by inoculation. Puerto Rico.

Oidium (see Erysiphe)

Pellicularia koleroga Cke.; thread-web blight. Puerto Rico, El Salvador, Panama, Honduras. Pellicularia sp. (aerial-type but not P. koleroga or P. filamentosa); leafspot with web on L. indica. C. America.

Pestalotia spp.; small delimited leafspots, sunken lesions on twigs (dryseason injury). C. America, W. Indies.

Phorodendron sp.; mistletoe-like parasite on L. speciosa. Puerto Rico.

Phthirusa seitzii Krug & Urb., P. squamulosa Eichl. and P. theobromae Eichl., all drooping mistletoe-like parasites on L. indica in Surinam, while P. theobromae was collected on L. speciosa in El Salvador and in Puerto Rico.

Phycopeltis lagerstroemiae Ell. & Ev.; leafspot disease on both hosts, but in addition tip blight only on L. indica. Subtropical USA, W. Indies.

Ragged-leaf on both host species, no single cause known, numerous weakly parasitic organisms found in withered tissues.   Puerto Rico.

Septoria sp.; leaf disease.   S. America.

Stomatochroon lagerheimii Palm; systemic algal leaf lesion. In Guatemala (on L. indica), in Puerto Rico (on L. speciosa).

Strigula complanata Fée; white lichen spot, only on L. speciosa. Puerto Rico, Costa Rica, Panama.

Trunk lesions called "caries," at base.   Basidiomycetous mycelium recovered, no sporulation for determination.   Puerto Rico.

Virus (undetermined) on L. indica in Puerto Rico.   Symptoms of streaking, typical mottle, leaf malformation.

Witches' broom of crape myrtle yielded Botryodiplodia sp. and other fungi, studies incomplete.   Puerto Rico.

LANTANA CAMARA L.   Native to tropical America, used as cover crop and soil erosion control, is both a weed and a prized ornament; "combará," "chumbinho" Po., "cariaquillo" Sp., "lantana" En.

Acanthostigma lantanae Theiss.; leaf disease.   Brazil.

Aecidium lantanae Mayor (synonyms: A. verbenicola Speg.; A. spegazzianum Sacc. & Trott.); rust. Brazil, Ecuador, Colombia, Nicaragua, Panama, probably Neotropical.

Alternaria sp.; rare leafspot.   Subtropical USA.

Asterina vagans Speg.; leaf disease.   S. America.

Asteridium radians Rehm (possibly a Meliola); black mildew with specially symmetrical hyphal spread.   Reported from Brazil on wild Lantana, very similar spot collected on L. camara in Costa Rica.

Botryosphaeria quercuum (Schw.) Sacc.; stem rot.   W. Indies, Florida.

Caeomurus (see Puccinia)

Cassytha filiformis L.; slender green dodder.   Puerto Rico.

Cephaleuros virescens Kunze; algal leafspot on unidentified species of Lantana.   Brazil.

Cercospora guianensis Stev. & Solh.; leafspot.   Brazil, nearby countries.   C. lantanae Chupp; leafspot.   Probably general.

Cuscuta americana L.; yellow dodder.   Puerto Rico.

Diatrypella lantanae Earle; stem disease.   Subtropical USA, W. Indies.

Epiphyma nervisquens (Chard.) Mill. & Burton; stem disease. Brazil.

Meliola ambigua Pat. & Gaill.; black mildew of leaf and stem spots.   W. Indies, Panama.   M. cookeana Speg., black mildew.   Florida.   M. lantanae Syd. & P. Syd.; black mildew.   Brazil.

Meloidogyne sp.; rootknot.   Scattered.

Mycena citricolor (Berk. & Curt.) Sacc.; leafspot.   Costa Rica.

Nectria lantanae Seav. (on L. odorata L.); stem disease. Bermuda.

Oidium sp.; white and scanty powdery mildew. El Salvador
(well up on side of Volcano Izalco).

Perisporina lantanae Stev.; leaf disease. W. Indies, C.
America, S. America.

Phyllachora sororcula Speg.; leaf disease. Paraguay, Brazil.

Phyllosticta lantanae Pass.; leafspot. Puerto Rico.

Phymatotrichum omnivorum (Shear) Dug. on wild Lantana;
dryland rootrot. Dry-hot Subtropical USA, Mexico.

Prospodium tuberculatum (Speg.) Arth. = Uredo tuberculata
Speg. W. Indies, Mexico, C. America.

Puccinia lantanae Farl. (synonyms: Uromyces lantanae Speg.,
Caeomurus lantanae (Speg.) Kuntze, Micropuccinia lantanae
(Farl.) Arth. & Jacks.); common rust. Neotropical.

Rosenscheldia paraguaya Speg.; stromatic stem disease.
Paraguay.

Septoria lantanae Gar.; small leafspot. Scattered.

Trabutia lantanae P. Henn.; leaf disease. Argentina.

Trochodium ipomoeae (Th.) Syd. (? = Uromyces), a rust.
Scattered in S. America.

Uredo (see Prospodium)

Uromyces dubiosus P. Henn.; rust. Brazil. Also U. lantanae
Speg. = Puccinia lantanae Farl.; rust. Brazil.

LAPAGERIA ROSEA Ruíz & Pav. From Chile and adapted to cool
tropics, an ornamental climber with showy long trumpet flowers,
becoming popular as a special plant; "copihue" Sp.

Cheilaria pulicaris Mont.; on leaves. Chile.

Cryptostictis lapageriicola Speg.; on stems and leaves. Chile,
Brazil.

Diaporthe eres Nits. and D. jaffueli Speg.; stem diseases.
Chile.

Hysterium chilense Speg.; stem disease. Chile.

Metasphaeria jaffueli Speg.; on leaves. Chile.

Pestalotia lapageriae P. Henn.; foliage lesions. Chile.

Phoma lapageriae P. Henn.; on stem. Chile.

Phyllosticta jaffueli Speg.; leaf spots. Chile.

Physalospora lapageriae Speg. = Glomerella lapageriae (Speg.)
Petr. & Syd.; stem disease. Argentina, Brazil.

Pleospora lapageriae Speg.; leafspot. Chile.

Sphaerella lapageriae Speg.; on dead foliage. Chile.

Trochila jaffueli Speg.; on foliage. Chile.

LATHYRUS ODORATUS L. Known for its fragrant flowers, it is of
Italian origin, grown in cool American tropics; "clarín" Sp.,
"ervilha-de-cheiro" Po., "sweet pea" En. (Has had only minor
attention to tropical diseases.)

Ascochyta lathyri Trail; stem lesion. Scattered reports in
C. America, S. America.

Botrytis cinerea Pers.; blossom blight, shoot graymold.
Scattered in moist-humid seasons. Widespread.

Cladosporium album Dows. (? - Erostrotheca multiformis
Mart. & Charles); whitemold. Cloudy area in Peru.

Colletotrichum gloeosporioides Penz. may deserve to be desig-
nated at least f. sp. pisi as isolates of it from Guatemala
did not infect ripe apple fruit, green banana, or young
squash; common anthracnose.   Peru, Guatemala.
Erwinia (? carotovorus): occasional market softrot of flower
stems.
Erysiphe polygoni DC.; white powdery mildew.   Chile, Argen-
tina, Peru.
Fusarium oxysporum Schl. f. sp. vasinfectum (Atk.) Sny. &
Hans.; wilt.   Florida.   F. solani (Mart.) Sacc. f. sp. pisi
(Jones) Sny. & Hans.; rot of stem and root.   Florida,
Ecuador.
Meloidogyne sp.; rootknot.   Florida.
Oidium sp. (? balsami); white powdery mildew disease (see
Erysiphe).
Peronospora trifoliorum dBy.; downy mildew.   Florida.
Phyllosticta sp.; leafspot.   Peru.
Polythrincium lathyrinum Syd.; black blotchspot.   Chile.   (Re-
ported as on Lathyrus sp., questionably L. odoratus.)
Pratylenchus pratensis Filipj.; root nematode.   Brazil.
Pythium sp.; in cool soils, damping-off and pre-emergence
seed decay.   Neotropics.
Rhizoctonia solani Kuehn the mycelial and parasitic state has
a basidial spore bearing state Thanatephorous cucumeris
(Frank) Donk = sometimes reported as Pellicularia filamen-
tosa (Pat.) Rogs.; the most common damping-off if soil is
moist and warm.   Scattered.
Rhizopus sp.; rot of petals on diseased flowers, market in
Guatemala.
Rusts are on Lathyrus spp. and should be watched for occur-
rence on sweet pea.   These rusts are generally species of
Uromyces.   S. America.
Sclerotinia rolfsii (Curzi) Sacc. = Pellicularia rolfsii (Sacc.)
West; ground-line rot.   Subtropical USA, ? symptoms seen
in Peru.   S. sclerotiorum (Lib.) dBy. = Whetzeliana
sclerotiorum (Lib.) Korf & Dumont; typical top rot with
sclerotia.   Guatemala.
Verticillium sp.; wilt.   Peru.
Virus mottlings seen (no determinations made) in C. America,
Peru.

LENS ESCULENTA Moench. (= L. CULINARIS Medic. and ERVUM
LENS L.)   A crop unknown as to precise origin but probably
man's oldest cultivated grain; grows with relatively small amounts
of moisture and fertility, has good quality seeds rich in protein,
crop called "lentilha" Po., "lenteja" Sp., "lentil" En.   (Intensive
disease studies in the American tropics just started.)   (Note:
This crop shows in the American tropics good adaptability, but
suffers from its Old World reputation as being the poor man's
food; in the Americas potentialities are considerable as a product
in protein-starved areas.   It should be pointed out that in Eurasia
this plant has remarkable variability under numerous types of

conditions, and newer kinds may eventually be secured for the tropical New World.)

Ascochyta pinodella L. K. Jones = Mycosphaerella pinodes (Berk. & Blox.) Vest. ; blight, footrot. Ecuador. A. pisi Lib. ; footrot. Chile.

Cladosporium sp. ; secondary invasion of old leaf. Chile.

Fusarium spp. ; wilts and root diseases. Scattered in S. America, some severe in Chile. Uncertain as to species, possibly many reports are F. roseum Lk. and forms of it. A "yellows" from a Fusarium is known in Peru.

Paratylenchus sp. ; nematode near roots. Chile.

Phyllosticta phaseolina Sacc. ; leafspot. Peru.

Pratylenchus sp. ; nematode near roots. Chile.

Rhizoctonia solani Kuehn = Thanatephorous cucumeris (Frank) Donk; damping-off. Ecuador.

Sclerotinia sclerotiorum (Lib.) dBy. = Whetzeliana sclerotiorum (Lib.) Korf & Dumont; plant rot. Chile.

Thielaviopsis basicola (Berk. & Br.) Ferr. ; root disease. Subtropical USA.

Uredo lentis Lagh. ; rust. Scattered.

Uromyces fabae dBy. ; rust. Peru, Chile, Brazil, Ecuador.

LEUCAENA GLAUCA Benth. A medium-sized shade tree used in plantation crops, acacia-like and important in some areas as browse for cattle and goats (horses lose their manes on eating it), valued soil protection; "acácia" Sp. Po. , "campeche, " "hedindilla" Sp. , "leucena, " "white popinac" En.

? Armilariella sp. (no sporophores, abundant shoe-string rhizomorphs on roots of dying trees); El Salvador.

Botryodiplodia theobromae Pat. ; common dieback. C. America.

Botryosphaeria dothidea (Moug.) Ces. & de Not. ; twig canker. C. America.

Camptomeris leucaenae (Stev. & Dal.) Syd. = Exosporium leucaenae Stev. & Dal. ; yellow leafspot. Venezuela, W. Indies.

Cassytha filiformis L. ; green dodder. Puerto Rico.

Cuscuta americana L. ; yellow dodder. Puerto Rico.

Ganoderma sp. ; tree base and stump decay. El Salvador.

Loranthus spp. ; bushy mistletoe-like parasites. El Salvador.

Oryctanthus spp. ; scandent phanerogamic parasites. Costa Rica, El Salvador.

Pythium sp. ; damping-off in seed bed. Guatemala.

Ravenelia sp. ; leaf rust. Guatemala, Mexico, El Salvador.

Strigula spp. , lichen milky white, through pink to gray spots on leaves and petioles. (No algae found as an associated leafspot.) El Salvador, Guatemala.

Thanatephorous cucumeris (Frank) Donk; abundant on debris near seedlings, the Rhizoctonia caused damping-off and strangle of seedlings. El Salvador, Guatemala.

Triblidiella rufula (Spreng.) Sacc. ; on die-back stems. Bermuda.

Virus symptom; mild leaf mottle and slight malformation. C. America.

LIBOCEDRUS CHILENSIS Endl.   A medium tall tree grown for ornament and wood, native of Chile, and L. DECURRENS Torr. much
   taller tree, valued for wood and ornament, origin western North
   America, was successfully introduced into S. America (much of
   the disease identifications in this section are from California and
   on the latter species).   Common names are such as "cedro
   incenso" Po., "cipres," "alérce" Sp., "incense cedar" En.
      Agrobacterium tumefaciens (E. F. Sm. & Towns.) Conn;
         crowngall.   California, Arizona.
      Anthostomella alcaluf Speg.; aging foliage disease.   Chile.
      Caeoma espinosae Syd.; rust.   Chile (S. America).
      Camaropycnis libocedri Cash; foliage disease.   California
      Camarosporium magellanicum Speg., on dead leaves of L.
         TETRAGONA Endl.   Chile.
      Chloroscypha jacksonii Seaver; twig disease.   California.
      Coryneum cardinale Wag.; branch canker.   California.
      Cytosporium fueguianum Speg.; foliage disease on L. tetragona.
         Chile, Argentina.
      Fomes pinicola (Sw.) Cke.; heartrot.   California.
      Gymnosporangium libocedri (P. Henn.) Kern; rust causing gall
         on leaf and branch, and witches' broom.   S. America.
      Herpotrichia nigra Hartig; feltblight.   California.
      Lenzites lepidus Fr.; trunk brown-heartrot.   General.
      Leptostroma magellanica Speg.; decaying branch and leaf of
         L. tetragona.   Chile, Argentina.
      Lophodermium juniperum (Fr.) deN.; needle-cast.   California.
      Marasmius ramealis Fr. = Marasmiellus ramealis (Bull.)
         Sing.; thread-cord disease.   Chile.
      Myzodendron sp.; dwarf mistletoe, report from Chile.
      Patinella xylographioides Rehm; lichen-like, leaf infection.
         Chile.
      Phoradendron spp.; mistletoe-like, woody and bushy branch
         parasites.   Widespread.
      Phymatotrichum omnivorum (Shear) Dug.; dry-land rootrot.
         Texas, ? Mexico.
      Polyporus spp., rots of dead trunks and large branches.
         General.
      Polystictus hirsutulus Schw.; woodrot.   Chile.
      Pseudopithyella minuscula Seaver; branch disease.   Bermuda,
         California.
      Sphaerella magellanicola Speg.; rot of foliage on L. tetragona.
         Southern S. America.
      Stereum hirsutum Willd., on dead branches.   California.
      Stigmatea sequoiae (Cke. & Harkn.) Sacc.; on foliage.   Cali-
         fornia.
      Trametes isabellina Fr.; pocketrot.   California.
      Tryblidiella macrospora Bonar & Cash; dieback.   California.
      Xylographa parallela Fr. (Lichen).   Chile.

LIGUSTRUM spp.   Ornamental bushes introduced from the Orient,
   used for hedges and windbreaks; "alheña," "ligustro" Sp.,
   "ligustrina," "alfeneiro" Po., "ligustrum," "privet" En.

Alternaria sp. (spores secured from scrapings off twigs with anthracnose dieback). El Salvador.

Armillaria mellea Vahl = Armilariella mellea (Fr.) Karst. ; rootrot. California.

Cephaleuros virescens Kunze; algal leafspot. Subtropical USA, Costa Rica, Argentina.

Cercospora liguistri Roum. ; large leafspot. Subtropical USA, C. America, S. America.

Chlorosis, apparently deficiency of minor elements. Causing both yellowing and stunt, bud death. Cuba.

Clitocybe tabescens Bres. = Armilariella tabescens (Scop.) Sing. ; rootrot. Subtropical USA, W. Indies.

Colletotrichum gloeosporioides Penz. (= C. ligustri Putt. or Gloeosporium ligustri P. Henn.) its perfect state is Glomerella (see below); anthracnose leaf and stem diseases. Brazil, Costa Rica, Colombia, Cuba, Subtropical USA, Puerto Rico.

Epicoccum lingustri P. Henn. ; on foliage, possible hyperparasite of other fungi. Occasional.

Glomerella cingulata (Stonem.) Sp. & Schr. ; identified as cause of stem canker. Puerto Rico. (Asco-state of Colletotrichum, see above.)

Lachnea hemisphaerica (Wigg.) Gill. (probably saprophytic); on dieback tissues. S. America.

Meloidogyne spp. ; mild rootknot effect. Subtropical USA.

Pellicularia koleroga Cke. ; thread-web blight. Scattered.

Pestalotia sp. ; leafspot, stem lesion. Neotropical.

Phyllosticta lingustri Sacc. ; small leafspot. Subtropical USA, S. America.

Phymatotrichum omnivorum (Shear) Dug. ; dryland rootrot. Texas, said to be in Mexico.

Pseudomonas garcae Am. , Teix. , & Pinh. ; bacterial leafspot. Brazil.

Pythium spp. ; cutting decay. C. America.

Strigula complanata Fée; lichen milky-spot. Common, but no reports from subtropical USA.

Virus: "lepra explosiva" on stems in Argentina.

Xenosporella berkleyi (Curt.) Lind. ; on foliage. S. America.

LINUM USITATISSIMUM L. An erect herbaceous annual, one of most famous fiber and oil plants of agriculture, was cultivated in the Old World since Egyptian antiquity, and was successfully introduced into Argentina and high-dry parts of other Neotropic countries, the moist-warm climates are not good for the crop; "lino, " "linaza" Sp. , "linho" Po. , "flax" En.

Alternaria sp. (A. alternata ?); leafspot, capsule blackening. Subtropical USA, Peru. A. solani (Ell. & Mart.) Sor. ; damping-off complex. Argentina.

Cercospora sp. ; leafspot. Brazil.

Cladosporium herbarum Lk. ; on dry leaf and dry field stalks. Brazil.

Colletotrichum lini Toch. (synonyms: C. lini Boll. and C.

linicola Peth. & Laff.); anthracnosis, blight.   Argentina,
Peru, Brazil, Chile, Uruguay.
Cuscuta indecora Choisy; dodder.   Peru.
Fusarium spp.   (Note: In this book the Fusarial parasites
    have been given in as simple a manner as possible.   In
    looking backward and forward in these pages it is to be
    remembered that Fusaria originated in warm lands, and
    their evolution deserves attention of mycologists living
    permanently in the tropics; in fact, major study should be
    dedicated to these exceedingly important and remarkably
    varied parasites.)   1. F. lateritium Nees; root and stem
    rot.   Argentina (southern S. America).   2. F. oxysporum
    Schl. f. sp. lini (Boll.) Sny. & Hans., damping-off, wilt,
    wherever susceptible types of flax are grown in subtropical
    or tropical Americas.   3. F. reticulatum Mont. = F.
    heterosporium Nees; flower and stem disease.   Argentina.
    4. F. scirpi Lamb. & Fautr. f. sp. acuminatum (Ell. &
    Ev.) Wr.; rootrot.   Argentina.   5. F. solani (Mart.)
    Sacc.; rootrot.   S. America.
Melampsora lini (Pers.) Lév. (sometimes given as M. lini
    Cast.); rust of leaf, stem, capsule.   Brazil, Argentina,
    Uruguay, Chile, Peru, Bolivia.
Mycosphaerella linorum (Wr.) Garcia-Rada (synonyms: M.
    linicola Naum., Sphaerella linorum (Wr.) Garcia-Rada,
    Septogloeum linicola Speg., Phylctaena linicola Speg.);
    causes pasmo, a variable spotting and blast of seedling
    cotyledons, mature plant leafspot, stem lesion and a cap-
    sule disease.   Subtropical USA, Peru, Brazil, Argentina,
    Uruguay (common in moist areas).
Oidium sp. (? = Erysiphe sp.); powdery white mildew.
    Argentina, Peru.
Olpidiaster radicis Pas.; root disease.   Argentina.
Olpidium brassicae Dang., root disease.   Uruguay.
Pasmo (see Mycosphaerella)
Phlyctaena (see Mycosphaerella)
Phoma sp.; black footrot.   Chile.
Polyspora lini Laff.; stem-break, discolored fiber.   Argentina.
Pythium ultimum Trow; damping-off.   Argentina.
Rhizoctonia solani Kuehn = Thanatephorous cucumeris (Frank)
    Donk; damping-off.   Wherever crop is grown.
Sclerotinia minor Jagger; flower disease.   El Salvador.   S.
    sclerotiorum (Lib.) dBy. = Whetzeliana sclerotiorum (Lib.)
    Korf & Dumont; blast and stemrot.   Subtropical USA, C.
    America, Peru.
Sclerotium rolfsii (Curzi) Sacc. = Pellicularia rolfsii (Sacc.)
    West; stem basalrot.   Subtropical USA.
Septogloeum (see Mycosphaerella)
Sphaerella (see Mycosphaerella)
Viruses: Aster Yellows Virus causing stunt and chlorosis in
    Mexico.   Beet Curly Top Virus causing stunt and much
    malformation in Subtropical USA.

LITCHI CHINENSIS Sonn. = NEPHELIUM LICHI Camb. Famed
fruit tree of China, appears increasingly grown in the American
tropics, most common name is "litchi."
 Cephaleuros virescens Kunze; algal leafspot. Costa Rica.
 Cladosporium sp.; leafedge disease (end of dryseason). Costa
 Rica.
 Clitocybe tabescens Bres. = Armilariella tabescens (Scop.)
 Sing.; rootrot. Florida.
 Colletotrichum gloeosporioides Penz. = C. litchis Gonz. Frag.
 & Cif.; anthracnose. Costa Rica, Florida, W. Indies.
 Eutypa erumpens Mass.; stem disease. Brazil.
 Loranthus sp.; (seedling on branch, removed by gardener before
 parasite had grown to maturity). Costa Rica.
 Meliola sp.; black mildew. Costa Rica.
 Mycena citricolor (Berk. & Curt.) Sacc.; leafspot. Costa Rica.
 Phyllosticta sp.; small leafspot. Costa Rica.
 Septobasidium sp., felt. Costa Rica.

LOLIUM spp. Perennial and annual pasture grasses from Eurasia,
and grown with some success in cooler-dryer parts of Neo-
tropics, has shown failure in warmer and low locations; "ballica"
Sp., "azevem" Po., "ryegrass" En.
 Chaetomium kunzeanum Zopf; on aging plants. Widespread.
 Claviceps purpurea (Fr.) Tul.; ergot. Colombia, Argentina,
 Brazil, Chile, Guatemala, El Salvador.
 Entyloma crastophyllum Sacc.; leafsmut. Colombia.
 Erysiphe graminis DC.; perfect state of Oidium (see below).
 Fusarium spp.; largely weak parasites. Widespread.
 Gibberella sp.; basal decay in cool areas.
 Oidium monilioides Desm. (also other species are named);
 imperfect states of Erysiphe. White powdery mildew.
 Common where the host is grown.
 Ophiobolus graminis Sacc. or O. cariceti Sacc.; footrot. S.
 America.
 Ovularia lolii Volk; on leaves report from S. America (?
 Brazil).
 Passalora punctiformis Sacc. (? = Cercospora with two celled
 spores); leafspot. Colombia.
 Pratylenchus sp.; nematode found in rhizosphere. Chile.
 Puccinia spp. rusts, all reported from S. America as follows:
 P. brachypus Speg., P. brachypus Speg. f. sp. loliiphila
 Speg. = P. recondita Rob., P. coronata Cda., P. coronata
 Cda. f. sp. lolii Erik., P. cryptica Arth. & Holw., P.
 graminis Pers., P. graminis Pers. var. avenae Erik. &
 E. Henn., P. graminis Pers. var. tritici Erik. & E. Henn.,
 P. striiformis Rest.
 Pythium sp.; cottony blight. Florida.
 Sclerotium clavus DC.; decay of plant. Brazil.
 Thanatephorous cucumeris (Frank) Donk = Rhizoctonia solani
 Kuehn; leaf and stem disease. Scattered in lower edges of
 hill pastures. C. America.
 Tylenchorhynchus sp.; nematode in root zone. Chile.

Ustilago striiformis (West.) Niessl.; headsmut, abortion.
    Colombia.

LONCHOCARPUS spp.   Several species of small trees and climbers
    indigenous to tropical Americas that are sources of fish stunning
    solutions put into rivers by native fishermen and also are em-
    ployed to kill rats, mice, and insects.   The plants are secured
    from jungle sources, but also multiplied by hand of man through
    stem cuttings.   They have numerous vernacular names: "bar-
    basco," "timbo," "cincho," "geno."   (Has had practically no
    serious study as to plant pathological problems.)
        Alternaria sp. (large spored) also reported as Macrosporium
            sp.; leafspot.   Colombia.
        Atelcauda (see Pileolaria)
        Cephaleuros virescens Kunze; algal leafspot.   Costa Rica.
        Cephalosporium deformans Crand.; on leaf and stem.   Peru.
        Cercospora atro-purpurescens Chupp; leafspot.   Colombia.
            C. gliricidiae Syd. & P. Syd. = C. gliricidiasis Gonz.
            Frag. & Cif.; leafspot.   W. Indies, S. America, C.
            America.   C. lonchicarpi J. A. Stev.; leafspot.   Brazil,
            Peru, British Guiana.
        Chlorosis (minor element deficiencies), roots shrunken and
            foliage yellowed.   Costa Rica, Nicaragua.
        Colletotrichum gloeospoioides Penz.; anthracnosis.   Neotropics.
        Dicheirinia archeri Cumm.; rust.   Surinam, Peru.   D.
            guianensis Cumm.; rust.   Northern S. America.   D.
            manaosense P. Henn.; brown leaf rust.   Brazil.
        Endothella lonchocarpicola (P. Henn.) Th. & Syd.; stromatic
            leaf lesions.   Brazil.
        Irenina lonchocarpi (Speg.) Stev.; black mildew.   S. America.
        Lenzites palisoti Fr.; woodrot.   C. America, S. America.
        Meliola bicornis Wint.; probably the most common black mil-
            dew.   Neotropical.   M. bicornis Wint. var. lonchocarpi
            Bat.; black mildew in Brazil.   M. lonchocarpi Speg. = M.
            lonchocarpicola Stev.; a common black mildew.   Argentina,
            W. Indies, Brazil, Nicaragua.   M. lonchocarpicola Stev.
            var. sancti-dominici Cif.; black mildew.   Dominican Re-
            public.   M. malacotricha Speg.; black mildew.   Argentina,
            Brazil.
        Microthyriella roupalae Syd.; small spots on foliage.   Brazil.
        Nectria theobromae Mass.; stem lesion.   S. America.
        Ophiodothella atromaculans (P. Henn.) Hoehn.; stromatic leaf-
            spots.   Brazil.
        Pellicularia filamentosa Rog. (an air-borne Rhizoctonia);
            target leafspot.   Peru.
        Phyllosticta sp.; small leafspot.   Colombia, Nicaragua.
        Pileolaria incrustans (Arth. & Cumm.) Thir. & Kern. =
            Atelcauda incrustans Arth. & Cumm.; crusty rust.   Panama.
        Porella sp.; leafy liverwort smother, in wet areas.   Nicaragua,
            Costa Rica, Panama.
        Puccinia arechavaletae Speg. = Caeomurus aeruginosus Kuntze;
            yellow leafrust.   Argentina, Uruguay, Brazil.   P. nephroidea
            Syd.; orange leafrust.   Brazil.

Ravenelia bakeriana Diet. ; brown leafrust.   Brazil.   R. lon-
    chocarpi Lagh. & Diet. ; a common brown leafrust. Brazil,
    Cuba, Guatemala, El Salvador (probably Neotropical).   R.
    lonchocarpicola Speg. ; leafrust.   Argentina.   R. mera
    Cumm.; leafrust.   C. America.   R. parahybana Viégas;
    leafrust. Brazil. R. tavaensis Viégas; leafrust.   Brazil.
Rhizoctonia solani Kuehn the mycelial parasitic phase of
    basidiospore-bearing Thanatephorous cucumeris (Frank)
    Donk; basal part of stem shrunken disease, cutting rot.
    El Salvador.
Trabutia amphigena Chard. and T. conica Chard.; both ir-
    regular shaped scabby leafspots.   W. Indies, S. America.
Virus: Mild leaf mottle from an unidentified mosaic-type
    virus in Nicaragua.

LUCUMA spp. [see also ACHRAS ZAPOTA]. (Perhaps most com-
    monly grown is L. NERVOSA DC.), this is a genus of many
    trees from tropical America and it has been planted and pro-
    tected in jungle growths among Indians for uncounted centuries
    as it is a welcome source of fruit; "caimito, " "lúcumo, " "mamey
    colorado, " "cánistel, " "abió. " (Except for a few disease collec-
    tions, no concerted plant pathological study has been developed
    on this important fruit genus.)
    Acrotelium lucumae (Arth. & Johnst.) Cumm. (= Uredo
        lucumae Arth. & Johnst.); powdery rust.   Cuba, Colombia,
        C. America.
    Ascochyta lucumae (P. Henn.) Wr. ; leafspot.   C. America.
    Aspergillus spp. ; fruit spot.   General.
    Botryodiplodia theobromae Pat. , = Diplodia sp. (see also
        Physalospora); dieback.   C. America.
    Catacauma cabalii Garces-O. ; leafspot.   Colombia.
    Cephaleuros virescens Kunze; algal leafspot.   Brazil.
    Colletotrichum gloeosporioides Penz. ; rootrot in Dominican
        Republic, fruitspot.   Widespread.
    Diaporthe eres Nits. = D. quilmensis Speg. ; fruitrot, dieback.
        S. America.
    Dityopeltis lucumae Bat. ; on leaves.   Brazil, C. America.
    Eutypa erumpens Mass. ; twig disease.   Trinidad.
    Fumagopsis triglifioides Speg. ; black fungus on leaves.   S.
        America.
    Ganoderma sp. ; base-of-trunk rot.   Honduras.
    Hydnum sp. ; dead branch woodrot.   Honduras.
    Irenina lucumae Cif. ; black mildew.   W. Indies, C. America.
    Loranthaceae are of widespread occurrence, often seen on
        caimito trees (also see Phoradendron).
    Macrophoma sp. (? = Diaporthe sp.); dieback.   C. America,
        S. America.
    Meliola lucumae Stev. ; black mildew.   Widespread.
    Nectria cainitonis P. Henn. ; twig canker.   S. America.
    Oidium sp. ; white mildew.   Specially reported in Peru, proba-
        bly widespread.
    Pestalotia lucumae Tehon; leaf and stem lesion.   Neotropical.

Phoradendron cheirocarpum Tul. and other species and genera of phanerogamic parasites on branches. Mexico, C. America.

Phyllosticta lucumae Syd.; leaf disease. Uruguay.

Physalospora obtusa Cke. = Botryosphaeria quercuum (Schw.) Sacc. is the asco-state of Botryodiplodia.

Poria sp.; dead branch rot. Honduras.

Puccinia lucumae Kern, Thur. & Whet.; leafrust. Scattered.

Scopella bauhiniicola (P. Henn.) Cumm., and S. lucumae Cumm.; yellow rusts on leaves. Reports especially from Brazil, (widespread?).

Strigula complanata Fée; common lichen spot on leaves. C. America, S. America. (Not seen in W. Indies or Florida.)

Symphaeophyma subtropicale Speg.; on foliage. Argentina, Brazil.

LUFFA ACUTANGULA Roxb. and L. AEGYPTIACA Mill. = L. CYLINDRICA Roem. Two squash-like crops introduced from Asia, producing an edible vegetable when young, and old fruits when retted in water provide sponge-like fibers for household use; "bucha" Po., "estropajo," "esponja" Sp., "sponge gourd" En. (Diseases listed common to both.)

Ascochyta (see Mycosphaerella)

Botryodiplodia sp. (small spored, tawny culture, not the common species called theobromae); vine lesion. El Salvador.

Cephaleuros virescens Kunze; algal leafspot. Scattered in W. Indies and C. America.

Cercospora citrullina Cke.; leafspot in Guatemala. C. cucurbitae Ell. & Ev.; leafspot in W. Indies, El Salvador, Brazil.

Colletotrichopsis luffae Bat. & da Sil.; anthracnose-like disease. Brazil.

Colletotrichum lagenarium (Pass.) Ell. & Halst. = C. orbiculare (Berk. & Mont.) Arx; common anthracnose. Everywhere grown.

Cuscuta americana L.; yellow dodder (by inoculation). Puerto Rico.

Diaporthe sp.; market stem-end blackrot. Guatemala.

Erwinia carotovora (L. R. Jones) Holl.; bacterial softrot. Widespread.

Erysiphe cichoracearum DC.; white powdery mildew. C. America in mountains.

Fusarium semitectum Berk. & Rav., F. roseum Lk., F. solani (Mart.) Sacc. f. sp. cucurbitae Sny. & Hans.; all three species recovered from root and stem rots. W. Indies, C. America.

Glomerella (? cingulata-like but with larger asci), asco-state of apparently a Colletotrichum on stems from highland grown plants in Costa Rica.

Guignardia momordica Rang.; spots on leaf and vine. Argentina.

Irenina? (not Meliola); black mildew. El Salvador.

Macrophoma sp.; stem disease. Colombia.

Meloidogyne sp.; rootknot. Florida.
Micropeltis luffaeicola Bat.; black mold-spots on leaf. Brazil,
    El Salvador.
Mycosphaerella citrullina (C. O. Sm.) Gross is the perfect
    state of Ascochyta citrullina Sm.; leaf and stem spot in
    cooler and dryer areas of C. America.
Pellicularia koleroga Cke.; leafrot and vine-end decay.   El
    Salvador, Guatemala, Puerto Rico, Colombia, Argentina.
Phytomonas cucurbitae Berg.; bacterial stem disease.
    Colombia, El Salvador.
Pseudoperonospora cubense (Berk. & Curt.) Rost.; mildew.
    Subtropical USA, C. America.
Pythium debaryanum Hesse (most likely in cool land), and P.
    ultimum Trow (most likely in warm land); damping-off.
    C. America, S. America.
Rhizoctonia solani Kuehn = Thanatephorous cucumeris (Frank)
    Donk; damping off. Widespread in warm locations.
Viruses. Cucumber Mosaic Virus in C. America, Florida.
    CMV Banana Race in C. America. CMV Celery-Comme-
    lina Race in Subtropical USA, Puerto Rico, El Salvador.

LUPINUS ALBUS L. and L. ANGUSTIFOLIUS L. Old World legu-
    minous crops introduced for soil enrichment, animal fodder, and
    seed (water soaked to remove bitterness) in human food; "lupine, "
    "tremoceiro" Po., "lupino, " "tarhuí" Sp., "lupin, " "lupine" En.
    Alternaria alternata (Fr.) Keiss. = A. tenuis Nees, blackening
        of old tissues. Rare, found in El Salvador and Guatemala.
    Botrytis cinerea Pers.; moldyblight. C. America in warm-
        moist areas.
    Ceratophorum setosum Kirch. = Pestalotia lupini Sor.; leaf
        brownspot, occasional stem lesion. W. Indies, Subtropical
        USA.
    Chaetoseptoria wellmanii J. A. Stev.; leafspot.   C. America.
    Chrysocelis lupini Lagh. & Diet.; leafrust. Subtropical USA,
        Costa Rica, Peru, Colombia, Argentina, Bolivia, Brazil,
        Ecuador.
    Colletotrichum dematium (Pers.) Grove; anthracnose.   Nica-
        ragua, Guatemala, Brazil.
    Cylindrosporium lini Ell. & Ev.; on leaf.   Subtropical USA.
    Erysiphe sp. is the ascospore state of Oidium.
    Fusarium avenaceum (Cda.) Sacc. asco-state is Gibberella
        avenacea Cook; seedling root and stem rot, wilt. Wide-
        spread. F. equiseti (Cda.) Sacc. = F. scirpi Lamb. &
        Fautr. asco-state G. intricans Wr.; pod disease, stemrot.
        C. America, S. America. F. oxysporum Schl. (no asco-
        state); wilt; isolated from rot of side roots. C. America.
        F. oxysporum Schl. f. sp. aurantiacum (Lk.) Wr., wilt.
        S. America. F. solani (Mart.) Sacc. with asco-state
        Nectria haematococca Berk. & Br. both states from old
        diseased stems in experimental plot. Guatemala. F.
        solani (Mart.) Sacc. f. sp. lupini Weimer; rootrot. C.
        America.

Helminthosporium sp.; leafspot. C. America.

Macrophomina phaseoli (Taub.) Ashby; stemblight.   Subtropical
    USA, El Salvador (Pacific slope), Puerto Rico.

Meloidogyne spp., rootknot nematodes. Florida, C. America.

Oidium sp. white powdery mildew, with occasional Erysiphe
    ascocarps in coolest dry areas. Scattered.

Pestalotia (see Ceratophorum)

Phomopsis rossianum Sacc.; on stem. Brazil, Bolivia.

Phyllosticta ferax Ell. & Ev. and P. lupini Bonar; shiny black
    leafspots. Bolivia, Colombia, W. Indies, Subtropical USA.

Phytophthora sp.; stem rot. C. America.

Rhizoctonia solani Kuehn the infective sterile mycelium of the
    basidial form Thanatephorous cucumeris (Frank) Donk;
    causing damping-off, stem lesion, occasional leafrot.   Neo-
    tropics but scattered.

Sclerotium rolfsii (Curzi) Sacc. = Pellicularia rolfsii (Sacc.)
    West which is the basidiospore state; bottom rot.   Sub-
    tropical USA.

Thecaphora sp.; seedsmut. S. America.

Uromyces spp. rusts; five species as follows: U. anthyllidis
    Schr. in Brazil, Argentina, U. elatus Syd. (synonym is
    Maravalia elata (Syd.) Mains) in Bolivia, Brazil, Peru,
    U. lupini Berk. & Curt., in Mexico, Uruguay, U. montanus
    Arth. in Mexico, Guatemala, and U. rugosus Arth. in
    Mexico.

Viruses: Abutilon Mosaic Virus in Argentina. Bean Yellow
    Mosaic Virus in Florida. Pea Mosaic Virus in Argentina.
    Brazilian Tobacco Streak Virus in Brazil. A mild to
    severe mottle from unidentified virus in C. American high-
    lands.

LYCOPERSICON ESCULENTUM Mill.   The somewhat variable annual
    to perennial fruit bearing herb (sometimes woody) of the Andean-
    Mexican mountains brought under cultivation by generations of folk
    horticulturists and producing the nutritionally very important
    tomato crops in all seasons; "tomáte," "tomatillo" Sp., "tomate,"
    "tomateiro" Po., "jito mate" in Mexico, "tomato" En., and
    numerous long unused and forgotten Indian names for many wild
    types. (Note: Small and large types have been grown for millenia
    by many Indians--South and Central American and West Indian.   I
    have seen wild types ranging from lush plants to those tough types
    hanging from rocks and clambering over semi-dry terrain in high-
    lands of Central America.   Foliage of tough types was fuzzy,
    gray-green, fruits irregular and acid, but sold in markets, and
    seemed distinctly not L. esculentum, the large fruit variety, or
    near it.   In Ecuador at a village in an outside market I saw (they
    had been torn away from the roots) a few ropy vines over six
    meters long, one of which had 45 stubby side branches, with
    fruits about the size of a pigeon egg that were crimson in color,
    tough skinned but tasting like good tomato; they were being bar-
    tered, bringing premium exchange for stringing on agave fiber
    for drying.   Tomato crops are increasing in importance and are

recognized by tropical institutions as needing study, and new
diseases, some not included in this list, are being reported.)

Actinomyces sp., scab on two to three year old underground
stems. Costa Rica, Honduras.

Aecidium cantensis Arth.; rust. Peru.

(? Aerobacter sp.); fruitrot complex. W. Indies.

Alternaria alternata (Fr.) Keiss. = A. tenuis Nees; benign
blackening of old tissues of leaf and calyx. Brazil, Argentina, Guatemala. A. solani (Ell. & Mart.) Sor.; foliage
and fruitspot, collar rot. Scattered but not severe in C.
America. A. tomato (Cke.) Brink. = A. tomato (Cke.)
Web.; nailhead fruitspot. Subtropical USA, El Salvador,
Puerto Rico, Mexico.

Aphalenchoides parietinus Stein.; root nematode. Brazil.

Ascochyta lycopersici (Plowr.) Brun. (easily confused with
Phyllosticta); mostly spots on lower leaves. Subtropical
USA, Ecuador, Mexico, Brazil. A. phaseolorum Sacc.;
leafspot. Brazil.

Aspergillus ochraceus Wilh.; yellow fruitmold. C. America.
A. tomartii Kita and A. terreus Thom cause fruit mold.
C. America.

(Bacterial species have been reported from unusual symptoms
at times as fruit decays on tomato, but designations are
uncertain. There is also a new bacterium which is related
to a round and raised leafspot, see note under Panagrolaimus.)

Blossom endrot of fruits, poor calcium balance. Scattered.

Botryodiplodia theobromae Pat. (synonyms: Diplodia theobromae (Pat.) Now. and D. natalensis P. Evans); stem and
fruitrot. C. America.

Botrytis cinerea Pers.; leaf-and-stem spots, fruitrots in
high moisture areas, and in wet season crops. Costa Rica,
Brazil.

Brachysporium tomato (Ell. & Barth.) Hir. & Wat.; fruitrot.
W. Indies, Subtropical USA (? El Salvador).

Cassytha filiformis L.; green dodder infection by inoculation.
Puerto Rico.

Catface, morphological distress. Scattered.

Cephaleuros virescens Kunze; algal leaf disease, stemspot.
El Salvador, W. Indies, Guatemala, Brazil.

Cephalosporium acremonium Cda.; ripe-fruit rot. Dominican
Republic.

Cercospora cruenta Sacc.; leafspot. Puerto Rico. C. diffusa
Ell. & Ev.; leafspot. Mexico. C. fuligena Rold.; leaf
disease. Mexico. C. nicotianae Ell. & Ev.; Texas. C.
physalidis Ell.; leafspot. Mexico. (Certain, if not all, of
these may be eventually treated as one species.)

Chaetomium bostrychodes Zopf; leaf and stem mold. Texas,
C. America.

Cladosporium fulvum Cke.; leafmold, fruit greenmold. Mostly
in moist areas, very severe on introduced host varieties.
Neotropics.

Cloudyspot on fruit, from insect feeding (in some cases seemed
associated with Nematospora). Scattered.

Colletotrichum gloeosporioides Penz. with asco-state Glomerella cingulata (Stonem.) Sp. & Schr. (synonyms: Colletotrichum phomoides Chester, Gloeosporium lycopersici
Krueg., G. phomoides Sacc., G. rufomaculans (Berk.)
Thue., and Hainesii lycopersici Speg.); anthracnose
diseases causing small spots on leaf and stem, fruit infections, sometimes recovered in isolations from stems
of plants wilted by Fusarium. Neotropical.
Corynebacterium michiganense (E. F. Sm.) Jens. = Aplano-
bacter michiganense (E. F. Sm.) Sw.; stem canker, leaf
disease, fruitspot with halo. California, Florida, Mexico,
Costa Rica, Panama (probably all C. America), Colombia,
Brazil.
Corynespora casiicola Wei (Cercospora-like spores); target
leafspot, stem lesions. In dry areas of Mexico, Guatemala.
Cuscuta spp. (many times these dodders are found on wild
and old neglected cultivated tomato plants in the American
tropics), a few tomato dodder species: C. americana L.,
C. campestris Yunc., C. indecora Choisy.
Cytospora sp.; from laboratory isolation in studying old infec-
tions from an apparent Orobranche secured in Cuba.
Diaporthe citri Wolf = D. medusae Nits.; fruitrot complex.
Trinidad. D. phaseolorum (Cke. & Ell.) Sacc.; fruitrot
with black stroma on surface. Dominican Republic, Sub-
tropical USA, C. America.
Ditylenchus dipsaci Filip.; rhizosphere nematode. Brazil.
Didymella lycopersici (Cke.) Hollos; on leaf and stem. Trini-
dad, other nearby islands.
Diplodia (see Botryodiplodia)
Diplodina destructiva Petr.; on foliage. Dominican Republic.
Dorylaimus krygeri Dit.; rhizosphere nematode. Brazil.
Erwinia carotovora (L. R. Jones) Holl. (synonyms: Bacillus
carotovorous Jones, Bacterium carotovorum Leh. & Neum.,
Erwinia aroideae Holl.); the common bacterial softrot in
storage and transport. Neotropics. E. lathyri Holl.;
saprophytic yellow bacterium in isolations from stems
diseased with fungi. Costa Rica.
Erysiphe (see Oidium)
Fasciation or flattened stems has been diagnosed as some
kind of virus, genetic effect, etc., is a rare and spectacular effect.
Fumago sp. (not Capnodium, not Meliola), under shade of
Inga and Erythrina trees in Mexico.
Fusarium spp., about nine occur in all countries growing
tomatoes but apparently are not everywhere in all farming
areas: 1. F. equisiti (Cda.) Sacc. = F. scirpi Lamb. &
Fautr.; isolated from fruitrot. Brazil, Chile. 2. F.
moniliforme Sheld. (but I did not encounter the asco-form)
isolations from stem wound infections. In warmest parts
of El Salvador and Costa Rica. 3. F. oxysporum Schl.;
from soil isolations the non-parasitic type. Puerto Rico,

Costa Rica. 4. F. oxysporum Schl. f. sp. lycopersici
(Sacc.) Sny. & Hans. races 1 and 2 in Florida, races 1,
2, and 3 in Brazil, common typical systemic wilts. In
infested slightly acid and medium warm, permeable soils,
almost any tropical country. 5. It has been suggested that
a different subspecific name be used for an especially virulent race that I secured in the United States of America and
in Costa Rica, as follows F. oxysporum Schl. f. sp. viru-
lenti. (In isolations it is relatively rare and the organism
does not sector to produce weakly parasitic races while in
the field, plant survival requires a type of resistance found
in Pan American tomato.) 6. Isolations were made repeatedly in Costa Rica, from browned root tips and from
soil at base of one plant in which was found a Fusarium
with multicelled macrospores very similar to, if not identical with, F. decemcellulare Brick. 7. In many cases I
secured F. roseum Lk.; market dry fruitrot. C. America
and I secured 8. F. roseum Lk. f. sp. retusum Wellman
in this case a non-wilting but stem-infecting fungus. El
Salvador, Guatemala. 9. F. solani (Mart.) Sacc.; isolated
in connection with blossom-end-rot lesions and secondary
leaf lesions. Mexico, Colombia, Jamaica, Argentina.
Geotrichum candidum Lk. (? possibly Oospora); in fruitrot
complex. C. America.
Guignardia sp.; minor infections on stem and leaf. Scattered.
Graywall, cause uncertain. El Salvador.
Hainseii (see Colletotrichum)
Hairyleaf from mites. Puerto Rico. (On wild tomato in El
Salvador.)
Haplocystis vexans Speg.; root disease. Brazil.
Helicotylenchus nannus Stein.; nematode near roots. Brazil.
Helminthosporium carposaprum Pol.; stem infection, leafspot.
Haiti, Panama.
Helosis mexicana Liebm.; balanophorous root parasite on wild
tomatoes growing under shade of urticaceous bushes. El
Salvador.
(Heterodera reports--see Meloidogyne)
Heterosporium sp.; diffuse leaf infections. Argentina, Panama.
Hollow stem (cause bacterial). Costa Rica, Panama.
Insect-related troubles, see Cloudyspot, Hairy leaf, Redring,
Russet, Nematospora, Yellows.
Internal browning of fruit, some believe physiological effect
from poor calcium supply, others think virus. Costa Rica,
W. Indies, Florida.
Leafcurl from physiological effects (? drought), leaf lamina
of normal thickness and green, possibly virus. Mexico,
El Salvador, Guatemala. (In one field in Guatemala it was
found only on large fruited cultivated tomatoes, not on
several plants of small fruited wild tomatoes that were
weeds in the field.)
Leafroll from physiological effect, leaf lamina thickened and
darker green, believed due to starch congestion (not in a

shaded field near the open field where it was widely present). El Salvador, scattered, more or less in small areas, C. America.

Lightning strike, plants torn at ground line, if young when struck some recover and stems show callus growth. Occasional.

Little leaf, no fruits set, possibly from zinc deficiency. Costa Rica.

Macrophomina phaseoli (Taub.) Ashby; stem charcoalrot. West coast El Salvador, Honduras, Puerto Rico.

Meloidogyne spp.; rootknot nematodes. Some species listed are M. javanica Chitw., M. exigua Goeldi, M. incognita Chitw., M. incognita Chitw. var. acrita Chitw. and various Heterodera spp. Rootknot found at least in Argentina, Brazil, Peru, Colombia, Chile, Venezuela, Panama, Costa Rica, Honduras, Nicaragua, Surinam.

Monilia platensis Speg.; leaf disease. Argentina.

Mycelium sp. (sterile hyphae on surface of dying roots). El Salvador, Costa Rica, ? California.

Mycena citricolor (Berk. & Curt.) Sacc.; rare leafspot. Costa Rica.

Myrotheciella catenuligera Speg.; leaf lesion. Argentina.

Myrothecium roridum Tode; fruit ringspot, stemcanker. Texas.

Nematospora coryli Pegl. (? = N. lycopersici Schn.); fruit disease, cloudy spot, seed blackening. Scattered occurrence in Subtropical USA, Cuba, Mexico, W. Indies, Peru.

Nigrospora oryzae (Berk. & Br.) Petch; blackened fruitrot. California, El Salvador.

Oedema; excrescences on leaf, following unexpected rains in the dryseason. Guatemala, Costa Rica.

Oidium spp., white powdery mildews given different specific names: O. balsamii Mont., O. erysiphoides Fr., O. lycopersici Cke. & Mass. and certain perfect state designations: Erysiphe cichoraceum DC., E. communis Wall., E. polygoni DC. These mildews are dry season diseases, and in fields where rain is infrequent. Special reports have been seen from Mexico, Puerto Rico, Ecuador, Peru, Bolivia, Argentina, Brazil, Chile, but disease will be found in other countries.

Oospore lactis (Fres.) Sacc. f. sp. parasitica Pritch.; in markets, sour yeastrot of fruit. C. America, Ecuador, Subtropical USA.

Ophiochaeta herpotrichia Sacc.; root disease. Brazil.

Orobranche ramosa L.; brownrape attack of roots. California, questionable in Cuba, (in El Salvador, believed to be this parasite.)

Panagrolaimus sp.; nematode inside leaf, carrying bacteria and causing raised leafspots. Honduras.

Penicillium spp. (several species on old tomato vines at end of wet season). El Salvador, Costa Rica. P. crustaceum Fr.; part of fruitrot complex. Chile.

Peronospora tabacina Adam; downymildew. Subtropical USA, W. Indies.

Pestalotia sp.; small stem-lesions, fruitrot isolates.   Texas,
    El Salvador west coast.
Peyronelia sp.; on foliage.   Puerto Rico.
Phoma destructiva Plowr.; black rot-spot of fruit, leafspot,
    sunken stem-spot.   Brazil, Chile, Ecuador, Colombia,
    Panama, Costa Rica, Puerto Rico, Santo Domingo, Trini-
    dad, Jamaica.
Phyllosticta hortorum Speg. (sometimes confused with Asco-
    chyta); spots almost always on lower leaves.   Occasional
    in both C. America and S. America.
Phymatotrichum omnivorum (Shear) Dug.; dryland rootrot.
    Arizona, Texas, Mexico.
Phytomonas vesicatoria Berg. et al.; fruitspot.   Argentina,
    Mexico.
Phytophthora spp., many reported in the Americas, here are
    eight species: 1. P. cactorum (Leb. & Cohn) Schr. as
    buckeye fruitrot, stem lesion.   2. P. capsici Leon. as
    zonate fruitrot.   3. P. infestans (Mont.) dBy. is by far
    the most common, causes the severe blight, fruitrot.
    (Note: In Brazil studies on P. infestans have indicated
    races 0, 1, 2, 3, 4 all present.   Races 3 and 4 most
    common.)   4. P. parasitica Dast. = P. terrestris Sher.
    is fruit buckeye rot, stem lesion.   5. P. mexicanum Hot.
    & Har. a fruit and foliage disease.   Countries reporting
    severe Phytophthora damage: Mexico, Ecuador, Puerto
    Rico, Venezuela, Colombia, El Salvador, Brazil, Argentina,
    Chile, Subtropical USA.   In less numerous reports: 6. P.
    drechsleri Tuck., 7. P. cinnamomi Rands, and 8. P.
    citrophthora (Sm. & Sm.) Leon. recovered in Argentina on
    tomato.
Pichia belgica (Lind.) Dek. and P. belgica (Lind.) f. sp.
    microspora Negr. & Fisch.; fruitrot complex.   Several
    places in S. America.
Pitting of fruit, physiological cause.   Subtropical USA, Puerto
    Rico.
Pleospora lycopersici Ell. & March. = P. herbarum Rab. (the
    probable asco-state of either a Heterosporium or a Clado-
    sporium); small fructifications at edges of shallow fruitrot
    lesions.   Costa Rica.
Pox-pit, fruit skin indentations.   Cause unknown.   W. Indies,
    Subtropical USA.
Pratylenchus pratensis (deMan) Filip.; root nematode.   Brazil,
    Peru.
Pseudomonas garcae Amar., Teix., & Pinh.; bacterial leafspot.
    Brazil.   P. solanacearum E. F. Sm.; bacterial wilt.   (In
    Brazil pathologists have identified races 1, 2, 3).   El Sal-
    vador, Jamaica, U.S. Virgin Islands, Mexico, Puerto Rico,
    Panama, Peru, Costa Rica, British Guiana, Brazil, Trini-
    dad, Ecuador.   P. tomato (Okab.) Alst.; bacterial speck of
    fruit.   Subtropical USA, C. America, Ecuador.
Puccinia pitteriana P. Henn.; brown leafrust (sometimes quite
    severe losses).   Costa Rica, Mexico, Panama, Colombia,
    Peru, Ecuador, Bolivia, Paraguay, Venezuela.

Puffs or pockets, blunt angled fruits, physiological causation.
Occasional.

Pullularia pullulans (dBy.) Berkh. = Aureobasidium pullulans
(dBy.) Arn.; market fruitspot. Guatemala.

Pythium spp. all cause damping-off and some are possibly
part of fruitrot complex. (At least these have been reported on tomato in the Neotropics: P. aphanidermatum
(Eds.) Fitzp., P. catenulatum Mat., P. debaryanum Hesse,
P. hydnosporium Schr., P. intermedium dBy., P. olygau-
drum Drechs., P. rostratum Butl., P. ultimum Trow.)

Red ring, insect injury of stem. Texas.

Rhizoctonia solani Kuehn = Thanatephorous cucumeris (Frank)
Donk causes damping-off, collar-rot, stem canker, fruit
soil-rot. El Salvador, Mexico, Ecuador, Jamaica, Argentina, Panama, Colombia, Puerto Rico, Costa Rica, Venezuela. (Probably in all tomato-growing countries, but some
soils more susceptible than others.)

Rhizopus nigricans Ehr. and R. stolonifer (Ehr.) Lind.; occasional market riperot. (I found it many times on unsprayed fruits if held beyond generally considered optimum
time for ripening.)

Rotylenchulus reniformis Lind. & Olive; root nematode.
Florida.

Rotylenchus queirozi Lorde.; root nematode. Brazil.

Russet of fruit and leaf. Mite causation. Texas, Mexico,
El Salvador.

(Scaptocoris talpa Cham. an interesting soil insect that, when
in the root zone, acts as a fungi-stat on Fusarium oxysporum lycopersici, and reduces injury from the rootknot
nematode as well in Honduras.)

Sclerotinia minor Jagg.; budrot and blossom disease, occasional
fruitrot. Subtropical USA, W. Indies, El Salvador. S.
sclerotiorum (Lib.) dBy. = Whetzeliana sclerotiorum (Lib.)
Korf & Dumont; leaf, stem, and fruit lesion and rot,
generally in old replanted fields. Uruguay, Brazil, Bolivia,
Chile, scattered in C. America.

Sclerotium rolfsii (Curzi) Sacc. (the basidial state, Pellicularia
rolfsii (Sacc.) West, was found once by me on old tomato
stems in El Salvador in a mountain farm), causes soil-line-
rot, may produce a confusing wilt, and fruitrot if fruits
touch soil with active fungus in it. Mexico, Costa Rica, El
Salvador, Colombia, Trinidad, Brazil, Peru, Venezuela,
Jamaica.

Septoria lycopersici Speg. = S. tomates Speg.; small stem and
leaf spots often confluent. Serious in non-sprayed fields in
moist areas, less severe in semi-arid locations. Practically general in distribution.

Spermophthora gossypii Ashby & Now.; part of fruitrot complex.
W. Indies.

Spyrotylenchus queirozi Lorde. & Cesn.; nematode on roots.
Brazil.

Stemphylium solani Web.; gray leafspot. Probably general in
distribution, rare in Honduras, seen in West Guatemala.

Stigmella sp.; on foliage.   Scattered in a few locations in S.
America.

Stomatochroon lagerheimii Palm; parasitic alga.   Guatemala
(moise area).

Sunscald of exposed fruits.   (Often after defoliation from
disease.)

Thielaviopsis basicola (Berk. & Br.) Ferr.; black rootrot.
Subtropical USA (dry parts), C. America (hill lands facing
the Pacific).

Trichoderma spp. (apparently 2 or 3 species); cause root and
stem base infections, also hyperparasites of various fungi.
Costa Rica.

Trichothecium roseum Lk.; fruitrot.   Guatemala, El Salvador.

Tylenchorhynchus sp.; nematode found in rhizosphere.   Peru.

Vascular discoloration in fruit.   (? lack of potash ?).
Florida, Costa Rica.

Verticillium alboatrum Reinke & Berth.; stemcanker, wilt.
Subtropical USA (dryer parts), El Salvador, Peru, Ecuador,
Brazil, Argentina.

Virus symptoms on tomato in the American tropics are sometimes bewildering to the observer but I have attempted to
cull from many reports authentic designations.   No doubt
some are missed, some confused, but most notable may be
here.   Those classed as mycoplasmas are considered as
viruses.   I have listed 26 of these diseases, all alphabeti-
cally, not in order of importance.   The word "Virus" is
not used as a name in the list: 1. Beet Curly Top (also
known in this case as Brazilian Curly Top) in Argentina,
Brazil, Bolivia.   2. Common Cucumber Mosaic scattered
in Subtropical USA, El Salvador, Cuba, Haiti, Peru, Costa
Rica.   3. Cucumber Mosaic strain Celery-Commelina in
Florida, Cuba, Puerto Rico, El Salvador, Guatemala, Costa
Rica, Haiti.   4. Eggplant Mosaic in Trinidad.   5. Euphorbia Mosaic in Brazil.   6. Potato Leaf Roll in Uruguay,
Brazil.   7. Potato X in Uruguay, Argentina.   8. Potato Y
in Brazil, Peru, Argentina.   9. Potato Y and Tobacco
Mosaic mixed in Brazil.   10. Sunflower (Argentine Virus)
in Argentina.   11. Tobacco Etch in Florida, Venezuela.
12. Tobacco Leaf Curl (Tomato Brazilian Curly Top) in
Brazil, Venezuela.   13. Tobacco Mosaic (Calico) in Brazil,
Puerto Rico.   14. Tobacco Mosaic (Common) (also known
as Common Tomato Mosaic) is of general distribution except
where too wet for insect survival.   15. Tomato Aucuba
Mosaic = Tobacco Mosaic AY in Brazil, Argentina.   16.
Tomato Big Bud in California, El Salvador, Puerto Rico,
Brazil.   17. Tomato Bunchy Top in Texas.   18. Tomato
Double Streak ("Riscas" of Brazil) = Potato X and Tobacco
Mosaic combined in Subtropical USA occasionally, Colombia,
Brazil, Argentina.   19. Tomato fruit mottle and ringspot,
from undetermined virus in Jamaica, Puerto Rico.   20.
Tomato leaf curl (? = Tomato Yellow Leaf Curl) in El
Salvador, Costa Rica, ? Peru.   21. Tomato purple top

(virus unidentified) in El Salvador, Costa Rica on West coast. 22. Tomato Spotted Wilt (= Tomato Brazilian Ringspot called "Corcova") in Uruguay, Brazil, Argentina, El Salvador, Puerto Rico. 23. Tomato Spotted Wilt var. Top Blight ("Peste negra," may be "Tisis") in Brazil, Bolivia, Puerto Rico, St. Croix. 24. Tomato rugose mosaic (undetermined virus) in Brazil, west coast of Costa Rica. 25. Tomato Yellow Leaf Curl (possibly "Tisis" may be a strain of this) in Puerto Rico, El Salvador. 26. Yellow chlorotic dwarf from chlorotic Sida carried by the white fly in Venezuela.

Water congestion.   Following unusual rainstorms leaves and
 stems show translucent injury, may end in temporary wilt.
Xanthomonas solanacearum (E. F. Sm.) Dows. = Pseudo-
 monas solanacearum E. F. Sm. ; bacterial wilt.   Widespread.
 X. vesicatoria (Doidge) Dows., bacterial leaf- and fruitspot.
 Subtropical USA, C. America, W. Indies.
Xiphinema sp. ; nematode ecto-parasite of roots.   Chile, Peru,
 Brazil.
Yellows due to the insect feeding of Psyllid sp.   Dry sub-
 tropical USA.

MACADAMIA TERNIFOLIA Muell.   A medium tall tree introduced
 from Hawaii, now being tested in a few select locations in Central
 America and the West Indies, and where grown long enough has
 given good crops of highly esteemed nuts.   Very few disease ob-
 servations made in the Americas.
  Arachnid injury of leaves.   El Salvador.
  Botryodiplodia theobromae Pat. (no asco-state seen), dieback.
   Puerto Rico, Costa Rica.
  Canker of leaf petioles (possibly insect, no fungus isolated).
   Costa Rica.
  Cephaleuros virescens Kunze; algal leafspot.   Puerto Rico,
   Costa Rica.
  Colletotrichum gloeosporioides Penz. ; anthracnose of leaves,
   flower petals and green pod.   Costa Rica.
  Phytophthora cinnamomi Rands; branch and trunk canker, root
   disease.   Costa Rica, Puerto Rico, California.
  Rosellinia bunodes (Berk. & Br.) Sacc. ; black rootrot.   Costa
   Rica.
  Strigula complanata Fée; lichen, white leafspot.   Costa Rica.
  Zinc deficiency effect (small leaves, short internodes, chloro-
   sis).   Puerto Rico.

MAGNOLIA GRANDIFLORA L.   From warm North America, a tall
 tree with handsome evergreen leaves and large flowers, an ornamental that is at times a considerable source of income to tropical florist and gardener; "magnolia" Sp. Po. En.
  Asterina comata Berk. & Rav. = Trichodothis comata Th. &
   Syd., imperfect and perfect forms of black mildew on
   leaves.   Subtropical USA.
  Botryodiplodia spp. also reported as Diplodia spp. =

Botryosphaeria quercuum (Schw.) Sacc.; perhaps most common as dieback. Neotropics. (Note: Dieback twigs are retained much longer on magnolia in the tropics than in the temperate zone. This gives imperfect states of this parasite much better chances of developing perfect states in lands where winter never comes.)

Botryosphaeria dothidea (Moug.) Ces. & deNot.; branch canker. Subtropical USA, W. Indies, Panama.

? Capnodium sp. ?; sooty mold (see Dimerosporium). El Salvador.

Cephaleuros virescens Kunze; algal leafspot (sometimes very large). Subtropical USA, W. Indies and probably scattered in many other locations.

Cercospora magnoliae Ell. & Harkn. = Isariopsis alba-rosella Sacc., the asco-state is Mycosphaerella milleri Hodges & Haasis; small leafspots. Subtropical USA, Argentina and probably other parts of S. America.

Cladosporium fasciculatum Cda.; leafspot. Subtropical USA. C. herbarum Lk.; leaf edge invasion. Puerto Rico.

Colletotrichum gloeosporioides Penz. (synonyms: Gleosporium magnoliae Pass., G. haynaldianum Sacc. & Roum., Colletotrichum magnoliae Sousa da Camara), the asco-state is Glomerella cingulata (Stonem.) Sp. & Schr.; on mature leaves infections result in striking large dark brown spots, on petioles are found relatively mild infections. Subtropical USA, Guatemala, Panama, probably widespread.

Coniothyrium olivaceum Bon.; leafspot. Dryer parts of Subtropical USA.

Criconemoides sp.; nematode recovered at roots. Florida.

Daedalea ambigua Berk.; trunk woodrot. Florida.

Dimerosporium magnoliae Tracy & Earle; sootymold. Subtropical USA, Panama, S. America.

Diplodia = Botryodiplodia.

Elsinoë magnoliae Jenk.; spot anthracnose. Subtropical USA.

Epicoccum nigrum Lk.; leafspot. Subtropical USA.

Exobasidium lauri Geyl.; leafgall. Florida found on Magnolia virginiana L.

Exophoma magnoliae Weed.; leafspot. Subtropical USA.

Fomes applanatus (Perk.) Gill.; basal rot of trunk, F. geotropus Cke.; brown heart rot, F. fasciatus (Sw.) Cke.; white heart rot. Observations and studies only in Subtropical USA.

Gloeosporium (see Colletotrichum)

Graphiothecium maculicolum Karst.; leafspot. S. America.

Helicotylenchus sp.; nematode in root zone. Florida.

Hemicycliophora sp.; nematode in root zone. Florida.

Heterosporium magnoliae Weed. (may be actually Cladosporium fasciculatum), leafspotting. Subtropical USA, something similar collected in Guatemala and Panama.

(Hypochnus rubrocinctus Ehr. = Chiodecton sanguineum Wain.; a lichen? reported on Magnolia sp. from S. America.)

Hypoxylon annulatum Mont.; on branch in Honduras.

Isariopsis (see Cercospora)
Leptothyrella longloisii (Ell. & Ev.) Sacc.; leaf disease. Sub-
    tropical USA moist areas.
Lophodermium maculare Fr.; leaf infections, subcuticular dark
    lesions. Florida, Puerto Rico, Honduras.
Macrophoma (where I checked reports) = Botryodiplodia.
Meliola amphitricha Fr.; black mildew on leaves. W. Indies,
    Subtropical USA, Venezuela. M. magnoliae Stev.; black
    mildew (fine hyphae). Puerto Rico, on leaf sent from St.
    Johns, Subtropical USA.
Meloidogyne spp.; rootknot. Subtropical USA, probably other
    regions.
Nectria sp.; secondary on dieback twig. Puerto Rico.
Pellicularia koleroga Cke.; thread-web blight. Panama, Hon-
    duras (probably in S. America), Subtropical USA.
Pestalotia guepini Desm.; secondary or at least following
    numerous leafspots. Subtropical USA, Puerto Rico, Costa
    Rica.
Phycopeltis sp.; algal plaques on leaves. Puerto Rico.
Phyllosticta cookei Sacc. in Subtropical USA. P. magnoliae
    Sacc. in S. America. Both small round leafspots.
Phymatotrichum omnivorum (Shear) Dug.; dryland rootrot.
    Texas.
Phytophthora citrophthora (Sm. & Sm.) Leon.; root disease.
    P. parasitica Dast.; stem canker and rootrot. (Both
    organisms in C. America and S. America.)
Polyporus spp.; wood rotters. Scattered.
Pratylenchus sp.; rhizosphere nematode. Florida.
Septobasidium tenue Couch; fungus felt-cover over insects.
    C. America, Subtropical USA, may be general.
Sirodesmium stictophyllum (Cke.) Sacc.; leaf disease. Sub-
    tropical USA.
Strigula complanata Fée; white-spot lichen. Well distributed
    but scattered.
Trichodothis (see Asterina)
Verticillium alboatrum Reinke & Berth.; in dead twigs. Sub-
    tropical USA.
Weak spot. Probably various fungi involved. Spots small,
    translucent at start, become necrotic on aging. Scattered
    in many countries.
Xiphinema sp.; nematode in rhizosphere. Florida.

MALUS SYLVESTRIS Mill. = PYRUS MALUS L. Origin temperate
    but not coldest region of Europe-Asia, and in the world it is one
    of the best known and one of the oldest cultivated fruit trees;
    was first introduced to American tropics by early Europeans and
    has become valuable export crop from Chile, Argentina, parts of
    Brazil and in addition there is scattered growing (as a more or
    less exotic) in occasional cool mountain plantings of other coun-
    tries; "manzano," "manzanero" Sp., "maciera" Po., "apple" En.
    (The list of tropical diseases given below no doubt is incomplete;
    many new problems are developing in new areas.)

Agrobacterium tumefaciens (E. F. Sm. & Towns.) Conn.;
    crowngall. Rare, probably introduced from temperate
    zone. Mexico, Argentina, Colombia.
Alternaria mali Roberts; leafspot, storage fruitrot. Subtropical
    USA, Argentina, .Uruguay, ? widespread. Alternaria (un-
    usual, larger spored than expected), given name Macro-
    sporium puttemansii P. Henn. (this name no longer tenable
    should be Alternaria puttemansii); leaf and fruit spots.
    Brazil.
Armillaria mellea Vahl = Armilariella mellea (Fr.) Karst.;
    spreading rootrot. Mexico, Subtropical USA, Colombia,
    Argentina.
Ascochyta sp. (Ascochytella sp.); leafspot usually small diameter.
    Southern parts of S. America.
Bitterpit, physiological fruit breakdown. Argentina.
Boletus fomentarius L. = Fomes fomentarius (L.) Kickx; root-
    rot, rot of trunk wood. Chile.
Boron deficiency causes drought spot, corky spot, rosette.
    Argentina, Guatemala.
Botryodiplodia sp. (Diplodia sp. see note immediately below);
    dieback. Widespread.
Botryosphaeria dothidea (Moug.) Ces. & de Not.; stem canker.
    S. America. B. quercuum (Schw.) Sacc. (synonyms:
    Physalospora obtusa Cke., Diplodia, sp., P. cydoniae Arn.,
    Sphaeropsis malorum Pk.); frog-eye leafspot, fruit black-
    rot and black mummy, branch canker. Colombia, El
    Salvador, Uruguay, Chile, Mexico, Guatemala, Peru.
Botrytis cinerea Pers.; fruit moldy-rot in Argentina, sucker
    tip mold in Colombia.
(? Capnodium spp.); sootymolds in all countries.
Cephaleuros virescens Kunze; algal leafspot. Colombia, Guate-
    mala.
Cephalothecium roseum Cda. = Trichothecium roseum Lk.;
    fruitspot, pink rot. Argentina.
Cercospora mali Ell. & Ev. and C. minima Tracy & Earle;
    leafspots in C. America and S. America. C. porrigo
    Speg. (the latter said to have been recovered on apple but
    was described on pear) = imperfect state of Venturia?
    Argentina.
Cladosporium sp.; spores recovered from dry leaf specimens
    from Guatemala, Panama, Colombia.
Colletotrichum gloeosporioides Penz. (synonyms: C. destruc-
    tivum O'Gara, C. piri Noack), asco-state is Glomerella
    cingulata (Stonem.) Sp. & Schr.; common anthracnose and
    bitter rot. Widespread.
Coniothecium chromatosporium Cda.; blister canker. Argentina.
Corky spot (see boron deficiency)
Corticium salmonicolor Berk. & Br.; branch wilt. Guatemala.
Dothiorella ribis (Fckl.) Sacc.; stem disease. Scattered in S.
    America.
Elsinoë piri (Wr.) Jenk. is the perfect state of Sphaceloma
    pirinum (Pegl.) Jenk. = Plectodiscella piri Jenk.; leafspot
    and scab spot. Argentina, Chile.

Entomosporium (see Fabraea)

Erwinia amylovora (Burr.) Winsl. et al.; fireblight of foliage.
Scattered in Mexico, W. Indies, C. America, S. America.

Fabraea maculata Atk. the asco-state of Entomosporium
maculatum Lév.; an uncommon leafspot.   Chile.

Fomes spp.; woodrot at base of old trees.   S. America.

Fruit cracking from season of abnormal rains after fruits are
green mature.   Scattered.

Fusarium solani (Mart.) Sacc. and F. roseum Lk.; both appear
to be secondary invaders following primary parasites.
Chile.

Fusicladium (see Venturia)

Ganoderma sessile Murr.; treerot at base.   S. America.

Gloeodes pomigena (Schw.) Colby; fruit sootyblotch.   Scattered
in Mexico, Guatemala, S. America.

Gymnosporangium clavariaeforme DC.; rust.   Mexico.

Hydnum sp.; secondary rot of old wood.   Scattered and relatively rare.

Jonathan spot, physiological fruit breakdown on several
varieties.   Scattered in Argentina.

Leptothyrium pomi (Mont. & Fr.) Sacc.; fruit flyspeck.   Scattered.

"Loco"; abnormal branch flowering while others on the tree
are dormant in certain varieties planted in high mountains
relatively near the equator.

Monilinia fructicola (Wint.) Honey; fruit brownrot.   C. America,
S. America.   M. laxa (Aderh. & Ruhl.) Honey; fruitrot.
Uruguay.

Myxosporium corticola Edg.; fruitrot and branch canker.   Chile,
Mexico, Peru.

Nectria ditissima Tul.; stem canker.   Mexico, Chile.   N.
galligena Bres.; canker.   Chile.   N. necatrix Berl.; canker.
S. America.

Nematodes, unidentified, found at times in rhizosphere but not
studied for pathology.   S. America.

Oidium spp. asco-stages are probably Podosphaera leucotricha
(Ell. & Br.) Salm., P. oxycanthae dBy., or P. clandestina
Lév.; white powdery mildew of leaves and immature twigs.
Seasonal in appearance.   Mexico, Brazil, Argentina, Colom-
bia, Peru, Uruguay, El Salvador.   (See also Sphaerotheca.)

Oryctanthus sp.; phanerogamic parasite of branch.   Guatemala,
El Salvador.

Pellicularia koleroga Cke.; web-thread blight.   Scattered occurrence.

Penicillium expansum Thom; storage softrot of fruit.   Peru,
Argentina, Chile (P. digitatum Sacc. reported in Chile).

Phoma pomi Pass.; leaf and fruit spot.   Chile, Argentina,
Brazil.

Phomopsis mali Rob.; bark canker.   Columbia, Chile.

Phyllosticta briardi Sacc. = P. mali (Prill. & Del.) Bria.;
leafspot.   Brazil, Peru, Dominican Republic.   P. solitaria
Ell. & Ev.; leaf and twig disease.   S. America.   A leafspot
said to be P. pruinicola Sacc. is reported from Uruguay.

Phytophthora boehmeriae Saw.; fruitrot. Argentina. P.
 cactorum Schr.; wilt and death of nursery plants. Ecuador.
Plectodiscella (see Elsinoë)
Podosphaeria (see Oidium)
Pratylenchus sp.; nematode found in rhizosphere. Chile.
Psittacanthus cuneifolius Bl.; large flowering parasitic bush
 on branches. Argentina.
Pythium debaryanum Hesse; seedling disease. Ecuador (cold
 area). P. oligandrum Drechs.; seedling disease. Argen-
 tina.
Rhizoctonia solani Kuehn (parasitic hyphal state) = Thanate-
 phorous cucumeris (Frank) Donk (spore bearing state);
 damping-off, seedling disease. Ecuador.
Rhizopus nigricans Ehr.; fruitrot. Colombia.
Rosellinia bunodes (Berk. & Br.) Sacc.; black rootrot. El
 Salvador, Colombia. R. necatrix (Hartig) Berl.; white root-
 rot. Chile, California, Argentina.
Rosette (see boron deficiency)
Scald (fruit skin disease induced in storage). Subtropical USA,
 Argentina, Chile.
Schizophyllum commune Fr.; disease of dying branches.
 Chile.
Sclerotium rolfsii (Curzi) Sacc.; = its basidial state is Pelli-
 cularia rolfsii (Sacc.) West.; rare soil-line rot of seedlings
 in abandoned lowland nursery. Guatemala.
Septobasidium spp.; brown felt-like growths over insects.
 Scattered.
Sphaerella pomi Pass.; leafspot. Chile.
Sphaeropsis (see Botryosphaeria quercuum)
Sphaerotheca pannosa (Wallr.) Lév.; asco-state of an Oidium.
 Reports of scattered occurrence, perhaps incomplete studies.
Sporotrichum malorum Kidd & Beaum.; fruitrot and stem
 canker. Colombia.
Stereum purpureum Pers.; silver-leaf, old tree woodrot, root-
 rot. Uruguay, Argentina.
Struthanthus sp.; scandent woody bush parasite on branch.
 Guatemala.
Trichothecium (see Cephalothecium)
Valsa leucostoma Pers.; twig canker, part of dieback complex.
 Chile.
Venturia inaequalis (Cke.) Wint. is the asco-state, its im-
 perfect state is Fusicladium dendriticum (Wallr.) Fckl.
 (Spegazzini studied it in his early years and named it
 Cercospora porrigo Speg.); scab disease of leaf, flower,
 fruit and twig. Argentina, Brazil, Uruguay, Chile, Colom-
 bia, El Salvador, Guatemala, Mexico.
Virus: Apple mosaic in Argentina.
Water core, physiological causation. Argentina.
Xylaria sp., rootrot. Chile.
Zinc deficiency causing little leaf, decline. Seen in Colom-
 bia, ? Guatemala.

MANGIFERA INDICA L. Originally from tropical India, a large
  tree which when properly grown produces good harvests sometimes twice a year of aromatic, delicious and popular fruit;
  "mango" Sp. En., "mangueira" Po.
    Aethanthus (? nodosus); parasitic shrub called "mucuma,"
      rare occurrence. Peru.
    Antennellopsis mangiferae Mend.; leaf scab-spot. Brazil.
    Aspergillus niger v. Tiegh.; market and storage fruitrot.
      Widespread.
    Botryodiplodia theobromae Pat. (Synonyms: Diplodia cacaoi-
      cola P. Henn., D. mangiferae Koord., D. recifensis Bat.,
      D. theobromae (Pat.) Now., Lasiodiplodia theobromae
      Griff.) the perfect state rarely collected is Physalospora
      rhodina Berk. & Curt. = Botryosphaeria quercuum (Schw.)
      Sacc.; twig disease, dieback, occasionally leafspot. Neo-
      tropical.
    Botryosphaeria dothidea (Moug.) Ces. & de Not.; stem canker.
      Brazil, Puerto Rico.
    Capnodium mangiferum Cke. & Br.; sootymold. General.
    Cephaleuros virescens Kunze; algal leafspot, leaf scurf.
      Florida, W. Indies, probably Neotropical in distribution.
    Ceratocystis fimbriata Ell. & Halst.; wilt and death, wood
      stain. Brazil.
    Cercospora (see Stigmina)
    Cladosporium herbarum (Pers.) Lk.; common leaf-edge disease
      of old leaves. Neotropics.
    Clasterosporium maydicum Sacc.; leafspot. Brazil.
    Colletotrichum gloeosporioides Penz. (sometimes given as a
      species of Gloeosporium) is the conidial and most commonly
      encountered, state of Glomerella cingulata (Stonem.) Sp. &
      Schr.; causes flower blight, fruitspot, withertip. Neo-
      tropics.
    Corticium salmonicolor Berk. & Br.; pink disease of branch.
      Brazil, Guatemala.
    Cryptosporium ipirangae Speg.; on leaves. Brazil.
    Cuscuta americana L.; dodder by inoculation on seedling.
      Puerto Rico.
    Dimerosporium mangiferum (Cke. & Berk.) Sacc.; black mil-
      dew. General.
    Diplodia (see Botryodiplodia)
    Elsinoë mangiferae Bitanc. & Jenk. = Sphaceloma mangiferae
      Bitanc. & Jenk.; spot anthracnose. Florida, Panama,
      Brazil.
    Hendersonia sp.; on foliage. S. America.
    Hexagonia tenuis (Hook.) Fr.; a polyporous rootrot. S.
      America.
    Lophodermium mangiferae Koord.; leaf disease. Puerto Rico.
    Marasmius sp.; cord blight disease. Brazil.
    Meliola mangiferae Earle; black mildew. W. Indies, C.
      America, Colombia, Peru.
    Microthyrium mangiferae Bomm. & Rouss.; twig disease.
      Scattered.

Mycena citricolor (Berk. & Curt.) Sacc. = Omphalia flavida
Maubl. & Rang.; leafspot. Mexico, W. Indies, C.
America, Northern countries of S. America.
Oidium mangiferae Berth. also O. mangiferae Av.-Sac. and
O. anacardii Noack., all conidial states of Erysiphe poly-
goni DC.; white powdery mildew. Brazil, Peru, Paraguay.
Oudemansiella canarii (Jungh.) Hoehn.; agaricaceous branch
decay. Brazil.
Pellicularia koleroga Cke.; thread-web foliage disease. Brazil,
C. America.
Pestalotia mangiferae P. Henn.; leaf and twig lesion. Florida,
Puerto Rico, El Salvador.
Phyllosticta mortoni Fairm.; common angular leafspot.
Florida, Puerto Rico, Cuba, Guatemala, Mexico, El Sal-
vador.
Phyllostictina mangiferae Bat. & Vit. = Phyllosticta mangi-
ferae Bat.; leafspotting. Brazil.
Phymatotrichum omnivorum (Shear) Dug.; dry-land rootrot.
Texas, ? Mexico.
Phytophthora sp.; rootrot. El Salvador.
Physalospora (see Botryodiplodia)
Polyporus sanguineus L.; sapwood rot. Florida.
Rosellinia bunodes (Berk. & Br.) Sacc.; black rootrot. W.
Indies.
Schizophyllum sp.; dieback of branch. Colombia, Costa Rica.
Sclerotinia sp.; flower blast. Brazil, W. Indies.
Sclerotium rolfsii (Curzi) Sacc. = Pellicularia rolfsii (Sacc.)
West; seedling disease at ground line, rot of fallen fruits.
Puerto Rico, Scattered in S. America.
Septobasidium lepidosoplus Couch, S. pilosum Boed. & St., S.
pseudopedicellatum Burt (and other spp.); brown felt that
is epiphytic and symbiotic with scale insects. Scattered.
Septoria sp.; leafspot. Florida.
Spegazzinia ornata Sacc.; leaf infections. Brazil.
Sphaerognomia mangiferae Bat.; leaf-edge spot. Brazil.
Sphaerostilbe repens Berk. & Br.; rather rare red rootrot.
S. America.
Stigmina mangiferae Ell. = Cercospora mangiferae Koord.;
small circular to somewhat angled leafspots. W. Indies,
Colombia, Bolivia, Brazil, Venezuela.
Trabutia mangiferae Char.; stromatoid leafspot. Brazil,
Puerto Rico.
Trametes corrugata Pers. and T. hydnoides Sw.; probably
secondary invaders of dead wood. Florida, Puerto Rico.
Trichoderma sp.; in moist country part of a dieback complex.
Widespread.
Tryblidiella fusca (Ell. & Ev.) Rehm and T. rufula (Spreng.)
Sacc.; on dead twigs and branches. Florida, El Salvador,
Puerto Rico.

MANIHOT UTILISSIMA Poh. (= M. ESCULENTA Crantz or JANIPHA
MANIHOT H.B.K.). From Brazil and the especially important

herbaceous to woody bush which feeds millions in the world's
tropics, grown for its tuberous roots; "manyóc," "mandioc,"
"mandioca" Sp. Po., "aipim," "maniçoba" Po. "cassava,"
"tapioca" En. (many forgotten Indian names).

Arthrobotrys superba Cda., on dying leaves.   W. Indies.
Aspergillus niger v. Tiegh.; market and storage rootrot.
   Widespread.
Asterina manihotia Syd.; dark colored mildew.   Costa Rica.
Botryodiplodia theobromae Pat. (synonyms: Diplodia theobromae (Pat.) Now., D. cacaoicola P. Henn., and D.
   natalensis P. Evans; its asco-state is Botryosphaeria
   quercuum (Schw.) Sacc.); dieback, stalk lesion, leaf
   disease, black root-lesion in markets.   Neotropical.
Botrytis sp.; blast of young leaf.   Peru.
Cercospora caribaea Chupp & Cif.; the common white-centered
   leafspot.   Neotropics.   C. henningsii Allesch. (Helminthosporium manihotis Rang.); the common brown leafspot and
   blotch.   Neotropics.   C. caricae Cif.; leafspot.   W. Indies,
   C. America.   C. manihobae Viégas; leafspot.   Brazil.   C.
   vicosae Muller & Chupp, leafspot.   Brazil.
Cercosporella pseudoidium Speg.; mildew-like leafspot.   S.
   America.
Colletotrichum gloeosporioides Penz. (C. janiphae Grove)
   sometimes listed as a Gloeosporium.   Perhaps deserves
   being given f. sp. designation as cultures appear darker
   and faster growing than most Colletotrichums.   Causes root
   disease, anthracnose lesions of leaf, petiole, branch, stalk.
   C. America, S. America.
Corallomyces jatrophae Moel.; stem infection.   Brazil.
Diaporthe manihoticola Viégas; stalk, and leaf, spot.   Brazil.
Dictyostomiopelta manihoticola Viégas; on foliage.   Brazil.
Dimerosporium pellicula Syd.; dark growth on leaves.   Brazil.
Eutypella manihoticola Viégas; on stems and branches.   Brazil.
Exidiopsis manihoticola Viégas; stem and root disease.   Brazil.
Fomes lignosus Kl.; white rot of root (sometimes a new-cleared-
   land disease).   C. America, S. America.
Fusarium spp.; rootrots, stem disease, branch infection.   (The
   species of Fusarium I encountered are of roseum and solani
   types, apparently no oxysporum.)   Neotropics.
Helminthosporium hispaniolae Cif.; leafspot.   W. Indies, Costa
   Rica.   (See also Cercospora henningsii.)
Macrophoma janiphae (Theum.) Berl. & Vogl. (? = Botryodiplodia), stem disease.   Brazil.   (Note: There is a
   Macrophoma manihotis P. Henn., described as occurring
   in East Africa; it could easily be that the two species are
   identical and spread from one hemisphere to the other by
   human intervention.)
Megalonectria pseudotrichia Speg.; on dead stalks.   Argentina
   and other countries of S. America.
Meliola manihoticola P. Henn.; common black mildew.   Widespread.
Meloidogyne sp.; rootknot.   Scattered.

Microsphaeria difusa Cke. & Pk. and as well M. euphorbiae
Pk.; conidial forms of widespread powdery mildews, most
often seen as an Oidium. Neotropics.
Microthyrium manihoticola Viégas; dark mildew. Brazil.
Mycosphaerella manihotis Syd. = Sphaerella manihotis Syd.;
brown leafspot. Brazil, Dominican Republic.
Nectria manihotis Rick; stalk disease. Brazil.
Oidium manihotis P. Henn.; white powdery mildew becoming a
brown blotch when disease ages. Asco-state is Micro-
sphaeria but ascocarps not commonly found in warmer lo-
calities. C. America, S. America.
Periconia manihoticola (Vinc.) Viégas; zonate leafspot. Brazil.
P. pycnospora Fres.; weak parasite causing leafspot under
adverse growing conditions. Panama, Puerto Rico, Brazil.
Phoma alterosae Viégas; leafspot. Brazil.
Phomopsis manihoticola Viégas; leaf and stem disease. Brazil.
Phyllosticta manihot Speg. (? = P. manihotae Viégas); a com-
mon black leafspot. Neotropics.
Physiological darkening of cut or broken tuberous root tissues,
common market and storage disfiguring effect. Neotropical.
Phytophthora parasitica Dast.; rootrot. Florida.
Pleophragmia manihoticola Viégas; leaf disease. Brazil.
Pseudomonas solanacearum E. F. Sm.; rot of plant base and
root. Colombia, Costa Rica.
Rhizoctonia microsclerotia Matz the mycelial state of basidial
state of the fungus Pellicularia microsclerotia (Matz) Rogs.;
web blight on lower foliage. W. Indies, C. America,
Brazil. R. solani Kuehn = Thanatephorous cucumeris
(Frank) Donk; rootrot. Scattered in old fields. (The bloom
from the Thanatephorous stage is seen especially in the
early mornings during moist weather on stalks, appears like
a thin powdery mildew.) Neotropical.
Rhizopus nigricans Ehr.; storage softrot of root. Widespread.
1. Rosellinia arcuata Petch; root disease. Brazil, Guatemala,
Nicaragua. 2. R. bunodes (Berk. & Br.) Sacc.; black
rootrot (most seen in C. America). C. America, S.
America. 3. R. pepo Pat.; a whiter rootrot (more com-
mon than R. arcuata). C. America, S. America.
Sclerotium rolfsii (Curzi) Sacc. (= Pellicularia rolfsii (Sacc.)
West is the basidiospore stage); sprout wilt, collar disease.
Costa Rica, Brazil, Colombia.
Uromyces janiphae Arth.; common black leafrust. W. Indies,
C. America, Brazil, Venezuela. U. manihotis P. Henn.;
brown to black rust on leaves, stems, petioles, flowers and
also causes witches' broom. Costa Rica, Colombia, Brazil.
U. striata Sehr.; leafrust. El Salvador. (U. carthagenensis
Speg., reported on Manihot carthagenensis Muell., possibly
may be on M. utilissima, a stem and leafrust. Argentina.)
Viruses (considered most damaging group of diseases in Brazil
and other parts of warm areas in South America): Cassava
Common Mosaic which causes witches' broom, mottle, stunt,
and very poor growth in Brazil, Paraguay, Colombia, El

Salvador, probably other countries.   Cassava Vein Mosaic
noted for vein-banding, mottle and stunt in Brazil, El Salvador.
Xanthomonas manihotis Starr; bacterial disease.   Panama,
    Costa Rica, El Salvador, Brazil.

MEDICAGO SATIVA L.   An introduction from the temperate zone at
first of questionable promise in the tropics until tests showed it
grew best in relatively cool and deep soils, requires special
treatment in some areas but where successful a valued perennial
forage and hay crop; "alfalfa" Sp. Po., "alfalfa, lucerne" En.
    Alternaria sp. ; from leafspot.   Colombia.
    Ascochyta imperfecta Pk. ; black stem, rootrot, blight.   Guate-
        mala, Ecuador.
    Asterocystis radicis Wild. (an olpidium); root invasion.
        Mexico.
    Cercospora spp. (cause leafspots and defoliation): C. helvola
        Sacc. in Mexico, C. America.   C. medicaginis Ell. & Ev.
        of special note in Puerto Rico, Santo Domingo, Guatemala,
        El Salvador, Venezuela, Uruguay, Peru, Ecuador, probably
        in all alfalfa growing countries.   C. zebrina Pass. in C.
        America, Peru, probably more countries.
    Chaetoseptoria wellmanii J. A. Stev. ; leafspots, may be con-
        fused with Cercospora medicaginis.   El Salvador, Costa
        Rica.
    Cladochytrium (see Urophlyctis)
    Colletotrichum spp. (all four are different from the common
        C. gloeosporioides which is almost on all crops): 1. C.
        dematium (Pers.) Grove with asco-state Sphaeria dematium
        Pers. ; leaf and stem infections.   Nicaragua, El Salvador,
        Guatemala.   2. C. destructivum O'Gara; anthracnose of
        stem, leaf and root.   Uruguay, Mexico, Nicaragua.   3. C.
        graminicola (Ces.) Wils. = C. graminicola (Ces.) Munt.;
        severe anthracnose.   Nicaragua, El Salvador.   4. C. tri-
        folii Bain & Essary; anthracnose, often serious losses.
        C. America, Uruguay, Mexico.
    Corynebacterium insidiosum (McCull.) Jens. ; bacterial wilt,
        rootrot.   Mexico, in Andean S. America.
    Criconemoides sp. ; nematode from near roots.   Peru.
    Cuscuta spp. (dodders): C. campestre Junc. in Argentina,
        C. epithymum Mur. in Argentina, C. indecora Choisy =
        C. racemosa Mart. in Argentina, C. suaveolens Seringe,
        in Argentina, and Cuscuta sp. in Venezuela, Peru, Guate-
        mala, Brazil, Chile.
    Darluca filum (Biv.-Bern.) Cast. ; hyperparasite on rusts.
        Scattered.
    Ditylenchus dipsaci (Kuehn) Filip. ; root feeding nematode.
        Peru, Argentina, Chile.
    Erysiphe polygoni DC. perfect state = Oidium balsamii Mart.
        or O. erysiphoides Fr. which are the names for the perfect
        state's conidial form; widespread white powdery mildew
        (also reported are E. communis (Wallr.) Fr. and E. martii

Lév.).   Mexico, C. America especially in the Pacific slope
lands, S. America.
1. Fusarium acuminatum Ell. & Ev. (its perfect state is
   Gibberella acuminata Booth); secured from diseased plant.
   Mexico.   2. F. culmorum (W. G. Sm.) Sacc. ; from
   diseased plant.   Mexico.   3. F. equiseti (Cda.) Sacc. =
   F. scirpi Lamb. & Fautr. (its perfect state is Gibberella
   intricans Wr.); from diseased plant.   Mexico.   4. F.
   oxysporum Schl. f. sp. medicaginis (Weim.) Sny. & Hans. ;
   root disease often ending in collapse in warm locations.
   Argentina, Peru, Ecuador, W. Indies (mountain plantings),
   California.   5. F. roseum Lk. = F. concolor Reink. ;
   disease of stem.   Chile, Guatemala.   6. F. sambucinum
   Fckl. var. minus Wr. (its perfect state is Gibberella pul-
   caris (Fr.) Sacc. ; from diseased plant.   Argentina.   7. F.
   solani (Mart.) Sacc. (its perfect state is Nectria haema-
   tococca Berk. & Br. = ? Hypomyces solani Reinke &
   Berth.); plantrot.   Mexico.
Helicobasidium (see Rhizoctonia crocorum)
Heterodera sp. ; nematode feeding on roots.   Chile, Peru.
Leptosphaeria circinans Sacc. ; on aerial parts of plant.   S.
   America.
Meloidogyne spp. ; rootknot.   Brazil, Argentina.
Minor element deficiencies may be main limiting factor in
   some areas where crop is a complete failure and when the
   needed trace elements are supplied growth is good.   Scat-
   tered especially in areas where seasonal rains may be
   heavy.
Nematodes of a number of types (undetermined) in soil around
   roots.   Guatemala.
Oidium (see Erysiphe)
Paratylenchus sp. ; nematode in soil around roots.   Chile,
   Peru.
Peronospora trifoliorum dBy. = P. aestivalis Syd. ; the com-
   mon downy mildew.   Argentina, Brazil, Peru, Uruguay,
   Bolivia, Chile, Ecuador, Guatemala, Costa Rica, Mexico,
   Colombia, Subtropical USA.
Phymatotrichum omnivorum (Shear) Dug. ; rootrot.   Texas,
   Arizona.   California.
Physoderma (see Urophlyctis)
Pleosphaerulina briosiana (Poll.) Hoehn. ; brown leafspot.
   Peru, Brazil, Colombia.   P. briosiana (Poll.) Hoehn. f.
   sp. brasiliensis Putt. ; severe leaf disease.   Brazil.
Pleospora herbarum (Pers.) Rab. = Stemphylium botryosum
   Wallr. ; leafspot.   Guatemala, Colombia.   P. hyalospora
   Ell. & Ev. ; leafspot.   Mexico.
Pseudomonas medigaginis Sack. ; bacterial blight, branch-rot.
   Ecuador, Argentina.
Pseudopeziza medicaginis (Lib.) Sacc. ; most common and
   severe black leafspot, defoliation.   In all alfalfa growing
   countries (for the fungus P. jonesii see Pyrenopeziza, and
   for P. trifolii see Pseudoplea).

Pseudoplea briosiana (Poll.) Hoehn. (maybe synonym is Pseudopeziza trifolii (Rostr.) Petr.); gray leafspot.   Brazil, Costa
Rica, El Salvador, Guatemala.
Pyrenopeziza medicaginis Fckl. (maybe synonym is Pseudopeziza jonesii Nanuf.); large yellow leafspots and blotch is
a common symptom. Colombia, Mexico, C. America.
Pythium spp.; damping-off, seedling disease.   Neotropical.
Rhizobium meliloti Dang.; nitrogen fixation nodules in roots,
if not present in soil naturally must be introduced.   Neotropics.
Rhizoctonia crocorum DC. (synonyms: R. violacea Tul., R.
medicaginis DC.), spore producing perfect state is Helicobasidium purpureum (Tul.) Pat.; purple or violet rootrot.
Colombia, Argentina, Ecuador, Dominican Republic, Peru.
R. solani Kuehn is the parasitic mycelial state, it has the
perfect spore-bearing state Thanatephorous cucumeris (Frank)
Donk; seedling damping-off, collar rot.   Scattered in alfalfa
growing countries.   Rhizoctonia sp. of aerial type; stem
bleach, leafrot. El Salvador, Dominican Republic.
Root sickness in small patches of plants, various fungi iso-
lated but pathogenicities not proved.   Occasional from
Argentina to Mexico.
Sclerotinia sclerotiorum (Lib.) dBy. = Whetzeliana sclerotio-
rum (Lib.) Korf & Dumont; stemrot, collapse.   Chile. S.
trifoliorum Eriks. = Whetzeliana trifoliorum (Eriks.) Korf;
rootrot, collar rot. Mexico.
Sclerotium rolfsii (Curzi) Sacc. = Pellicularia rolfsii (Sacc.)
West; collar rot in warm and moist area.   Subtropical USA,
Guatemala, Nicaragua.
Septoria medicaginis Rob. & Desm.; leafspotting.   Argentina.
Stagonospora sp.; leafspot.   Argentina, Ecuador.
Stemphylium (see Pleospora)
Tylenchorhynchus sp.; nematode recovered from rootzone soil.
Chile.
Uromyces striatus Schroet. (synonyms: U. medicaginicola
Speg., U. medicaginis Pass. note also U. lupulina Speg.
was reported on alfalfa by Spegazzini in Argentina); stem
and leaf rust, very severe in warm localities.   Santo
Domingo, C. America, Chile, Peru, Argentina, probably
all Neotropics. Also reported is U. striatus Schroet. var.
medicaginis (Pass.) Arth.; severe rust, sometimes called
"rainy season rust." Mexico, C. America, S. America.
Urophlyctis alfalfae (Lagh.) Magn., synonyms: Cladochytrium
alfalfae Pat. & Lagh., Physoderma alfalfae (Pat. & Lagh.)
Karst.; crownwart, wart on stem and leaf, root affected.
Mexico, Peru, Chile, Argentina, Ecuador.
Xanthomonas alfalfae (Ricker, Jones, & Davis) Dows.; bac-
terial leaf and stem spot.   Subtropical USA, Guatemala,
El Salvador, Nicaragua.
Xiphinema sp.; nematode from rhizosphere.   Peru, Chile.

MELICOCCA BIJUGA L.   Tree of the West Indies, Central and

South America, grown for its fruit with pulp eaten fresh and seed roasted; "mamoncillo" Sp., "jungle lime" En.

Ascochytella melicoccae Gonz. Frag. & Cif.; leafspotting. W. Indies.

Cephaleuros virescens Kunze, algal leafspot. C. America.

Colletotrichum gloeosporioides Penz. (no asco-state encountered); anthracnose of fruit, leaf, stem. C. America, Colombia, El Salvador.

Fruitrot, dark color (not Colletotrichum). Honduras.

Hydnum-like sporocarps on log wood. Guatemala.

Leaf spot, unusually large and black. Isolations and microscopic examinations disclosed no fungus or bacterial bodies in host tissue. El Salvador.

Meliola cruciferae Starb.; black mildew. W. Indies. M. sapindacearum Speg.; black mildew. Panama, Colombia, Venezuela, El Salvador, Puerto Rico, Honduras.

Pestalotia sp.; lesions of leaf and petiole. Honduras.

Phoradendron sp.; phanerogamic parasite, mistletoe-like. El Salvador.

Phthirusa sp.; phanerogamic parasite. Costa Rica, W. Indies.

Septoria melicoccae Gonz. Frag. & Cif.; leafspot. W. Indies.

Strigula complanata Fée; lichen spotting on stem. Honduras.

Tricholoma pachymeres (Berk. & Br.); mushroom on decayed tree trash. S. America.

Virus: Mottle and leaf malformation like typical mosaic in El Salvador.

MICONIA spp. Numerous, tropical American, wild and cultivated species of valued shrubs, some tree-like, with handsome foliage, planted for hedges and background masses, used for soil conservation, wind-breaks, short-lived agricultural shade, and a few for medicines by indigenous "curanderos." The common names for these bushes in Spanish are such as "camasey" or "quina blanca," but there are many vernacular names that are not recorded. Diseases are numerous and a large list of black mildews is reported. (Note: An emphasis on collecting in village yards, in wild areas, and by work in selected mycological herbaria and libraries, would result in upwards of a thousand items for a disease list.)

Asterina spp. (about 20); black mildews, tropical America.

Auerswaldia fiebrigii P. Henn. and A. miconiae P. Henn.; scab-like leaf diseases. S. America.

Aulographum culmigenum Ell.; black spot. Puerto Rico.

Bagnisiopsis spp. (about 8); dieback, leaf and twig lesions. Widespread.

Blastotrichum miconiae Stev.; leaf disease. W. Indies.

Borinquenia miconiae Stev.; stem disease. W. Indies.

Botryosphaeria multipunctata Syd.; granular erumpent lesions. S. America.

Capnodium spp.; sootymolds. Scattered in the Neotropics.

Cassytha filiformis L.; green dodder. Puerto Rico.

Catacauma miconiae Th. & Syd. = Catacaumella miconiae

Th. & Syd. and C. nitens Petr. = C. miconiae or Cata-
caumella pululahuensis Th. & Syd.; scabby tarspot diseases
on foliage. Brazil, and other countries in S. America.
Cephaleuros (? minima?); algal leafspot. Costa Rica.
Cephaleuros virescens Kunze; algal leafspot. Puerto Rico,
Costa Rica.
Cercospora erythrogena Atk.; leafspot. Scattered in Neo-
tropics. C. miconiae Gonz. Frag. & Cif. and C. miran-
densis Chupp, both these latter are common leafspots.
General in occurrence.
Chaetostigmella parasitica Toro; leaf disease. S. America.
Coscinopeltella montavoae Chard.; leaf disease. S. America.
Cronartium egenulum Syd.; leafrust. Brazil, nearby
countries.
Cryptocline andina Petr.; anthracnose-like disease. Moun-
tainous S. America.
Cuscuta americana L.; coarse yellow dodder. Puerto Rico.
C. domingensis Urb.; dodder. Dominican Republic.
Cyclotheca miconiae Th.; leafspot. Brazil.
Dothidina fiebrigii (P. Henn.) Th. & Syd.; leafspot. Paraguay.
D. miconiae Th. & Syd.; leafspot with black bodies on it.
Puerto Rico, Brazil. D. peribebuyensis (Speg.) Chard.;
leafspot studded with black fungus fruiting bodies. Puerto
Rico, Panama, Brazil, probably widespread (also reported
as Bagnisiopsis see above).
Echidnodella miconiae Ryan; blackmold and black mildew. W.
Indies.
Elsinoë miconiae (Stev. & Weed.) Limb. & Bitanc.; spot
anthracnose. Brazil.
Guignardia atropurpurea Chard., G. melastomataceae (P. Henn.)
Th., G. punctiformis Chard.; leaf and stem diseases. Co-
lombia and other countries.
Helminthosporium insuetum Petr.; on leaves. S. America.
Hemidothis pitterii Syd.; leafspot. Brazil and other nearby
countries.
Hyalosphaera miconiae Stev.; leafspot. W. Indies.
Hypocrella tamoneae Earle; leaf disease with stroma. W.
Indies.
Hysterostomella costaricensis Syd. in C. America, and H.
miconiae P. Henn. (= Hysterostomina miconiae); leaf
disease with spots having on them thick stromata. C.
America, S. America (? in W. Indies I saw symptoms but
did not confirm by microscope).
Irene melastomacearum (Speg.) Th. & Syd. = Irenina melas-
tomacearum (Speg.) Th. & Syd. a black mildew. W. Indies.
Irenopsis miconiae Stev. (synonyms: Meliola miconiae Stev.,
I. miconiaecola Stev.); black mildew. W. Indies, C.
America.
Lasiosphaeria miconiae Viégas; stem disease. Brazil.
Lasiostemmella sydowii Petr.; on foliage. Widespread.
Lembosia diffusa Wint., L. melastomatum Mont., L. sclerolo-
bii P. Henn.; on leaf all are dark mildew patches. W.
Indies.

Marasmius sp.; black cordblight (only parasitic vegetative stage seen).   Costa Rica.

Meliola (see Irenopsis)

Microclava miconiae Stev.; probably hyperparasite on fungus. Puerto Rico, Dominican Republic.

Monogrammia miconiae Stev.; attack of senescent leaf.   W. Indies, (possibly scattered in general).

Morenoella spp.; black patches on leaves.   W. Indies, C. America.

Mycena citricolor (Berk. & Curt.) Sacc.; leafspot, round with discrete edges.   Puerto Rico, Costa Rica.

Mycosphaerella miconiae Syd.; leafspot.   S. America.

Napicladium fumago Speg.; black and sooty leafspot.   Argentina, Puerto Rico.

Nematodes (undetermined) rhizosphere inhabitants.   (Note: Nematodes were found when making microscopic examinations of Rosellinia and a root disease of uncertain cause; no attempt was made to determine if the nemas were parasitic.)   Puerto Rico, Costa Rica.

Pellicularia koleroga Cke.; thread-web blight.   W. Indies, C. America.

Phaeofabraea miconiae Rehm; leaf disease.   Brazil, W. Indies.

Phyllachora miconiicola Speg.; black leafspot.   Costa Rica.

Phyllachorella multipunctata (Wint.) Petr.; roughened black spot on leaf and stem.   Brazil.

Physalospora melastomicola Speg., P. miconiae Sacc., and P. miconicola Gonz. Frag. & Cif.; are irregular leafspots, stem infections.   Brazil, W. Indies (probably widespread).

Phytophthora-Pythium (rootrot isolates, uncertain identity as fungi had both Phytophthora and Pythium characters). Costa Rica, in Puerto Rico more nearly Phytophthora.

Placoasterina antioquensis Toro; black mildew, leaf disease. Colombia.

Polystigma melastomatum Pat.; on foliage.   S. America.

Pseudomeliola perpusilla Speg.; black mildew, on leaves.   S. America.

Pseudopeziza subcalycella Rehm; leafspot.   S. America.

Rosellinia bunodes (Berk. & Br.) Sacc.; rootrot.   C. America, W. Indies.

Scoriadopsis miconiae Mend.; on foliage.   S. America.

Septoria miconiae Rangel (? = S. miconiae Garm.); leafspot. Brazil, Puerto Rico.

Sorokina blasteniospora Rehm; on leaves.   Brazil.

Stemphylium dichroum Petr.; leafspot.   S. America.

Strigula spp. (white, gray, tan colors); W. Indies, C. America.

Tryblidaria subtropica Wint.; leaf and stem disease.   S. America.

Xenostomella meridensis Toro; on leaf.   Colombia.

MONSTERA DELICIOSA Liebm.   Famed ornamental from Mexico, an epiphytic vine having very large leaves handsomely sculptured,

fashioned with spectacular perforations.   Plant is multiplied by cuttings, clambers in trees or on old walls, grows in great pots or in patio gardens, and has had a minimum of attention by plant pathologists.   Vine is furnished with remarkable rope-sized aerial roots of exotic appearance, and the vine bears large elongate yellowish fruits relished by many; "piñanóna" Sp., "monstera" Sp. Po. En.

  Bacterial diseases (undetermined species found in diseased tissues examined microscopically): 1. Leafspot starting small but expanding, in Cuba.   2. Petiole and aerial-root lesions, in Cuba and in Panama.   3. Softrot of fallen leaf almost anywhere.

  Black, sunken lesion on aerial root.   Phoma-like fungus.   El Salvador.

  Botryodiplodia theobromae Pat.; stem and aerial root dryrot. Panama, probably widespread.

  Cephaleuros virescens Kunze; algal leafspot.   Costa Rica, Cuba, Puerto Rico.

  Colletotrichum gloeosporioides Penz. with often numerous bodies of asco-state Glomerella cingulata (Stonem.) Sp. & Sch.; anthracnose.   C. America.

  Diplodia = Botryodiplodia.

  Dothiorella vagans Speg.; leaf disease.   Brazil.

  Fusarium sp.; "roseum type" semi-parasitic type isolated from leafspots.   El Salvador.

  Leaf lamina collapse between large veins, unknown cause. Panama.

  Macrophoma philodendri Pk.; leafspot.   Florida.

  Mycena citricolor (Berk. & Curt.) Sacc.; leafspot.   Costa Rica.   (This fungus also attacked Monstera gigantea Eng. and M. pittieri Engl. in Costa Rica.)

  Neohenningsia brasiliensis P. Henn.; leaf disease.   Brazil.

  Phyllosticta (? philodendri?); leafspot.   C. America.

  Pythium sp.; rootrot.   Cuba.

  Strigula complanata Fée; white lichen spot.   General?   Found especially abundant in Cuba and Puerto Rico and observed in jungles of Panama, El Salvador and Costa Rica on wild piñanónas.

  Sun scald.   Costa Rica (west coast).

  Uredo monstericola Kern, Thur., & Whet.; leafrust.   S. America.

MONTEZUMA SPECIOSISSIMA Moc. & Ses.   A spectacular hibiscus-like flowered tree, not widely grown, from Puerto Rico.   Common names are "magda" Sp., "tree hibiscus" En.

  Botrytis sp.; on dying flower petals.   Puerto Rico.

  Capnodium (see Microxyphiella)

  Colletotrichum gloeosporioides Penz. as well as (on fallen leaves) the asco-state Glomerella cingulata (Stonem.) Sp. & Schr.; anthracnose.   Puerto Rico.

  Loranthus sp.; mistletoe-like parasite.   Puerto Rico.

  Microxyphiella sp., its conidial state can well be called a

Capnodium, the lesion is restricted to the base of the leaf.
Puerto Rico.
Phakopsora desmium (Berk. & Br.) Cumm. = Cerotelium
  desmium Arth.; leafrust. Puerto Rico.
Phyllachora minuta P. Henn.; leaf tarspot. Puerto Rico.

MORUS ALBA L., and M. NIGRA L. Adapted (in early years for
  silkworm food) to the American tropics from temperate zone
  Asia and Europe; medium sized often popular ornamental fruit-
  bearing trees; "morera" Sp., "amoreira" Po., "mulberry" En.
    Acrosporium sp. (= ? Oidium) sometimes applied to white
      mildew. Dry sides of W. Indies islands, C. America.
    Armillaria sp. = Armilariella sp.; rootrot. W. Indies.
    Botryodiplodia theobromae Pat. = Diplodia natalensis P.
      Evans; dieback. Neotropics.
    Capnodium spp. = Antennaria spp.; fumagina or sootymold.
      Widespread.
    Cephaleuros virescens Kunze; algal leafspot. Brazil, Puerto
      Rico, probably widespread.
    Cercospora mori Speg., C. mori Cke., and C. moricola Cke.
      (may be synonymous); cause markedly large leafspots.
      Subtropical USA, W. Indies, C. America, S. America.
    Ciboria carunculoides (Siegl. & Jenk.) Whet.; fruit disease,
      hardberry. Subtropical USA.
    Cylindrosporium mori Berl. its asco-state is Mycosphaerella
      mori (Fckl.) Lind.; the russet leafspot. Argentina, probably
      in Brazil and other parts of south S. America.
    Cytospora sp.; twig canker. Dry parts of subtropical USA.
    Diaporthe orientalis Sacc. & Speg.; branch disease. Argen-
      tina, other parts of S. America.
    Diatrypella amorae Viégas; twig and branch disease. Brazil.
    Dothiorella mori Berl.; branch canker. Subtropical USA, W.
      Indies.
    Epicoccum sp.; leaf disease. Chile.
    Hemicycliophora sp.; rhizosphere nematode. Peru.
    Kuehneola looseneriana (P. Henn.) Jacks. & Hollw.; rust. S.
      America.
    Meloidogyne spp.; rootknot nematodes. Scattered occurrence.
    Mycena citricolor (Berk. & Curt.) Sacc.; leafspot. Costa
      Rica.
    Mycosphaerella (see Cylindrosporium)
    Nectria cinnabarina Tode; twig and branch canker. Widespread.
    Oidium (see Acrosporium)
    Phleospora mori Sacc. (possibly Cylindrosporium); leafspot.
      Brazil.
    (? Phyllactinia sp.; leaf disease. Seen in a mountain garden
      in Puerto Rico.)
    Phyllosticta concors Sacc.; brown leafspot. Mexico.
    Phymatotrichum omnivorum (Shear) Dug.; dryland rootrot.
      Texas, Arizona.
    Physalospora sp.; branch canker. W. Indies.
    Polyporus spp.; woodrots of old trees. W. Indies, C. America.

Pratylenchus sp.; nematode found in rhizosphere.   Peru.
Pseudomonas mori (Boy. & Lamb.) Stev.; bacterial gumming.
  Brazil.
Rhytidhysterium ruflum (Spreng.) Speg. = Tryblidiella rufula
  Spreng.; on dieback wood.   W. Indies, S. America.
Rosellinia aquila (Fr.) deNot.; rootrot.   W. Indies.
Schizophyllum alneum (L.) Schr., and S. commune Fr.;
  branch rots.   W. Indies, S. America.
Septobasidium saccardinum (Rang.) Mar.; felt growth over
  insects.   Argentina, Brazil.
Septogloeum mori Berk. & Br.; leafspot.   El Salvador, Vene-
  zuela.
Stomatochroon sp.; a typical systemic algal leaf infection.
  Puerto Rico.
Strigula complanata Fée; white lichen spot.   Puerto Rico.
Tryblidiella (see Rhytidhysterium)
Ustilago haesendockii West.; rootsmut.   Argentina.
Virus: An undetermined mosaic mottle in Puerto Rico.

MUCUNA spp. = STIZOLOBIUM spp. of Asiatic origin.   Brought in
  as plants to protect fragile neotropical lowland and hot country
  soils, they are vigorous fast creeping legumes; "mato," "frijol
  de terciopelo" Sp., "velvudo" Po., "velvet bean" En.
  Cercospora cruenta Sacc. (perfect state is Mycosphaerella
    cruenta (Sacc.) Lath.); leafspot irregularly spread in old
    lamina.   Bolivia, W. Indies, Subtropical USA, C.
    America.   C. mucunae H. & P. Syd. (see Mycosphaerella);
    leafspot.   Puerto Rico.   C. stizolobii H. & P. Syd. (see
    Mycosphaerella); leafspot.   Antilles, Subtropical USA, S.
    America, C. America.
  Chaetoseptoria wellmanii J. A. Stev.; large round, destructive
    leafspot.   C. America.
  Fusarium sp.; podspot.   Puerto Rico, Texas.
  Meloidogyne sp.; nematode rootknot.   Subtropical USA (may
    be much more widespread).
  Mycosphaerella mucunae Stev. (? = the asco-state of Cerco-
    spora mucunae); on decayed spotted leaves.   Puerto Rico.
    M. stizolobii (see C. stizolobii above).
  Pellicularia filamentosa Rog. (aerial type, bearing small
    sclerotia but basidiospores not found); toprot.   C. America.
  Penicilliopsis brasiliensis Moell.; senescent leaf and pod
    disease.   Brazil, spread to much of S. America.
  Phyllosticta antilleana Cif.; small leafspots.   W. Indies.   P.
    mucunae Ell. & Ev.; small leafspots.   Subtropical USA.
  Phymatotrichum omnivorum (Shear) Dug.; dry country rootrot.
    Subtropical USA (probably down into Mexico).
  Physiological chlorosis: iron deficiency in Costa Rica, and
    zinc deficiency in Florida.
  Phytophthora parasitica Dast.; root and stem diseases.   Guate-
    mala.
  Rhizoctonia solani Kuehn its spore bearing state is Thanate-
    phorous cucumeris (Frank) Donk.; stem and root disease.
    Guatemala.

Sclerotium rolfsii (Curzi) Sacc. = Pellicularia rolfsii (Sacc.)
West; rootrot and collar rot. Nicaragua.
Uredo famelica Arth. & Cumm.; leafrust. S. America.

MUSA PARADISIACA L. An extremely important food source is a
sturdy plant, productive, widely grown, its fruits are starchy
but seedless and much eaten (mostly cooked) but not often grown
for long distance shipping; names used: "plátano" Sp. Po.,
"plantain," "cooking banana" En., M. SAPIENTUM L. (= M.
PARADISIACA L. var. SAPIENTUM Kuntze), a highly lucrative
fruit crop for long distance shipment in special conditions; it
is the most commonly eaten fruit in the Western tropics, its
fruits are seedless and of dessert character; names used:
"banana" Sp. Po. En., "cumbres," "guinéo" Sp., "bananeira"
Po., M. TEXTILIS Née, a tough plant grown for its strong and
excellent fiber from the pseudostem; common name is "abacá"
Sp. Po. En., "manila hemp" En. From the Orient, the three
species are large herbs ("plants," not "trees"); have a subterranean corm ("head") on which stands the enwrapped leaf-sheath
structure that suggests a strong trunk, known as the pseudostem;
at the top are the palm-like outspread leaves and an inflorescence which pushes up through the pseudostem to become the
"fruit bunch" or "fruit stalk." The fruits ("fingers") are closely
set in whorls ("hands"). Plantains and bananas have resulted from
chance crosses between the seedbearing wild diploids M. ACU-
MINATA L. (designated AA) and M. BALBISIANA L. (BB); the
result: true bananas are triploid AAB or AAA and some the
tetraploid AAAA; plantains are either triploids AAB or ABB.
There are many varieties of bananas, plantains, and abacas.
Some have descended from ancient folk selections and some out
of experimental breeding studies that are still in progress. (Unless otherwise specified any disease organism given is understood to attack all three host species.)

  Acrothecium lunatum Wakk.; part of banana fruit disease com-
    plex. General.
  Alternaria sp.; rare find in leaf-edge disease complex. Hon-
    duras.
  Aposphaeria musarum Speg.; on dying leaves of banana and
    plantain. Argentina and other parts of S. America.
  Armillaria mellea Vahl = Armilariella mellea (Fr.) Karst.;
    corm and root rot. Florida, Brazil.
  Aspergillus luchuensis Inui; part of banana and plantain fruit
    rot complex. C. America.
  Asterostroma musicolum Mass.; leathery semi-saprophytic
    growth on fallen banana trash. W. Indies.
  Belondira parva Thorne; banana root nematode. Puerto Rico.
  Botryodiplodia theobromae Pat. (conidial state only); fruit
    peduncle and finger rot, corm blackhead, stalk rot. Neo-
    tropics.
  Calostilbe striispora Seav. (? = Sphaerostilbe musarum Ashby);
    on banana, occurrence is associated with corm and fruit
    stalk failure. W. Indies.

Capnodium musae Viégas; pellicle of sootymold on Musa
  foliage and on maturing fruit stalk.   Brazil, El Salvador,
  Costa Rica.   (This is apparently not Fumago.)
Cephalothecium sp.; in banana and plantain fruit decay complex. Widespread.
Ceratocystis paradoxa (Dade) Mor., synonyms are: Thiela-
  viopsis paradoxa Hoehn., Ceratostomella fimbriata Dade;
  banana and plantain stalk and finger rot.   General, probably
  also on abacá.
Cercospora hayi Calp.; secondary on leaves, on fruit as brown-
  spot occurs as minor on abacá, mild on plantain, but under
  some conditions severe on banana.   Mexico, Honduras,
  Cuba, Costa Rica, Panama, Ecuador, Colombia, Surinam,
  Guatemala.   C. musae and C. musarum see Mycosphaerella.
Chaetothyrina musarum (Speg.) Th. = Chaetothyrium musarum
  Speg.; the common sooty blotch on fruits of banana and
  plantain. Brazil, C. America.
Chloridium musae Stahel; leaf speckle.   Surinam, Guatemala,
  Jamaica.
Cladosporium musae Mason, C. (? atriellum), C. herbarum
  Lk.; all three part of sooty moldyness of banana bagged
  fruit. C. America, Jamaica.
Clasterosporium maydicum Sacc.; on banana a weak plant
  parasite. Trinidad.
Clitocybe tabescens (Scop.) Bres. = Armilariella tabescens
  (Scop.) Sing.; banana cormrot.   Florida.
Coccomyces musae Sacc.; disease of banana leaf.   C. America.
Colletotrichum musae (Berk. & Curt.) Arx (not C. gloeo-
  sporioides) its perfect state is Gloeosporium musarum Cke.
  & Mass.; the most common fruitspot on any Musa spp.,
  banana fruit neckrot and scald.   Neotropical.
Cordana musae (Zimm.) Hoehn. = Scolecotrichum musae
  Zimm.; leafspot that is most severe on plantain and abacá,
  milder on banana.
Corynespora casiicola Wei; a banana fruitspot, occasional as
  a leafspot. Honduras, W. Indies.
Criconema sp.; banana rhizosphere nematode.   C. America.
Cuscuta americana L.; yellow dodder.   C. America, W.
  Indies.
Cylindrocolla musicola Speg.; on banana fruit skin.   Argentina
  and other countries.
Cyphella musaecola Berk. & Curt.; plantain pseudostem
  disease. Cuba.
Deightoniella torulosa (Syd.) Ell. Synonyms: Helminthosporium
  torulosum Ashby., Brachysporum torulosum Syd.; on
  banana and plantain fruit causes speckle and blacktip, and
  on leaf blackspot, on abacá seedling damping-off, leafspot,
  pseudostem deepspot.   Neotropical.
Diamond spot of banana fruit and leaf is a complex of Fusarium
  spp. and Cercospora hayi.   C. America, S. America, W.
  Indies.
Drechslera gigantea Ito = Helminthosporium giganteum Heald &
  Wolf; fruit spot, leaf eyespot.   Widespread.

Erwinia carotovora (L. R. Jones & Dows.) Holl. = Pectobacterium carotovorum Wal.; bacterial softrots, part of
  transit and storage fruitrot complex.   Neotropical.
Finger-drop of banana.   Mechanical pedicel injury invaded by
  Botryodiplodia, Colletotrichum, bacteria, Fusarium, and
  other organisms.   Neotropics.
Fuligo septica Gmel.; rare saprophyte on moist banana trash.
  Puerto Rico.
Fumago vagans Pers.; on banana apparently a special sootymold on bagged fruit bunches under protective bags.   Neotropics.
1. Fusarium moniliforme Sheld. the asco-state is Gibberella
  fujikuroi (Saw.) Ito, often followed by secondary bacteria
  and a Cephalosporium; on all Musas, causes leaf-sheath
  disease, associated with a heartrot of pseudostem, predis-
  poses plants to wind damage.   Widespread.   2. F. oxysporum Schl. (wild type); only infects roots of Musa balbisiana.
  Honduras, Colombia, Panama, Costa Rica.   3. F. oxysporum Schl. f. sp. cubense (E. F. Sm.) Sny. & Hans.; the
  panama disease, systemic and spectacular fusarial wilt.
  In many countries, in more or less acid tropical soils.
  In abacá varieties some susceptible and some moderately
  to fully resistant, certain bananas resistant (gros michel
  banana susceptible, red banana is moderately resistant, a
  dwarf cavendish is resistant), plantains are mostly resistant.   (Some fungus race differences: race 1, kills gros
  michel banana AAA, silk banana AAB, some abacás, and
  balbisiana BB; race 2, severe on plantains such as bluggoe
  and chato ABB, mild on balbisiana BB, severe on acuminata AA; race 3, variable in disease on balbisiana BB,
  non-pathogenic on chato and bluggoe ABB.   It is notable
  that some dwarf bananas are susceptible to Fusarium oxysporum cubense in the Orient but resistant to it in the
  Occident.)   4. F. (roseum-type); part of a rot complex of
  banana and plantain fruit in storage and transport.   5. F.
  semitectum Berk. & Rav.; banana fruit peduncle disease.
  S. America.   6. F. solani (Mart.) Sacc.; rootrot (non-
  systemic).   W. Indies.
Gnomonia veneta Kleb. (? = Gloeosporium nervisequum Sacc.);
  on plantain and banana the leafvein anthracnose disease,
  leaf tip browning.   Argentina.
Haplolaimus sp.; nematode found in rhizosphere of banana.
  Dominica.
Helicotylenchus multicinctus Gold.; sting root nematode on
  plantain.   Cuba.   Helicotylenchus spp.; banana sting root
  nematodes.   Guyanas, Surinam, Trinidad, Jamaica, Bolivia,
  Dominican Republic, Virgin Islands, Barbados.
Helminthosporium (see Deightoniella and see Drechslera)
Helotium musicola Speg.; leaf lesion of banana and plantain.
  C. America, S. America.
Hemicycliophora sp.; nematode found in banana rhizosphere.
  C. America.

Hendersonula toruloidea Natt.; banana fruit tiprot.   Jamaica
    (at high elevations).
Hormodendrum cladosporioides (Fr.) Sacc. = Cladosporium
    cladosporioides (Fr.) deVries; banana leaf and fruit moldy
    disease.   Trinidad, Puerto Rico, Cuba, Jamaica.
Leptosphaeria musicola (Speg.) Sacc. & Trott. = Leptosphae-
    rella musicola Speg.; banana leaf ringspot.   S. America.
Leptothyrium musae Cif. & Frag.; banana leaf disease.
    Dominican Republic.
Loranthaceae, occasional seeds dropped by birds on Musa
    leaves.   Seeds germinate, pierce leaf lamina but finally
    exhaust the endosperm, and seedling dies.   Uncommon,
    but are traceable from nearby jungles of parasitized dico-
    tyledonous trees.   (Several times observed.)
Macrophoma (see Phyllostictina)
Marasmius stenophyllus Mont. (M. semiustus Berk. & Curt.)
    = Marasmiellus inoderma Sing.; common rootrots.   Neo-
    tropics.
Melanospora pampeana Speg. = Spicaria mucoricola Speg.;
    found as a banana leaf disease.   Southern S. America.
Meliola musae (Kunze) Mont.; on abacá leaf, fine-woven black
    mildew.   Costa Rica.
Meloidogyne spp.; rootknot.   Jamaica, Surinam, Guyanas,
    Trinidad, Virgin Is., Barbados, Dominica, Dominican Re-
    public.
Microxyphium sp. (ascomycete state of) sootymold on banana
    fruit under protective bag.   C. America.
Mucor caespitulosis Speg.; infection of dead flower ends on
    Musa ensete in Brazil and Guatemala and Costa Rica (con-
    firmed by moist chamber cultures in C. America).
Mycena citricolor (Berk. & Curt.) Sacc.; leaf infections.
    Guatemala, Costa Rica.
Mycogone sp.; (? hyperparasite) on diseased banana leaf.
    Colombia.
Mycosphaerella musae Speg. with short ascospores is asco-
    state of Cercospora (? musarum); a minor leafspot of
    banana.   S. America.   M. musicola Leach with longer
    ascospores is asco-state of the common and devastating
    Cercospora musae Zimm.; the severe sigatoka leaf disease
    on bananas and some plantains, not on abacá.   Neotropics.
    (M. fijiensis Mered. = M. musicola Leach var. fijiensis
    Stov.; cause of banana black leafstreak in Orient thus far
    not in west tropics but can be expected sooner or later.)
Nectria suffulta Berk. & Curt.; on plantain dead pseudostems.
    St. Thomas, Honduras.
Nectriella musicola (Speg.) Sacc. & Trott.; on plantain and
    banana leaf.   S. America.
Nigrospora oryzae Petch; on senescent leaf.   Widespread.
Papularia sphaerosperma Hoehn.; banana leaf disease.
    Trinidad.
Paratylenchus sp.; rhizosphere nematode.   C. America.
Pectobacterium (see Erwinia)

Penicillium glaucum Lk. and P. musae Weid.; part of sooty-
mold complex on bagged banana fruit bunches. Widespread.
Periconiella musae Stahel = Ramichloridium musae Stahel;
banana leafspeckle and smoky blotch. Surinam.
Pestalotia leprogena Speg.; on probably all cultivated species
of Musa, causing fruitspot and rot of main stalk in bunch.
Widespread. P. palmarum Cke. = Pestalotiopsis palmarum
Stey.; banana and plantain circular spot following injuries.
Jamaica, C. America, northern S. America.
Peziza livida Schum.; and P. musicola Sacc. & Sacc.; on
dying plantain pseudostems. S. America.
Phaeosaccardinula musaecola Bat.; banana leaf disease.
Brazil.
Phoma musaecola Speg.; plantain fruit freckle. S. America.
P. musarum Cke.; banana fruit freckle. C. America, W.
Indies.
Phomopsis sp.; a banana fruit stalk disease. Scattered but
widespread.
Phyllosticta gastoni Tassi and P. musae-sapienti Gonz. Frag.
& Cif.; spots only on banana leaf. W. Indies.
Phyllostictina musarum Petr. (= Macrophoma musae (Cke.)
Berl. & Vogl. or Phoma musae Cke. including other
synonyms in such genera as Dothidea, Phyllachora,
Sphaeropsis, and Guignardia); freckle leaf disease on
banana and plantain. W. Indies, C. America, Brazil.
Physiological diseases (that is, no apparent parasites involved), examples of: Blow down or lodging in wind prone
areas. Chlorosis in soils with excess alkalinity. Corm
navel rot, degeneration of central "eye." Corm root inhi-
bition. Corm water-soaked tissue (internal effect, not
from poor soil drainage). Fruit black spot. Tissue sterile,
no apparent virus. Leaf streak, not virus, not insect,
tissue sterile.
Polyporus sapurema Moel.; on banana root and plant base
disease complex. Brazil.
Pratylenchus coffeae Filip. & Stek.; a root nematode that is
severe on plantain and abacá but mild on banana. Surinam,
Dominica, Puerto Rico, Jamaica, Honduras, Cuba, Guatemala, Costa Rica. P. musicola Filip.; root nematode of
abacá. Costa Rica, Panama, Honduras.
Pseudomonas solanacearum E. F. Sm. = Xanthomonas solanacearum (E. F. Sm.) Dows.; bacterial wilt. This bacterial
species is composed of races, those important to Musa wilt
are: Race 1, attacks several diploid species including M.
acuminata AA, and infects numerous other hosts, tomato,
miscellaneous Solanaceae and is often called the "tomato
strain." Race SFR (pure culture appearance of colonies
is "smooth, fluid and round") is markedly adapted to insect
distribution by wasps, bees, flies that feed on ooze from
inflorescences and visit healthy buds. Is specially severe
on plantain. Race 2, most common on triploid banana.
Widely distributed.

Pseudoplea trifolii (Rostr.) Petr.; banana fruitrot.   Trinidad.
Puccinia heliconiae (Diet.) Arth. = Uredo heliconiae Diet.;
    rare banana and abacá leafrust.   Guatemala, W. Indies.
Pyricularia grisea Sacc. = Pyriculariopsis parasitica Ellis;
    banana fruit pitting, semi-parasitic on dying leaves.
    Mexico, Panama, Colombia, C. America, Ecuador, Brazil,
    and occasional but rare in Surinam and Caribbean islands.
Radopholus similis Thorne; burrowing root and corm nema-
    tode, on Musa spp.   Widespread.   (Note: This nematode
    when accompanied with Fusarium oxysporum cubense will
    cause increased disease, it acts "synergistically," as there
    is less disease if the nematode is used alone or if the
    fungus is used alone.   However, this nematode is by no
    means a prerequisite to Fusarium infection of roots.)
Rhizoctonia grisea Matz; aerial-adaptation causing leaf infec-
    tions on young plantain and banana plants.   W. Indies,
    Costa Rica, (? Colombia).   R. solani Kuehn = Thanate-
    phorous cucumeris (Frank) Donk, the soil inhabiter that
    causes occasional rootrot and seedling disease of any Musa,
    will grow only superficially on healthy mature pseudostem
    surfaces of banana and abacá, but will invade outside cells
    of plantain.
Rhizopus nigricans Ehr.; market fruitrot of banana.   Trinidad,
    Peru.
Rosellinia pepo Pat.; base-of-plant rot.   S. America.
Rotylenchulus reniformis Lind.; root attack on all three
    Musa species.   Probably in all banana growing countries.
Sclerotinia sclerotiorum (Lib.) dBy. = Whetzeliana sclerotiana
    (Lib.) Korf & Dumont; banana fruitrot where cool.   Ber-
    muda.
Sclerotium rolfsii (Curzi) Sacc. = Pellicularia rolfsii (Sacc.)
    West; on various Musa seedlings in breeding gardens.
    Trinidad.
Sphaerella musae Speg.; a banana leaf disease.   Dominican
    Republic.
Sphaeroderm argentinensis Speg.; on banana leaf.   Argentina.
Sphaeropsis paradisiaca Sacc. var. minor Gonz. Frag. &
    Cif.; banana and plantain leaf fungus.   Dominican Republic.
Sphaerostilbe (see Calostilbe)
Sporormia minima Avers.; on banana leaf.   Trinidad.
Sporotrichum musarum Torr.; hyperparasite of insect.   S.
    America.
Stachylidium theobromae Turc.; banana and plantain fruit
    cigar-endrot, flower infection.   Mexico, Jamaica, Trinidad,
    Bermuda, El Salvador, Honduras, Costa Rica, Colombia,
    Ecuador, Peru, Venezuela, Brazil, Bolivia.
Stictis musae Seav. & Wat.; on banana trash in Bermuda.
    S. tropicalis Speg.; on plantain trash in Brazil.
Stomatochroon sp.; systemic leaf blade invasion of alga.
    Costa Rica.
Thielaviopsis (see Ceratocystis)
Trichodorus sp.; banana rhizosphere nematode.   C. America.

Tylenchus musicola Cobb; banana root attack. W. Indies,
   Colombia.
Uredo (see Puccinia)
(A rust, Uromyces musae P. Henn., is on banana in the
   Orient and should be considered as an eventual arrival in
   the Neotropics.)
Verticillium theobromae Turc.; banana and plantain corm rot
   and plant wilt. Guayanas, Puerto Rico, Jamaica, Cayman
   Island. Verticillium sp.; part of banana cigar endrot complex. Neotropics.
Viruses of Musa spp. need more attention and a comprehen-
   sive study. Some obvious examples: On abacá, plantain
   and banana the Cucumber Mosaic Virus causes leaf mottle,
   mild stripes and is scattered. On banana and plantain the
   virulent strain of Cucumber Mosaic from Commelina causes
   much more severe mottle, dwarfing, often causing heartrot
   and lodging in the W. Indies, Florida, Panama. Something
   very similar to if not identical with Banana Bunchy Top
   occurs on banana in Brazil. On banana the Rayadilla is
   possibly the same as Stripe or Streak in Colombia, Venezuela. A severe Mosaic on plantain is found in Brazil.
   (The insect vectors for first and second viruses are:
   Aphis gossypii Glover and A. correopsidis Thomas in con-
   trolled conditions whether in Florida or in Puerto Rico.
   In the case of the third virus the vector is one aphid:
   Pentalonia nigronervosa Coq. that occurs in Brazil.)
Xanthomonas (see Pseudomonas)
Xiphinema sp.; banana rhizosphere nematode inhabitant. C.
   America.
Zygophiala jamaicensis Mart.; banana leafspeckle. Jamaica.
(N.B.: Several items included above are rare in occurrence
   and some of questionable nature; a number of fungi are not
   listed even though described as possible disease associates.
   However, no pathologist, least of all the present one, can
   have any certainty that through genetic accident a seemingly
   minor parasite might not contribute a race of major sig-
   nificance. Finally, the list I give lacks some 30 or more
   reported "diseases" that are actually saprophytes and
   epiphytes.)

MUSA SAPIENTUM; M. TEXTILIS see MUSA PARADISIACA

MYRISTICA FRAGRANS Houtt. A medium-sized tree (origin the
   Mollucas) that is the source of spices. "Nuez moscada" Sp.,
   "nozmoscada" Po., "mace, " "nutmeg" En.
   Botryodiplodia sp.; dieback. Costa Rica.
   Cephaleuros virescens Kunze; algal leafspot. Puerto Rico.
   Ceratocystis sp.; branch death and wood-stain. Costa Rica.
   Colletotrichum sp.; anthracnose spots on leaf and fruit. C.
      America.
   Corticium salmonicolor Berk. & Br.; branch pink-disease.
      C. America, W. Indies.

Eutypa erumpens Mass.; woodrot, bark disease.   Trinidad,
    Granada, Cuba.
Fomes spp.; wood rotting.   C. America.
Marasmius equirinus Muell.; cord or wire blight.   W. Indies,
    C. America.
Meliola sp.; black mildew of leaf and fruit.   Costa Rica.
Oidium sp.; white powdery mildew.   Guatemala (dry season).
Pellicularia koleroga Cke.; leafrot.   Trinidad, Guatemala,
    Panama.
Phoma-like fungus isolated from leafspot.   El Salvador.
Phomopsis sp.; fruitspot.   W. Indies.
Phyllosticta sp.; leafspot.   W. Indies.
Phytophthora sp.; flush-shoot rot, root disease.   Costa Rica.
Rosellinia bunodes (Berk. & Br.) Sacc.; black root, and R.
    pepo Pat; white rootrot.   C. America (wet areas).
Stem canker, undetermined cause.   Panama.
Strigula complanata Fée; lichen leafspot.   Puerto Rico.
Trentapohlia sp.; algal tufts on leaves.   Costa Rica, Puerto
    Rico.

MYRTUS spp. (the classic M. COMMUNIS L. is the "myrtle" of
    song and story, not to be confused with the crape myrtle).
    Several species of bushes and small trees planted for sentiment
    and the attractive, fragrant leaves and fruit; "arrayán" Sp.,
    "myrtle" En., "mirto" Po. (Note: Some, if not all, diseases
    on them may be expected to be found on much more economically
    important tree relatives, some of which are guava, rose-apple,
    eucalyptus, etc. Lists I secured are only names of fungi and
    not diseases in Ecuador, Chile, Brazil, and Argentina; I have
    seen only five sources, and some of it is fragmentary.)
    Antennaria scoriadea Berk.
    Asterina pellicularia Berk.
    Asterinella cylindrotheca (Speg.) Th.
    Aulographum inconspicuum Rehm
    Clypeosphaeria chilensis Speg.
    Crinipellis myrti Pat.
    Cylindrocladium candelabrum Viégas
    Dacromyces deliquescens Bull.
    Deparzea myrticola Klo.
    Didymosphaeria eugenicola Speg.
    Dothidea granulosa Klo.
    Fomes senex Cke.
    Hypoxylon marginatum Berk.
    Leptosacca lumae Syd.
    Limacinia fernandeziana Neg.; L. scordiadea Keiss.
    Marasmius isabellinus Pat.; M. myrti Sacc.
    Melanomma australe Speg.
    Meliola valdiviensis Speg.
    Melophia superba Speg.
    Merulius chilensis Speg.
    Microcyclus labens Sacc. & Syd.
    Mycosphaerella lumae Syd.; M. myrticola Speg.

Myiocopron valdivianum Speg.
Napicladium fumago Speg.; N. myrtaceum Speg.
Negeriella chilensis P. Henn.
Pestalotia decolorata Speg.
Pezomela lumae Syd.
Phoma tularostoma Fr.
Phomopsis lumae Syd.
Phyllachora conspurcata Sacc. = Dothidea conspurcata Berk.
Phyllosticta costesi Speg.
Physalospora philippiana Wint.
Pilidium myrtinum Dur. & Mont.
Polystomella granulosa Th. & Syd. = Asterina granulosa
      Hook. & Arn.
Puccinia psidii Wint.
Rehmiodothis myrtincola Th. & Syd.
Sclerotium rolfsii (Curzi) Sacc.
Seynesia chilensis Speg.
Sphaerella myrticola Speg.
Sphaeria mammaeformis Pers.
Stictis myrti Pat.
Stomiopeltella patagonica Th. = Microthyrium patagonicum
      Speg.
Stomiopeltis chilensis Syd.
Trichothyrium chilensis Speg.

NEPHELIUM see LITCHI

NEPHROLEPIS EXALTATA (L.) Schott var. BOSTONIENSIS Davenp.
   A cultivated fern common in almost all communities of tropical
   America is the spectacularly variable, easily grown ornamental
   fern; "helecho común," "helecho bostón" Sp., "samambaia,"
   "samambaia-boston," "feto-boston" Po., "house fern," "boston
   fern" En.   (Note: Two centuries ago whaling vessels needing
   drinking water visited Central American ports where captains
   and seamen saw plant-loving peasant people who grew frilly
   leaved ornamental ferns in pots inside rooms and under eaves
   of houses.   Plants were bought to grow in ship cabins at sea
   and sailors took plants to New England, often near Boston, to
   grow in houses.   A century later Central Americans on visiting
   New England took back to the tropics the so-called "boston fern.")
      Botrytis cinerea Pers.; leaf tipmold.   Costa Rica (Caribbean
         Coast area).
      Capnodium spp.; epiphytic sootymold (invariably traceable to
         honeydew from nearby sucking insects on citrus or other
         tree).   Scattered.
      Cassytha filiformis L.; green dodder by inoculation.   Puerto
         Rico.
      Cercospora sp.; uncommon leafspot.   C. America.
      Cuscuta americana L.; a yellow dodder by inoculation.
         Puerto Rico.
      Cylindrocladium pteridis Wolf; leafspot.   Florida.
      Glomerella nephrolepis Faris (conidial state is an unidentified

Colletotrichum with small curved spores); leaflet disease
and plant base decay.   El Salvador in a mountain village.
Meliola sp. (on Nephrolepis rivularis Mett.); black mildew.
Widespread.   Also M. (probably amphitricha); black mildew.
Costa Rica, collected in a mountain village.
Milesia columbiensis Diet.; a rust on wild Nephrolepis.
Colombia, Puerto Rico.   M. insularis Faull on boston
fern.   Puerto Rico.
Nigrospora sp.; secondary infections on edges of old leaves.
Puerto Rico.
Pellicularia sp. (an aerial form, no sclerotia); leaflet rot.
Puerto Rico (rainforest ecology).
Pestalotia sp.; secondary infections on midrib of old leaves.
Guatemala.
Phyllosticta sp.; leafspots.   Florida.
Rhizoctonia solani Kuehn = Thanatephorous cucumeris (Frank)
Donk; rot at plant base.   C. America, W. Indies, Florida.

NERIUM OLEANDER L.   Originating in Mediterranean lands, this
woody bush-like tree is planted in tropical America for its
handsome growth characters and its fragrant flowers; "alhelí,"
"adelfa" Sp., "oleandro," "espirradeira" Po., "rose bay,"
"oleander" En.
Alternaria sp. (Macrosporium nerii Cke.); leafspot.   Sub-
tropical USA.
Armillaria mellea Vahl = Armilariella mellea (Fr.) Karst.;
mushroom rootrot.   Subtropical USA (dry area).
Ascochyta oleandri Sacc. & Speg.; leafspot.   Southern S.
America.
Capnodium elongatum Berk. & Desm.; sootymold.   Subtropical
USA, C. America, S. America.
Cassytha filiformis L.; green dodder (by inoculation).   Puerto
Rico.
Cephaleuros virescens Kunze; algal leafspot.   Puerto Rico.
Cercospora neriella Sacc.; leafspot.   Neotropics.
Clitocybe tabescens (Scop.) Bres. = Armilariella tabescens
(Scop.) Sing.; rootrot.   Florida, Puerto Rico.
Colletotrichum gloeosporioides Penz. = Gloeosporium oleandri
Sacc. (perfect state is Glomerella cingulata (Stonem.) Sp.
& Schr.); anthracnose.   Widespread.
Cuscuta americana L.; fast-growing yellow dodder.   Puerto
Rico.   C. indecora Choisy; slow-growing more slender
yellow dodder.   Florida.
Diaporthe nerii Speg. = D. eres Nits.; twig disease.   W.
Indies, Subtropical USA, Panama.
Diatrypella inflata Rick; stem canker.   Brazil.
Diplodia nerii Sacc. (= Botryodiplodia sp.); twig dieback and
part of complex causing the fungus-induced witches' broom.
California, Puerto Rico.   (See Physalospora.)
Erwinia carotovora (L. R. Jones & Dows.) Holl.; softrot of
cuttings, flower softrot in market.   El Salvador.
Flower petal disease.   Fungus-induced early fading of flower

in moist area of Puerto Rico. Isolations gave indefinite results.

Glomerella (see Colletotrichum)

Haplosporella nerii Sacc.; branch canker, erumpent leafspot. Florida, El Salvador.

Hypocrella orbicularis Syd.; stem and leaf infections. C. America.

Macrophoma oleandrina Pass.; stem swellings. S. America.

Mycena citricolor (Berk. & Curt.) Sacc.; leafspot. Costa Rica.

Oryctanthus ruficaulis Eichl.; parasitic scandent bush on host branches. Surinam.

Pellicularia koleroga Cke.; leafrot and web-blight. Argentina.

Pestalotia versicolor Speg.; lesions on leaf, petiole and stem. W. Indies, C. America, S. America.

Phthirusa sp.; parasitic bush. El Salvador.

Phyllosticta glaucispora Del. and P. nerii West.; light colored leafspots. Subtropical USA, C. America.

Phymatotrichum omnivorum (Shear) Dug.; dryland rootrot. Texas (? Mexico--tourist report).

Physalospora sp. = Botryosphaeria sp., asco-state of Diplodia; dieback and witches' broom (black). Puerto Rico.

Pseudomonas savastanoi E. F. Sm. var. nerii C. O. Sm. and P. tonelliana (Ferr.) Burkh.; bacterial branchknots. Dry subtropical USA.

Root failure of cuttings from mother plants that were grown on soils deficient in boron or zinc. Subtropical USA, W. Indies.

Rosellinia bunodes (Berk. & Br.) Sacc.; black rootrot. Moist area in Costa Rica. R. subiculata Sacc.; rootrot. S. America.

Schizophyllum commune Fr.; basal bush rot and old branch rot. Puerto Rico.

Septoria oleandrina Sacc.; leafspotting (small lesions). Widespread.

Sphaeropsis sp.; stem canker, witches' broom. Florida.

Strigula sp.; small greenish lichen spots on foliage. Puerto Rico, Virgin Islands.

Sunscald of leaves and new stems after shade tree was cut. El Salvador.

Virus. Undetermined but caused symptoms of leaf mottle and aborted flowers with mild witches' broom-like growth. Puerto Rico.

Woodrots. Undetermined, soft rots, some firm, light and dark colors, in old trees at base. No sporocarps. C. America, W. Indies.

NICOTIANA TABACUM L. An important crop produced for its pleasure-giving value, native of Mexico, Central America and part of South America is this large plant, herbaceous at first but in a year becoming woody, its fragrant leaves containing a mild narcotic recognized by the aboriginal users who chewed and

smoked it, and who gave it to earliest explorers who introduced it into Europe and the rest of the world; "tabaco" Sp., "tabaco," "fumo" Po., "tobacco" En. and local names seldom recorded. (Note: There are some 40 species of Nicotiana; some have poisonous properties, and a few are utilized instead of, or in addition to, common tobacco. For example N. longiflora Cav. is smoked and additionally favored for its flowers, and N. sylvestris Speg. & Comes found in Argentina by Spegazzini, with its strongly fragrant stimulating leaves, is used by Indians to reinforce common tobacco.)

Algae, green scum on seedbeds and in low grounds, results in unthrifty growth. Scattered.

Alternaria alternata (Fr.) Keiss. = A. tenuis Nees (small spores); not common, causes blackened curing and "pole-rot." Subtropical USA, Cuba, Brazil. A. longipes (Ell. & Ev.) Mason (large spores); brown-spot. Subtropical USA, Cuba, Brazil, Ecuador. A. solani Sor., A. tabacina Hori, and even A. brassicae Sacc. sometimes listed as causing brown-spot.

Ascochyta nicotianae Pass.; (in W. Indies), and A. phaseolorum Pass.; (in Brazil), causing leafspots.

Aspergillus spp.; about 5 spp. cause or add to shed burn and similar curing and storage problems. Widespread.

Botryobasidium (see Rhizoctonia)

Botryodiplodia cacaoicola (Pat.) P. Henn. = B. theobromae Pat.; wound parasite on old stems. General but scattered.

Botrytis cinerea Pers.; brown rot especially of harvested leaves. Widespread.

Cassytha filiformis L.; green-slender dodder (in a two-year-old abandoned field). Puerto Rico.

Ceratocystis fimbriata Ell. & Halst. (Ceratostomella fimbriata Ell.); systemic stain disease in old stems. W. Indies, Venezuela.

Cercospora nicotiana Ell. & Ev. (C. apii Fres.); frog-eye spot, burn on lower leaves. W. Indies, Subtropical USA, Nicaragua, Colombia, Ecuador, Venezuela (possibly nearly general).

Cincinnabolis cesstii dBy.; hyperparasite of white mildew. Reported in Colombia, seen by me in mildewed leaves sent from Venezuela.

Cladosporium (? herbarum); spores on spot diseased old leaf. Cuba.

Cold injury, buds became chlorotic and dropped after 3 hours at freezing. Mountain garden in Guatemala.

Colletotrichum gloeosporioides Penz. deserves f. sp. nicotiana. (synonyms: Colletotrichum nicotianae Av.-Sacc. or C. nicotianae Boem.); anthracnose. Brazil, Costa Rica, El Salvador. (I believe it to be Neotropical.)

Criconemoides sp.; nematode feeding on root-tip. Subtropical USA.

Cuscuta americanum L.; fat-yellow dodder. Puerto Rico. Cuscuta dodder species. Several seen, undetermined. C. America, Cuba, Puerto Rico.

Cytospora nicotianae Av. -Sacc.; stem disease.   Brazil.
Dark cure, apparently because of potash deficiency.   C.
America, Venezuela.
Distortion, strap leaf, etc., injury from 2, 4-D.   Scattered.
Drought spot (large and numerous sterile lesions coupled with
curved leaf edges).   Guatemala, El Salvador.
Drowning collapse.   Following flood and water-logging of soil
lasting over 24 hrs.   El Salvador.
Erwinia sp. (? aroidea); stalkrot.   Ecuador.   E. carotovora
(L. R. Jones & Dows.) Holl.; soft stalk, seedling rot.
El Salvador (in field where soil was thin and moist).
Erysiphe cichoracearum DC. = Oidium tabaci Th.; white
powdery mildew.   Special reports from Jamaica, Guate-
mala, Brazil but widespread.
False broomrape (? bacterial cause).   Nicaragua, Subtropical
USA.
Fleck (see Weather)
Frenching, undetermined cause, known also as "macho."   C.
America, Venezuela, Jamaica.
1. Fusarium episphaeria (Tode) Sny. & Hans. = F. dimerum
Penz. (small spores, 2-3 cells); weakly parasitic on es-
caped tobacco plants on hillside.   El Salvador.   2. F.
moniliforme Sheld. the perfect state is Gibberella fujikuroi
(Saw.) Ito; damping-off, collar rot of young plants.   Puerto
Rico.   3. F. oxysporum Schl. f. sp. nicotianae (Johns.)
Sny. & Hans.; systemic wilt.   At one time much in Argen-
tina and Chile and in W. Indies, but now uncommon from
use of resistant strains of host.   4. F. solani (Mart.)
Sacc. its asco-state is Nectria haematococci Berk. & Br.;
damping-off, stem lesions, is isolated from borders of
leafspots.   Widespread, from W. Indies to Argentina.   5.
F. tabacivorum Del. = F. oxysporum nicotianae; disease
of large seedlings.   Argentina.
Hail injury, rare but occasional.
Insecticide (DDT), injury causes yellowing and leaf malformations, from excesses in application.
Lightning strike, round-shaped killed area in field and the
bases of live plants at periphery are torn open.   Scattered.
Macrophoma tabaci Av. -Sacc.; gray stemspot.   Brazil.
Macrophomina phaseoli (Taub.) Ashby (which probably deserves
f. sp. tabaci); collar rot, charcoal rot.   Guatemala hot dry
area, Venezuela, El Salvador.
Meloidogyne spp.: at least hapla Chit., incognita Chit., and
javanica Chit.; rootknot nematodes.   Widespread.   M.
inornata Lord.; rootknot.   Brazil.
Mineral deficiencies, symptoms:
Boron lack causes distortion, growing point killed, stunt.
Calcium lack causes dead leaf margins, apical necrosis,
intense green color.
Copper lack causes sudden loss of stiffness.
Iron lack causes marked chlorosis, severe at top.
Magnesium lack causes leaf chlorosis between veins.

Manganese lack causes general chlorosis, poor growth.
Nitrogen lack causes yellow-green chlorosis, stunt.
Phosphorus lack causes leaf curl, slow growth, small
   leafspots.
Potassium lack causes mottling, necrotic spotting of leaf.
Zinc lack causes chlorosis of old leaves, necrotic areas
   in younger leaves.
Mucor spp.: molds of cured tobacco. Where storage is
   poorly handled.
Nectria cinnabarina Tode f. sp. solanicola Av.-Sacca = N.
   balansae Speg.; branch canker on old plants. C. America,
   S. America.
Nematodes (see generic names of)
Oidium (see Erysiphe)
Olpidium brassicae (Wr.) Dang.; in young roots (benign but
   vector of a necrotic virus). S. America, Subtropical USA.
Orobranche ramosa L.; broomrape parasite. Cuba (probably
   scattered in other countries).
Ozone damage, white cloudy leafspots and fleck, least severe
   under shade. Most likely near large centers of industrial
   development.
Penicillium spp.; green, reddish, white, brown molds of
   cured leaf found in poor storage.
Peronospora nicotianae Speg. (described 1898), = P. tabacina
   Adam (1933); bluemold. First found in Argentina in 1891,
   in Texas in 1906. Neotropics.
Phoma solanicola Prill. & Del. = Plowrightia solanicola Av.-
   Sacc.; stem and branch disease. Brazil.
Phyllosticta hainensis Gonz. Frag. & Cif.; leafspot. Domini-
   can Republic, Puerto Rico, Jamaica, Brazil. P. nicotiana
   Ell. & Ev.; seedling leafspot. W. Indies. P. nicotiani-
   cola Speg.; a whitish leafspot. Argentina. P. tabaci Pass.;
   leafspot. Brazil.
Phymatotrichum omnivorum (Shear) Dug.; rootrot. Texas
   (? Mexico).
Physarum spp.; slime mold, smothering epiphyte. Subtropical
   USA.
Phytophthora parasitica Dast. var. nicotianae Tuck.; black
   shank. Neotropics.
Placosphaeria nicotianae Av.-Sacc.; stem blackrot. Brazil.
Pleospora infectoria Fckl.; on old leaves. Brazil. P.
   nicotianae Av.-Sacc. (asco-state of Alternaria?); fruiting
   bodies in old diseased leaves. C. America, Brazil, Venezuela.
Pratylenchus spp.; root-lesion nematodes. At least the following species: P. brachyurus Fil. & Stek., P. minyurus
   Sher. & All., P. pratensis Fil., P. zeae Grah. Scattered,
   cause of root brown-rot, and these nemas often have associations with Fusarium, Macrophomina, Pythium, Rhizoctonia, Thielaviopsis, and certain bacteria.
Pseudomonas angulata (From. & Murr.) Holl.; angular leafspot, blackfire. (Occurrence varies: severe some years,
   recedes others.) Subtropical USA, Puerto Rico, Colombia,

Brazil, Argentina (probably widespread). P. solanacearum
E. F. Sm. ; bacterial wilt, slime disease. Argentina,
Brazil, Ecuador, Colombia, Chile, Dominican Republic,
Puerto Rico, Cuba.
Puccinia nicotianae Arth. ; rust. (On Nicotiana tomentosa
   Ruiz. & Pavón in S. America.) P. substriata Ell. &
   Bar. ; rust (has several grass hosts). Honduras.
Pythium aphanidermatum Fitz. , P. debaryanum Hesse, and
   P. ultimum Trow. ; all cause damping-off, reported from
   various countries. The latter often accused in warm wet
   seasons and warm soils. Neotropics.
Pythium-Phytophthora (certain isolates not separable on study
   in culture); damping-off, rootrot, stem base cortex rot.
   El Salvador, Costa Rica.
Rhizoctonia solani Kuehn = Thanatephorous cucumeris (Frank)
   Donk, (synonym: Botryobasidium solani Donk); damping-off,
   soreshin. Of common occurrence, special studies in Vene-
   zuela, Cuba, Dominican Republic, Argentina, Puerto Rico,
   Nicaragua, and El Salvador. (Certain soils "will recover
   from disease in subsequent years of planting. ") Rhizoc-
   tonia sp. (aerial type); top-rot. S. America.
Rhizopus sp. ; storage decay. Probably common when leaf is
   inadequately dried.
Rotylenchus sp. and Rotylenchulus sp. ; root zone nematodes.
   S. America.
Sclerotinia sclerotiorum (Lib.) dBy. = Whetzeliana sclerotio-
   rum (Lib.) Korf & Dumont; storage-rot. Widespread, and
   it is also reported as cause of damping-off in Venezuela.
Sclerotium rolfsii (Curzi) Sacc. = Pellicularia rolfsii (Sacc.)
   West which is the basidial state. Soil-rot, stem collapse.
   Scattered in warm moist areas. Subtropical USA, Trinidad,
   Venezuela, and other countries.
Septomyxa affinis (Sherb.) Wr. ; blotch and scabby lesions.
   Florida.
Septoria nicotianae Pat. ; frog-eye, zonate leafspot. W. Indies,
   C. America, Brazil, Ecuador. S. diversa Sacc. & Syd. ;
   reddish leafspot. Brazil. S. palan-palan Speg. (on Nico-
   tiana glauca Grah.); leafspot. Southern S. America.
Sphaerulina hainensis Gonz. Frag. & Cif. ; leafspots (some
   confluent). Dominican Republic, Brazil.
Stemonitis sp. ; slime mold developing as a disfiguring epiphyte
   on leaf. Moist country occurrence. W. Indies, C. America.
Sunburn or scald, dead areas of leaf. Appears at times on
   plants after long period of cloudy weather followed by sudden dry time and continued winds.
Thielavia basicola Zopf; damping-off, root attack. Scattered
   but widespread.
Thielaviopsis basicola (Berk. & Br.) Ferr. (Distinct from
   the above). Causes black stem and black root disease of
   older plants. Subtropical USA, Jamaica, Costa Rica,
   Colombia, Brazil.
Toxicity from excess chemicals:

Boron causes spotted leaves, stunting.

Chlorine causes leaves thick, brittle, cupped upwards.

Manganese causes leafspots that are sterile and necrotic, stunt, chlorosis.

Nitrogen causes yellow seedlings, extremely green older plants, browned roots.

Trichodorus spp.; stubby-root nematode.   Florida.

Tylenchorhynchus claytoni Stein.; stunt nematode.   Subtropical USA.

Uredo nicotianae Anast. & Splend.; leafrust.   Brazil (see also Puccinia above).

Verticillium albo-atrum Reinke & Berth.; a fungus wilt.   Uncommon, may be largely in Argentina.

Viruses are numerous and some common (tobacco is a notable experimental plant for many viruses) and I list here 18: 1. Abutilon Mosaic in Brazil.   2. Beet Curly Top (Brazilian) in Brazil.   3. Common Cucumber Mosaic = CMV in Subtropical USA, W. Indies, possibly S. America. 4. CMV Strain Celery-Commelina (which is more virulent) in Subtropical USA, Cuba, Puerto Rico, Honduras.   5. Malva Yellows that comes from the weed source Sida in Brazil, Venezuela, seen lately in C. America.   6. Potato Y = Chili Pepper Mosaic in Venezuela, Puerto Rico, Trinidad.   7. Sunflower Mosaic (Argentine) in Argentina.   8. Tobacco Etch in Puerto Rico, Venezuela, Subtropical USA. 9. Tobacco Leafcurl in Puerto Rico, Brazil, Colombia, Panama, Venezuela.   10. Common Tobacco Mosaic = TMV in all countries.   11. Tobacco Mottle (mild) in Argentina, Puerto Rico, Peru.   12. Tobacco Necrosis (vector is an Olpidium), occurrence is rather rare in Brazil, may be elsewhere.   13. Tobacco Rattle = Potato Corky Ring Spot in Florida.   14. Tobacco Ring Spot in Brazil.   15. Tobacco Rosette = TMV with Tobacco Vein Distortion in Puerto Rico, Venezuela.   16. Tobacco Streak in Brazil, Venezuela. 17. Tobacco Spotted Wilt = "Corcova" in S. America.   18. Tobacco Vein Banding = ? Potato Virus X ? in Subtropical USA, Venezuela, Puerto Rico.

Weather spots, fleck, glaze on recently mature leaves.   Uncommon but brought occasionally for clinical determination.

White leaf-veins in cigar tobacco.   Too rapid drying of leaf. Panama.

Xiphinema sp. largest of the root zone nematodes.   Appears to transmit Ring Spot Virus but this has not been reported up to date in the American tropics.

NOTHOFAGUS spp. by some FAGUS spp.   Trees native to the dreary, cool in summer and cold in winter, climate of Tierra del Fuego at the very southern tip of South America.   The area is not strictly tropical, but fungi on the trees must have held special interest for Spegazzini as his fruitful mycological studies on them indicate.   The large display of the so-called higher fungi collected under Fuegian conditions, must have struck delight

in the hearts of mycologists who worked mostly in warm countries where insects and microorganisms quickly erode and con-
sume sporocarps. It is believed all organisms listed here were
in Chile. The trees from which they came vary in size from
large to small; "haya de antarctica" Sp., "faia do antarctica"
Po., "fuegian beech," "dwarf beech," "beech of the antarctic"
En.

Acanthostigma imperspicuum Speg.; felt disease.

Agaricus spp. from woodrots: A. brunswickianus Speg., A.
   gossypinulus Speg., A. variabilis Pers.

Agyrium antarcticum Rehm; part of a lichen.

Ameghiniella australis Speg.; woodrot.

Antennaria antarctica Speg., A. pannosa Berk., A. scoriadea
   Berk.; on leaves (all are sootymolds).

Anthostoma ollantosporum Speg. and A. urodelum Speg.; on
   foliage.

Anthracoderma selenospermum Speg.; leaf disease.

Anthracophyllum discolor Sing.; agaracaceous parasite.

Aposphaeria clausa (Wint.) Speg. and A. freticola Speg.;
   Phoma-like foliage disease.

Ascomyces fuegianus Speg.; leaf blister disease.

Auricularia sambucinum Mart.; ascomycetous woodrot.

Bolbitius aleuriatus (Fr.) Sing.; woodrot.

Boletus loyo Phil. f.; conk on trunkrot.

Calocera cornea Batsch.; branch and woodrot. Tremella-like
   fruiting body.

Calosphaeria antarctica Speg. and C. chilensis (Speg.) Petr.
   & Syd.; branch and twig decay.

Cantharellus tarnensis (Speg.) Sing.; agaricaceous woodrot.

Cenangium australe Pat.; dieback.

Chlorosplenium auruginosum (Oed.) deNot.; on dead wood.

Clitocybe pleurotus Sing. and C. subletoloma Sing.; rootrot
   agarics.

Corticium spp. dark colored resupinate hyphae with basidia,
   on foliage, stems, roots: C. lacteum Fr., C. leve Pers.,
   C. majusculum Speg., C. tremellinum Berk., and C.
   triviale Speg.

Cortinarius darwinii Speg. and C. tarnensis Speg.; both are
   agarics occurring on decayed wood.

Coryne sarcoides (Jacq.) Tul. and C. urnalis Sacc.; wood
   decay fungi.

Crepidotus brunswickianus (Speg.) Sacc., C. nephrodes
   (Berk. & Curt.) Sacc., C. phalligens Mont., and C.
   xerotoides Speg.; all are agarics on rotted tree debris.

Cucurbitaria antarctica Speg.; foliage lesions.

Cylindrium fuegianum Speg.; leaf infection.

Cyphella vitellina (Lév.) Pat.; part of a lichen.

Cytospora antarctica Speg.; twig attack.

Cytosporium fuegianum Speg.; stem and twig lesions.

Cyttaria spp. as follows: C. berteroi Berk., C. darwinii
   Berk., C. espinosae Lloyd, C. hariotii Fisch., C. hookeri
   Berk., C. hookeri Berk. f. candida March., C. hookeri

Berk. f. typica March., C. intermedia Palm, C. johowii
Esp. (Note: These remarkable fungi were seen, and some
collected, by Charles Darwin in 1834, during his famous
voyage on the Beagle. They are Ascomycetes, similar to
Morchella, and occur in abundance on Nothofagus trees
which they parasitize. Darwin wrote of C. darwinii: "It
is a globular bright-yellow fungus, which grows in vast
numbers on [Fuegian] beach-trees.... [T]he fungus in its
tough and mature state is collected in large quantities...
and it is eaten uncooked.... [T]he natives eat no other
vegetable food besides this fungus.... [It is for them]...
a staple article of food." He noted that these sweetish
ascocarps are harvested regularly in Tierra del Fuego and
that they are the only fungus known in the world which is
a main and dependably used staple food for man.)
Dascyscypha cerina (Pers.) Fckl. and D. dusenii Rehm; both
cause stem cankers.
Diderma antarctica (Speg.) Sturgis; an epiphytic, occasional
myxomycete.
Durella bagnisiana Sacc.; on foliage.
Eriosphaeria vulgaris Speg.; woodstain fungus.
Eurotium herbariorum (Wigg.) Lk. var. fuegianum Speg. (Its
conidial state is a Penicillium); leaf infection of old tissues.
Exidia auricula-judae Fr.; yellowish sporocarp on rotten wood.
Faoydia dusenii Speg.; agaric on rotted wood surface.
Fistulina antarctica Speg., F. endoxantha Speg., F. hepatica
(Schaef.) Fr.; all cause brown woodrots, sporocarps are
often conspicuous in temperate zone, but not commonly
observed in tropics.
Flammula inopoda (Fr.) Sacc.; woodrot.
Fomes adamantinus Berk., F. leucosphaeria Mont., F. live-
scens Speg., and F. marginatus Fr.; these are associated
with log-rots, heart-rots of standing trees, and stumprots.
Fumago pannosa Speg.; sootymold on leaves.
Gnomonia magellanica Speg.; leafspot.
Harknessia antarctica Speg.; erumpent leafspots.
Helminthosporium subolivaceum Speg.; on foliage.
Helotium spp. causing disease of twigs and branches which
exhibit cup-like sporocarps on bark: H. citrinum (Hedw.)
Fr., H. epiphyllum Fr., H. gregarium Boud., H. mega-
losporum Speg., H. ochraceum Boud., and H. pergracile
Speg.
Heterobotrys antarctica Speg.; black mildew on leaf.
Hirneola antarctica Speg. and H. vitellini Lév.; woodrots
with Tremella-like sporocarps.
Humaria antarctica Speg.; branch disease.
Hymenochaete tenuissima Berk. and H. ferriginea (Bull.)
Mass.; resupinate parasitic fungi.
Hypocrea gelatinosa (Tode) Fr. and H. rufa Fr.; tree-base
wood and root rots, with surfaces bearing stroma.
Hypoxylon spp. cause cankers on stems, twigs, and trunks:
H. annulatum Mont., H. bovei Speg., H. coccineum Bull.,

H. creoleucum Speg., H. diatrypelloide Speg., H. magellanicum Speg., H. pseudopachyloma Speg., H. rubricosum Fr.

Hysterographium spp., wood infections: H. cylindrosporium Rehm, H. cylindrosporium Rehm var. intermedium Rehm, H. fuegianum Speg., H. fuegianum Speg. var. intermedium Rehm, H. magellanicum Speg.

Inocybe cerasphora Sing. and I. mariluanensis (Speg.) Sing.; both agarics associated with woodrot.

Irpex obliquus (Schrad.) Fr.; woodrot.

Kuehneromyces cystidiosus Sing.; parasitic agaric.

Laestadia diffusa (Crié) Sacc. & Trav.; on foliage.

Leptothyrium nothofagi Speg.; small leafspots.

Lophiostoma pseudomachostoma Sacc.; on foliage.

Lophodermum hysterioides (Pers.) Sacc.; leaf disease.

Melanconis antarctica Speg.; probably dieback.

Melanconium stromaticum Cda.; leaf and stem lesions.

Melasmia antarctica Speg.; leafspot.

Melittosporium caeruleum Rehm; leafspot.

Microdiplodia mafilensis Speg.; dieback.

Micronegeria fagi Diet. & Neger = Melampsora fagi Diet. & Neger; rust with pustules that tend to be covered.

Mollisia caesia (Fckl.) Sacc. f. brachycarpa Speg. and M. cinerea Karst.; woodrot.

Mycena citrinella (Pers.) Quel. and M. obtusiceps Speg.; cause wood decay.

Mycenella funebris Sing.; black agaric on dead wood.

Myzodendron antarcticum Gand. and M. oblongifolia DC. with other species of these phanerogamic parasites; cause development of "wood rose" and kill branches.

Odontia barba-jovis Fr.; woodrot.

Ophiobolus antarcticus Speg.; root disease.

Patellaria eximia Syd.; P. lecideola Fr.; on dead twigs.

Phaeomarasmius ciliatus Sing.; branch disease.

Pholiota megalosperma Sing.; bark rot.

Phoma spp. on foliage, twigs and roots: P. antarctica Speg., P. clausa Wint., P. desolationis Speg., P. hariotiana Wint., P. subantarctica Speg.

Pleonectria vagans Speg.; on dead and dying branch tissues.

Pleosphaeria patagonica Speg.; on foliage.

Pleurotus portegnus Speg. and P. sutherlandii Sing.; agarics (on rotted branches).

Pluteus defibulatus Sing. and P. spegazzinianis Sing.; agarics on dead wood.

Polyporus spp., rots of trunk-heart, stumps, standing trees: P. adustus (Willd.) Fr., P. albus (Huds.) Fr., P. australis Fr., P. fomentarius Fr., P. fuegianus Speg., P. gayanus Lév., P. marmoratus Berk., P. senex Nees & Mart., P. sulfureus (Bull.) Fr., P. vaporarius Fr.

Propolis pulchella Speg.; on foliage.

Pterula campoi Speg.; rot at tree base.

Pyrenopeziza chilensis Speg.; foliage disease.

Rhizomorpha subcorticalis Pers.; internal rot (? of root).
Rosellinia magellanica Speg., R. pulveraceae (Ehr.) Fckl.,
    R. reticulospora Greis & Greis-Deng.; all rootrots.
Schizostoma vicinissimum Speg. and S. vicinissimum Speg.
    var. fagi Rehm; on foliage.
Scolecosporium fagi Lib.; on leaves.
Septoria fagicola Speg.; leafspot.
Sphaerella spp.; foliage mildews: S. antarctica Speg., S.
    nothofagi Speg., S. subantarctica Speg.
Sphaeronaema conicum (Tode) Sacc. and S. stictum Berk.;
    leaf spotting.
Sporocybe antarctica Speg.; stem canker.
Stereum rugosum Fr. and S. sarmienti Speg.; woodrots.
Taphrina entomospora Thaxt.; leafcurl.
Teichospora perpusilia Speg.; on foliage.
Trametes albida-rosea Bromm. & Rouss.; woodrot.
Tremella mesenterica Retz.; on rotted wood.
Tympanis antarctica Speg.; probably from dieback.
Uncinula magellanica Thaxt. and U. notofagi Thaxt.; are leaf
    mildews.
Xerotus discolor Mont.; on diseased wood.
Xylaria antarctica Speg. and X. hypoxylon Grev.; rootrot.
Zignoella patagonica Speg.; on foliage.

OCOTEA spp.   A number of tropical American, small- to medium-
    sized, nut-bearing trees, both protected in the wild and planted
    in rough forest areas, that have hard, muskily fragrant wood.
    The timber is used in construction and cabinet work, and logs
    when chipped-up, or broken nuts when used, are distilled to
    produce the perfumers "wood oil" or "guyana linaloe oil" of
    commerce.
    Asteridiellina portoricensis (Speg.) Seav. & Toro; black leaf-
      spot. Puerto Rico.
    Cassytha filiformis L.; slender green dodder by inoculation.
      Puerto Rico.
    Catacauma ocoteae Stev.; scabby leafspot. W. Indies.
    Cephaleuros virescens Kunze; algal leafspot. Puerto Rico
      (probably scattered in general).
    Cephalosporium tumefaciens Wint.; on foliage. Brazil.
    Clinoconidium farinosum (P. Henn.) Pat.; gall disease. S.
      America.
    Cuscuta americana L.; yellow dodder. W. Indies.
    Didymaria rostrata Speg.; on foliage. S. America.
    Dothidella arechavaletae Speg. = Oligostroma arechavaletae
      (Speg.) Th. & Syd.; tarspot. Uruguay.
    Drepanoconis brasiliensis Schr. & P. Henn., D. larviformis
      Speg., D. tumefaciens (Wint.) Viégas (compare with
      Clinoconidium above); all are gall producing. S. America.
    Echidnodes microspora (Chard.) Seav. & Toro (? Lembosia);
      black mildew. Mexico, Puerto Rico.
    Helicomyces larviformis Speg. = Actinopeltis lauricola (Speg.)
      Petr.; on living leaves. C. America, S. America.

Helminthosporium ocoteae Stev.; apparently this is a hyper-
  parasite of Meliola on the host.   Puerto Rico.
Irenina calva (Speg.) Stev. = Meliola calva Speg.; typical
  black mildew.   C. America.   I. glabroides Stev.; black
  mildew.   Puerto Rico.
Lembosia microspora Chard. = Morenoella portoricensis
  Speg.; spreading black leafspot causes defoliation.   W.
  Indies.
Leptothyrium ampullulipedum Speg.; small spots on leaves
  and stems.   Widespread.
Lichens coupled with mosses and algae of several kinds and
  no identifications attempted, causing seriously overburdened
  branches.   C. America.
Loranthaceae (of at least four genera: Loranthus, Phthirusa,
  Psittacanthus, and Oryctanthus) phanerogamic attacks on
  branches of mature and old trees.   C. America.
Meliola ocoteae Stev.; ? = M. ocoteicola Stev.; black mildew.
  C. America, W. Indies.
Micropeltis leptosphaerioides Speg.; small black leafspots.
  C. America.
Microphyma microsporium Speg.; leafspot.   S. America.
Microthyriella ocoteae Bat.; small star-like leafspots.   Brazil,
  probably other countries in S. America, maybe in C.
  America.
Morenoella (see Lembosia)
Mycena citricolor (Berk. & Curt.) Sacc.; leafspot.   Costa
  Rica.
Napicladium laurinum Speg.; scab-like leafspot.   Neotropics.
Phragmopeltis melanoplaca Petr. & Cif.; leafspot.   W.
  Indies.
Phyllachora spp., rounded black spots on leaves, some spots
  become angular: P. catesbyana Chard. in Florida, P.
  nectandrae Stev. & Dal. in Virgin Islands, P. ocoteicola
  Stev. & Dal. in Puerto Rico and Brazil, and P. perplexans
  Chard. in Puerto Rico.
Phytophthora cinnamomi Rands; root disease.   Scattered.
Rosellinia bunodes (Berk. & Br.) Sacc.; black rootrot.   C.
  America.
Strigula complanata Fée; white lichen-leafspot.   Neotropical.
Uredo farinosa P. Henn.; rust deforming inflorescenses.   W.
  Indies, S. America.
Verticillium heterocladium Penz.; confusing spore deposit
  from hyperparasite of insect.   Puerto Rico.
Zukalia lauricola Speg.; leaf disease.   S. America.

OLEA EUROPEA L.   An exceedingly tough low-growing fruit tree
  taken to the New World from Europe in the 1600s, brought from
  Mexico to California in 1769 and somewhat later to Argentina,
  it is a subtropical tree killed by cold at about minus 12°C., but
  it needs a chilling season to fruit, thus is a failure in warm
  even-tempered tropical areas.   Some countries where most
  serious tests have been grown are: Mexico, Guatemala, Peru,

Chile, Uruguay, Brazil, California and Argentina. In the last
two it has become an important agricultural production; "olivo, "
"aceituno" Sp. , "oliveira, " "azeitona" Po. , "olive" En.
Alternaria alternata (Fr.) Keiss. = A. tenuis Nees; mild
    fruitspot and leaf tipspot. Argentina.
Antennaria elaeophila Mont. (= ? Capnodium sp.); sootymold,
    fumagina. Argentina.
Armillaria mellea Vahl = Armilariella mellea (Fr.) Karst. ;
    rootrot. California, Texas.
Asterina oleina Cke. ; black leafspot. Florida, Puerto Rico.
Botryosphaeria dothidea (Mong.) deNot. = B. berengerina
    deNot. ; branch canker, dieback. Argentina.
Calospora oleicola Speg. ; contributes to dieback. Argentina.
Capnodium (see Antennaria)
Ceratocystis sp. ; woodstain, branch disease. Chile.
Cercospora cladosporiodes Sacc. ; fruit and leafspot. Chile,
    Peru.
Cladosporium herbarum Lk. ; edge-of-leaf disease. Chile,
    Argentina.
Colletotrichum (see Gloeosporium)
Cuscuta indecora Choisy; smother and causes stemgall.
    California.
Cycloconium oleaginum Cast. ; eyespot and defoliation. Chile,
    Argentina, Peru, Ecuador, California.
Dasyscypha fasciculata Seav. & Wat. ; stumprot. Bermuda.
Dematium-like, black web fungus on leaf surfaces. Chile.
Dieback, physiological from deficiency of soil organic matter.
    California.
Fruit dryrot, physiological from excessive fertilization. Cali-
    fornia.
Fruit pitting, physiological from lack of boron. California.
Fusarium sp. ; associated with dieback. Chile, Argentina.
Gloeosporium olivarum Alm. = Colletotrichum gloeosporioides
    Penz. probably deserves a special f. sp. such as olivarii
    (asco-state is Glomerella cingulata (Stonem.) Sp. & Sch.);
    anthracnose infections of flower, peduncle, fruit, leaf,
    twig. Argentina, Peru, Puerto Rico, El Salvador, Cali-
    fornia.
Hainesia oleicola Speg. ; on foliage. Scattered.
Heartrot of stump and standing trunk. Scattered. (Note:
    Near Tarsus, Turkey, a "tree" was pointed out to me
    which men claimed furnished shade for St. Paul in his
    preaching. This "tree" consisted of several trunks all
    bearing good crops, and these subsidiary trees were dis-
    persed around an opening which was evidently at one time
    filled by a large tree trunk. An olive tree under the right
    condition is so resistant to animals, insects and disease
    microorganisms that it may be almost a permanent feature
    like a rock. In some Old World countries ancient olive
    trees are owned, harvested, and even sold separately from
    the very land where they grow, which land is cultivated for
    different crops, or used for grazing by lesser owners.)

Hendersonia oleae (Speg.) Sacc. & Trott.; on foliage. Argentina.

Hendersonulina oleae Speg.; leaf infection. Chile.

Massaria argentinensis Speg.; on foliage. Argentina.

Meloidogyne sp.; nematode rootknot. Peru, California.

Myriophysella chilensis Speg.; small irregular leafspots. Chile.

Oidium sp.; white mildew. Argentina.

Pestalotia scirrofaciens Brown; tumor on branch. Florida.

Phyllachora mayepeae Stev. & Dalb.; leafspot. Venezuela.

Phyllosticta sp.; minor leafspot. Argentina.

Phymatotrichum omnivorum (Shear) Dug.; rootrot. Texas.

Pratylenchus musicola (Cobb) Filip.; root nematode. California.

Pseudomonas garcae Amar., Teix., & Pinh.; bacterial leafspot. Brazil. P. savastonoi (E. F. Sm.) Stev.; bacterial knot of root, stem, flower, fruit (this is an accidental spread from Europe). Subtropical USA, Peru, Mexico, Guatemala, California, Colombia, Argentina, Chile, Brazil.

Psittacanthus cuneifolius Bl. = Loranthus cuneifolius Ruiz & Pav. (including other members of the Loranthaceae); attacks by phanerogamic parasites on tree branches. Argentina and other countries.

Rhizoctonia solani Kuehn (basidial state not reported); damping-off of seedlings. Peru.

Rosellinia necatrix (Hart.) Berl.; rootrot. Argentina, Chile.

Septobasidium guararicticum Mar.; following insect work. Argentina.

Septoria serpentina Ell. & Mart.; leafspot. Colombia.

Tylenchulus semipenetrans Cobb; nematode in rhizosphere. California.

Virus: Olive Partial Paralysis Virus in Argentina, and an undetermined graft-transmissible apparent virus mottle disease in Peru.

OPUNTIA spp. There are several needle-studded cacti of this genus, some selections are spineless, and all are adapted to marginal lands of desertic tropical America where they are valued and planted for soil retention, grazing, edible fruits, and the commercially exported cochineal dyestuff used in textiles, food color, and cosmetics. Plants are fibrous rooted, tops are of segments (cladodes) or pads, one fitted on top of another; "túna," "chumbo," "nopal," "west indian fig," "prickly pear."

Aecidium (see Puccinia)

Alternaria (see Macrosporium)

Anthostomella cacti (Schw.) Sacc.; cortex or "skin" disease. California, Puerto Rico.

Aspergillus alliaceus Thom & Church; cladode rot. Texas, Guatemala, Mexico, Puerto Rico.

Botryodiplodia opuntiae Sacc. (= Diplodia); tip disease. Widespread.

Broomeia congregata Speg. f. sp. argentiensis Speg.; puffball basalrot. Argentina.

Cassytha filiformis L.; green dodder.   Puerto Rico.
Ceratocarpia wrightii Toro (perfect state of an Aspergillus);
   Puerto Rico.
Ceriospora bonaerensis Speg.; on cladodes.   Argentina.
Colletotrichum gloeosporioides Penz. = Gloeosporium cac-
   torum Ston. and/or G. opuntiae Ell. & Ev. (asco-state
   is Glomerella cingulata (Stonem.) Sp. & Sch.); anthrac-
   nose, sunken spot, zonate spot on cladodes.   W. Indies,
   C. America and probably General.
Criconemoides sp.; rhizosphere nematode.   Peru.
Cuscuta americana L.; yellow dodder, deformed wrapped
   cladodes.   W. Indies.
Didymella acanthophila Speg.; spine disease.   Argentina.
Diplotheca tunae (Spreng.) Starb.; cladode blackspot.   Puerto
   Rico.
Erwinia aroideae (Town.) Holl.; bacterial rot of market fruits.
   Guatemala.
Fomes robustus Karst.; (on very old plants) basal decay with
   large bracket fungus.   C. America, Mexico.
Fusarium lateritium Nees f. sp. majus Wr.; cladode disease.
   Scattered occurrence.   F. opuntianum Speg. ? = F. lateri-
   tium majus; branch wilt.   Argentina.
Hendersonia bonariensis Speg. and H. opuntiae Ell. & Ev.;
   scorch.   Dry Subtropical USA, Mexico, C. American
   Pacific side deserts.
Hyponectria cacti (Ell. & Ev.) Seav.; cladode disease.   Moist
   parts of Subtropical USA.
Lembosia cactorum Tracy & Earle; black mildew.   Florida.
Leptosphaeria opuntiae Dodge; cortex spots.   Texas.
Macrophoma opuntiicola (Speg.) Sacc. & Syd.; spots on cladodes.
   S. America.   M. opuntiicola (Speg.) Sacc. f. sp. tunae Cif.
   & Gonz. Frag.; on cladodes.   W. Indies.
Macrophomina phaseoli (Taub.) Ashby = Sclerotium bataticola
   Taub.; charcoal-rot.   Texas, Mexico.
Macrosporium opuntiicola Cif. & Gonz. Frag.; cladodespot.
   (Genus name no longer tenable, should be Alternaria.)   W.
   Indies (spores of an Alternaria secured by me from scrap-
   ings off a blackened and withered túna fruit in Guatemala).
Meloidogyne sp.; rootknot.   Peru.
Montagnella opuntiarum Speg. f. sp. minor Speg.; on wilted
   cactus branches.   Paraguay.
Mycosphaerella opuntiae (Ell. & Ev.) Dearn.; black rot-spot.
   Florida, Texas, Puerto Rico.
Myriangium tunae (Spreng.) Petr.; scabspots.   Scattered.
Pestalotia opuntiicola Cif. & Gonz. Frag.; purple lesions.
   W. Indies.
Phyllosticta opuntiae Sacc. & Speg. (= P. cacti Arch. and P.
   concava Seav.) and P. opuntiae Sacc. & Speg. var. micro-
   scopica Cif.; causing several symptoms from white to brown
   cladode spots, to sunken spots and dryrot.   Neotropics.
Physalospora obtusa (Schw.) Cke. and/or P. rhodina Cke.;
   blackrot.   Florida, W. Indies. (= Botryosphaeria quercuum
   (Schw.) Sacc.)

Phytophthora parasitica Dast.; stemrot.   Puerto Rico.
Pratylenchus sp.; nematode from rhizosphere.
Puccinia opuntiae (Magn.) Arth. & Holw. = Aecidium sp.;
    rust.   S. America.
Pythium debaryanum Hesse in California.   P. ultimum Trow.
    in El Salvador, causing stemrot.
Rosellinia opuntiicola Speg.; woodrot.   Argentina.
Sclerotinia opuntiarum Speg. is perfect state of Sclerotium
    opuntiarum Speg. (= Whetzeliana sclerotiorum (Lib.)
    Korf-Dumont var. opuntiarum (Speg.) Alip.); plant col-
    lapse.   Argentina, Chile, Colombia.
Septoria fici-indicae Vogl.; cladode spot.   Puerto Rico, Texas.
Stagonospora opuntiae Speg.; killing of branches.   Argentina.
Stevensea wrightii (Berk. & Curt.) Trott.; blackspot.   Florida,
    Puerto Rico, Texas.
Teichospora mammoides Ell. & Ev. var. opuntiae Dearn. &
    Barth.; cortex rot.   California.
Tretopileus opuntiae Dodge; cladode disease.   Florida.
Venturia sp.; cladode infection.   Colombia.

ORCHIDS, Domesticated (those of mainly "botanical types" not included).   Included here are orchids grown in epiphytic manner
as moist country plants in partial shade and adapted to garden
and inside patio management by being established in pots of
mixed charcoal bits, bark, and moss; grown also on top of
sections of tree branches as well as tops of posts and in crotches
of growing trees specially preserved.   In the Neotropics there are
at least 200 orchid genera, all with diseases, much unstudied and
not reported, and in my experience the more commonly encountered orchid genera were: ANGRACEUM (1), BRASSAVOLA (2),
CATASETUM (3), CATTLEYA (4), CYCNOCHES (5), CYMBIDIUM
(6), CYRTOPODIUM (7), DENDROBIUM (8), EPIDENDRUM (9),
HABENARIA (10), IONOPSIS (11), LAELIA (12), LYCASTE (13),
MAXILLARIA (14), MILTONIA (15), ODONTOGLOSSUM (16),
ONCIDIUM (17), PHALAENOPSIS (18), SOBRALIA (19), STAN-
HOPEA (20), VANDA (21).   The hundreds of species have many
local names depending on localities and interest of people growing
them; the main common names of orchids are "orquidie" Sp.,
"orquídea" Po., "orchid" En.   (In the disease list below I have
indicated under each disease the host genera I have found the
disease on or seen reported as attacked.   Diseases of these
plants have not been a main responsibility of tropical American
pathologists.)
    Acremoniella sp.; leafspot.   Brazil (4).
    Algae (green types that are mostly single-celled); occasional
        on pseudobulb.   Costa Rica (4, 6, 9).
    Alternaria sp.; rare in leafspot complex.   Dominican Re-
        public (on wild epiphytic orchid).
    Amerosporium orchidearum Speg.; a leaf disease fungus.
        Argentina (on wild epiphytic orchid).
    Aphelenchoides ritzmabosi (Schw.) Stein. & Buh.; yellow bud.
        Florida (21).

Asterinella epidendri (Rehm) Th.; smudge leafspot.   Brazil
    (9), Panama (9).
Botryosphaeria quercuum (Schw.) Sacc. (synonyms: Botryo-
    diplodia theobromae Pat., Diplodia theobromae (Pat.)
    Now., Diplodia sp., Physalospora rhodina Berk. & Curt.);
    leafspots, flowerstalk lesions, stem cankers.   Costa Rica,
    Puerto Rico (4, 11, 16), Ecuador (4), Panama (2, 3, 9,
    15, 17, 21).
Botrytis cinerea Pers.; fading-flower petal disease.   Neo-
    tropical (probably all genera are susceptible but disease
    was determined on 4, 5, 6, 9, 13, 16, 17, 20, 21).
Capnodium citri Berk. & Desm. and other species; super-
    ficial but ugly fungus deposits subsisting on insect exu-
    dates dripped by aphids and other sucking pests living on
    the shade and fruit trees that are above the orchids.   Neo-
    tropical (on all genera).
Cephaleuros virescens Kunze; the well known common scurfy
    algal spot.   El Salvador, Costa Rica, Guatemala, Honduras,
    Panama, Puerto Rico, probably more widespread (1, 4, 6,
    8, 9, 12, 15, 17, 18, 21).
Ceratostomella (see under Mycorrhizal fungi)
Cercospora dendrobii Burn.; leafspot.   Florida (8).   C.
    odontoglossi Prill. & Delacr.; leafspot.   Florida (2, 4,
    12).   Cercospora sp.; large leafspot.   Florida (2), Puerto
    Rico (4, 6, 9), El Salvador and Costa Rica (3, 9, 11,
    14, 17, 21).
Chaetodiplodia sp. (? = Botryodiplodia); leafspot and pseudo-
    bulb injury.   W. Indies, Brazil (4).
Ciliélla epidendri (Rehm) Sacc. & Syd.; leaf disease and
    collapse.   Brazil (9).
Colletotrichum gloeosporioides Penz., its asco-state is
    Glomerella cingulata (Stonem.) Sp. & Schr.; common an-
    thracnose on leaf, flower, spathe, pseudobulb, and aerial
    root.   The symptoms of infections vary from small discrete
    spots to large spreading and livid diseased areas and rotted
    flowers.   Neotropical, and on all host genera observed.
    (It should be noted that there are over 25 reported fungus
    synonyms for the orchid anthracnose organism, some
    genera of these confusing synonyms are Colletostroma,
    Discella, Gloeosporium, Gloeosporiopsis, Hainesia, Hypo-
    dermium, Macrophoma, Marsonia, Myxosporium, Vermicu-
    laria, and Volutella.)
Corticium (see under Mycorrhizal fungi)
Cuscuta sp.; dodder.   Florida (10), El Salvador (21).
Diplodia natalensis P. Evans, D. laeliocattleyae Sib., D.
    paraphysaria Sacc.; are usually all stem diseases, often
    a slowly progressing rot following primary infection by
    a fungus or other injury.   The first: Florida (3, 9),
    Puerto Rico (4, 13, 17, 7, 12, 9), California (4), Costa
    Rica (1, 21, 17, 8), Brazil (13), Colombia (4, 9).   The
    second: Florida (4, 12).   The third: Dominican Republic
    on several orchids.

Erwinia carotovora (L. R. Jones) Holl.; bacterial softrot.
 Florida (2, 4, 5, 6, 16, 17, 18), Costa Rica (4, 6, 9,
 12, 15, 18, 19, 20, 21), Guatemala in market (4).
Fusarium oxysporum Schl. f. sp. cattleyae Foster; systemic
 stem disease. Florida (2, 12, 17), California (4, 8, 9,
 17). F. roseum Lk.; leaftip disease. Costa Rica (1).
 F. solani (Mart.) Sacc.; rootrot. C. America (4, 7, 9,
 12, 16, 18, 19, 21).
Gloeodes pomigena (Schw.) Colby; leaf sooty blotch. Florida
 (2, 4, 9).
Gloeosporium affine Sacc., said to be the "European anthracnose" developing smaller spore pustules in Florida (4),
 Costa Rica (4, 6, 12). G. maxillariae Alles. abnormally
 large spore pustules in Nicaragua (13). G. orchidearum
 Alles., G. orquidearum Karts & Hart., are both from the
 typical organism in Colombia, Florida, and Puerto Rico
 (4). The first organism probably deserves a subspecific
 designation that I suggest might be Colletotrichum gleosporioides Penz. var. affine (Sacc.), and the latter
 organisms should all be C. gloeosporioides, see also note
 on genus synonymy in Colletotrichum.
Hemileia americana Mass. = Uredo americana (Mass.) Arth.;
 lesions on flower petal, scape, leaf, brown to yellow red
 rust pustules. C. America, W. Indies (4). H. oncidii
 Griff. & Maubl.; leafrust. Brazil (9, 13, 17), C. America
 (9, 17).
Hendersonia epidendri Keiss.; leaf disease. American Virgin
 Islands (9).
Leafblade spots (from isolation studies no fungus or bacterium
 involved). Puerto Rico and Costa Rica (4, 6).
Lichens, various (but not Strigula) on pseudobulbs of old
 orchids. Costa Rica (4, 6, 13, 14, 17).
Loranthus sp. seeds placed on pseudobulb of Cattleya sp.
 where they germinated, but no infection. Costa Rica.
Macrophoma oncidii P. Henn. (not anthracnose); leafspot and
 leaf collapse, lesions with discrete borders and perithecia
 at the center. Mexico (4, 17), Costa Rica (4, 18), El
 Salvador and Honduras (12), Brazil (17).
Marsonia (see Colletotrichum synonymy)
Melampsora repentis Plowr.; leafrust. Colombia (17), Guiana
 (10, 18).
Meliola sp.; black mildew. Mexico (12), Puerto Rico (18),
 Costa Rica (on numerous small wild orchids).
Memnoniella echinata (Riv.) Gaill.; leaf destruction. S.
 America (4).
Micropeltis bakeri (Syd.) Cash & Wats. and M. orchidearum
 P. Henn.; both black mildews. Dominican Republic,
 Brazil (various orchids).
Microthyriella rubi Petr. (? = Leptothyrium pomi (Mont. &
 Fr.) Sacc.) is the asco-state of Zygophiala jamaicensis
 Mason; leaf flyspeck. Florida (2, 4, 8, 9).
Moelleriella epidendrum Rehm; stem canker. Brazil (9).

Morenoella calmi Rac.; black mold of leaf.   Puerto Rico (9).
Mycena citricolor (Berk. & Curt.) Sacc.; leafspot.   Costa
 Rica (2, 8, 9, 17) and on wild so-called "botanical types"
 growing on coffee tree branches that bore leaves diseased
 with this parasite.
Mycorrhizal fungi, apparently non-pathogenic but symbionts
 with orchids, are Basidiomycetes, in my experience many
 of them belonging to species of Corticium-Rhizoctonia.
 (Note: In Costa Rica, Honduras and El Salvador I have
 grown or had access to some 300 species of orchids.
 Whenever examined closely with microscope, supplemented
 with 20X hand lens, I found fine networks of fungus hyphae
 attached or closely adjacent to orchid aerial roots and in
 two host genera (4, 17) microscopic study revealed slender
 fungal hyphae inside root cells.   Basidiomycete genera
 reported to be orchid symbionts are: Armillaria or
 Clytocybe [Armilariella], Ceratocystis also given as Cera-
 tostomella, Corticium, Fomes, Marasmius, Pellicularia,
 Pleurotus, Ptychogaster, Rhizoctonia [special type?], and
 Xerotus.
Myiocopron corrientinum Speg. = Microthyrium corrientinum
 (Speg.) Cke.; on dead stems and leaves.   Argentina (17).
Nectria bulbicola P. Henn.; disease and dryrot of pseudobulbs.
 Widespread (14).   Nectria sp.; dry lesion of stem and
 flower stalk.   C. America (various orchids).
Ophiodothella orchidearum Cash. & Wats.; foliage disease.
 Brazil (4).
Pestalotia sp.; leaf and stem lesions.   Colombia (4), Costa
 Rica (21), Florida and Puerto Rico (8).
Phoma corrientiana Speg. = Sphaeropsis corrientina (Speg.)
 Kuntze; leaf disease.   Argentina (17).   P. oncidii Speg. =
 Sphaeropsis oncidii (Speg.) Kuntze; on dry scapes.   Argen-
 tina (17).   (Phoma oncidiisphacelati Tassi = ? Sphaeropsis
 sp.; leaf collapse.   Mexico (17).
Phomopsis orchidaphila Cash & Wats.; leafspot, stem black-
 canker, flower peduncle canker, flower calyx spot.
 Mexico, El Salvador, Costa Rica, Honduras, Trinidad,
 probably in S. America (various orchid genera, specially
 studied on Cattleya).
Phyllosticta laelia Keissl.; leaf and leaf sheath spots.   Mexico
 (11, 12).   P. nigromaculans Sacc.; leafspot and on becoming confluent leaf section may collapse.   Brazil (4?),
 Costa Rica (17, 8, 15); Panama (1, 2, 8, 19).   P. valparadisiaca Speg.; leafspotting.   Chile (on various genera
 of orchids).
Phyllostictina pyriformis Cash & Wats.; leaf infection.
 Florida (8, 9).
Physalospora camptospora Sacc.; leafspot, stemrot.   Brazil
 (4), Costa Rica (4).   P. cattleyae Maub. & Lasn.; stem
 and leaf disease.   Brazil (4).   P. orchidearum P. Henn.;
 leafspot, stem dryrot, flower peduncle foldover disease.
 Costa Rica and probably other countries in C. America

(4, 8, 9, 17, 18). P. wildemaniana Sacc.; disease spreading in large parts of leaf. Brazil and W. Indies (various unspecified orchids).

Physiological disease from lack of minor elements, not reported in garden-grown orchids.

Phytophthora cactorum (Leb. & Cohn) Schroet.; a severe blackrot. Florida (2, 4, 8, 9, 17), Costa Rica (17), Surinam (4), Puerto Rico (4).

Pleospora sp.; ? secondary on leaf lesions. Mexico and El Salvador (4).

Pratylenchus pratensis (De Man) Filipj.; root nematode on greenhouse cultivated orchid from use of contaminated growing media, almost impossible to expect to find in outside gardening growths in most of Neotropics. Florida (4).

Pseudomonas cattleyae (Pavar.) Savul.; bacterial brownspots of leaf and stem. Florida (4, 8, 9, 18).

Puccinia cinnamomea Diet. & Holw.; rust. Mexico, Dominican Republic (on undetermined epiphytic orchid).

Pythium ultimum Trow.; the most common and serious blackrot, most destructive of all orchid diseases. Neotropical in distribution (all host genera believed susceptible).

Rhizoctonia solani Kuehn (its perfect state is Thanatephorous cucumeris (Frank) Donk); rootrot. Florida (various orchids) all are probably susceptible, Colombia (4, 9). (In some cases is mycorrhiza?)

Rooting failure, little studied. Noted especially in orchids planted with moss on branches of a smooth barked tree, found twice on a rough barked branch where green moss was bound around orchid pseudobulbs. Costa Rica (4, 8, 15, 20).

Sclerotinia fuckeliana (dBy.) Fckl.; petal blight. Florida (4, 8, 17, 18, 20).

Sclerotium rolfsii (Curzi) Sacc. (perfect state is Pellicularia rolfsii (Sacc.) West); rare stemrot. Florida (5), Puerto Rico (4).

Scolecopeltis ionopsidis Toro; black mildew. Puerto Rico (11).

Septobasidium sp.; felt disease. S. America (12).

Septonema intercalare Cash & Wats.; inflorescence disease interception. ? From some area in Neotropics on Laelia sp. S. orchidophilum Speg.; inflorescence disease. Argentina (17).

Septoria codonorchidis P. Henn.; leaf and scape spotting. Chile (7). S. orchidearum West.; leafspotting. S. America (12). S. selenophomoides Cash & Wats.; leafspot. Florida (8), Chile (6), C. America (12, 16). Septoria sp.; leafspot. Colombia (4).

Sphaeropsis orchidearum Cif. & Gonz. Frag.; canker, leafspot. Dominican Republic (17, and other undetermined orchid).

Sphenospora kevorkianii Lind; rust. Florida (9, 20, 11), Panama (17, 19, 20), Nicaragua (17, 19, 9, 20); S. mera Cumm.; rust. Florida (5). S. saphena Cumm. (possible

synonym of Uredo guacae); rust.   Florida, Puerto Rico,
Jamaica (17).
Stigmatea sp.; irregularly spreading leafspot.   Costa Rica (4).
Strigula complanata Fée; lichen spotting of leaves, pseudo-
bulbs, aerial roots.   C. America (probably on all orchids
but in Costa Rica noted as specially abundant on Cattleya
and Dendrobium genera).
Sunscald, discoloration, chlorosis and tissue death on plants
suddenly exposed to full sunlight in dry season after long
growth in shade and lush development during the wet season.
Occurs anywhere on probably all orchids.
Trentepohlia sp.; brownish-green hairy algal epiphyte.   Costa
Rica (21).
Trichobelonium epidendri Rehm; hyperparasite on black mildew.
S. America (9).
Tubercularia orchidearum Speg.; pseudobulb disease.   Argen-
tina (17).
Uredo spp., all are leafrusts as follows: U. behnickiana P.
Henn. in Florida (9, 12, 17), Jamaica (4), Panama (17),
Brazil (17).   U. darnosa Speg. in Paraguay and Brazil
(various orchids) but in Brazil especially on Catasetum.
U. epidendri P. Henn. in Florida (1, 9, 11, 12, 17),
Honduras (17), Mexico (17), Brazil (9, 17).   U. cyrtopodii
Syd. in Cuba, Puerto Rico, Brazil (7).   U. guacae Mayor
in Ecuador, El Salvador, Puerto Rico, Florida, Brazil (9).
U. gynandrearum Cda. in Puerto Rico, Florida, Trinidad,
Jamaica, Cuba (10).   U. incognita Speg. in Paraguay (on
several orchids).   U. nigropunctata P. Henn. in Puerto
Rico (believed all orchids susceptible), Florida (4, 20, 7,
9), Brazil (20, 7), C. America (16).   U. oncidii P. Henn.
in California, Guatemala, Mexico, Brazil (17).   U. orchidis
Mart., or U. orchidis Wint. in Trinidad (intercepted there
on orchid from another country).   U. wittmackiana P. Henn.
& Klitz. in Mexico (9).
Venturia hariotiana Speg.; leafrot.   Argentina, Brazil (17).
Volutella albido-pila Boud. possibly Colletotrichum but without
certainty; severe leafspot.   (An interception in USA temperate zone of diseased Phalaenopsis sp. plant from tropics.)
Viruses: Cattleya Color Break in Florida (2, 4, 6, 12, 16,
17), in Venezuela (6).   Common Cucumber Mosaic in Brazil
(in an orchid).   Cymbidium Mosaic in Florida (1, 4, 2,
6, 9, 12, 8, 15, 17, 18, 12), in Puerto Rico (4, 9, 6,
12), in Costa Rica (11).   Oncidium Ring Spot in Florida (17),
in a country of S. America (16).   Common Tobacco Mosaic
Virus in Florida (4, 21), in Puerto Rico (2, 4, 6, 9, 12).
"Mosaic symptoms" in Venezuela (4).
Zygophiala (see Microthyrium)

ORYZA SATIVA L.   Its seed is a famous staple world food; the
crop originated in the Oriental tropics, was brought to the New
World two centuries after the time of Columbus, and is grown
widely as a dryland crop becoming one of four most important

crops in Brazil, and a major grain planted in subtropical USA.
This crop plant has the largest list of diseases known, and in
the whole world tropics, counting its weak parasites, the straw
and grain problems, epiphytic to semiparasitic associates, the
saprophytes, and clearly severe and definite disease organisms,
the total number is much over 600.   There are a few less,
strictly in the Neotropics; "arróz" Sp. Po., "rice" En.   (The
Latin genus name comes from the Arabic.)

Achlya sp.; seedling rot.   California.

Acrothecium lunatum Wakk.; leaf and culm disease.   S.
    America.

Albinism is a genetic defect.   Occasional.

Algae, green types.   Seen smothering seedlings.   Puerto
    Rico, Guatemala.

Alternaria alternata (Fr.) Keiss. (small spores) = A. tenuis
    Nees; black headmold, leafspot, grainspot.   Costa Rica,
    Mexico, Subtropical USA, Argentina where it follows Pyricu-
    laria.   Also Alternaria sp. (large spores); leaf disease.
    Costa Rica, Panama.

Aphelenchoides besseyi Christie; nematode at roots.   Sub-
    tropical USA, El Salvador.   A. oryzae Yak.; white tip.
    Panama, Venezuela, Peru, Subtropical USA, Costa Rica.
    A. parientinus Stein.; semi-parasite.   Brazil.

Aposphaeria sp.; black spot disease.   Brazil.

Ascochyta oryzae Catt.; glume and leaf disease.   Brazil (and
    probably many other countries in S. America).

Aspergillus niger v. Tiegh. and A. oryzae Cohn; storage musty
    grain.   Occasional.

Bacterial disease, in poorly dried stored grain, called pink
    wet-rice.   Panama.

Balansia oryzae (Syd.) Nar. & Thir. (= Ephelis sp. which is
    the conidial state); black-ring disease, sterility.   Subtropi-
    cal USA.

Briosia sp.; leaf decay.   Scattered occurrences.

Capnodium spp.; sooty molds.   Widespread.

Cassytha filiformis L.; green dodder.   Puerto Rico.

Cephalothecium (see Trichothecium)

Ceratosphaeria grisea Heb. is the asco-state of Pyricularia
    grisea Sacc. which is the synonym of P. oryzae Cav.;
    riceblast.   Neotropical.

Cercospora oryzae Miy.; linear spots on leaf and glume.
    General.

Chlorosis, physiological from alkaline soils.   Subtropical
    USA, W. Indies.

Cladosporium herbarum Pers., C. maculans Cast.; on old
    and dying leaves.   Common, widely distributed.   (Also
    sometimes given as C. maculans Boed.)

Clasterosporium putrefaciens Sacc.; leafrot.   Mexico.   C.
    punctiforme Sacc.; leafrot.   Brazil.

Cochiobolus (see Helminthosporium)

Corticium sasakii (Shira) Matsu.; sheath spot, banded leaf,
    rootrot.   Louisiana, Texas, C. America.

Criconemoides morgense Taylor; rhizosphere nematode.
    Brazil.
Curvularia brachyspora Boed. noted especially in Argentina.
    C. falcata (Tehon) Boed. in Subtropical USA, Puerto Rico,
    C. America.   C. lunata (Wakk.) Boed., of nearly general
    distribution.   C. pallescens Boed., especially in Argen-
    tina.   These closely related fungi all cause a blight of
    seedlings, grain and glume disease, stem disease.
Cuscuta spp.; dodders.   El Salvador, Puerto Rico, Guate-
    mala, Costa Rica.
Didymella glumicola Speg., glume disease.   S. America.
Entyloma oryzae H. & P. Syd. (? = E. dactylidis Cif.);
    leafsmut.   Mexico, Dominican Republic, Nicaragua, Guate-
    mala, Costa Rica, Panama, Peru, Surinam, Colombia,
    Venezuela, Bolivia, Subtropical USA.
Epicoccum neglectum Desm.; kernel discoloration.   Widespread.
Fusarium moniliforme Sheld., this species identified in Costa
    Rica, Surinam, Panama, Puerto Rico, and Trinidad and
    causing seedling disease.   (It is uncertain that it produces
    the "bakanae" symptoms characteristic of it in the Orient,
    thus this in the New World may be a race deserving of a
    subspecific designation).   The ordinary fusarial organisms
    recovered designated as Fusarium spp. are semi-parasitic,
    some mixed in a fungus complex resulting in seedling leaf
    blights, occasional foot disease, wilting (in Colombia and
    Guatemala), and grain attack.
Gibberella fujikuroi (Saw.) Ito (asco-state of F. moniliforme),
    identified in Surinam, causing footrot, and a Gibberella sp.,
    causing a scab-spot of grain in Brazil and Bolivia.
Haplographium olivaceum Cke. & Mass.; leaf attack.   Argen-
    tina, cool parts of Brazil.
Helicoceras oryzae Linder & Tullis; sheathrot, grainspot.
    Subtropical USA.
Helminthosporium oryzae deHaan; occurs widespread, causes
    inflorescence abortion, seedling death, brown leaf and seed
    lesion (has resulted in famines in the Orient), the perfect
    state of this fungus is Cochliobolus miyabeanus Drechs.
    H. sigmoideum Cav. = the asco-state is Leptosphaeria sal-
    vinii Catt. but the latter seldom found in the American
    tropics; culm rot, leaf veinspot, brown leafspot.   Found at
    least in the Texas and Louisiana states of Subtropical USA,
    Trinidad and Tobago, Jamaica, Peru, Ecuador, Costa Rica,
    Honduras, Nicaragua, Panama, Cuba.
Hendersonia sp.; leaf disease.   Venezuela.
Heterosporium echinulatum Cke.; leafmold.   Chile.
Leptosphaeria (see Helminthosporium)
Magnesium deficiency, numerous symptoms: chlorosis, white
    leaf tips, distinct leaf lamina invasion by saprophytic fungi,
    added susceptibility to nematodes.
Meloidogyne spp.; root attack.   Surinam, Jamaica, Brazil,
    Panama, Subtropical USA.
Monascus purpureus Went.; red kernel.   Scattered.

Monhystera stagnalis Stein.; rootzone nema. Brazil.
Mycena citricolor (Berk. & Curt.) Sacc.; leafspot by inoculation. Costa Rica.
Mycosphaerella spp. (these following are perhaps synonyms), in any case the disease produced is leafspot and sheath infection and it occurs rather generally: M. malinverniana Miy., M. oryzae (Catt.) Miy., M. shiraiana Miy., M. usteriana (Speg.) Hara = Sphaerella usteriana Speg.
Neovassia horrida (Tak.) Pad. & Khan (N. barclayana Bref.) = Tilletia horrida Tak.; smut. General but scattered.
Nigrospora oryzae (Berk. & Br.) Petch; grainrot, spikemold, disease of old leaf. Neotropical occurrence.
(Ophiobolus spp., given as perfect states of Helminthosporium are the following: O. cariceti Sacc., O. graminis Sacc., O. oryzae Miy., O. oryzinus Sacc. (some may be synonyms); cause such diseases as take-all, footrot, culm lesions, black sheatrot, lodging. Widespread.)
Penicillium spp.; grain bluemold, straw mold, musty grain. C. America, Subtropical USA. (Note: Experience in work with isolates and by moist chamber studies show that old straw used in₀matting, braided rope, sandals, and bedding, aged field stubble, or new field straw and the like will yield a very long list of numerous species of epiphytes and saprophytes such as algae, scavenger nematodes, bacteria, yeasts, myxomycetes, Penicillium, Rhizopus, Aspergillus, Monilia, Oospora, Gliocladium, Curvularia, Fusarium, Chaetomium, and others. Many such occurrences can be found but they are not included in the present list, unless indicated as a special problem. There is a danger that some present saprophytes may evolve into parasitic races but until this happens there is little value in trying to cover all contingencies, and prophetic choice is impossible by me.)
Phaeoseptoria oryzae Miy.; leafspot. Mexico, Guatemala.
Phaeosphaeria oryzae Miy., synonyms: Trematosphaerella oryzae Pad., Phyllosticta oryzae Hori; brown leafspot. Colombia, Nicaragua, Brazil, Mexico, Costa Rica, Bolivia.
Phoma glumarum Ell. & Tracy = Phyllosticta glumarum (Ell. & Tracy) Miy.; kernel disease, glume- and leaf-spot. Subtropical USA, Brazil. P. glumicola Speg.; disease of culm and glume. Subtropical USA, C. America. P. necatrix Th. = Phyllosticta necatrix (Th.) Miy.; black leafspots. Subtropical USA, W. Indies, C. America. P. oryzae Hori reports from Colombia, Brazil, El Salvador, Nicaragua. P. usteriana Speg.; blackculm and leaf lesions. S. America.
Phyllosticta oryzae Hori = Phoma oryzae Sacc. which is the conidial state and has as its perfect state Phaeosphaeria see above.
Pleospora sp.; leaf disease. Mexico, Costa Rica, Panama.
Pratylenchus pratensis Filip.; root nematode. Brazil.
Pseudomonas (see Xanthomonas)
Puccinia graminis Pers. f. sp. oryzae Frag.; rust. Scattering in Costa Rica, Panama, W. Indies.

Pyricularia (see Ceratosphaeria [in the field, the conidial
　　state of Pyricularia is the one always encountered])
Pythium arrhenomanes Drechs.; rootrot. Brazil. P.
　　rostratum Butl.; damping-off, plant wilt, rootrot, root
　　tip disease. Guatemala, Panama, California.
Red streak, unidentified cause. Panama, Guatemala.
Rhizoctonia spp. (the most definitive studies of these are still
　　needed in the American tropics) provisionally four species
　　are recognized on rice as follows: R. grisea (J. A. Stev.)
　　Matz (? = Corticium sasakii Matsu); banded sheathrot.
　　Panama, Puerto Rico, Subtropical USA. R. solani Kuehn
　　(the basidial state I believe not reported on rice in the
　　Americas); stemrot, damping-off, rot of flagged leaves.
　　Probably scattered in general, special notes show occur-
　　rences in Brazil, Subtropical USA, W. Indies, Venezuela,
　　Panama, Costa Rica, Mexico, Guatemala. R. oryzae
　　Ryker & Gooch; bordered sheathrot. Dominican Republic,
　　Guatemala, El Salvador, Panama, W. Indies, Guiana,
　　Ecuador. R. zeae Voorh.; huskspot, sheathrot. Florida,
　　Panama, El Salvador, W. Indies, Guiana, Ecuador.
Rhynchosporium oryzae Has. & Yok.; scald, brown leaf-tip-
　　blight. El Salvador, Guatemala, Peru, Costa Rica, Colom-
　　bia, Bolivia.
Sclerotinia sclerotiorum (Lib.) dBy. = Whetzeliana sclerotio-
　　rum (Lib.) Korf & Dumont; plant wateryrot. Florida.
Sclerotium coffeicola Stahel; leafspot. Surinam. S. oryzae
　　Catt.; base-of-stem disease. Guiana, Louisiana, Colombia,
　　Costa Rica, Brazil. S. rolfsii (Curzi) Sacc. = the basidial
　　state is Pellicularia rolfsii (Sacc.) West; seedling blight.
　　Subtropical USA, C. America, Mexico, Colombia.
Septoria oryzae Catt.; leaf and sheath disease. Brazil,
　　Colombia, Jamaica, Florida. S. oryzae (Catt.) f. sp.
　　brasiliensis Speg.; leaf disease. Brazil. S. poae Catt.,
　　leaf disease. Brazil.
Spermospora subulata Spr.; gray mold on leaves. Dominican
　　Republic.
Sphaerella (see Mycosphaerella)
Sphaeropsis oryzae Sacc.; leaf disease, sheathrot. Brazil.
Sterile head, cause is physiological. Panama, Cuba, Uruguay,
　　Subtropical USA.
Teichosporella oryzae Bat. & Gay.; on leaves. Brazil.
Tilletia (see Neovassia)
Trematosphaerella (see Phaeosphaeria)
Trichochonis caudata (App. & Str.) Clem.; pink kernel. Sub-
　　tropical USA, Panama, Mexico. T. padwickii Gan., leaf-
　　spot. Peru, Surinam.
Trichoderma viride Pers.; part of sheathrot complex. Sub-
　　tropical USA, Costa Rica.
Trichotecium roseum Lk. = Cephalothecium roseum Cda.;
　　pink disease. Chile.
Tylenchorhynchus sp.; rhizosphere nematode. Surinam,
　　Jamaica.

Ustilaginoidea virens (Cke.) Tak.; false smut on spike. Widespread.

Variegation, genetic anamoly. Scattered.

Viruses: Hoja Blanca Virus causing stunt, white leaf. (It was noted as early as 1935 in Colombia, started slow increase and by 1962 became widespread) in Florida, Texas, Mexico, Honduras, Cuba, Colombia, Surinam, Venezuela, Guiana, Dominican Republic, Ecuador, Panama, in Costa Rica by 1969. White stripe, still to be carefully studied reported in Panama. (A severe stripe is also known in the Orient.) Yellow Dwarf, a Mycoplasma disease, in Colombia.

Wind burn. Scattered, depends on seasonal accidents.

Xanthomonas itsana Dows. ? = X. oryzae Dows. (also reported under Pseudomonas sp.); grainspot, blight. Panama (probably brought in on Oriental seed). Nicaragua.

Xiphinema americanum Cobb and X. campinense Lorde.; root ecto-feeding nematodes. Brazil.

OXALIS TUBEROSA Mol. Grown for its slightly elongated, nutritious, flavorsome tubers, it is the second most important root crop of the Mexican Cordillera and in the Andes, where it is planted at high elevations, 3000 meters and above, is grown as an annual herb, a heavy producer on marginal lands, and is popular since very ancient Inca times to the present and is an indispensable food but not much studied for diseases. (Countries where most cultivated: Mexico, Venezuela, Bolivia, Colombia, Brazil, Peru, Chile, Argentina.) Common Indian names: "oká," "quimbu," "cuiba," "papa estranjera," and the processed whole dried tubers are called "chuño."

Aspergillus sp. on dead leaf sent to clinic from Argentina.

Botrytis sp.; leaf blight. Ecuador, Peru (low elevations).

Cercospora oxalidiphila (Speg.) Chupp & Muller; leafspot. Widespread. C. oxalidis Muller & Chupp; leafspot. Brazil.

Colletotrichum gloeosporioides Penz.; anthracnose spots on leaves and stems. Venezuela (probably widespread).

Dryrot of tuber, unidentified cause. Ecuador.

Erwinia sp.; bacterial softrot. In markets, Ecuador, Mexico, Peru, (? other countries).

Heterodera rostochiensis Wr.; root nematode. Peru.

Mycosphaerella sp.; large leafspot. Colombia, Brazil, Argentina.

Oidium sp.; white mildew. Mexico, Colombia, probably more widespread.

Phyllosticta oxalidicola Speg.; leafspot. Argentina, Venezuela.

Puccinia oxalidis Diet. & Ell.; leafrust. Bolivia, Colombia, Ecuador, Peru.

Rhizoctonia sp. (not R. solani) hyphae on tuber surfaces in market. Ecuador.

Rhizopus nigricans Ehr.; market softrot of tubers. Argentina, Ecuador.

Rosellinia bunodes (Berk. & Br.) Sacc.; black rot. Ecuador, and on test introduction planted in Costa Rica.

Septoria sp.; leaf disease from coalescing spots.   Bolivia.
Urocystis oxalidis Paz.; smut.   Peru, Brazil.
Virus: Mosaic mottle of leaves in Peru.

PACHYRHIZUS AHIPA Par., P; EROSUS Urb., and P. TUBEROSUS
    Spreng. All leguminous vines native to medium-high parts of
    tropical America that bear pods which are cooked when young,
    have edible seeds, and about two years after planting the vines
    have large, swollen, nutritious roots and for subsistence farmers
    these are an important food when certain other crops fail.   These
    are peasant products from Mexico to Brazil and Argentina, these
    crops have had little scientific attention (I know of no published
    disease studies, the list of diseases below is from my own inci-
    dental notes); "jícama," "ajipa" Sp., "jacatupe" Po., "yam bean"
    En.
        Black mildew, undetermined organism(s), observed under shade.
            C. America.
        Chaetoseptoria wellmanii J. A. stev.; leafspot. El Salvador.
        Colletotrichum probably C. gloeosporioides Penz.; anthracnose
            spots on leaves and pods. El Salvador.
        Erwinia carotovora (L. R. Jones) Holl. (typical symptoms);
            typical market bacterial softrot of tuberous roots. Guate-
            mala, probably widespread.
        Fusarium sp. (spores identified in scrapings from tuber lesions
            of a market dryrot). Panama.
        Isariopsis griseola Sacc. = Phaeoisariopsis griseola (Sacc.)
            Fur.; angular leafspot. El Salvador.
        Meloidogyne sp.; mild rootknot reported by students. Brazil.
        Nigrospora sp.; from microscopic determination of spores
            from black lesions on old vine. Guatemala.
        Oidium sp.; white powdery mildew. El Salvador, Guatemala.
        (? Phyllosticta sp. ?); small black leafspots. Ecuador.
        Phytophthora sp. (microscopic identification); rootrot. El
            Salvador.
        Rhizoctonia (? solani); seedling rot. Panama, Guatemala.
        Rhizopus (? niger); softrot of tuberous roots in markets. El
            Salvador, Costa Rica and is probably general.
        Rosellinia pepo Pat.; rootrot in field. Costa Rica.
        Virus, mild mottle. C. America.

PALMS, Ornamentals. Where grown in the Neotropics, they are
    considered the most elegant and romantic of trees (of course any
    palm is an ornament but there are a few in the American tropics
    that are planted, cultivated and harvested specifically for the
    major agricultural, money-making, crops for which they are
    known, and these are found elsewhere in this book under the
    genus names to which they belong), the genera of ornamentals on
    which the disease list below is based are as follows: ACRO-
    COMIA (1), ARECASTRUM (2), BACTRIS (3), BUTIA--including
    YATAY (4), CARYOTA (5), COCCOTHRINAX (6), COCOS--ex-
    cluding the coconut, which is listed elsewhere (7), COPERNICIA--
    including the wild Brazilian wax palm (8), INODES--certain wild

species used for hat and fan making (9), LEVISTONA (10),
PHOENIX--excluding the date, found elsewhere (11), ROYSTONEA
(12), SABAL (13), THRINAX (14), WASHINGTONIA (15), and
unidentified wild specimens (16). There are literally hundreds
of common names used locally for palms; general names are
"palma" Sp. Po., "arbol de palma" Sp., "palmeira" Po.,
"palm, " "palm tree" En.

Acrotheca acrocomiae Gonz. Frag. & Cif.; leafspot. W.
    Indies (1).
Agaricus gardneri Berk.; mushroom on palm trash. S.
    America (16).
Alternaria sp.; in leafspot complex. Florida (12).
Amerosporium sabalinum Ell. & Ev.; leaf decay organism.
    Subtropical USA (13).
Anthostoma versicolor Starb.; stem disease. S. America (7)
    and A. yatay Speg.; stem and trunk infection. Argentina
    (4, 7).
Anthostomella hemileuca Speg. = Entosordaria hemileuca
    (Speg.) Hoehn.; rot of leaves. Paraguay, Argentina (4, 7).
    A. leucobasis (Ell. & Mart.) Sacc., A. melanosticta Ell.
    & Ev., A. minor Ell. & Ev., A. sabalensioides (Ell. &
    Mart.) Sacc.; these four on dead leaf blades and stalks.
    Subtropical USA (13). A. phoenicola Speg.; petiole disease.
    Argentina (11).
Apiosporium australe Speg.; associated with trunk decay. S.
    America (6, 8, 16).
Aposphaeria jubaea Speg.; leaf disease. Chile (16).
Armillaria mellea Vahl = Armilariella mellea (Fr.) Karst.;
    root rot. California (2, 11).
Asterina sabalicola Earle; leaf blotch. Subtropical USA (13),
    Puerto Rico (16).
Auerswaldia guilielmae P. Henn.; leafspots with stroma. S.
    America, C. America (4). A. palmicola (Speg.) Petr.
    with the synonyms: Bagnisiopsis palmicola (Speg.) Petr.,
    Coccostroma palmicola (Speg.) Arx & Muller, Dothidina
    palmicola (Speg.) Th. & Syd.; small black stromatic leaf-
    spots. W. Indies, C. America, S. America (4, 7, 14,
    16), California (15). A. rimosa Speg.; stromatic spots on
    palm leaves. Noted specially in Paraguay but in other
    countries adjacent (4, 7, 1).
Bagnisiopsis = Auerswaldia
Bagnisiopsis bactridis (Rehm) Th. & Syd. also given in early
    name Bagnisiella bactridis Rehm; stromatic leaf disease.
    Brazil (3). B. diplothemii (Rehm) Th. & Syd. Paraguay
    (16).
Botryodiplodia theobromae Pat. = B. palmarum (Cke.) Petr.
    & Syd.; often on old leaves it causes infection and disease
    of leaflet, petiole, midrib, enlargement of leafspots from
    other parasites. Neotropical (on all palm genera, some
    specially noted: 2, 3, 4, 7, 11, 12, 15, 16).
Botryosphaeria dothidea (Moug.) deNot also given as B. ribis
    Gross. & Dug.; trunk lesions. Scattered in Neotropics

(7, 11, 15, 16). B. palmigena (Berk. & Curt.) Bomm. &
Rouss.; leaf disease. Only in Costa Rica (16).
Botryosporium palmicolum Speg.; leafrot. Argentina (4).
Bulgariella foliacea Starb.; on fallen palm debris. Scattered
  (16).
Calocera cornea Fr.; on fallen palm debris. Scattered (16).
Calonectria ferruginea Rehm; lesions on palm frond midribs.
  W. Indies, C. America (16).
Camarotella astrocaryae (Rehm) Th. & Syd.; leaf disease.
  Brazil (16).
Capnodium sp.; sootymold. Puerto Rico (5).
Catacauma mucosum (Speg.) Th. & Syd. with synonyms:
  Phyllachora capitata Speg., P. cacaoicola P. Henn., P.
  mucosa Speg.; leaf tarspot. Brazil and nearby countries
  (2, 4, 7, 16). C. sabal Chard.; black leafspot. Subtropi-
  cal USA, Mexico, W. Indies (13).
Cenangium sabalidis (Ell. & Mart.) Sacc.; leafstalk disease.
  Florida (13).
Cephaleuros virescens Kunze; algal leafspots. Puerto Rico
  (5).
Cephalosporium lecanii Zimm.; parasite of palm scale. Puerto
  Rico (16).
Ceratostoma australe Speg.; raceme disease. Argentina,
  Brazil, Paraguay (4, 7, 116).
Cercospora acrocomiae J. A. Stev. = Exosporium palmarum
  Sacc.; leafspot. Puerto Rico, C. America (1). C. palmi-
  cola Speg. sometimes given as Helminthosporium sp.;
  elliptical leafspot. C. America (4), Paraguay, Brazil
  (4, 7). Cercospora sp.; leafspot. Florida, Puerto Rico
  (5).
Cercosporella sp.; irregular leafspots. C. America (7).
Chaetomium fieberi Cda. var. macrocarpi Speg.; on fallen
  leaf. Argentina (4).
Chaetosphaeria iquitosensis (P. Henn.) Th. = Meliolina iquito-
  sensis (P. Henn.) Stev.; black leafblotches. S. America
  (3).
Chlorosis often is angular patches in leaf lamina, stunt. Neo-
  tropical but in scattered areas (on all palms).
Chromosporium albo-roseum Speg. = C. argentinum Sacc. &
  Syd.; rachis decay. Argentina (4).
Ciferria coccothrinax Gonz. Frag.; on leaf. W. Indies (6).
Circinatrichum maculiforme Nees; leaf disease complex. S.
  America (16).
Cladosporium herbarum (Pers.) Lk.; saprophytic on senescent
  leaflets, on fallen and undisturbed piles of cut leaves, on
  palm thatch. El Salvador, Honduras, Guatemala, perhaps
  Neotropical (5, 6, 9, 11, 12).
Clitocybe tabescens (Scop.) Bres. = Armilariella tabescens
  (Scop.) Sing.; rootrot. Subtropical USA (8, 11, 15).
Clusia rosea Jacq.; epiphytic smothering tree. W. Indies, C.
  America (16). (In a few cases Loranthaceae have been
  seen living on the epiphyte Clusia, thus giving the incorrect
  impression of attack on palms.)

Cocconia sparsa (Pk. & Cke.) Sacc. ; leafstalk disease.
Florida (13).

Coccostromopsis palmigena Plunk. (? = Coccostroma sp.);
leaf disease. Trinidad (various palms).

Colletotrichum gloeosporioides Penz. (= Gloeosporium coccoes
Allesch.), its perfect state is Glomerella cingulata (Stonem.)
Sp. & Schr. not often seen or reported; common anthracnose diseases of any part of palm frond, flower infection,
spot of fruit pedicel, fruitspot. Neotropics, some special
studies: Subtropical USA (12, 15), Puerto Rico, Virgin
Islands (2, 12), Costa Rica (2, 5), Chile (7, 11).

Coniosporium crustaceum Speg. ; leaf decay.   Argentina (14).
C. myiocoproide Speg. ; on dead leaf.   Paraguay (4).

Corallomyces mauritiicola P. Henn. ; disease of leaf midrib.
S. America (16).

Cylindrocladium macrosporium Sherb. ; leafspot.   Florida (15).

Cyphella mauritiae Pat. & Gaill. ; woodrot.   S. America (16).
C. paraensis P. Henn. ; woodrot.   Brazil (3).

Cytospora palmarum Cke. ; spathe and leaf infections.   Florida
(several palms).   C. yatay Speg. ; spathe disease.   Argentina (4).

Daedalea elegans Spreng. ; accompaniment of wood decay.   W.
Indies and more widespread (1, 16).

Dictyotheriella guianensis Stev. & Maub. ; frond mold.   W.
Indies, S. America (16).

Didymella jaffueli Speg. ; frond leaflet spots often circular and
sunken.   Chile (16).   D. phacidiomorpha (Ces.) Sacc. ;
leaflet lamina spots.   Florida (6).

Didymosphaeria astrocaryi Hoehn. ; leaf and frond midrib
disease.   Guiana (16).

Dimerium bactridicola P. Henn. ; hyperparasite of black mildews.   Scattered (3).

Diplodia (see Botryodiplodia)

Diplosporium macrosporum Speg. ; accompanying trunk decay.
Argentina (4).

Discosiella acrocoma-maculiforme Bat. ; leaf infection.   Brazil
and other countries nearby (1).

Dothidina (see Auerswaldia)

Dothiorella gregaria Sacc. ; canker and gummosis.   Subtropical
USA (2).

Ellisioidothis crustacea (Speg.) Bat. & Peres (see Myiocopron);
leafspots.   Brazil (16).   E. inquinans (Ell. & Ev.) Th. ;
leaf disease.   Subtropical USA (13).   E. pitterii Syd. ; leafspots.   Brazil (16).

Erwinia carotovora (L. R. Jones) Holl. ; isolations of heartrot
from well placed clandestine rifle shots.   Cuba (12).

Euryachora neowashingtonianae Dearn. ; leafstalk disease.
Florida (15).

Exosporium acrocomiae Lk. = Cercospora acrocomiae J. A.
Stev. ; large gray leafspots.   Puerto Rico (1).   E. palmi-
vorum Sacc. ; brown leafspots.   Subtropical USA, Dominican
Republic, Puerto Rico (2, 11).

Ficus spp. (so-called "strangling figs"); epiphytic, smothering, causing break-over of palms. C. America (9, 12).

Flammula palmarum Speg.; mushroom-like sporocarp on fallen trunks. Brazil (16).

Frizzle leaf with necrosis, tree stunt and little leaf, caused by manganese deficiency. Scattered in Neotropics (2, 10, 11, 12, 16 and other palms).

Ganoderma lucidum (Ley.) Karst.; trunk-base-and-log-rot. Cuba (12). G. sulcatum Murr.; on dead trunks. Subtropical USA (13).

Gemma infection body (not Mycena citricolor); leaf disease. Puerto Rico (5).

Geotrichum coccophilum Speg. = G. candidum Lk.; scale insect hyperparasite causing possible confusion in disease diagnosis. Brazil (7).

Glenospora microspora Speg.; rot of fallen palm frond. Paraguay (16).

Gloniopsis multiformis Starb.; accompanying wood decay. S. America (8). G. verbasci (Schw.) Rehm; trunk rot with surface perithecia. Brazil (2).

Gnomonia sabalicola Earle; leafstalk rot. Subtropical USA (13).

Gorgoniceps confluens Seav. & Waterst.; stem lesion. Bermuda (13).

Grallomyces portoricensis Stev. = Phialetea aerospora Bat. & Nasc.; on leaf surfaces. W. Indies (16).

Graphiola congesta Berk. & Rav.; small raised scab lesions. Subtropical USA (13, 16). G. phoenics (Moug.) Poit. = Phacidium phoenicis Moug.; scab-like small rough leaf and scape lesions (the well known so-called "false smut"); Subtropical USA, Puerto Rico, Cuba (2, 6, 11, 13, 15), C. America (11, 12, 13), Uruguay (11, 16), Brazil (11, 16). G. thaxteri Fisch.; "false smut." Florida (13).

Graphium sp.; on dead palm. Costa Rica (3).

Haplosporella asterocaryi P. Henn.; leaf disease. Brazil (16). H. brasiliensis Speg. = Sphaeropsis brasiliensis (Speg.) Sacc.; petiole disease. Brazil (6, 14).

Helicomyces roseus Lk. = Ophionectria cylindrothecia Seav.; stalk canker. Bermuda (13).

Helminthosporium bactridis P. Henn.; leafspot. Brazil (3). H. spiculiferum Ell. & Ev.; leafspot. Subtropical USA (6, 13). (? Helminthosporium sp. = Cercospora sp. ?); linear leafspot. S. America, C. America (7, 12, 16).

Hendersonia sabaleos Ces.; leaf diseased condition. Subtropical USA. (13).

Humarina waterstonii Seav.; seed disease. Bermuda (10).

Hyaloderma bakeriana P. Henn.; hyperparasite of leaf fungi. C. America, S. America (3, and others, 16).

Hypocrella glaziovii P. Henn. and H. semen Bres. (probably others of these species); hyperparasites of insects. Widespread (16, and other genera).

Hypoxylon spp.; dryrot of frond midribs and in some cases on trunks. Widespread (many palms).

Langsdorffia hypogaea Mart. (member of Balanophoraceae);
root parasite called "siego." From Mexico to southern
Brazil (on several palm genera).
Laschia auriscalpina Charles; attacking wood of weakened
trees. Cayenne, Panama, Cuba, Florida (13).
Lembosia diplothemii P. Henn.; black leafspot. Brazil (16).
L. cocoes Rehm and L. drymidis Lév. = Leveillella dry-
midis (Lév.) Th. & Syd.; both black moldy leafspots.
Brazil (4).
Leptosphaeria coccothrinacis Cif. & Gonz. Frag.; leafspotting.
Dominican Republic, nearby locations (6).
Leptostroma jubaeae Speg.; leaf disease. Chile (16).
Leptothyrium bactridis P. Henn.; small leafspots. C.
America (3). L. trithrinacis Speg.; small leafspots.
Argentina (14).
Lightning strike; killing top and therefore the tree which may
stand a long time dead. Scattered (any palm genus).
Linospora palmetto Ell. & Ev.; leafspot. Subtropical USA
(13).
Little leaf and some necrosis, from manganese deficiency.
Florida (5).
Macrophoma sp.; leafspot. Brazil (5).
Melanconium palmarum Cke.; prickly looking fungus growths
on leaf stalk, on stem. Puerto Rico (1), Subtropical USA
(13). M. sabal Cke.; leaf stalk disease. Florida (12).
M. yatay Speg.; on rachis and spathe. Argentina (4).
Meliola spp. (all are black mildews on leaves): M. amphi-
tricha Fr. = M. trichostroma (Kunze) Toro is widespread
(3, and several other genera); M. bidentata Cke. in
Florida (13); M. decora Syd. in Subtropical USA (3); M.
denticulata Wint. in Puerto Rico (12); M. furcata Lév. in
Subtropical USA (13), Puerto Rico (6, 16), S. America
(6); M. furcata Lév. in Subtropical USA (13), Puerto
Rico (6, 16), S. America (6); M. furcata Lév. var.
coccothrinacis Cif. in Dominican Republic (6); C. furcata
Lév. var. coperniciae Speg. in Argentina, Paraguay (8);
M. iquitosensis P. Henn. in Subtropical USA (13), S.
America (3); M. kerniana Cif. in W. Indies (3); M. mauri-
tiae Stev. in S. America (16); M. palmicola Wint. in Sub-
tropical USA (13), Panama (1, 13), Puerto Rico (13), Ar-
gentina (8); M. palmicola Wint. var. coperniciae Speg. in
Argentina (8); M. sabalidis Sacc. in Florida (13).
Meliolinopsis hyalospora (Lév.) Beeli = Meliola hyalospora
Lév.; black mildew. S. America (16). M. palmicola
Stev.; black mildew. Trinidad (3).
Melittosporiopsis pachycarpa Rehm; leaf disease. Chile (16).
Meloidogyne sp.; rootknot nematode. Subtropical USA (11,
12, 15).
Mendogia manaosensis (P. Henn.) Th. & Syd.; on leaf.
Brazil (16).
Metasphaeria cocoes Speg. = M. spegazzinii Sacc. & Trott.;
spathe rot. Argentina (2, 7). M. palmetta (Cke.) Sacc.;

leaf stalk rot. Subtropical USA (13). M. washingtoniae
Earle; leaf disease. Subtropical USA (15).
Microthyrium caaguazense Speg. f. coperniaciae Rehm; large
  leafspot. S. America (8). M. caludovicae P. Henn. ;
  dark brown spreading leafspot. Brazil (10).
Mitopeltis chilensis Speg. ; on leaf. Chile (16).
Mucor funebris Speg. (? = Rhizopus sp.); fruitrot. S.
  America (various palm genera).
Mycosphaerella advena Syd. ; leafspot. S. America (16). M.
  bambusae Pat. var. coccoes Rehm; leafspot. Brazil (7).
  M. palmae Miles; leafspots often becoming confluent.
  Puerto Rico (4, 15, 16). M. serrulata (Ell. & Ev.) Diehl;
  leafspot. Florida (13). M. washingtoniae Rehm; leafspot
  and blade collapse. California (15). Mycosphaerella sp. ;
  severe leafspot. Cuba, Puerto Rico (12).
Myiocopron crustaceum Speg. = Microthyrium crustaceum
  (Speg.) Cke. or Ellisiodothis crustacea (Speg.) Bat. &
  Peres; leaf spotting on senescent leaf. Paraguay (16).
Myringinella sabaleos (Weed.) Limb. & Junk. ; black speck.
  W. Indies, Subtropical USA (13).
Myxomycetes of many kinds, saprophytes and epiphytes, on
  both seedlings and on palm debris. Widespread (on any
  palm genera).
Myxosporium balmoreanum Speg. ; rachis disease. Argentina
  (16).
Nectria sp. ; bark disease. Florida, W. Indies (12).
Omphalia pigmentata Bliss; seedling rootrot. California (15).
  Related O. tralucida Bliss; seedling rootrot, tree decline.
  California (11, 15).
Ophiobolus veroisporus Ell. & Mart. ; leaf-stalk rot. Florida
  (12).
Ophionectria cylindrotheca Seav. ; on palm debris. Bermuda
  (13).
Oxydothis poliothea Syd. ; leaf disease. S. America (16).
Parmulinea rehmii (Maubl.) Th. & Syd. ; leafspot. Brazil (3).
Patellina subconoides Speg. ; petiole rot. Argentina (11).
Penicillium vermoeseni Biou. ; in California, canker (3, 11)
  and budrot (15).
Pestalotia coperniciae Speg. synonyms: Pestalotiopsis coper-
  niciae (Speg.) Stey. , P. coperniciae (Speg.) Bat. & Peres. ,
  P. microspora Speg. ; on fruits. Argentina, Brazil (8).
  (P. funerea Desm. is probably incorrect for:) P. palmarum
  Cke. ; the common and most well-known leaf edge invasion.
  C. America, W. Indies (1). Leafspot and occasional sys-
  temic spread in main vein of leaf. C. America, W. Indies,
  Subtropical USA (1, 2, 6, 7, 11, 12, 13, 15, 16).
Pestalotiopsis acrocomiarum Bat. ; brown leafspot. Brazil (1).
Phaeochora acrocomiae (Mont.) Th. & Syd. , P. indaya Viégas
  and P. mauritiae (Mont.) Hoehn. ; leafspots with stroma on
  surfaces. S. America (1, 7, 16). P. neowashingtoniae
  (Shear) Th. & Syd. ; leafstalk infection. California (15).
Phoma coccoes Allesch. , P. coccophila Speg. ; petiole lesions.

Brazil (7). P. maculata (Cke. & Harkn.) Sacc.; leaf
spreading-spot. Widespread (on many palm genera). P.
palmicola Wint.; leafspot disease. Florida (15).
Phomatospora kentiae Speg.; on petioles. Argentina (16).
Phomopsis pritchardiae (Cke. & Harkn.) Sacc.; on leaf and
   petiole. Dry parts of Subtropical USA (15).
Phycopeltis sp.; algal plaques on leaves and petioles. Puerto
   Rico (1, 5, 6, 11, 13, 16).
Phyllachora acrocomiae (Mont.) Sacc.; tarspot. C. America,
   Argentina, Brazil (1). P. palmicola Speg.; tarspot. Para-
   guay (8, 14). P. roystonea Johnst. & Brun.; midrib tar-
   spot. Cuba (12).
Phyllosticta cearensis Bat.; leaf lamina spot. Brazil (8).
   P. cocoes Allesch.; oblong leafspots. Brazil (7). P.
   palmetta Ell. & Ev.; leafspot. Subtropical USA (13).
   Phyllosticta sp.; leafspots of medium size. Florida,
   Puerto Rico (2).
Phymatotrichum omnivorum (Shear) Dug.; rootrot. Texas (15).
Physalospora (see Botryodiplodia)
Phytophthora palmivora Butl.; budrot and wilt. Subtropical
   USA (2, 13, 7, 15), Puerto Rico (12, 13, 16).
Placostroma diplothemii Syd.; rough black tarspot. Brazil
   (16).
Plowrightia diplothemii Rehm; petiole disease. Brazil (16).
Polyporus nivosellus (Murr.) J. A. Stev.; woodrot. Puerto
   Rico (3, 4, 7, 12). P. tulipifera (Schw.) Overh.; on fallen
   trunk. Florida (13).
Poria buttneri P. Henn.; on fallen leafstalks. S. America
   (7). P. cocos Wolf; root infection (producing giant sclero-
   tium). Florida (13). P. heteromorpha Murr.; leaf stalk
   disease. Florida (13).
Pseudographis cocoes P. Henn.; leaf and leafstalk rot. S.
   America (2).
Pseudomonas washingtoniae (Pine) Elliott; bacterial leafspot.
   Arizona (15).
Pucciniopsis guaranitica Speg.; called "leaf rust" but see note
   with heading Rusts below. Paraguay, Argentina (4).
Pythium sp.; wilt. Florida (2, 15).
Rhabdospora sabalensis Cke.; leaf stalk-rot. Subtropical
   USA (13).
Rhizoctonia solani Kuehn, its basidial state seldom reported on
   palms is Thanatephorous cucumeris (Frank) Donk also re-
   ported as Pellicularia filamentosa (Pat.) Rogers; root decay.
   Florida, W. Indies (7, 12, 16).
Rusts, there apparently are no true rusts on palms; the in-
   correctly called "leaf rust" on certain palms is either
   Pucciniopsis or the algal leafspot from Cephaleuros.
Saccardomyces bactridicola P. Henn.; leaf mold. Argentina,
   Brazil (3).
Schizophyllum commune Fr.; leaf stalk-rot, trunk rot with
   sunken sporocarps rifting outer tissues. Puerto Rico, C.
   America, Panama, Colombia (specially seen were on 5, 7,
   11, 13 but perhaps most, if not all, genera are susceptible).

Septobasidium sabalis Couch and S. sabal-minor Couch; fungus
    felts.   Subtropical USA (13).
Septoria cocoina Ell. & Ev.; large sometimes gray leafspot.
    Subtropical USA, Puerto Rico (1, 2, 7, 13, 15, 16).
Sphaerella bambusae Pat. var. cocoes Rehm; leafspot.
    Brazil (3, 7). S. frenumbensis Speg.; leafspot. Brazil
    (3, 7, 11).
Sphaerodothis acrocomiae Chard. & Baker; black leafspot.
    W. Indies (1). S. neowashingtoniae Shear; black leafspot.
    California (15).
Sphaeropsis sabalicola Ell. & Carver; leaf stalk rot problem.
    Subtropical USA (13).
Spiculostilbe dendritica Morris; leaf disease.   Panama (16).
Stagonospora desmonci P. Henn.; leafspot and irregular
    splotch.   S. America (16).
Steirochaete mcchellandii Toro; yellowish leafspots.   Puerto
    Rico (16).
Stilbella melanotis Syd.; leaf sheath-rot.   S. America (3).
Stilbum aurantio-cinnabarinum Speg.; trunk canker, petiole
    rotted spot.   Argentina, Panama (4, 16).   S. aurantio-
    cinnabarinum Speg. subsp. fusiceps Speg.; accompanying
    trunk and leaf petiole decay.   Argentina (7, 11).
Strigula complanata Fée; common white lichen spotting of
    leaves and petioles.   C. America, S. America (1), Costa
    Rica, Cuba, Puerto Rico (2, 12), Puerto Rico, Virgin
    Islands (5, 11, 16).
Stunt, along with chlorosis and some necrosis in leaflets as
    well as mild little leaf.   Cause is manganese deficiency.
    Scattered in Neotropics (all palms probably susceptible, es-
    pecially noted on 10, 11, 16).
Stylina disticha (Ehr.) Syd.; smut.   Puerto Rico interception.
    (11).   This disease reported from the Orient many years ago.
Tapesia succinea Rehm; on leaf.   Brazil (3).
Thielaviopsis paradoxa (de Seyn.) Hoehn.; rootrot, darkened
    vasculars of stem, trunk bleeding.   W. Indies, subtropical
    USA (16), Brazil (5).
Trabutia mauritiae (Mart.) Sacc. (= Sphaeria mauritiae Sacc.)
    and T. atroinquinans (Wint.) Th. & Syd.; both are black
    and shiny leafspots.   Brazil and contiguous countries (16).
Trametes cubensis (Mont.) Sacc.; on decayed trunk.   Florida,
    Cuba, C. America (13, 16).
Trentepohlia sp.; tufts of brown algal growths, benign.   Puerto
    Rico (5).
Trullula olivascens Sacc.; on dead stalks.   Brazil (7).
Typhula trailii Berk. & Cke.; accompanying decay of fallen
    leaves.   S. America (16).
Uleopeltis manaosensis P. Henn.; leaf disease.   Brazil (16).
Valsa sabalina Cke.; leaf stalk disease, isolated from leafrot
    of a few palm genera.   Subtropical USA (13), Costa Rica
    (3, 7, 11, 14).
Venturia sabalicola Ell. & Ev.; leaf disease.   Subtropical
    USA (13).

Xanthomonas vasculorum (Cobb) Dows.; bacterial wilt, gumming. W. Indies, Trinidad, Brazil, probably widespread
(12).

PANDANUS UTILIS Bory = P. VEITCHI Dall. A tall exotic Oriental tree that has prominent prop roots, edible fruits, and along
its branches, arranged in spiral or screw-like pattern, are the
broad-based spine-tipped leaves much used in weaving baskets,
hampers, house walls; "sabután," "palma de tirabuzón" Sp.,
"pandano" Po., "pandanus," "screw-pine" En.
　Botryodiplodia theobromae Pat.; leaf disease.   Puerto Rico,
　　C. America (probably widespread).
　Botryosphaeria dothidea (Moug.) Ces. & de Not.; stem canker
　　and proproot canker (identification from moist chamber
　　studies).   Panama, questionable symptoms in Puerto Rico.
　Cephaleuros virescens Kunze; algal leafspot.   W. Indies in
　　moist locations.
　Ceratocystis sp.; stain of leaf vein (such attacked leaves
　　sometimes used decoratively in woven material).   W.
　　Indies.
　Colletotrichum omnivorum Halst. = C. gloeosporioides Penz.
　　(synonyms: C. dematium Grove, Vermicularia herbarum
　　Westend., C. zingiberis Butl. & Bisby); large and spreading
　　leafspot.   Puerto Rico, Subtropical USA.
　Cytospora palmarum Cke.; leaf-end disease.   Florida.
　Dothidea pandanae Schw.; stromatic spots on leaves.   S.
　　America.
　Epicoccum pandani P. Henn.; on leaves.   Panama.
　Macrophoma pandani (Lév.) Berk. & Vogl.; leafspot.   Sub-
　　tropical USA.
　Melanconium pandani Lév.; black leafspot, branch-tip disease,
　　various branch lesions.   Subtropical USA, W. Indies, C.
　　America.
　Meliola musae Mont. (and other unidentified related fungi);
　　black mildews.   W. Indies, C. America.
　Pestalotia palmarum Cke.; leaf lesions.   Common and wide-
　　spread.
　Phyllosticta pandanicola Young; leaf scorch (some cases with
　　stippled markings), diseased leaftips.   Puerto Rico, Panama,
　　Colombia, El Salvador.
　Physalospora pandani Ell. & Ev.; on old diseased leaves.
　　Dominican Republic.
　Rhizoctonia microsclerotia Matz; web-blight of lower leaves.
　　Puerto Rico.
　Schizophyllum commune Fr.; basal decay of old leaves.
　　Puerto Rico.
　Strigula complanata Fée; white lichen spot.   W. Indies, C.
　　America.
　Tremella sp.; on decayed stump.   Honduras.

PANICUM spp. [but see P. BARBINODE and P. MAXIMUM, following two entries].   About 70 or more of these grasses are in

mixed tropical American pastures.   Diseases result naturally in
disease rotations: killing off or weakening of some species,
ascendency of others, these succumbing, finally original kinds
"coming back, " and thus repeated.   On certain of these hosts
occur many as yet undescribed parasitic organisms.   However,
species of Balansia are frequently seen, Claviceps is not uncommon, Phyllachoras are almost everywhere, black mildews are
widespread, Monilias both strong and weakly parasitic are found,
collections are easily made of semi-saprophytic Fusariums,
Cladosporiums, Spegazzinias, as well as others, and there are
described rusts some unique to these pasture grass species, and
a number of smuts are also present.

PANICUM BARBINODE Trin. = P. PURPURASCENS Rad. [see also
    PANICUM spp., preceding entry, and P. MAXIMUM, following
    entry].   One of the two most used species is native of Brazil
    where it is one of the excellent tropical pasture grasses; "zacate
    de pará, " "palojillo" Sp., "pasto do pará, " "capím" Po., "para
    grass" En.
        Cephaleuros virescens Kunze; algal leafspot.   Brazil.
        Cerebella andropogonis Ces.; on the sphacelia state of Clavi-
            ceps.   Widespread.
        Claviceps purpurea Tul. (possibly C. uleana P. Henn.); ergot.
            Puerto Rico, C. America and widespread.
        Fusarium sp.; weakly parasitic.   C. America.
        Fusarium oxysporum Schl. f. sp. cubense (E. F. Sm.) Sny.
            & Hans.; rootrot.   Honduras.
        Helminthosporium sp.; leaf disease.   Subtropical USA, W.
            Indies.
        Marasmius sacchari Wakk.; culmrot.   Puerto Rico.
        Myriogenospora paspali Atk.; black-crust, tangle-top.   Sub-
            tropical USA, C. America.
        Nigrospora oryzae (Berk. & Br.) Petch; leafmold.   W. Indies,
            C. America.
        Perisporium zeae Berk. & Curt.; sootymold.   In many
            countries.
        Puccinia panicicola Arth.; rust.   Trinidad, Peru and probably
            widespread.   P. puttemansii P. Henn.; rust.   S. America.
        Uromyces leptodermus Syd.; rust.   El Salvador, Barbados,
            Puerto Rico, Virgin Islands, Panama, Florida.
        Viruses: Sugarcane Mosaic Virus is widespread, and Sugar-
            cane Ratoon Stunting Virus in C. America.

PANICUM MAXIMUM Jacq. [see also PANICUM spp. and P.
    BARBINODE, preceding two entries].   A vigorous introduction
    from Africa that was quickly accepted in the American tropics;
    "gramalote, " "zacatón, " "graminea guinea" Sp., "pasto do
    guine, " "pastagem guine" Po., "guinea grass" En.
        Balansia spp. (external symptoms are numerous spectacular
            heads sometimes almost geometrically arranged) B. clavi-
            ceps Speg. (synonyms: B. trinitensis Cke. & Mass., B.
            claviceps Speg. f. sp. fasciculata Speg., Ephelis mexicana

Fr.); sclerotial disease of spike. Mexico to southern S.
America. Other species occur, e.g. B. diadema Moel.,
B. strangulans (Mont.) Diehl = Dothichloe nigricans
(Speg.) Chard., B. strangulans (Mont.) Diehl f. sp. dis-
coidea Diehl, and B. subnodosa Atk.

Balansiella orthocladae P. Henn. (? Claviceps) seed infection.
S. America.

Calonectria geralensis Rehm; on dead tissues. C. America.

Cercospora fusimaculans Atk.; typical elongated leafspots (so
common it is accepted as "normal" by farmers and a
characteristic of this pasture grass). Widespread.

Cerebella andropogonis Ces. and C. panici Tracy & Earle;
on inflorescence. S. America.

Cladosporium (probably the species is herbarum); ragged leaf
edges. W. Indies, C. America.

Claviceps maximensis Theis, C. purpurea (Fr.) Tul., and C.
uleana P. Henn.; ergots (of poisonous nature to grazing
animals). Widespread.

Colletotrichum graminicola (Ces.) Wils. the asco-state is
Glomerella tucumanensis (Speg.) Arx & Moel.; leaf, stem
inflorescence and seed injury. Neotropical.

Coniothyrium panici Syd.; brown elongated lesions. Colombia,
Guatemala, Puerto Rico, Venezuela, probably more widespread.

Crozalsiella argentina Hirsch.; smut. Argentina.

Cytosporella cereina Speg.; leaf & culm disease. Scattered
occurrence.

Dothichloe nigricans (see Balansia strangulans)

Entyloma speciosum Schr. & P. Henn.; swollen white leafspots. Argentina.

Ephelis (see Balansia)

Epicoccum nigrum Lk.; on leaves where it may resemble a
smut. W. Indies, C. America.

Eriospora ambiens (Cda.) Sacc. and E. Leucostoma Berk. &
Br.; on culms. Southern S. America.

Fusarium avenaceum Sacc. = F. pseudonectria Speg. (the
perfect state is Gibberella avenacea Cook); on dead culms.
Argentina and nearby countries.

Gamosporella hysterioides Speg. (considered by some mycologists as a "nomen confusum"); diseased culms. Paraguay.

Glomerella (see Colletotrichum)

Helminthosporium flagelloideum Atk.; leafspot. Nicaragua,
Subtropical USA, Puerto Rico, El Salvador.

Himantia stellifera Johnst.; collar rot, culm lesion. Puerto
Rico, Cuba.

Meliola amphitricha Fr., M. panici Earle, and Meliola spp.;
black mildews. Widespread.

Mycosphaerella panicicola P. Henn. = Sphaerella panicicola
(P. Henn.) Sacc. & D. Sacc.; rounded leafspot. Widespread.

Myriogenospora paspali Atk.; crusty leaf disease. Subtropical
USA.

Ophiodotis vorax (Berk. & Curt.) Sacc.; black stroma on leaf
and culm. Brazil.
Otthia panici Stev.; rifted lesion on culm. Puerto Rico.
Phyllachora spp.; black and usually shiny leafspots. May be
encountered in all countries, eight of the species are: P.
acutispora Speg., P. apiculata Speg., P. chardoni Orton,
P. cynodontis (Sacc.) Niessl., P. graminis (Pers.) Fckl.,
P. heterospora P. Henn., P. paniciolivacei Chard., P.
pazschkeana Syd.
Phyllosticta panici Young; leafspotting. Subtropical USA, W.
Indies, (probably in C. America), S. America. P. panici-
maximi Cif. & Gonz. Frag.; leafspotting. Dominican Re-
public, Brazil.
Physalospora panici Rehm; on leaves. S. America.
Puccinia spp. Rusts (the host appears remarkably tolerant
to these rusts and some of the names may be synonyms)
are found widespread on this pasture grass in the American tropics, names as follows: P. atra Diet. & Holw.,
P. faccida Berk. & Br., P. huberi P. Henn., P. Levis
(Sacc. & Bizz.) Magn., P. negrensis P. Henn., P. panici
Diet., P. puttemansii P. Henn., P. recondita Rob., P.
substriata Ell. & Barth. (= P. tubulosa Arth. = P. paspalicola Arth.). (Note: It is sometimes difficult to dif-
ferentiate Puccinia from Uromyces on this grass; however, teliospores of the former have two cells and such
spores of the latter are only one-celled, and by microscopy
is the only sure way to make this determination.)
Rhizoctonia solani Kuehn as the non-sporuliferous state causing
minor damage but abundant. Puerto Rico.
Sclerotium compactiusculum Speg.; sclerotial disease. Paraguay (see Ustilagopsis).
Septoria tandilensis Speg.; spots on leaf and sheath. Argentina, other parts of S. America.
Spherella (see Mycosphaerella)
Tolyposporium minus Schr., T. reticulatum Speg.; both spike
disease. S. America.
Uromyces graminicola Burr., U. leptodermus Syd., U. putte-
mansii Rang.; rusts. Widespread. (See note under
Puccinia.)
Ustilago spp.; panicle smuts, widespread but given most at-
tention in South America. Examples: U. balansia Speg.,
U. bonariensis Speg. (= Sphacelotheca bonariensis (Speg.)
Cif. = Sorosporium bonariensis (Speg.) Zund.), U. brasi-
liensis Zund., U. vesiculosa P. Henn.
Ustilagopsis bertoniensis Speg., U. compactiuscula Speg. (it
is suggested by some that these are imperfect states of
Claviceps); disease of flower. Paraguay.
Viruses: Sugarcane Mosaic Virus is widespread. Sugarcane
Ratoon Stunting Virus found in S. America, C. America;
Severe stunt, virus is unidentified. El Salvador.

PASPALUM spp. There are over 50, all native in the American

tropics serving as valuable pasture, ground-cover, and lawn
grasses of which a few most popular are P. CONGUGATUM Ber.,
"horquetilla, " "spreading paspalm"; P. DILATATUM Poir,
"yerba dalis, " "dallis grass"; P. LAXUM Lam. , "matojo, "
"bunch grass"; P. NOTATUM Flu. , "yerba bahia, " "bahia lawn
grass"; P. VIRGATUM L. , "matojo blanco, " "tall papalum. "
In the list below, the diseases appear to be able to attack almost any Paspalum.

Aecidium (see Puccinia)

Angiospora (see Puccinia)

Balansia strangulans (Mont.) Diehl; strangles.   Scattered.

Cassytha filiformis L. ; green dodder.   Puerto Rico, Florida.

Cephaleuros virescens Kunze; algal leafspot.   Brazil.

Cerebella paspali Cke. & Mass. ; associated with ergot es-
    pecially on P. dilatatum.   Subtropical USA.

Chaetospermum tubercularioides Sacc. ; on dead culms.
    Florida.

Cladosporium herbarum Lk. ; headmold and disease of sene-
    scent leaves.   Neotropics.

Claviceps deliquescens (Speg.) Haum. and C. paspali Stev.
    & Hall; both ergots.   Widespread.   C. lutea Moel. (long
    curved sclerotia in fruiting spikes); ergot.   Brazil.   C.
    purpurea (Fr.) Tul. ; common ergot.   Neotropics.

Colletotrichum graminicola (Ces.) Wils. ; anthracnose spots
    of leaf and spike.   (Is also a hyperparasite of Physoderma
    paspali.)   Subtropical USA, Puerto Rico, C. America,
    Brazil (probably general).

Curvularia sp. ; leafspot, culm lesion.   Florida, Nicaragua.

Cuscuta americana L. ; yellow fast growing dodder.   Puerto
    Rico.   Cuscuta spp. ; various dodders.   C. America,
    Florida, W. Indies.

Darluca (see Eudarluca)

Diplodia (? theobromae); on old culms of the host P. virgatum.
    Puerto Rico.

Endodothella platensis (Speg.) Th. & Syd. = Dothidella platen-
    sis Speg. ; elongated black leaf-stromata.   Argentina.

Erineum lorentzii Speg. ; leaf disease.   Argentina.

Eudarluca australis Speg. is the perfect state of Darluca
    filum (Biv. -Bern.) Cast. which is a hyperparasite on
    Puccinias.   Widespread.

Fusarium spp. (often isolated and obtained in moist chambers);
    weak parasites causing secondary enlargement of leaf
    lesions from primary parasites.   F. heterosporium Nees;
    scab and headmold.   Subtropical USA, Puerto Rico, C.
    America.   F. oxysporum Schl. f. sp. cubense (E. F. Sm.)
    Sny. & Hans. ; root invasion.   Honduras.   F. roseum Lk. ;
    headmold, present with ergot.   Widespread.

Gibberella zeae (Schw.) Petch (imperfect state is a Fusarium);
    basalrot.   Brazil.

Helminthosporium brevipilum Cda. , in W. Indies and Sub-
    tropical USA; H. mayaguezense Miles only in Puerto Rico;
    H. micropus Drechs. in Florida; H. penicillosum Speg.

in S. America; and H. rostratum Drechs. in Subtropical
USA.   All cause leafspot and blast.
Hendersonia rhizomatophila Speg.; on dying rhizomes.   Argen-
tina.
Himantia stellifera Johnst.; rhizome and collar disease.
Puerto Rico.
Isariopsis paspali Syd.; on leaves.   S. America.
Leptosphaeria proteispora Speg.; on decaying culms.   Argen-
tina.
Lophiotrema paspaicola Speg.; possibly saprophyte, found on
dead culms.   Argentina.
Meliola amphitricha Fr., M. panici Earle, M. stenotaphri
Stev., also other species; black mildews.   Neotropical.
Mycena citricolor (Berk. & Curt.) Sacc.; leaf infection.
Costa Rica.
Myriogenospora bresadoleana P. Henn. and M. paspali Atk.;
cause tangletop and black streak.   Nicaragua, Grenada,
Trinidad, Subtropical USA.
Odontia saccharicola Burt; secondary parasitism on dying
plants.   Brazil, Argentina, Puerto Rico.
Ophiobolus sp.; leaf and sheath disease.   S. America.
Phyllachora spp.:   P. acuminata Starb., P. apisculata Speg.,
P. brevifolia Chard., P. congruens Rehm, P. cornispora
Atk., P. cornispora-necrotica Chard., P. eriochloae
Speg., P. graminis (Pers.) Fckl., P. graminis (Pers.)
Fckl. f. sp. tupi Speg., P. guianensis Stev., P. infuscans
Wint., P. leonardi Chard., P. microspora Chard., P.
molinae Chard., P. murilloi Chard., P. ortonii Chard.,
P. oxyspora Starb., P. paspalicola P. Henn., P. paspali-
virgati Chard., P. quadroaspora Tehon, P. wilsonii Orton,
P. winkleri Syd.; leaf tarspots that range from small to
large, from round to elongate, and from dull to shiny.
Tarspots were found on Paspalum leaves in all countries I
visited, many are very widespread in occurrence.
Physoderma paspali J. A. Stev. (on Paspalum plicatum)
brown streak, brownspot.   W. Indies (probably elsewhere).
Pleospora culmicola Speg.; stem and leaf sheath disease.
Chile.
Puccina spp.   (Note: Puccinia substriata--no. 11--deserves
special mention here as its alternate hosts are several
dicotyledonous hosts such as tobacco and various Solanums.)
About 11 named: 1. P. atra Diet. & Holw. = P. esclaven-
sis Diet. & Holw.   2. P. chaetochloae Arth. = P. mau-
blancii Rang.   3. P. compressa Arth. & Holw. (synonyms:
Aecidium compressa Mains, Angiospora compressa Mains,
Uredo stevensii Arth.).   4. P. dolosa Arth. & Fromme.
5. P. huberi P. Henn.   6. P. levis (Sacc. & Trott) Magn.
= P. paspali Tracy & Earle.   7. P. panicophila Speg.   8.
P. paspalicola Arth. said to be the same as P. substriata.
9. P. paspalicola Kern, Thur. & Whet.   10. P. pseudoatra
Cumm.   11. P. substriata Ell. & Barth. (synonyms:
Aecidium tubulosum Pat. & Gaill., P. tubulosa Arth. and

possibly P. paspalicola Arth.). All cause rust. Many are general in occurrence.

Pyrenophora chaetomioides Speg.; semi-parasitic on culms. Argentina, may be widespread and at times only a saprophyte.

Pyricularia grisea Sacc. (see Pyricularia under ORYZAE); blight and leafspot. Subtropical USA.

Ramularia sp.; leafspot. Nicaragua.

Rhizoctonia (? solani) non-sporuliferous type; causing brown patch in lawns. W. Indies.

Sclerotinia sp.; causing lawn-spot. Subtropical USA, W. Indies.

Sootymolds (possibly Capnodium sp.) that occur widespread and appear most likely during dryseasons.

Spermoedia stevensii Seav. = Claviceps sp.; fruiting spike disease. Subtropical USA, Puerto Rico.

Sphacelia deliquescens (Speg.) Haum.; on leaves. S. America notably in Argentina.

Sphacelotheca cordobensis (Speg.) Jacks.; inflorescence smut. Brazil. S. paspali-notati Clint.; head smut. Florida, Puerto Rico.

Sphaerulina subtropica Speg.; leaf lamina attack. Argentina.

Stagonospora paspali Atk.; leafspot. Brazil, Subtropical USA.

Telimena dominguensis Chard.; leaf scab-spot. W. Indies.

Tilletia paspali Lind., T. rugispora Ell. = T. ulei Schr. & P. Henn.; both smuts. Subtropical USA, Brazil (and probably countries in between).

Uredo (see Puccinia)

Uromyces sp.; questionably reported, probably one of the Puccinias.

Ustilago spp., all nine are panicle and seed smuts: 1. U. carbo Tul., in S. America. 2. U. garcesi Zund., in S. America. 3. U. microspora Schr. & P. Henn., in S. America. 4. U. paspali Speg., in S. America. 5. U. paspali-dilatati P. Henn., in Brazil. 6. U. schroeteriana P. Henn., in Subtropical USA, Brazil, Panama, Puerto Rico. 7. U. subnitens Schr. & P. Henn., in Brazil. 8. U. venezuelana Syd., in Brazil, ? other parts of S. America. 9. U. verrucosa Schr., in Brazil.

Virus: Sugarcane Mosaic virus is widespread.

PASSIFLORA spp. A number of fruit-bearing vines native to tropical America, not often widely publicized but very important for needed fresh fruits, mostly found in local markets and in backyard gardens. Examples: P. EDULIS Sims, P. LIGULARIS L., P. QUADRANGULARIS L., P. RUBRA L. Some of many common names; "granadilla," "tacsa," "galipa," "curubá" Sp., "granadillo," "maracujá," "maracuja melão" Po., "grenadilla," "sweet calabash," "passion vine fruit" En. The pathology of these plants is not well studied. In the disease list here given, items are from occurrences on both wild and cultivated species.

Alternaria sp.; rare fruitspot. Colombia.

Arcyria nodulosa Macr.; probably epiphytic. Colombia.

Asterella balansae Speg. var. macrospora Speg.; dark mildew.
  Argentina, nearby countries.

Asterina spp.; all black mildews, smudge spots: A. arnaudi
  Ryan in W. Indies, A. cubensis Sacc. & Syd. in W. Indies,
  A. megalospora Berk. & Curt. in Puerto Rico, Cuba,
  Colombia, Ecuador, Peru, Brazil, A. mulleri Stev. in
  Brazil, A. tacsoniae Pat. var. passiflorae Ryan in Puerto
  Rico.

Asterostomella gregariella Petr. & Cif.; disease at base of
  aging plants. Dominican Republic, Puerto Rico.

Botrytis cinerea Pers.; fruitrot and branch-tip disease. Colombia.

Cephalosporium sp.; isolation from diseased leaves. El
  Salvador.

Cercospora spp. all causing typical leafspots: C. biformis
  Peck, in Puerto Rico, Subtropical USA, C. calospilea
  Syd. in Colombia, C. fusco-virens Sacc. in Brazil, Trini-
  dad, C. passiflorae Muller & Chupp in Brazil, C. regalis
  Th. in El Salvador, Guatemala.

Cladosporium maracuja Viégas; leaf disease. Brazil, and
  other parts of S. America.

Clypeolum hieronymi Rehm; on leaf (report from S. America).

Colletotrichum gloeosporioides Penz. (perfect state is
  Glomerella cingulata (Stonem.) Sp. & Schr.); stem, fruit,
  and leaf anthracnose spots, also occasional seedling blight.
  Neotropics.

Didymosphaeria innumerabilis Wint.; leafspot. Brazil.

Dimerosporium passiflorae Pat. = D. passiflorae Speg.;
  Argentina.

1. Fusarium oxysporum Schl. (f. sp. passiflorae?), swelling
   and bark-split at base of stem, vascular discoloration,
   top chlorosis followed by wilt and death. Puerto Rico
   (inoculation studies showed my isolates did not cause
   disease in coffee or tomato in Puerto Rico). 2. F. gloeosporioides Speg. = F. semitectum Berk. & Rav. 3. F.
   roseum Lk.; both associated with fruit disease. S.
   America, C. America. (Note: In Puerto Rico isolations
   were made from crown rot and collapsing branches on
   three plants, and typical Fusarium spores obtained in isolates were identical in growth, morphology, and measurements with F. retusum Wellman, a tomato parasite I
   studied, believed now to be a form of F. oxysporum Schl.
   var. redolans (Wr.) Gordon.)

Helminthosporium stahlii Stev.; leafspot. Puerto Rico,
  Dominican Republic.

Irenina glabra (Berk. & Curt.) Stev. and Irenopsis moelle-
  riana (Wint.) Stev.; black mildews. W. Indies, C. America.

Meliola aristata Toro; black mildew. Dominican Republic.

Meloidogyne sp.; rootknot nematode. Florida, Brazil,
  Puerto Rico.

Mycena citricolor (Berk. & Curt.) Sacc.; leaf lesion.   Costa
   Rica.
Mycosphaerella passiflorae Rehm; leafspot.   S. America.
Oidium sp.; white mildew.   Colombia.
Ovulariopsis passiflorae Syd.; white mildew.   S. America.
   (? = Oidium.)
Phaeodimeriella tachirensis Toro; leaf mildew.   Scattered in
   W. Indies.
Phomopsis sp.; leaf and stem lesions.   Colombia.
Phyllosticta superficiale Stev.; leafspot.   Puerto Rico.
Puccinia scleriae (Paz.) Arth.; rust.   Widespread.
Rhizoctonia (? solani); ground-level lesion.   Puerto Rico.
Schiffnerula pitteriana Syd.; mildew.   Scattered in S. America.
Sclerotinia sclerotiorum (Lib.) dBy. = Whetzeliana sclerotiorum
   (Lib.) Korf & Dumont; collar rot.   Subtropical USA, W.
   Indies.
Septoria passiflorae Syd.; small leafspot.   Colombia.
Seynesia balansae Speg. and S. megalospora (Berk. & Curt.)
   Rehm, on leaves and green vines.   S. America.

PELARGONIUM HORTORUM Bail.   Popular perennial ornamental,
  becomes woody with age and of considerable economic value for
  florists, its origin South Africa; "geranio" Sp., "gerânio" Po.,
  "geranium" En.
    Alternaria (? brassicae); rather rare dryseason leafspot.
     Puerto Rico.
    Aspergillus sp.; stemrot.   Subtropical USA, Puerto Rico.
    Botryosphaeria berengeriana deNot., B. dothidea (Moug.) Ces.
     & deNot.; both stem cankers.   First one reported in S.
     America, latter in Subtropical USA.
    Botrytis cinerea Pers.; rainyseason graymold, leafspot,
     blossom blight, cutting disease.   Widespread but not seen
     by me in dry areas.
    Cercospora brunkii Ell. & Gall.; leafspot, leaf burn.   Sub-
     tropical USA, Guatemala, Venezuela, Puerto Rico, Panama.
    Colletotrichum gloeosporioides Penz. = Gloeosporium pelar-
     goni Cke. & Mass. (perfect state is Glomerella cingulata
     (Stonem.) Sp. & Schr.); anthracnose leaf and stem disease.
     Puerto Rico, St. Thomas, S. America (? general).
    Criconemoides sp.; root zone nematode inhabitant.   Peru.
    Diaporthe meduseae Nits.; stem disease.   California.
    Fusarium sp.; part of complexes causing basal rot and root
     failure.   W. Indies.
    Leaf scorch, lamina collapse from physiological cause, some-
     times edge crumples, followed by invasion with weak para-
     sites.   Puerto Rico.
    Leptosphaeria pelargonii Rehm; spreading leafspot.   Brazil.
    Meloidogyne sp.; rootknot.   Florida.
    Nectria pelargonii Speg.; disease on old stems.   Scattered in
     Argentina and S. America.
    Phytophthora drechsleri Tucker; cutting disease, stem rot.
     W. Indies.

Pseudomonas erodii Lewis; bacterial leafspot.   Subtropical
   USA.
Puccinia callaquensis Neg.; rust.   Chile, Argentina.   P.
   geraniphila Lindq. (on wild geranium); rust.   Argentina.
   P. leveillei Mont.; rust.   Bolivia.
Pythium spp.; stem rots, cutting rots.   Subtropical USA, C.
   America, reported from Peru, probably scattered in Neo-
   tropics.
Rhizoctonia solani Kuehn = Thanatephorous cucumeris (Frank)
   Donk, which is the basidial stage that is rarely reported;
   stem disease, rot at ground line.   Widespread.
Solenia villosa Fr.; on old woody stems.   S. America.
Stilbum aurantio-cinnabarinum Speg. = Sphaerostilbe rosea Kal.;
   stem disease.   Argentina.
Tubercularia pelargonii Speg.; on old dying branches.   Argen-
   tina.
Viruses: Beet Curly Top in California, the common Cucumber
   Mosaic Virus in Florida, the Cucumber Mosaic Virus viru-
   lent strain Celery-Commelina in Florida, Guatemala, El
   Salvador, Cuba.   By inoculation Sunflower virus (Argentine)
   in Argentina.   Tomato Spotted Wilt in Texas.

PENNISETUM CLANDESTINUM Choi.   (1) An introduction from the
   mountains of east Africa, used much in dairy farms in certain
   cooler parts of the American tropics, a reasonably short grass;
   "kikuyo," "kukuyu" Sp. Po. En.   P. PURPUREUM Schum. (2)
   came from the African plains and was planted in the American
   tropics in the warmer medium to low pastures; "elefante" Sp.
   Po., "napier," "elephant grass" En.   Pathology of these two
   valued grasses is as yet sketchy and incomplete.
   Balansia claviceps Speg. = Ephelis trinitensis Cke. & Mass.;
      inflorescence blight.   Argentina, Puerto Rico (2).
   Beniowskia sp.; small red spots on leaf.   Ecuador (2).
   Cassytha filiformis L.; green dodder.   Puerto Rico (2).
   Cochliobolus sp.; on diseased leaf.   Mexico (2).
   Curvularia sp.; leafmold.   General (1, 2).
   Cuscuta americana L.; yellow dodder.   Puerto Rico (2).
   Dorylaimellus labiatus Thorne; root nematode.   Puerto Rico
      (2).
   Dorylaimoides sp.; root nematode.   Puerto Rico (2).
   Ephelis (see Balansia)
   Fusarium graminearum Schwabe; leaf disease.   S. America
      (1).   F. (? roseum); culm disease.   C. America (2).
   Fusicladium sp.; leafspot.   (2).
   Helminthosporium ocellum Faris; leafspot.   C. America (2).
      H. sacchari Butl.; large eyespot of leaf, rainyseason
      disease.   Subtropical USA, W. Indies, C. America (1, 2).
      Helminthosporium sp.; leafspot.   Widespread.   (1 in low
      plantings.)
   Leaf disease (coalescing spots and collapse), poorly formed
      perithecia, possibly a Didymella.   Guatemala (1).
   Meloidogyne sp.; rootknot nematode.   Panama (2).

Nigrospora sacchari (Speg.) Mason; leafspot, leaf edge
   disease. C. America (1, 2).
Oxydirus tenuicaudus Thorne; root nematode. Puerto Rico (2).
Phyllachora (? minutissima); small tarspots. Costa Rica (1).
Puccinia sp.; rust. El Salvador, Guatemala (2).
Pyricularia grisea Sacc.; leafspot, blast. W. Indies, C.
   America (1, 2). For perfect stage of Pyricularia see
   under ORYZA.
Pythium sp.; root disease. Highland wet season disease.
   Costa Rica, Guatemala (1).
Septoria penniseti Gonz. Frag. & Cif.; leafspot. W. Indies
   (2).
Sorosporium chardonianum Zund.; smut. El Salvador (2).
Viruses: Chlorotic streak, unidentified, in test in lowland
   grown field of Puerto Rico (2). Mosaic mottle and slight
   leaf malformations in hot land planting in Panama (1).
   Sugarcane Mosaic Virus in Subtropical USA, Panama,
   Argentina (2).

PERSEA AMERICANA Mill. synonyms P. GRATISSIMA Gaertn.
and P. DRYMIFOLIA Cham. & Schl. (1). A handsome fruit
tree, its origin is in northern sector of American tropics.
Aside from its well-established commercial renown, it satisfies
hunger for millions in the subtropics or warm tropics, bears
crops of fruits that have in them richly nutritive pulp; "aguacate,"
"palto" Sp., "abacate," "abacateiro" Po., "avocado," "alligator
pear" En., "palta," "curo," "aoacatl," "ahuacatl" Ind. P.
LINGUE Nees (2); the closely related central South American
tree, noted for its unique wood used in fine furniture and carved
ware, and for the specially commercially valuable tan-bark,
known as "chilean tan-bark" or "lingue."
   Abaphospora (see Melanopsamma)
   Acrothecium lunatum Wakk.; on leaves. S. America (1).
   Alternaria sp.; fruitrot. California (1). So far as known,
      not reported in other parts of the American tropics.
   Antennaria aequatorialis Speg.; black leafmold on wild Persea.
      Costa Rica.
   Anthostomella lingue Speg. (2) and A. perseicola (Speg.) Sacc.
      & Trott. (2); both leaf diseases, both in Chile.
   Armillaria mellea Vahl. = Armilariella mellea (Fr.) Karst.;
      rootrot. California, Ecuador (1).
   Aspergillus niger v. Tiegh.; fruitrot. S. America, C.
      America, W. Indies (1).
   Asterina delitescens Ell. & Mart.; black leafspot. Puerto
      Rico (1).
   Asteromella gratissima Petr. & Cif.; on leaf and twig. Do-
      minican Republic (1).
   Belonium valdiviensis Speg.; branch canker and on decorticated
      logs. Chile (2).
   Botryodiplodia theobromae Pat. also reported as Diplodia theo-
      bromae (Pat.) Now. (asco-state is Botryosphaeria quercuum
      (Schw.) Sacc.); stemend rot, fruit mummy, dieback.

spread (1).   (Asco-state also given by some as a Physa-
lospora.)
Botryosphaeria dothidea (Moug.) de Not. ; branch disease, fruit
    canker, gumming on bark.   Puerto Rico, Florida, Ecuador,
    Chile, Trinidad, Argentina (1).
Botrytis cinerea Pers. ; mold on leaves and flowers.   Chile
    (1).
Calothyrium apiahynum (Speg.) Bat. ; black leaf mold.   Brazil
    (1, 2).
Capnodium sp. ; fumagina or sootymold.   Scattered in general
    (1, 2).
Cephaleuros virescens Kunze; algal leafspot, jelly spot.
    General (1, 2).
Cercospora lingue Speg. ; fruitspot on (1), angular leafspot
    on (1, 2).   Peru, Chile.   C. perseae Ell. & Mart. ; leaf-
    spot, fruit and leaf blotch.   Peru, Venezuela, Ecuador,
    Bolivia, Subtropical USA, C. America (1).   C. purpurea
    Cke. = Mycosphaerella sp. (is the asco-state); angular
    small leafspot, fruitspot.   Bolivia, Chile, Nicaragua,
    Puerto Rico, Cuba, Guatemala, El Salvador (1).
Ceuthospora monocarpa Mont. and C. perseae Petr. & Syd. =
    perfect state is Physalospora chilensis Speg. ; on foliage,
    a wet season disease.   Chile (2).
Circinotrichum maculiforme Nees; leaf disease, Chile (2).
Cladosporium herbarum (Pers.) Lk. ; leaf-edge disease, fruit-
    spot.   Chile, Brazil, Bolivia, Guatemala (1).
Clasterosporium maydicum Sacc. ; leafspot.   S. America (1).
Clathrus cibarius Fisch. ; tan bark rot.   Chile (2).
Claudopus variabilis Pers. ; decay of bark, and wood.
    Chile (2).
Clitocybe tabescens (Scop.) Bres. = Armilariella tabescens
    (Scop.) Sing. ; rootrot.   Subtropical USA (1).
Colletotrichum gloeosporioides Penz. (sometimes reported as
    a Gloeosporium sp.), the asco-state is Glomerella cingu-
    lata (Stonem.) Sp. & Schr. ; withertip, dieback, leafspot,
    fast riperot of fruit.   Neotropical (1).   No doubt this
    plurivorous organism attacks (2).
Corticium salmonicolor Berk. & Br. ; branch pinkdisease.
    W. Indies, Costa Rica (1).
Criconemoides sp. ; rhizosphere nematode.   Chile (1).
Diaporthe citri Wolf (? = Phomopsis), fruitrot.   W. Indies (1).
Diatrype valdiviensis Speg. ; eruption disease of bark.   Chile
    (2).
Diplodia (see Botryodiplodia)
Dothiorella spp. ; trunk canker, fruitrots, leaf diseases.
    Widespread but scattered (1).
Ellisiela chilensis Speg. = Colletotrichopsis chilensis Bat. ;
    leaf disease.   Chile (2).
Endothia havanensis Bruner; branch and twig disease.   Cuba
    (1) (of interest as it is reported also in Japan).
Entsordaria perseicola Speg. = Anthostomella perseicola
    (Speg.) Sacc. ; leaf disease, see under Anthostomella (2).

Epicoccum neglectum Desm. ; on foliage.   Chile (1).
Fomes applanatus (Pers.) Gill. ; log and stump rot.   Chile
     (2).   F. lamaoensis Murr. ; honey-comb old tree woodrot.
     C. America (1).   F. leucophaeus Mont. ; log rot.   Chile (2).
Fracchiae heterogena Sacc. ; branch and twig disease.   Chile
     (1, 2).
Fusarium lateritium Nees, F. roseum Lk. , F. sambucinum
     Fckl. ; all appear to cause fruitrot, dieback, root-tip infec-
     tion, enlargement of leaf spots, branch canker.   Wide-
     spread (1).   F. oxysporum Schl. (there may be a special
     avocado race); rootrot, vascular invasion and wilt of large
     trees, gum extrusion at bark cracks, wilt.   Costa Rica
     (dry Pacific slopes), Dominican Republic (1).
Ganoderma australis Fr. ; root and trunk-base infection.
     Honduras (1), Chile (2).   Also G. pseudoferreum Ov. &
     Stein. ; root disease.   Scattered (1).
Glomerella (see Colletotrichum)
Glonium valdivianum Speg. = Psiloglonium valdivianum (Speg.)
     Urr. ; log-rot.   Chile (2).
Guignardia lingue (Speg.) Sacc. & Trott. = Laestradia lingue
     Speg. ; leaf, fruit, and twig infection.   S. America (1, 2).
     G. perseana Rang. ; leafspot.   S. America (1).
Helicotylenchus sp. ; nematode at root.   Surinam (1).
Helminthosporium fumosum Ell. & Mart. ; leafspot.   Florida,
     California, in El Salvador of rare occurrence (1).
Helotium leucopus Mont. ; on dead branch.   Chile (2).
Hendersonia sp. ; twigrot, leaf disease.   Argentina, Brazil (1).
Hydnum sp. ; root and stump rot.   C. America, S. America
     (1).
Hymenochaete subiginosum Lév. ; tan-bark decay.   Chile (2).
Hypocrea atrovirens Mont. ; on decorticated logs.   Chile (2).
Hypoxylon crocatum Mont. and H. valsarioides Speg. = H.
     sassafras (Schw.) Curt. ; both on decorticated logs, branch
     rot.   Chile (2).   H. subffusum Speg. ; on dead branch.
     Puerto Rico (1).
Irene perseae (Stev.) Toro; black mildew of leaf.   Florida,
     California, Puerto Rico, Jamaica, Colombia, may be in
     nearly all countries (1).
Irenopsis martiniana (Gaill.) Stev. ; black mildew.   Subtropical
     USA, C. America (1).
Lambottiella chilense Speg. ; leaf disease.   Chile (2).
Leaf bronzing from red spider attack.   Scattered in dry
     countries, and as in El Salvador may appear and then
     disappear depending on conditions (1).
Leptosphaerella lingue Speg. ; = Leptosphaeria lingue (Speg.)
     Sacc. & Trott. ; on foliage.   Chile (2).
Lophodermium hysterioides (Pers.) Sacc. ; on leaf and petiole,
     defoliation.   Chile (2).
Loranthus robustissimum Trel. ; large bushes parasitizing old
     tree tops.   El Salvador (1).
Macrophoma sp. ; gall-like swellings on twig and leaf.   Scat-
     tered in C. America (1).

Marasmius capillipes Speg. f. sp. chilensis Speg. and M.
 eburneus Speg. f. sp. chilensis Speg.; both are thread fungi,
 leaf attack.   Chile (2).
Melanopsamma microscopicum Speg.; wood discoloration.
 Chile (2).   M. valdiviensis Speg. = Abaphospora valdivien-
 sis (Speg.) Kirsch.; both wood disease and discoloration.
 Chile (2).
Meliola amphitrica Fr.; black mildew of leaf.   El Salvador,
 Guatemala (1).   M. perseae Stev.; black mildew.   Domini-
 can Republic (1).   M. perseae Stev. f. setulifera Speg.
 (synonyms: M. setulifera (Speg.) Stev., M. martiniana
 Gaill.); black mildew of leaves.   Florida (1).
Meloidogyne sp.; rootknot nematode.   Florida (1).
Metasphaeria valdiviensis Speg.; branch canker disease.
 Chile (2).
Microcyclus scutula (Berk. & Curt.) Sacc.; twig disease.   :
 Scattered (1).
Microthyrium microscopicum Speg. f. sp. minor Speg.; on
 leaves.   Chile (2).
Mucor sp.; on moist fallen leaves, occasional in fruit-rot
 complex.   Widespread (1).
Mycena citricolor (Berk. & Curt.) Sacc.; leafspot, defoliation.
 W. Indies, C. America, Bolivia, Peru (1).   M. galericu-
 lata Quél.; agaric disease at root.   Chile (2).
Mycosphaerella perseae Miles is asco-state of Phyllosticta
 perseae--see below.   M. tulasnei Jaez.; leafspot, occasional
 fruitspot.   Chile (1).
Nectria theobromae Mass.; twig canker.   Scattered (1).
Nigrospora oryzae (Berk. & Curt.) Petch; mild leaf and fruit-
 spots.   Widespread (1).
Oidium sp.; white powdery mildew.   Neotropics (1, 2).
Otthiella brenesii Petr.; on foliage.   S. America (1).
Paranthostomella valdiviana Speg.; leaf decay.   Chile (2).
Pellicularia koleroga Cke. (aerial adaptation); web-blight,
 leafrot.   Honduras, Costa Rica (1).
Penicillium expansum Lk., P. digitatum Sacc., P. italicum
 Wehm.; all cause fruit rots and molds in markets and
 transportation.   General (1).
Pestalotia guepini Desm.; fruitspot, leafspot, and seedling
 blight.   Florida, California, Texas, Puerto Rico (1).   P.
 leprogena Speg.; bud and graft union failure.   Peru, Brazil,
 (1).   P. neglecta Th.; leafspot.   Peru, Bolivia (1).
Phoma valdiviensis Speg.; branch lesion and rot.   Chile (2).
Phomopsis sp.; branch-tip disease, fruit stem-end rot.
 Ecuador, Trinidad, Puerto Rico (1).
Phoradendron dimidiatum Eichl., in Surinam and P. rensoni
 Trel., in C. America and Mexico; both are spectacular
 phanerogamic parasites on tree branches (1).
Phthirusa pyrifolia Eich.; drooping vine-like phanerogamic
 parasite on tree branches (1).
Phyalosporella chilensis Speg. = Ceuthospora chilense (Speg.)
 Petr.; leaf disease.   Chile (2).

Phycopeltis sp.; algal plaque on leaf.   Costa Rica, Puerto
  Rico (1).
Phyllachora gratissima Rehm; tarspot.   Venezuela, Brazil,
  Bolivia, Ecuador, Colombia, Guatemala, Honduras, Puerto
  Rico, Costa Rica, Dominican Republic (1).   P. maculicola
  Chard.; tarspot.   Brazil, Venezuela (1).
Phyllosticta micropuncta Cke.; small leafspots.   In moist
  areas of Subtropical USA, W. Indies, C. America (1).
  P. perseae Ell. & Mart. is the conidial state of the ascus
  bearing Mycosphaerella perseae Miles; large leafspot, leaf-
  blotch. Florida, Mexico, W. Indies, C. America (1).   P.
  persearum Bat. & Vit.; leafspot.   Brazil (1).
Phymatotrichum omnivorum (Shear) Dug.; dryland rootrot.
  Texas (1).
Physalosporella chilensis Speg. ? = Ceuthospora perseae Petr.
  & Syd., is synonym of Physalospora perseae Doidge; leaf,
  fruit, stem disease.   Widespread (1).
Physiological diseases occur under special conditions, are
  often seen, sometimes incorrectly diagnosed, have been
  studied only on (1), and have wide range of effects
  examples of which are: copper deficiency causes dieback,
  dieback caused by tree exhaustion after extra heavy crop;
  fruit skin or carapace apparently effected by a hardened
  bruise, fruitcracks are from unseasonable weather.   Fruit-
  end drying from overmaturity.   Iron deficiency causes
  chlorosis and weakened growth.   Leaftip burn from accumulation of salts occurs in poorly drained fields.   Leaf sterile-
  spots from numerous still undetermined causes.   Mottling of
  leaf from several physiological disturbances.   Nitrogen deficiency causes yellowing while nitrogen toxicity from heavy
  fertilizer applications may cause leaftip necrosis in young
  leaves and develops into excessive green color and arrested
  growth.   Russetting (not from insects) of leaves and fruits
  from undetermined physiological disturbance at time fruits
  are over half grown.   Salt toxicity near seashore may
  cause ragged leaf and chlorosis and necrosis.   Sulfur
  toxicity in volcanic regions causes leaf drop.   Water-logging
  injury results in root failure and wilting.   Zinc deficiency
  causes little-leaf and rosette.
Phytophthora cactorum (Leb. & Cohn) Schr.; trunk canker,
  collar rot.   Mexico, California, Puerto Rico, Surinam,
  Guatemala, Costa Rica (dry area), Grenada, Honduras (1).
  P. cinnamomi Rands; decline, canker, rootrot, slow death,
  often most severe in crop planted on soil with hard pan.
  Scattered in many countries.   Both (1) and (2) susceptible.
  P. palmivora Butl.; leaf infection, seedling blight, occa-
  sional branch canker by infection through side twig.   Scat-
  tered in Mexico, Subtropical USA, C. America (1).   P.
  parasitica Dast.; collar rot, branch canker.   Florida,
  Puerto Rico (1).
Polyporus hirsutus Wulf. and P. mutabilis Berk. & Curt.;
  both may cause stump- and heart-rot of old standing trees.
  W. Indies, C. America (1).

Polystictus aculeifer Berk. & Curt., P. versicolor Fr.; seem
    equally the cause of branch-and-log rot. Chile (2).
Porella sp. (leafy liverwort); leaf smother. Under very moist
    conditions in Puerto Rico and Guatemala (1).
Poria sp.; branch decay. C. America (1). P. vaporaria
    Sacc. causes tan-bark, branch and log decay. Chile (2).
Pratylenchus sp.; rhizosphere nematode inhabitant. Chile,
    Surinam, Florida (1).
Pseudomonas syringae Van Hall; fruitspot. Colombia, Cali-
    fornia (1).
Pseudoplea trifolii (Rastr.) Petr.; rare leafspot. C. America
    (1).
Psittacanthus calyculatus Don; large bush parasite on main
    branches of host tree. El Salvador (1).
Pyrenopeziza araucania Speg.; branch and bark rot. Chile (2).
Pythium ultimum Trow., and P. vexans dBy.; both cause
    rootrot. California, Costa Rica (1).
Radopholus similis Thorne; root zone nematode. Florida,
    Guyana, Trinidad, Tobago (1).
Rhizoctonia solani Kuehn is non-sporulating state of the
    basidial Thanatephorous cucumeris (Frank) Donk, but the
    sporeless stage causes seedling stem disease and as well
    infects the seed fleshy endosperm in germination seedbed.
    Guatemala, El Salvador, Costa Rica, Colombia, California,
    Puerto Rico, and probably other countries on (1).   Rhizoc-
    tonia sp., aerial web-blight on young shoots in nursery.
    El Salvador (1). (This second Rhizoctonia was much coarser
    in hypha character than the so-called microsclerotia type.)
Rhizopus nigricans Ehr.; fruitrot. Chile, but scattered in C.
    America and found on one fruit in Florida (1).
Rosellinia arcuata (Hartig) Petch; rootrot. Costa Rica,
    Panama (1). The more common R. bunodes (Berk. &
    Curt.) Sacc.; a black rootrot. Jamaica, Puerto Rico,
    Guayana, St. Vincent, Haiti, Peru, Cayman Island (1).
    R. necatrix Prill. (has been reported as Dematophora sp.);
    rootrot. Guatemala, Honduras, Puerto Rico, Panama (1).
    The restricted R. pepo Pat.; a white rootrot. C. America,
    Ecuador (1). R. valdiviensis Speg.; branch decay. Chile
    (2).
Schizophyllum sp.; weakly parasitic fungus causing secondary
    expansion of branch decay following other organisms. C.
    America, S. America (1).
Sclerotinia sclerotiorum (Lib.) dBy. = Whetzeliana sclero-
    tiorum (Lib.) Korf & Dumont; seedling collar-rot. Cali-
    fornia (1).
Sclerotium rolfsii (Curzi) Sacc. = Pellicularia rolfsii (Sacc.)
    West which is the spore-producing form; seedling collapse,
    ground blast. El Salvador, California, Peru, Florida (1).
Spegazzinia ornata Sacc.; leaf disease. Argentina, Brazil (1).
Sphaceloma perseae Jenk. (reported by some as Sporotrichum
    citri Butl.); spot anthracnose, fruit scab, leaf scab-spot.
    Distribution is widespread (1).

Sphaeropsis sp.; fruit-skin disease.   Brazil (1).
Sphaerostilbe cinnabarina Tul.; fruit canker.   Brazil (1).
Stereum hymenoglium Speg.; tan-bark and branch infectious
     fungal growth.   Chile (2).
Stilbella sp.; secondary invasion of branch-tip following die-
     back.   Bolivia (1).
Stomatochroon lagerheimii Palm; algal systemic leaf disease.
     Guatemala (1).
Stomiopeltis sp.; leaf-smudge.   Brazil (1).
Strigula complanata Fée; lichen-caused milkyspot on leaf.
     Not reported in Subtropical USA but otherwise probably
     distribution is about Neotropical (1), and probably also on
     (2).
Trichoderma lignorum (Tode) Harz; rootrot, part of branch
     disease and dieback complex.   Scattered (1).
Trichopeziza punctiformis (Fr.) Fckl.; canker.   Chile (2).
Triposporium perseae Speg.; leaf infection.   Chile (2).
Trochila perseae Speg.; leaf disease.   Chile (2).
Tryblidiella rufula (Spreng.) Sacc. = Rhytidhysterium rufulum
     (Spreng.) Petr.; following anthracnose dieback.   Wet area
     in Puerto Rico, moist hillside in El Salvador (1).
Tylenchorhynchus sp.; found as a rhizosphere nematode.
     Chile (1).
(Uromyces placentula Berk. and Uredo fruticola P. Henn. are
     rusts of tree relatives of the hosts in the present list but
     rust has not been recorded thus far on either avocado or
     lingue.)
Valsa sp.; death of seedlings in two nurseries.   Costa Rica (1).
Verticillium albo-atrum Reinke & Barth.; one-sided wilt and
     then quick death, discolored vasculars.   California, Chile,
     Ecuador, Guatemala, Bolivia (1).
Virus: Avocado Pear Sunblotch Virus in California, Mexico,
     Puerto Rico, C. America, Peru, probably elsewhere (1).
Xiphinema sp.; rhizosphere nematode collection.   Chile (1).
Xylaria sp.; root disease.   C. America (1).

PETROSELINUM CRISPUM (Mill.) Nym.   An herb of European intro-
     duction noted for its attractive finely divided, flavorsome leaves,
     used in cooking and garnish, also grown as ornamental pot
     plants; "perejíl" Sp., "salsa" Po., "parsley" En.
     Alternaria dauci (Kuehn) Groves & Skolko; leafspot, blight.
          Puerto Rico.
     Capnodium (? citri); sootymold on old pot plant near orange
          tree.   Guatemala.
     Cercospora apii Fres.; leafspot.   El Salvador, Guatemala,
          Costa Rica, Puerto Rico, Colombia, Venezuela.
     Chaetomella atra Fckl.; weak parasite.   (From a locality in
          S. America.)
     Ditylenchus dipsaci Filip.; stem nematode.   Florida.
     Erwinia aroideae (Town.) Holl., and E. carotovora (L. R.
          Jones) Holl.; both bacterial market softrots.   C. America,
          probably Neotropical.

Erysiphe polygoni DC, its conidial state is Oidium erysiphoides Fr.; common white powdery mildew.　On old plants
protected from rains or in dry season.　Widespread.

Fusarium (? roseum); damping-off and plant base-rot.　El
Salvador, Costa Rica.

Galera crystallophora Speg.; questionably parasitic, seen on
soil under host leaves.　Argentina.

Meloidogyne sp.; rootknot nematode.　Jamaica, Subtropical
USA (not much reported since host root is not often examined in tropics, probably somewhat well distributed).

Oidium (see Erysiphe)

Plasmopara nivea (Ung.) Schr.; downymildew.　Argentina.

Puccinia petroselini Lindr.; leafrust.　Argentina (? nearby
countries).

Pythium mastophorum Drechs., and P. ultimum Trow.;
damping-off.　Widespread.

Rhizoctonia solani Kuehn the parasitic non-sporulating stage
of the basidial Thanatephorous cucumeris (Frank) Donk;
causes damping-off, bottom rot.　Texas, Florida, El
Salvador, Panama (probably many other countries).

Sclerotium sclerotiorum (Lib.) dBy. = Whetzeliana sclerotiorum (Lib.) Korf & Dumont; leaf and stem rot.　Subtropical USA.

Septoria petroselini Desm.; leaflet blights.　Barbados, California, Texas, Puerto Rico, Mexico, Chile, Argentina,
Bolivia, Uruguay.

Stemphylium radicinum Neerg.; leafspot.　Argentina.

Viruses: Aster Yellows Virus in California.　Virulent
Cucumber Mosaic Virus race Celery-Commelina in Florida,
El Salvador, Cuba.　Beet Curly Top Virus in California.
Sunflower Virus (Argentine) in Argentina.

PETUNIA HYBRIDA Vilm.　This is a plant, mostly of Argentinian
origins, of a complex nature from artificial crosses of wild
species, thereafter selected and purified, it is a much used
profusely flowering annual; common names "petúnia" Sp. Po.,
and "petunia" En.

Alternaria sp.; blight of leaf and flower stalks.　Costa Rica.

Ascochyta petuniae Speg.; leafspot.　Scattered in many
countries.

Botrytis cinerea Pers.; damping-off, young plant basal rot.
Puerto Rico.

Cassytha filiformis L.; green and pungent smelling dodder.
Puerto Rico, St. Johns.

Cercospora canescens Ell. & Mart.; leafspot.　S. America.
C. petuniae Chupp & Muller; leaf blotchy-spot.　Venezuela,
Guatemala, Florida, Puerto Rico, Brazil.

Choanephora conjuncta Couch; flower petal disease.　Costa
Rica.

Corynebacterium fascians (Tilf.) Dows.; fasciation, bacterial
systemic invasion.　California, Guatemala.

Cuscuta americana L.; yellow brittle-dodder.　C. America.

Entyloma petuniae Speg. ; leafsmut, sori make pale leafspots. Argentina.

Erysiphe polygoni DC. = Oidium

Fusarium roseum Lk. ; damping-off. (Isolations made in Costa Rica and Puerto Rico, but it is much more widespread.)

Meloidogyne sp. ; nematode rootknot. Subtropical USA, Puerto Rico.

Mycena citricolor (Berk. & Curt.) Sacc. ; leafspot (by inoculation). Costa Rica.

Oidium erysiphoides Fr. (= the perfect state is Erysiphe polygoni DC. , but this is seldom seen); white powdery mildew. Costa Rica, Guatemala, Florida, Brazil (probably is widespread).

Physiological scorch of leaf, cause not understood. Puerto Rico.

Puccinia aristidae Tracy; leafrust. Dry parts of Subtropical USA, said to be in Mexico but no confirmation.

Pythium? - Phytophthora? (borderline morphology). Damping-off of large seedlings in rainforest area. Puerto Rico.

Rhizoctonia solani Kuehn is the sterile but parasitic mycelial state of the basidial Thanatephorous cucumeris (Frank) Donk; damping-off and stemrot. Probably Neotropical, reports and seen: Texas, Florida, Costa Rica, St. Thomas, El Salvador, Guatemala, Panama, Colombia.

Thielavia basicola Zopf; black rootrot. Subtropical USA.

Viruses are limiting factors in seed production. A number of these diseases have been determined and reported: Beet (Argentine) Curly Top Virus in Argentina. Cucumber Mosaic Virus causes relatively mild mottle, rather common distribution, but scattered. The virulent Cucumber Mosaic Virus, strain of Celery-Commelina, that on petunia is significantly most severe, results in stunting and with marked malformation in Cuba, Florida, Puerto Rico. Leafroll (unidentified virus) in Puerto Rico, Cuba. Potato Virus Y in Peru. Common Tobacco Mosaic Virus, has wide Neotropical distribution. Tomato Ring Spot Virus (Brazilian) in Argentina.

PHASEOLUS VULGARIS L. A native of the American tropics, is an annual bush to vine type plant (of which there are hundreds of varieties), that produces under a remarkably wide range of climatic conditions and is the best known, most important single source of vital protein for millions of human beings in the American tropics; examples of common names: "frijól, " "poroto, " "caraota, " "habichuela" Sp. , "feijoeiro, " "feijáo, " "gráo, " "vagem" Po. , "bean, " "dry-bean, " "green-bean, " "kidney bean" En. , and literally hundreds of local Indian names. (My notes and observations developed information on difference in diseases on beans under a wide range of tropical conditions. In addition E. Echandi has correlated, with notable success, bean diseases as related to recognized ecologic zones as defined by soils and forestry scientists. It is possible now to indicate

climatic effects on some diseases and these effects are given
within brackets under the disease item.)

Alga, green and non-filamentous disfiguring and spoiling exposed seed in old ripe but not dry pods lying on soil.   El
Salvador.

Alternaria alternata (Fr.) Keiss. = A. tenuis Nees; grain
spot and flavor change.   Mexico, C. America, Venezuela,
Colombia.   [Grain spot increases with unexpected rains at
harvest.]   A. brassicae Sacc. f. sp. phaseoli Bruner
(sometimes reported as A. fasciculata Jones & Grout);
rare leafspot.   Mexico, Chile, Florida, Brazil.

Aphelenchulus radicicolus Steiner; root nematode.   Subtropical
USA.

Arcyria sp.; epiphytic myxomycete giving confusion to growers
who feared disease on crop.   El Salvador.

Aristostoma oeconomicum (Ell. & Tracy) Tehon; leafspot.
Honduras.

Ascochyta boltshauseri Sacc.; pod disease, leafspot with concentric pattern.   W. Indies in hills, C. America.   [Most
severe in wet and cool areas.]   A. phaseolorum Sacc.;
leafspot, pod disease.   El Salvador, Panama, Guatemala,
Bolivia, Colombia, Peru, Brazil.   A. pisi Lib.; leaf
disease.   Venezuela.

Bacteria from soil, secondary rot of planted bean seed.   Neotropical.

Basidiomycete (unidentified, with white mycelium in pure culture); parasite on leg and petiole causing branch collapse.
El Salvador, Guatemala.

Belonolaimus longicaudatus Rau; root nematode.   Florida.

Boron deficiency, stunt and flower abortion.   Scattered.

Botryodiplodia natalensis Evans = B. theobromae Pat. with
the asco-stage Botryosphaeria quercuum (Schw.) Sacc.;
attack of old main stalk, tissue blackening.   Mexico, C.
America.   [Apparently in rather dry areas.]

Botrytis cinerea Pers.; gray-mold blight of shoot tips and
flowers.   Subtropical USA, Peru, Trinidad, El Salvador.
[In El Salvador seen as specially severe and common
late in the rainy seasons in hillside fields.]

Calcium deficiency result is bleached chlorosis, slight necrosis, bud drop.   Florida.

Calonectria sp.; stem-disease in old field area.   El Salvador.

Cephalosporium gregatum Alling. & Chamb.; stemrot.
Ecuador.

Cercospora spp. (a number reported, nine given here, all
cause leafspots and a few attack pods) as follows: 1. C.
canescens Ell. & Mart.; very common leaf and pod spotting.
Subtropical USA, Puerto Rico, Trinidad, Jamaica, Brazil,
in Venezuela in its central provinces.   2. C. caracallae
(Speg.) Chupp is also given as Cercosporina caracallae
Speg.; leafspot.   Puerto Rico, Venezuela, Brazil, Argentina.   3. C. cruenta Sacc. its perfect state is Mycosphaerella cruenta Lath.; leafspot.   Widely distributed.   4. C.

griseola Sacc. = C. columnaris Ell. & Ev.; a gray leafspot. C. America. 5. C. phaseolina Speg. (sometimes
spelled C. faseolina); a severe leafspot. Argentina, Paraguay. 6. C. stizolobii Syd.; leafspot. W. Indies, El
Salvador. 7. C. stuhlmannii P. Henn. reported sometimes,
is probably C. columnaris or its synonym C. griseola.
8. C. vanderysti P. Henn. is believed to be the most
severe gray-blotch organism. C. America, Colombia,
Venezuela, Brazil, probably Neotropical. [Most severe in
C. America in areas with lowland moist ecology, in Brazil
reported most severe in the warm moist season.] 9. C.
zonata Wint.; leafspot, sometimes involving large sections
of leaf lamina. Brazil.

Chaetomium sp.; on weathered bean straw. El Salvador.

Chaetoseptoria wellmanii J. A. Stev.; round gray leafspot.
Mexico, all C. America, Panama, W. Indies, symptoms
seen in Venezuela, ? Colombia. [In C. America it is
most severe in locations of from medium low to medium
high elevations and worst when rains are broken by short
dry periods.] Probably all species of Phaeolus are susceptible to this organism.

Chlorosis problems, physiological causes, dependent on soil
content: chlorosis from deficiencies of copper, iron,
nitrogen, zinc and manganese are examples. Study will
separate such effects from chloroses due to viruses or
microorganisms.

Cladosporium herbarum Pers.; podspot and at times severe
leaf disease. Fairly general distribution but especially in
Chile, Brazil, Peru, and C. America. [This fungus often
at best a weak parasite on many crops is sometimes severe
on beans, and especially in C. America when pods are on
vines but kept in piles for one to three weeks before
thrashing. Has been at times a major severe disease in
Brazil.]

Cold injury symptoms are stunt and stem swelling. Guatemala.

Colletotrichum lindemuthianum (Sacc. & Magn.) Briosi & Cav.,
also listed as Gloeosporium linemuthianum Sacc. & Magn.
(the perfect state not reported?); leaf, stem, pod and seed
anthracnose lesions. Occurrence Neotropical. [Using contaminated and infected seed, the disease is most severe in
moist-warm areas, rare under dry conditions where irrigation is used, in C. America plantings most susceptible
are at medium low elevations.] C. truncatum Andrus &
Moore = C. dematium (Pers.) Grove f. truncata (Schw.)
Arx; blotch, stem anthracnose. Nicaragua, Venezuela,
Puerto Rico.

Copper deficiency causes elongated leaves, slender and brittle
stems. Florida.

Corticium salmonicolor Berk. & Br. (by inoculation) plant rot.
Costa Rica.

Corynebacterium flaccumfaciens (Hedges) Dows.; rather rare
bacterial wilt. Mexico, C. America.

Criconemoides sp.; found as a rhizosphere nematode. Chile.
Curvularia sp.; secondary leafspots that are enlarged after
  start by anthracnose, and following insects. Venezuela,
  El Salvador.
Cytospora-Valsa, isolation from old vines. El Salvador.
Diaporthe phaseolorum Sacc. is the perfect state of the conidial organism, Phoma subcircinata Ell. & Ev.; pod, leaf
  and stem disease. Widespread. [Rarely as severe epi-
  demic, cool-moist conditions favor disease, but occurs in
  fields at both low and high elevations.] D. sojae Lehm.;
  on pod and stem. Weak parasite. Honduras.
Diplodia (see Botryodiplodia)
Entyloma sp.; rounded angular leafspot, smut inside the leaf
  tissues. W. Indies, Guatemala, Honduras.
(? Entyloma sp., not the above.) Smut sori buried in leaf
  lamina with symptoms similar to above but spots more
  rounded, spores different, mycologists undecided about
  disposition, may require a new genus. Collected by E.
  Echandi in Costa Rica.
Erwinia carotovora (L. R. Jones) Holl.; green-bean slimy
  softrot. Common under poorly conducted transport and
  market conditions. Neotropics.
Erysiphe (see Oidium)
Eudarluca australis Speg. = E. caricis (Fr.) O. Eriks., is
  perfect state of the rust parasite Darluca filum (Biv.-
  Bern.) Sacc. noted at times on bean rust.
Eutypa phaseolina Mont.; parasitic on old dying stems.
  Trinidad, Guatemala-El Salvador brought to clinic from
  field at national boundary.
Fusarium spp. (about nine of these bean parasites) on vari-
  ously diseased plants from warm, not necessarily hot,
  areas. Certain of these organisms determined as follows:
  1. F. culmorum (W. G. Sm.) Sacc.; from dying plant
  stem. El Salvador. 2. F. equiseti (Cda.) Sacc.; special
  type of damping-off. Colombia, Peru. 3. F. lateritium
  Nees; stem canker. Costa Rica. 4. F. oxysporum Schl.
  f. sp. phaseoli Kendr. & Sny. (apparently no perfect state);
  yellows-type wilt. C. America, Panama, Colombia. 5. F.
  oxysporum Schl. f. sp. vasinfectum (Atk.) Sny. & Hans.;
  a vascular invader often fatal in small field areas. Mexico,
  Honduras, W. Indies, Subtropical USA, Peru, Guatemala,
  Nicaragua. 6. F. roseum Lk. is often recovered and reported, a variable and complex organism; secondary leafspot enlargement, insect feeding lesion infection, pod spot,
  seed spot, change of bean flavor, damping-off, stem and
  branch canker-like lesions, rootrot, flower petal brown
  rot, from dry bean straw months after harvest. El Salvador, Honduras, Costa Rica, Peru, Colombia, Argentina,
  probably in all countries and islands. 7. F. semitectum
  Berk. & Rav.; from tip-decay of mature pods recently
  harvested. Honduras, El Salvador. 8. F. solani (Mart.)
  Sacc. with its asco-state Nectria haematococca Berk. &

Br. also given as Hypomyces solani Reinke & Berth.; dry-rots of stem and pod. Chile, Peru. 9. F. solani (Mart.) Sacc. f. sp. phaseoli (Burkh.) Sny. & Hans. with its perfect state N. haematococca as above; one of the most serious diseases, wilt from stem and root rot, discrete dryrot canker on stem, pod infection, has been recovered as spores on seed surfaces. Special isolation studies and reports from Subtropical USA, W. Indies, Costa Rica, Mexico, Chile, Peru, Ecuador, Venezuela, Panama, El Salvador, Nicaragua, Brazil. [Losses most severe in warm, distinctly friable, slightly acid, low-lying but not wet areas.]

Gloeosporium (see Colletotrichum)

Helminthosporium sp.; secondary enlargement of other leaf-spots. Venezuela, Subtropical USA (not seen in C. America or W. Indies).

Heterosporium sp.; confusing but almost a superficial leaf fungus. Chile, C. America.

Iron deficiency causes very pale leaf chlorosis with green veins, poor set of pods. Scattered in C. America.

Isariopsis griseola Sacc. (an apparent segregate is named Phaeoisariopsis griseola Ferr.), in S. America it has been some times reported incorrectly as Cercospora columnaris Ell. & Ev. = C. griseola Sacc.; on mature leaves typical gray angular leafspot, attacks pods and stem, severe defoliation. Mexico, W. Indies, C. America, Panama, Colombia, Peru, Ecuador, Venezuela, Brazil, probably in nearly every country of the warm American tropics. [Most notable and severe on medium low but moist hillsides, seen by me occasionally on high wet hills, found in C. America and W. Indies most on warm mountain sides and on table lands. Disease reduced undershade. For some reason, mild in occurrence in Subtropical USA.]

Leptosphaeria phaseolorum Ell. & Ev.; stem disease. Puerto Rico.

Macrophomina phaseoli (Taub.) Ashby = Sclerotium bataticola Taub.; root and stem charcoalrot. Subtropical USA, Mexico, C. America, Peru, Brazil, Colombia, Venezuela, probably more widespread. [It is considered sporadic, but is almost confined to areas of high temperature, dry soils and high air humidity. In C. America it is severe at times in dry forest type areas at sea level and up to about 700 meters elevation.]

Magnesium deficiency causes leaf bleach and thin lamina. Puerto Rico, Costa Rica.

Manganese deficiency causes yellowish chlorosis of whole plant, bud drop, killing most severely affected plants. Florida, Honduras.

Meliola sp.; hyperparasite on Chaetoseptoria wellmanii and on Colletotrichum gloeosporioides. Costa Rica.

Meloidogyne sp. sometimes reported as M. incognita Chitw.; rootknot nematode. Brazil, Subtropical USA, Costa Rica, Jamaica, Chile, Peru.

Mycena citricolor (Berk. & Curt.) Sacc. leafspot. [Scattered,
within splash-drop distance of other diseased hosts in shady
conditions.] By inoculation I caused leafspot in Costa Rica,
but also found later in field, natural field occurrence in
rainforest area of Puerto Rico and Costa Rica.

Mycosphaerella spp., these are considered perfect states of
Cercospora spp. deserve much intensive research in the
tropics; cause large leafspots and blotches, M. cruentha
Lath., in Florida, Texas and M. diversa Rangel, in Brazil.

Myrmaecium roridum Tode; pod disease.   El Salvador,
Guatemala.

Nematospora coryli Pegl. = N. phaseoli Wing.; rare bean-
seed yeast-spot follows insect pod punctures.   Florida, W.
Indies, Costa Rica, Ecuador, Peru, Brazil.

Nitrogen deficiency causes poor growth and general yellowing.
Scattered in Neotropics.

Oidium balsamii Mart. or O. erysiphoides Fr. is the readily
found conidial state, its perfect state is Erysiphe polygoni
DC.; white powdery mildew.   Is probably Neotropical in
occurrence.   Reports from: Subtropical USA, W. Indies,
C. America, Ecuador, Chile, Peru, Colombia, Uruguay,
Argentina, Brazil, Paraguay, Venezuela. [In C. America
and W. Indies, is in fields from low to high elevations, in
wet through moist to dry. Most severe effect seen in El
Salvador and Guatemala in semi-dry areas in low fields
during dry season.   In Costa Rica the disease is absolutely
absent in chronically wet area, heavy rains washed away
spores.   In Brazil more severe at medium high elevations,
in Argentina it has been noted that beans are most severely
injured when mildewed at their most vegetative stage.]

Ovularia phaseoli Drum. = Ramularia phaseolina Petr. or R.
phaseoli Deigh.; mealy spot, white mealy leafspot.   Brazil,
Colombia, Venezuela, Ecuador, Honduras, Panama, Guatemala, Dominican Republic, Nicaragua.

Parodiella perisporioides (Berk. & Curt.) Speg. is also reported as Dimerium grammodes (Kuntze) Garman; a black
mildew.   Scattered reports in Puerto Rico, Mexico, C.
America, S. America.

Peduncle weakness, severe pod drop, physiological but not
much studied.   Scattered, noted as specially severe in dry
season in El Salvador.

Penicillium spp.; certain of them are definitely parasitic on
senescent leaves and on grains.   Storage spoilage.   Scattered occurrences, under moist primitive conditions.

Periconia pycnospora Fr.; pod disease.   El Salvador.

Phoma (see Diaporthe)

Phosphorus deficiency causes crop failure, also in some soils
leaves are small and brittle.   Costa Rica.

Phyllachora phaseoli (P. Henn.) Th. & Syd.; tarspot.   Puerto
Rico, Dominican Republic, scattered in C. America,
probably in other countries.

Phyllosticta noackiana All. (sometimes as P. noackianum All.);
leafspot.   Brazil, Argentina, Paraguay.   P. phaseolina

Sacc.; a common leafspot, blotch, podspot in Subtropical
USA, Costa Rica, Nicaragua, El Salvador, Guatemala,
Peru, Brazil, Argentina, Puerto Rico.
Phymatotrichum omnivorum (Shear) Dug.; dryland rootrot.
Texas, Arizona (? possible spread to Mexico).
Physarum sp.; epiphytic myxomycete on living plant stems and
on debris. W. Indies, C. America.
Phytophthora nicotianae de Haan var. parasitica (Dast.)
Waterh.; common preemergence damping-off, normal
damping-off, podrot. W. Indies, S. America. P. para-
sitica Dast.; stem canker, pod diseases, downy mold on
leaves especially during flowering. Rare in Mexico, com-
mon in Cuba, especially noted in El Salvador, Venezuela,
Puerto Rico, Peru, Panama, probably widespread. P.
phaseoli Thaxt.; downy mildew. Mexico, C. America, W.
Indies, Subtropical USA, Brazil, Paraguay, Argentina. P.
terrestris Sherb.; podrot. Panama.
Potash deficiency causes weak growth and buddrop. Costa
Rica.
Pratylenchus sp.; nematode feeding in root zone. Chile.
Pseudomonas garcae Am., Teix., & Pinh.; large leafspot
(by inoculations?). (P. phaseoli, see Xanthomonas.) P.
phaseolicola (Burkh.) Dows. = P. medicaginis Sacc. var.
phaseolicola Stapp & Kotte; common halo blight, blast,
greasy spot was severely increased by presence of thrips.
Widespread. [One of the rainyseason diseases, most
severe at warm but not hot temperature levels, not important in dry, cool lands.] P. solanacearum E. F. Sm.;
wilt, occasionally in subtropical USA, Puerto Rico, C.
America, S. America. [Invariably found where, in the
season or year before, a crop has been grown that had the
disease. Most common and severe in warm areas, under
best growing conditions for beans.] P. syringae Van Hall
= P. viridifaciens Tisd. & Williams; bacterial leafspot
and podspot, black-stem, rot of mature pod where in contact with soil. [Occurs in many countries but is seldom
if ever spread over whole fields.]
Pythium spp., although general disease symptoms are readily
recognized, organisms are not easily determined as to
species but reports of five are: 1. P. aphanidermatum
(Edson) Fitz. = P. butleri Subr.; a rootrot and wilt,
green-pod rot. 2. P. debaryanum Hesse; rootrot and wilt.
3. P. helicoides Drechs.; rootrot. 4. P. oligandrum
Drechs.; rootlet and mainroot rot. 5. P. ultimum Trow.;
rootrot, wilt podrot. Species 3 and 4 are specially described from Florida but it is still impossible to know
whether any of these five species are limited in distribution, and they may all be widespread. [Generally speaking,
pre-emergence and seedling damping-off is more severe in
a wet and cool climate. Possibly 5 is more serious as a
root parasite in warmer-moist areas. Where beans are
harvested and sold as "green beans" the market podrot

known as "nesting" from 1 and 5 is increased when harvest
is in the rainy season or picking is done with dew on the
plants.]
Ramularia (see Ovularia)
Rhizobium phaseoli Dang. ; nodule-forming bacterium is a
  "disease" in the true sense but it is much desired.
  General in distribution. If it does not occur it must be
  introduced.
Rhizoctonia dimorpha Matz; a true plant rot.   Puerto Rico
  and apparently confined to that island.   R. microsclerotia
  Matz; the often severe web-blight under certain conditions;
  Florida, scattered in W. Indies and C. America, Colombia,
  some areas in S. America.   [Most severe in moist medium-
  low mountain valleys, in exposures partly shaded in a por-
  tion of the day, it is a rainyseason disease.]   R. solani
  Kuehn, perfect state given is Corticium solani (Prill. &
  Del.) Bourd. & Galz., or Pellicularia filamentosa Rog. or
  Thanatephorous cucumeris (Frank) Donk; the destructive
  and common damping-off, stem and pod canker, zonal leafspot.   Widespread.   [In low fields to medium low table
  lands, less severe in slightly acid soils.]
Rhizopus stolonifer (Ehr.) Lind; market-rot of greenbeans.
  Neotropics.
Sclerotinia sclerotiorum (Lib.) dBy. = Whetzeliana sclerotiorum (Lib.) Korf & Dumont; market water-rot, white-
  mold and occasional seedling damping-off, attack at times
  in field on shoots, flowers, pods.   Subtropical USA, Mexico,
  C. America, Panama, Colombia, Ecuador, Peru, Chile,
  Brazil, probably widespread.   [Most common in warm-wet
  fields, at medium high to lower moist elevations, a rainy-
  season disease.]
Sclerotium rolfsii (Curzi) Sacc. = the perfect state is Pelli-
  cularia rolfsii (Sacc.) West; seedling blast and soil-line
  rot or footrot.   Subtropical USA, Mexico, Honduras, Costa
  Rica, El Salvador, Nicaragua, Surinam, Venezuela, Peru,
  Colombia, Brazil.   [Medium low to low lowlands, rains
  increase severity.]
Septoria phaseoli Maubl. ; leafspot.   Brazil.
Sootymold (undetermined organism) on insect honeydew.   Costa
  Rica.
Stemonitis sp. ; epiphytic fungus at base of plants.   Added confusion to disease diagnosis with bacterial wilt and web
  blight.   Costa Rica.
Stemphylium botryosum Wallr. ; leafspot.   Peru.
Stunt, caused by injected toxin associated with feeding of an
  insect, the Empoasca.   El Salvador, Puerto Rico, probably
  more widespread.
Sun scorch when shade is suddenly removed as in weeding,
  most severe killing of leaves on western slope field exposures, small spots develop in less severely sun scorched
  leaves in El Salvador.
Thielaviopsis basicola (Berk. & Br.) Ferr. ; black rootrot.
  Subtropical USA.

Tylenchorhynchus sp. ; found as a rhizosphere nematode.
  Chile.
Uromyces phaseoli (Pers.) var. typica Arth. which has some
  synonyms: U. appendiculatus (Pers.) Fr. , U. phaseolorum
  dBy. , Puccinia phaseoli Rebent. , U. phaseolicola Speg. ;
  this is the common and important leafrust with often some
  pustules on shoots and pods. Neotropics. [Most severe
  in not cold but cool regions, during the moist season. In
  C. America and W. Indies at medium high elevations on
  hillsides facing the Atlantic or the Caribbean.]
Vermicularia polytricha Cke. and V. truncata Schw. (? =
  Colletotrichum sp.); podrot. El Salvador, reported to me
  by fungicide agents in Guatemala and Panama.
Verticillium albo-atrum Reinke & Berth. ; root and shoot
  disease. El Salvador, Peru, Chile.
Viruses [There are about 20 virus diseases, in some instances
  they are disseminated by seeds, but mostly are carried by
  insects. Dry areas with irrigation or where rainfall is
  moderate, is where insects are most plentiful. In wet
  areas, or where planting is so timed that the crop develops
  in its first stages during rains, the insects are practically
  eliminated and with them the viruses. Beans as a whole
  are quite susceptible to virus diseases, although there are
  many locally grown varieties that are tolerant]: 1. Abutilon
  Mosaic (by white fly inoculation) in Brazil. 2. Bean Common Mosaic (severe, it is seed and insect transmitted) is
  widespread. [Most prevalent in low, semi-dry areas.]
  3. Bean Common Mosaic var. Bean Curly Leaf (leaf hopper
  transmitted) in El Salvador, Ecuador. 4. Bean Curly Top,
  unidentified (white fly transmitted) in El Salvador. 5. Bean
  Leaf Mottle, unidentified (white fly transmitted, possibly
  from cotton, perhaps the same virus as immediately above)
  in El Salvador. 6. Bean Severe Mosaic in Mexico, W.
  Indies, Costa Rica, Honduras, Nicaragua. [Most prevalent
  at lower elevations, in tropical dry-forest ecology.] 7.
  Bean Yellow Mosaic (scattered, is spectacular but often of
  minor importance) in Florida, Dominican Republic, Cuba,
  Guatemala, Mexico, Nicaragua, Costa Rica, Ecuador,
  Brazil, is probably Neotropical in distribution. 8. Bean
  Yellow Mosaic var. Sweet Pea Streak (by inoculation) in
  Argentina. 9. Citrus Crinkly Leaf (by inoculation) in
  Florida. 10. Citrus Infectious Variegation (by inoculation)
  in California. 11. Cucumber Mosaic (by aphid transmis-
  sion and found in field) in Puerto Rico, probably in C.
  America, and W. Indies. 12. Cucumber Mosaic var.
  Celery-Commelina (in field and by inoculation) in Florida,
  Cuba, Costa Rica, and nearby countries. 13. Euphorbia
  Mosaic in Brazil. 14. Pea Enation Mosaic (by inoculation)
  in Argentina. 15. Pea Mosaic in Argentina. 16. Rhyn-
  chosia Mosaic (white fly transmitted) in Puerto Rico. 17.
  Sida Yellow Mosaic, also called Bean "Golden Mosaic" in
  C. America, Venezuela, Brazil. 18. Sunflower Mosaic,

Argentine, (by inoculation) in Argentina.   19. Tobacco
Streak = Brazilian Tobacco Streak in Brazil.   20. Tomato
Black Ring var. Bean Ring Spot (a rare occurrence) in
Honduras.
Xanthomonas phaseoli (E. F. Sm.) Dows. synonyms are
   Bacterium phaseoli E. F. Sm. and Pseudomonas phaseoli
   E. F. Sm.; common systemic bacterial blight.   Probably
   Neotropical in distribution but with notable reports in:
   Puerto Rico, Mexico, Dominican Republic, El Salvador,
   Guatemala, Nicaragua, Costa Rica, Trinidad, Ecuador,
   Peru, Brazil, Argentina. [In W. Indies and C. America
   it is the most common at flood plain elevations to low
   foothills, in areas where rainyseason is followed with warm
   temperatures.]
Xiphinema sp., nematode recovered in root zone soil, uncer-
   tain if it feeds on or injures bean roots.   Chile.
Yeast, wild type, recovered from rotting green pods in
   market. El Salvador.

PHILODENDRON spp. All handsome-leaved ornamental vines, grown
   in village huts and city houses, and rooted healthy plants are
   sold in metropolitan temperate zones, of economic importance to
   florists and plant shippers; "paisaje," "bejuco de casa" Sp.,
   "vinha de viagem," "filodendro" Po., "philodendron" En.
   Actinopeltis philodendri Bat.; leaf disease. Brazil.
   Ascochyta philodendri Bat.; leafspotting. Brazil.
   Ascopolyporus moellerianus Moell.; leaf and stem canker.
      Brazil, Argentina, nearby countries.
   (Bacteria, unidentified but proved common and epiphytic in
      leaf creases. Puerto Rico.)
   Botryodiplodia theobromae Pat. (and occasionally its perfect
      state: Botryosphaeria quercuum (Schw.) Sacc. Cke. is
      found); die-back and stem lesions. Panama, Puerto Rico,
      Costa Rica, Guatemala.
   Capnodium sp. 1. A coarse, characteristic, common sooty-
      mold growing on insect droppings. Puerto Rico.   Capno-
      dium sp. 2. a fine grained sootymold, slow growth and
      subsisting, without insects present, on leaf exudations that
      are normal for healthy plants at certain periods. Puerto
      Rico.
   Cassytha filiformis L.; green dodder. Puerto Rico.
   Cephaleuros virescens Kunze; algal leafspot. Puerto Rico.
   Colletotrichum gloeosporioides Penz. (conidial state); common
      anthracnose. Probably on all host species and widespread,
      of special note in Brazil, Paraguay, Argentina, Puerto
      Rico, Florida, and seen in Trinidad. C. philodendri Speg.
      or C. philodendri P. Henn. (considered by some as synony-
      mous with C. gleosporioides but the symptoms are dif-
      ferent); large leafspot, stem and petiole sunken lesions with
      purple edges. This organism is perhaps at the very least
      a special form of C. gloeosporioides. Reports from
      Puerto Rico, Panama, El Salvador, Guatemala, Colombia,
      Brazil.

Cuscuta americana L. ; common dodder (by inoculation).
    Puerto Rico.
Dendrophoma vagans Syd. ; leaf lesions.   Southern countries
    in S. America.
Dictyothyriella guianensis Stev. & Man. , and D. philodendri
    Stev. & Man. ; black mildew spots on living leaves.   C.
    America, northern part of S. America.
Diplopeltis sp. ; leaf disease.   Brazil.
Erwinia chrysanthemi Burkh. et al. , (philodendron strain); a
    common bacterial disease.   Florida, W. Indies, C. America.
Gloeosporium philodendri Speg. , its synonym is Colletotrichum
    gloeosporioides.
Hymenopsis paraensis Syd. ; plant disease.   Brazil.
Leafspot, from exudates out of stomata at warm temperatures
    on specially vigorous plants.   Exudates cause cell injury,
    and in the exudates develop weakly parasitic bacteria and
    fungi.   California, Florida, Puerto Rico, Costa Rica,
    Ecuador.
Lembosia patouillardii Sacc. & Syd. ; leaf smudge that seriously
    disfigures surfaces.   Argentina.
Meliola musae Mont. and M. philodendri Stev. ; common black
    mildew spots.   W. Indies, C. America (? General).
Micropeltis aroidicola Stev. & Man. and M. macromera Syd. ;
    colored fungus webs on leaf surface.   W. Indies.
Mycena citricolor (Berk. & Curt.) Sacc. ; leafspot (by inocula-
    tion).   Puerto Rico, Costa Rica.
Mycosphaerella philodendri (Pat.) Lind. ; leafspot.   Venezuela.
Phyllachora chloridicola Speg. and P. philodendronis Pat. =
    P. engleri Speg. ; both cause tarspot on leaf, first one re-
    ported from Argentina and W. Indies, the second from W.
    Indies, C. America, S. America.
Phyllosticta cearensis Bat. , P. imbe Bat. , and P. philodendri
    All. ; leafspots that may coalesce.   Brazil.
Phytophthora sp. ; leafspot and collapse, stem lesion.   Florida.
Pycnothyrium philodendricola Bat. (? = Diplopeltis sp.); on
    living leaves.   Brazil.
Rooting failure in alkaline soil without organic matter.   Guate-
    mala.
Sclerotium rolfsii (Curzi) Sacc. , its basidial state is Pellicu-
    laria rolfsii (Sacc.) West; base-rot and wilt, occasional but
    rare in rooting benches.   Florida, Puerto Rico, Guatemala,
    Costa Rica, Panama.
Scolecopeltella microcarpa Speg. ; leaf disease.   Puerto Rico.
Scolecopeltidium multiseptatum Stev. & Man. ; leafspot.   S.
    America.
Septoria guembe Speg. ; leafspot.   Paraguay.
Strigula complanata Fée; milky leafspot.   Puerto Rico, but
    probably of general distribution.
Trichopeltis reptans (Berk. & Curt.) Speg. ; black mildew.
    Puerto Rico, El Salvador.
Tripospermum gardneri Hug. ; on leaves part of a disease
    complex.   Brazil.

Uredo monsterae Syd.; leafrust.   Mexico.
Viruses: Dasheen Mosaic, causes leafmottle malformation of
   lamina and stunt in El Salvador, Costa Rica, U. S. Virgin
   Islands, Cuba, Panama (Note: This virus disease is very
   mild in some Philodendrons.   It may well be common and
   possibly holds over in these ornamental hosts, to then be
   actively effective on the food producing colocasias).   Philodendron Selorum Mosaic in Florida.
Xanthomonas dieffenbachiae Dows.; bacterial leafspot, spots
   sometimes confluent result in scorch-like symptom, vine
   tip infection fatal to branch.   Florida (symptoms seen in
   Guatemala but no isolations made).

PHOENIX DACTYLIFERA L.   Famed in desert Egypt and Arabia,
   the tall, handsome palm tree was introduced into the Neotropics
   by Spain for fruit and ornament (for fruit requires desert climate,
   specially skilled exact irrigation, pruning, hand pollenation);
   "dátil," "datylis" Sp., "datil," "tamariera" Po., "date palm" En.
   Alternaria citri Ell. & Pierce; fruit spot.   Dry Neotropical
      USA, A. stemphylioides Bliss; fruit rot.   California.
   Alternaria sp.; leaf spot.   Dry Subtropical USA.
   Aspergillus niger v. Tiegh.; fruit rot at calyx end.   Cali-
      fornia.
   Auerswaldia palmicola Speg. synonyms are: Bagnisiopsis
      palmicola (Speg.) Petr., Dothidina palmicola (Speg.) Th.,
      Sphaerodothis palmicola (Speg.) Shear and Coccostroma
      palmicola (Speg.) Arx & Muell.; leaf spots and blast with
      stromatic growths.   Puerto Rico.
   Black nose of fruit from humidity during ripening.   California,
      W. Indies.
   Botryosphaeria dothidea (Moug.) Ces. & de Not. = B. ribis
      Gross. & Dug.; leaf stalk rot.   W. Indies.   B. quercuum
      (Schw.) Sacc. its most used synonyms are: Botryodiplodia
      theobromae Pat., Diplodia theobromae (Pat.) Now.,
      Physalospora rhodina Berk. & Curt.; leaflet spot, midrib
      rot following injury, spathe spot.   W. Indies, S. America.
   Brachysporiella gavanus Bat.; on rotting leaves.   Brazil.
   Catenularia fuliginea Saito; fruit rot.   California.
   Cephaleuros virescens Kunze; algal leafspot.   Brazil.
   Ceratocystis fimbriata Ell. & Halst., leaf blight, stem
      bleeding, vascular discoloration.   W. Indies, Florida.
      C. paradoxa Dade; bud-rot, black scorch.   California.
      C. radicicola (Bliss) Mor. = Ceratostomella radicicola
      Bliss; rootrot, wilt.   California.
   Cladosporium sp.; on herbarium specimen of leaf.   Costa Rica.
   Colletotrichum gloeosporioides Penz. which is the imperfect
      state; leaf disease.   Special note in Texas, Puerto Rico,
      probably wherever this palm is grown.
   Diplodia phoenicum (Sacc.) Fawc. & Kl.; leafstalk rot, shoot
      blight, fruit rot from stem end.   California, Puerto Rico.
   Dothidina (see Auerswaldia)
   Exosporium palmivorum Sacc.; brown leafspot.   Subtropical
      USA, W. Indies.

Fusarium sp. (of one type); inflorescence rot, fruit rot. Subtropical USA, W. Indies. Fusarium sp. (of a different
kind, unidentified); rotting of roots, stem disease, wilt and
death. Peru.
Graphiola phoenicis (Moug.) Poit.; leaf blight, false smut,
raised scab-like spot. Peru, Jamaica, Trinidad and To-
bago, Dominican Republic, Puerto Rico, C. America,
Venezuela, Argentina, Peru.
Helminthosporium molle Berk. & Curt.; brown spot on fruit,
leafspot (superficial injury). California, Puerto Rico.
Leptonchus granulosus Cobb; root nematode. California.
(Loranthaceae, so common on many trees cannot parasitize
palms.)
Marasmius eburneus Th.; rootrot. S. America.
Meliola furcata Lév.; black mildew. Puerto Rico.
Meloidogyne sp.; rootknot nematode. California.
Omphalia tralucida Bliss and O. pigmentata Bliss; rootrot,
decline. California.
Paurodontus densus Thorne; nematode associated with roots.
California.
Penicillium roseum Lk.; fruitrot. California.
Pestalotia palmarum Cke. (? = P. phoenicis Vize); leafspot.
California, C. America, Puerto Rico.
Phomopsis phoenicola Trav.; fruitrot. California.
Phymatotrichum omnivorum (Shear) Dug.; rootrot. Arizona,
California (? Mexico, unconfirmed report).
Pleospora herbarum (Pers.) Rab. perfect state of Stemphylium
sp; leafspot, fruitspot. California, Mexico on fruit sample.
Poria ambigua Bres., P. ravenelae (Berk. & Br.) Sacc., P.
versipora (Pers.) Rom.; rots of old trunk wood and leaf.
California (apparently one of these fungi but immature on
fallen leaf base in Puerto Rico).
Pythium sp.; seedling rot. C. America, W. Indies.
Sebacina obscura Mart.; on trunk. S. America.
Stilbum aurantio-cinnabarium Speg. = Sphaerostilbe rosae
Kal.; canker and rot of trunk and leaf petiole. Argentina.
Strigula complanata Fée; white lichen leafspot. Brazil.
Thielaviopsis paradoxa (De Sey.) Hoehn.; black scorch, bitten-
leaf. Widespread occurrence.
Yeasty flavor, unidentified organism. Unacceptable fruit
disease. Puerto Rico.

PHYSALIS IXOCARPA Brot. (= P. EDULIS Hort.), P. PERUVIANA
L. (= P. EDULIS Sims), and P. PUBESCENS L. Of over 50
species these three herbaceous annual plants, native to tropical
America, bear crops of fruits very much relished and most used
by "huerta" or home-yard peasant gardeners, some fruits sold
in village markets but of little major commercial monetary importance and in consequence have had only minor disease attention; "tomatillo," "capulí," "uchuba" Sp., "físalis," "tomate
silvestre" Po., "physalis," "husk tomato" En.
Aecidium (see Puccinia)

Capnodium spp.; sootymolds (epiphytic black fungus growths
on insect exudate from pests on trees shading Physalis
plants. Widespread.
Cephaleuros virescens Kunze; algal leafspot.   Puerto Rico.
Cercospora diffusa Ell. & Barth.; serious leafspot that be-
comes systemic.   Puerto Rico, C. America.   C. physa-
licola Ell. & Barth.; round discrete leafspot.   Puerto Rico.
C. physaloides Ell.   Puerto Rico, C. America, Brazil.
C. physalidicola Speg. = Cercosporina physalidicola Speg.;
leafspot and rot.   Argentina.
Cladosporium sp.; disease on senescent leaves and fruit husks.
Subtropical USA, C. America.
Colletotrichum sp.; small anthracnose leaf and stem lesions,
found systemic in old stems.   C. America, S. America.
Corynespora casiicola (Berk. & Curt.) Wei; leafspot.   Hon-
duras.
Cylindrosporium (see Entyloma)
Entyloma australe Speg. = E. physalidis Wint. (= Cylindro-
sporium australe Speg.); white swollen smut spots on leaves.
Puerto Rico, Subtropical USA, Brazil.
Fumago vagans Pers. (different from Capnodium); sootymold.
S. America.
Gloeosporium australe Speg. (small spored, as a hyperpara-
site of Entyloma australe).   Argentina.
Meloidogyne sp.; rootknot nematode.   Florida.
Phyllosticta calycicola Speg.; husk spots.   Argentina.   P.
physaleos Sacc.; leaf spotting.   Argentina.   P. physaleos
Sacc. var. calycicola Speg.; fruit and hust spotting.   Ar-
gentina.
Phymatotrichum omnivorum (Shear) Dug.; dry-land rootrot.
Texas.
Phytophthora infestans dBy.; wet-season blight.   Mexico, C.
America.
Puccinia physalidis Peck = Aecidium physalidis Burr.; leaf-
rust.   Subtropical USA (probably more widespread).
Rhizoctonia solani Kuehn; rare but can occur as seedling and
young plant disease.   El Salvador, Guatemala.
Septoria lycopersici Speg.; mild small to confluent leafspots.
Widespread.
Stemphylium solani Web.; leafspot.   Florida.
Viruses (found in the field); Cucumber Mosaic causing mild
mottle in C. America, Cucumber Mosaic of the Celery-
Commelina virulent strain causing severe malformation and
marbling of leaves in Florida, Cuba, Puerto Rico, Honduras, Mosaic mottle still unidentified causes a mild mottle
with bronzing in W. Indies, and Tobacco Mosaic causing
typical mottling in Neotropics.

PIMENTA DIOICA Merr. [see also P. RACEMOSA, following entry].
Tropical American small trees that produce the allspice of commerce.   They have not been given much plant disease study;
"pimento" Sp. Po., "allspice tree" En.

Botryodiplodia theobromae Pat.; dieback.   Panama, Costa
   Rica.
Cephaleuros virescens Kunze; algal leafspot.   Puerto Rico.
Loranthaceae (at least three genera according to student ob-
   server, but no collections or determinations made by me).
   Panama.
Meliola amomicola Stev.; black mildewed leaves.   Puerto
   Rico.
Mycena citricolor (Berk. & Curt.) Sacc.; leafspot near
   diseased coffee trees.   Costa Rica.
Puccinia psidii Wint.; leafrust.   Jamaica (said to be in nearby
   islands).
Rhizoctonia (aerial type); on foliage, web-type thread blight.
   Puerto Rico.
Rostrella coffeae Zimm. = Ceratocystis sp.; branch and tree
   disease resulting in woodstain.   Jamaica, Panama.
Strigula spp.; lichen leafspots, of various colors and types.
   General.
Valsa sp.; killing of small branches.   Panama.

PIMENTA RACEMOSA Moore [see also P. DIOICA, preceding
   entry].   A small tree, somewhat tolerant of shade, grown in
   groups and protected in the wild, from which are plucked the
   leaves.   On drying these are shipped and then distilled for the
   bayrum oil.   This is still practically a forest product, of great
   help and advantage to the gatherers; "malagueta" Sp., "bayrum
   tree" En.
   Calothyriolum apiahynum Speg. (? = Asterina sp.); a black
      mildew.   Brazil, Puerto Rico, Jamaica, Costa Rica,
      Panama, Brazil.
   Capnodium coffeicola Pat.; sootymold, spoils leaves for ex-
      traction process, acts as a smother on branch tips.   W.
      Indies.
   Cercospora sp.; leafspot, defoliation in newly sun-exposed
      trees.   Puerto Rico, Costa Rica.
   Colletotrichum gloeosporioides Penz.; leaf anthracnose lesions.
      Puerto Rico.
   Cuscuta sp.; slender vine type of dodder.   Puerto Rico.
   Drepanoconis sp.; on dying branches.   Costa Rica.
   Meliola amomicola Stev.; black mildew on leaves, requiring
      rubbing off of blemishes before leaves are dried prepara-
      tory to sale.   Puerto Rico, Jamaica.
   Pestalotia sp.; lesions on twigs from which leaves have been
      plucked.   Costa Rica.
   Phyllosticta sp.; leafspots.   Puerto Rico.
   Porella sp.; leafy liverworts on old leaves left on for more
      than one year.   Costa Rica.
   Puccinia psidii Wint.; leafrust.   Jamaica.
   Rosellinia bunodes (Berk. & Curt.) Sacc., and R. pepo Pat.;
      two rootrots.   Puerto Rico.
   Strigula complanata Fée; white lichen-leafspot.   Puerto
      Rico.

PINUS spp. Numerous valuable species of forest and planted timber
trees being used in some areas and newly tried in others in the
cool tropics, where it is found they grow much more rapidly than
in the temperate zone; "pino" Sp., "pinheiro" Po., "pine" En.
Alternaria alternata (Fr.) Keiss = A. tenuis Nees (small
    spored); blackening of needles. Guatemala, Honduras.
Antennaria sp.; needle smoky-mold. Colombia.
Aphelenchoides fragariae Christy; root nematode. Florida.
Aphelenchus xylophilis Steiner & Buhrer; appears at times to
    be connected with bluestain of wood. Subtropical USA.
Arceuthobium spp. = Razoumofskya spp.; about 12 species of
    dwarf mistletoes that cause witches' brooms of branches
    and terminal trunk growth. They seem to be confined to
    Subtropical USA, Mexico, Guatemala, Honduras, Haiti,
    and Dominican Republic.
Armillaria mellea Vahl = Armilariella mellea (Fr.) Karst.;
    rootrot. California, Brazil.
Atropellis tingens Loh. & Cash; twig and branch canker.
    Subtropical USA.
Belonopsis tropicalis Rick; leaf disease. Brazil.
Bifusella striiformis Dar.; needle cast. Subtropical USA.
Boron deficiency, severe dieback. Chile.
Botryodiplodia theobromae Pat. (the asco-stage is Botryo-
    sphaeria quercuum (Schw.) Sacc.); twig and stem lesions,
    die-back. Scattered, more or less generally. (Also re-
    ported as Diplodia megalospora Berk. & Curt., D.
    natalensis, P. Evans, P. pinea Kickx.)
Botryosphaeria pinicola Speg., on cone scales and B. ribis
    Gross. & Dug. = B. dothidea (Mong.) Ces. & deNot.;
    branch cankers. Both in Argentina.
Botrytis cinerea Pers.; seedling blight. Especially studied
    in Colombia and Chile, but widespread.
Cenangium ferruginosum Fr.; twig disease. Subtropical USA.
Ceratocystis sp.; bluestain of sapwood. Subtropical USA,
    and found as severe following bark beetle attack in Guate-
    mala in 1929 killing trees.
Chytridium citriforme Spar.; on pollen. Cuba.
Clitocybe tabescens Bres. = Armilariella tabescens (Scop.)
    Sing.; rootrot. Florida.
Coleosporium spp.; 14 needle rusts, examples: C. apocyna-
    ceum Cke., in Subtropical USA. C. solidaginis (Schw.)
    Th. in Mexico. C. vernoniae Berk. & Curt., in Florida,
    Texas, W. Indies, Colombia.
Collybia aquosa Bull.; an agaric on dead wood. Brazil.
Corticium amorphum (Pers.) Fr.; seedling disease? S.
    America.
Cronartium spp. (It is to be noted that these rusts on pro-
    ducing tongue-like aperculate peridia, become a new form
    genus which is known as Peridermium, see below), which
    cause hypertrophy, examples: C. cerebrum Hedgc. &
    Long; the gall rust. Florida, Texas. C. coleosporioides
    Arth.; blister rust with swellings on trunk and branch.

California, Guatemala, (probably also in Mexico and Honduras). C. conigenum Hedgc. & Hunt; cone rust, spectacularly swollen and yellow. Guatemala. C. fusiforme
Hedgc. & Hahn; fusiform swellings on branches and trunk.
Subtropical USA. C. quercuum (Berk.) Miy.; cone and
branch galls. Mexico, Guatemala, Brazil. C. ribicola
Fisch.; blister rust. Texas. C. harknessii Mein.;
globoid stem gall. C. America, S. America. C. strobi-
linum Hedgc. & Hahn; cone rust. Subtropical USA, Mexico.
Cryptoporus volvatus Hubb. = Polyporus volvatus Peck; trunk
and stump rot. Mexico.
Cylindrocladium scoparium Morg.; seedling damping-off.
Brazil.
Diplodia (see Botryodiplodia)
Dothiostroma (see Scirrhia)
Flammula sp.; occasional rot on fallen branches in S. America.
Fomes annosus (Fr.) Cke.; buttrot and rootrot. Special reports from Guayana, Jamaica, seen in Mexico, may be
general in distribution. F. pini (Brot.) Karst.; pocket rot
and redringrot. S. America, Subtropical USA. F. marginatus Fr.; stump and log rot. S. America.
Fracchiaea cucurbitarioides Speg. f. sp. pini-insignia Speg.;
disease of trunk and branch. Argentina.
Fusarium spp,, several have been isolated from infections of
pine seedlings, branch disease specimens, twig lesions,
and in connection with needle cast studies. Some of the
organisms cause damping-off and some are only weakly
parasitic: F. roseum Lk. and F. solani (Mart.) Sacc.,
including various formae of these and F. oxysporum Schl.
which acts as a wilt organism in seed beds and nursery
plantings. W. Indies, Guatemala, Chile, Colombia, Argen-
tina, Brazil.
Geotrichum sp.; pink wood-stain. Subtropical USA.
Glonium stellatum Muhl.; on dead branch (sometimes die-back).
S. America.
Haplosporella sp.; die-back. Chile.
Hydnum mucidum Pers.; woodrot. Mexico.
Hypoderma hedgcockii Dearn.; needle blight. Subtropical USA,
W. Indies. H. lethale Dearn.; gray leafspot. Subtropical
USA.
Hypodermella cerina Dar., and H. montana Dar.; needle tar-
spot, needlecast. Ecuador and California.
Kneiffia subsulphurea (Karst.) Bres.; on branchlets. Southern
S. America.
Lecanosticta (see Scirrhia)
Lepiota sp.; agaric under trees (probably not parasitic). C.
America.
Lophodermium nitens Pers.; needle-cast. California, Ecuador.
L. pinastri (Schrad.) Chev.; needle-cast. Widespread.
Macrophoma sp.; twig disease. Jamaica, Argentina.
Meloidogyne sp.; seedling rootknot nematode. Florida, Chile.
Monochaetia pinicola Dearn.; needle disease. Florida.

Mycorrhizas, there may be a complex of such fungi. Where
pines have been absent and seeds are introduced, without
proper mycorrhizas present the pine seedlings fail to make
trees.

Naemacyclus niveus Sacc. ; drop of needles. Uruguay.

Oidium tomentosum Lind. ; white mildew. Bahamas.

Omphalia campanella Batsch and O. hydrogramma Fr. ; rootrots. Brazil. (? also other countries in S. America.)

Paratylenchus sp. ; nematode rhizosphere inhabitant. Chile.

Peridermium spp. (see note under Cronartium); P. cerebrum
Peck = Cronartium quercuum and P. filamentosum Peck =
Cronartium coleosporioides. A special needle rust in
Florida is known as P. floridanum Hedgc. & Hunt, and a
needle rust in Guatemala (which also occurs at least in
Florida, Mexico, and Brazil) is known as P. guatemalense
Arth. & Kern.

Pestalotia funerea Desm., P. lignicola Cke., P. hartigii Tub.,
and P. macrochaeta (Speg.) Guba; all attack needles and
twigs and are found widespread from W. Indies to Argentina.

Phaeociboria brasiliensis Syd. ; on foliage. Brazil.

Phoma sp. ; on twigs and on peduncles of cones. Scattered.

Phymatotrichum omnivorum (Shear) Dug. ; basal-rot of young
trees. Texas.

Phytophthora cinnamomi Rands; rootlet disease, decline. Subtropical USA, Argentina, Brazil. P. citrophthora (Sm. &
Sm.) Leon. ; rootrot. Brazil. P. parasitica Dast. ; stem
canker of young trees. Brazil.

Polyporus schweinitzii Fr. ; brown woodrot, butt rot. Noted
especially in Mexico, Subtropical USA, W. Indies. (Also
see Cryptoporus.)

Poria cocos Wolf; root disease. Florida. P. mollusca Fr. ;
woodrot, root infection. Scattered in S. America.

Pratylenchus sp. ; rhizosphere nematode inhabitant. Chile.

Psittacanthus calyculatus Don; bush-like parasite on branch
and trunk, both old infections and new attacks. El Salvador.

Pythium (? debaryanum); seedbed damping-off at high elevations. P. rostratum Butl. ; seedbed damping-off. Scattered
notices. P. ultimum Trow. ; seedbed damping-off and basal
attack of seedling trees. Chile, Argentina, Colombia.

Razoumofskya (see Arceuthobium)

Rhizoctonia solani Kuehn (basidial stage is rare); damping-off
and brown superficial mycelium on old seedling roots.
Dominican Republic by personal communication, Puerto
Rico, Mexico, Chile, Argentina, Brazil.

Rhizopagon honorati Speg. ; an ectotropic mycorrhiza. Chile.

Rosellinia spp. ; rootrots. C. America.

Sarcoscypha concatenata Rick; on decayed wood. (? Chile.)

Sarcosoma godronioides Rick; on decayed wood. (? Chile.)

Scirrhia sp. (sometimes spelled "Scirria"), synonyms:
Dothiostroma sp., Lecanosticta sp., Septoria sp. ; causes
brown needlespot. Subtropical USA, Mexico, W. Indies,
Chile, Brazil.

Sclerotinia fuckeliana Fckl.; moldy attack of seedlings.   Chile,
  Colombia.
Septoria (see Scirrhia)
Sphaerocolla citrina Speg.; root disease.   Argentina.
Sphaeropsis ellisii Sacc.; branch-tip disease.   Uruguay.   S.
  pinicola Speg.; on pine debris.   S. America.   S. puttemansii P. Henn.; on branch material.   Brazil.
Stereum sp.; trunk disease.   Mexico, Guatemala.
Struthanthus orbicularis Blume; a scandent but woody hyper-
  parasitic bush with original infection on Psittacanthus that
  was killing pine and the vigorous Struthanthus was drooping
  down to reach and then attack the Pinus (host) that was
  originally infected by Psittacanthus.   El Salvador.
Thelephora (? palmata); a smothering coarse non-sporulating
  fungus on seedlings (sporocarps on debris).   Widespread.
Trametes sp.; woodrot.   Argentina.
Trichodorus californicus Allen; root disease nematode.   California.
Verticillium albo-atrum Reinke & Berth.; seedling wilt.   Chile.
Zinc deficiency; little leaf, followed by stunt and growth
  failure.   Scattered occurrence.

PIPER NIGRUM L., since very ancient times one of the most, if
  not the most, valued spice plant of the tropics.   It was intro-
  duced years ago to warm-moist areas in the Americas now
  slowly increasing as a specialty production; "pimiento negro"
  Sp., "piment," "pimentlira" Po., "pimiento," "black pepper"
  En.
Capnodium sp.; sootymold.   Puerto Rico, Brazil.
Cephaleuros virescens Kunze; algal leafspot, berry infections.
  Puerto Rico.
Cercoseptora piperis (Stev. & Dalb.) Petr.; leafspot.   Puerto
  Rico.
Cercospora piperis Pat.; leafspot.   Puerto Rico, El Salvador,
  Cuba, Haiti, Panama, Brazil.
Colletotrichum gloeosporioides Penz. (= the reported asco-
  state is Glomerella cingulata (Stonem.) Sp. & Schr.); the
  former found as anthracnose lesions of root and climbing
  stem, the latter related to leaf disease in moist weather.
  Puerto Rico, Jamaica, Brazil.
Cyclodothis pulchella Syd.; black leafspot.   Puerto Rico.
Didymothozetia mimisoensis Rangel; on leaves.   Peru, Brazil,
  Argentina.
Diplodia theobromae (Pat.) Now. = Botryodiplodia theobromae
  Pat.; associated with footrot.   Puerto Rico.
Erwinia carotovora (L. R. Jones) Holl.; slimy softrot of
  leaves and fruit.   Puerto Rico.
Fusarium oxysporum Schl.; wilt.   Puerto Rico.   F. solani
  (Mart.) Sacc. f. sp. piperis Alb.; footrot.   Brazil.
Guignardia pipericola Stev.; leafspot.   Puerto Rico, Virgin
  Islands.
Irenina glabra (Berk. & Curt.) Stev. and I. tortuosa Stev.;
  both black mildews and both in Puerto Rico and in Panama.

Meliola contorta Stev. and M. pancipes Stev. ; both smudge
  black mildew, both in Puerto Rico.
Mycena citricolor (Berk. & Curt.) Sacc. ; leafspot next to
  diseased coffee. Puerto Rico.
Pellicularia koleroga Cke. ; leafscorch, branchrot. Puerto
  Rico, Brazil.
Phytophthora palmivora Butl. ; root disease, footrot, leafspot.
  W. Indies, Brazil.
Podosporium effusum Pat. ; leaf infection. Puerto Rico.
Pseudomonas syringae v. Hall; bacterial leaf disease. Brazil.
Radopholus similis (Cobb) Thorne; yellows. Widespread.
Rhizoctonia solani Kuehn; damping-off in seedbeds. Brazil.
Rosellinia bunodes (Berk. & Br.) Sacc. ; rootrot. Puerto
  Rico.
Sclerotium rolfsii (Curzi) Sacc. (perfect state is Pellicularia
  rolfsii (Sacc.) West.); cutting disease, wilt. Florida.
Stigmatea piperis Rehm; leaf infection. Virgin Islands.
Strigula complanata Fée. ; white lichen leafspot. Puerto Rico.
Uredo piperis P. Henn. ; leafrust. Puerto Rico, Peru, Brazil.
Viruses, perhaps more than one, reported as most serious
  limiting disease in 1960-70s in Brazil.

PISCIDIA ERYTHRINA L. Trees protected and planted by indigenes,
  who for generations have been employing in rivers the emulsion
  from bruised young branches and leaves for stunning fishes,
  bark extract materials are used in human medicine for hypnotic
  effect; "barbasco" Sp. Po. Ind.
    Cercospora piscidiae P. Henn. ; leafspot. Jamaica.
    Colletotrichum sp. ; leafspot. Costa Rica.
    Die-back of young branches. Complex of unstudied insects
      and fungi, severe in trees growing in lowlands. Costa
      Rica.
    Marasmius sarmentosus Berk. ; horse-hair (black cord)
      blight. Costa Rica, Panama.
    Matsuchotia complens Moell. (Hydnum-like sporocarp); rot
      at tree base. S. America.
    Ravenelia piscidiae Long; leafrust. Florida, W. Indies.

PISTACIA VERA L. An introduced Mediterranean nut tree now
  being grown in several areas; the nut well received is of unusual type and relished; "pistacia" Po. , "pistacho, " "alfóncigo"
  Sp. , "pistacio" En.
    Botryodiplodia sp. ; twig disease, dieback. Costa Rica.
    Colletotrichum gloeosporioides Penz. ; anthracnose. Florida.
    Meloidogyne sp. ; rootknot (nematode). Costa Rica, California.
    Pellicularia koleroga Cke. ; thread-web blight. Costa Rica,
      Subtropical USA.
    Phyllosticta lentisci (Pass.) Alles. ; leafspot. Texas.
    Phymatotrichum omnivorum (Shear) Dug. ; dryland rootrot.
      Texas, California, Arizona, (? Mexico).
    Pleurotus ostreatus Jacq. ; sapwood rot. California.
    Schizophyllum commune Fr. ; woodrot, branch disease.
      Mexico, C. America.

Septoria pistaciarum Caracc. ; leafspot. Texas.
Tryblidiella sp. ; dieback. Subtropical USA, W. Indies.

PISUM SATIVUM L. An herbaceous annual from the Old World
that under tropical conditions is fairly sensitive and grows best
in near neutral and slightly alkaline good soils in the cooler
somewhat dryer parts. It is eaten mostly as a cooked green
garden seed; "arveja, " "alverja, " "guisante" Sp. , "ervilha, "
"ervilhas verdes" Po. , "pea, " "garden pea, " "green peas" En.
   Aphanomyces eutiches Drechs. ; rootrot. Ecuador.
   Ascochyta pinodella L. K. Jones; footrot. Florida, Ecuador.
      A. pinodes Berk. & Bloz. (its perfect state is Mycosphae-
      rella pinodes (Berk. & Blox.) Vest.); blight, footrot favored
      by rainy season and heavy dews (true of all these Ascochyta
      diseases). Brazil, Haiti, Argentina, Costa Rica, Jamaica.
      A. pisi Lib. ; leaf-, pod-, stem-spot. Mexico, Costa Rica,
      Haiti, Panama, Colombia, Venezuela, Ecuador, Bolivia,
      Chile, Guatemala, Peru, Brazil, Argentina.
   Cercospora pisi-sativae Stev. ; leafspot. Subtropical USA,
      Cuba, Puerto Rico, Santo Domingo, Guatemala, Ecuador.
   Chlorosis, physiological from various deficiencies: iron,
      manganese, nitrogen, zinc. Scattered.
   Choanephora conjuncta Couch; leaf and stem disease. Sub-
      tropical USA.
   Cladosporium herbarum Lk. ; leaf decline. California, Brazil.
      C. pisicola Sny. ; scab, pod and foliage blackspot. Cali-
      fornia, Mexico.
   Colletotrichum dematium (Pers.) Grove = Sphaeria dematium
      Pers. ; anthracnose injuries. Mexico. C. dematium
      (Pers.) Grove f. sp. truncata Arx = Vermicularia truncata
      Schw. ; leafspots. Mexico. C. pisi Pat. synonyms: C.
      gloeosporioides Penz. and C. gloeosporioides (pea strain);
      common anthracnose. Widespread.
   Cuscuta sp. ; dodder. Peru.
   Erysiphe polygoni DC. is the perfect state of Oidium, see
      below. (E. cichoracearum DC. is believed erroneously
      included in a few reports.)
   Fusarium oxysporum Schl. f. sp. pisi (v. Hall) Sny. & Hans.
      (= F. orthoceros Ap. & Wr.); wilt. California, Texas,
      Haiti, Ecuador. F. solani (Mart.) Sacc. f. sp. pisi (L.
      K. Jones) Sny. & Hans. ; rootrot, death. Florida, Brazil.
   Leptostroma pisi Gonz. Frag. & Cif. ; black leafspot. Do-
      minican Republic.
   Macrophomina phaseoli (Taub.) Ashby; stem blight. Texas,
      Costa Rica, Guatemala.
   Meloidogyne sp. ; rootknot nematode. Florida, Panama, Brazil.
   Mycosphaerella (see Ascochyta)
   Oidium erysiphoides Fr. ; common white powdery mildew pro-
      ducing abundant conidia on pea, its asco-stage, not thus
      far reported in most warm tropical countries, is Erysiphe
      polygoni DC, but the Oidium is of general distribution.
   Peronospora viciae (Berk.) Casp. = P. pisi Syd. ; downy

mildew on underside of leaf, moist season disease.   Ecuador, Jamaica, Brazil, Colombia, Argentina.
Phymatotrichum omnivorum (Shear) Dug.; rootrot.   Texas, Arizona.
Phytophthora spp.; wet season rootrot, podrot, leaf collapse. C. America, Brazil (doubtless other countries).
Pleospora sp.; leafspot disease.   Brazil.
Pratylenchus sp.; nematode in rhizosphere.   Peru, Chile.
Pseudomonas pisi Sac.; bacterial blight.   In C. America found at lower elevations.
Pythium spp.; damping-off, podrot, tip blight, rootrot.   Subtropical USA, C. America, Brazil.
Rhizobium leguminosarum Baldw. & Fred; root nodule (much desired infections, if the bacterium is absent must be brought in).
Rhizoctonia solani Kuehn = Thanatephorous cucumeris (Frank) Donk (its basidial state); rootrot, damping-off, stem and pod canker.   Guatemala, Colombia, Peru, Brazil.
Rhizopus stolonifer (Ehr.) Lind.; black podmold in fresh produce markets.   Guatemala, Panama, Ecuador.
Sclerotinia sclerotiorum (Lib.) dBy. = Whetzeliana sclerotiorum (Lib.) Korf & Dumont; pod and stem rot.   Chile, Subtropical USA, Peru, scattered in C. America and W. Indies, Brazil, Argentina.
Sclerotium rolfsii (Curzi) Sacc.; soil blight.   Florida, Brazil.
Septoria papilionacearum Cif. & Gonz. Frag.; small and round dark leafspots.   Brazil, Dominican Republic.   S. pisi Westd.; yellow and light colored leafspots, blotch. Ecuador, Chile, Brazil, Bolivia.
Tylenchorhynchus sp.; nematode found in rhizosphere.   Chile.
Uromyces fabae (Pers.) dBy. and U. pisi (DC.) Otth.; rusts. Chile, Ecuador, Texas, California, Peru.
Vermicularia (see Colletotrichum)
Viruses: Abutilon Mosaic (by inoculation) in Argentina.   Pea Enation Mosaic in Argentina.   Pea Mosaic, the most common pea virus disease is widespread.   Pea Streak of Argentina in Argentina.   Pea Spotted Wilt in Subtropical USA.   Sweetpea Streak in Argentina.

PITCAIRNIA see PUYA

PITHECOLOBIUM spp.   Consist of over 200 small to large trees that are native to the world tropics, but in the Americas only about ten are notably used as shade in pastures, for hedges, in ornamental planting, for posts and for construction.   Some Spanish names: "guamá," "tamarinda," "escambrón," "uña," "iña," "guaymachíl," "rolón."   (The "Saman," as can be seen in pages below, is not included in this treatment, although some botanists consider it a species of the genus Pithecolobium.   However, pathologically at least it is quite different.)   Diseases as I have listed them have given me some idea of the range of pathological problems found in wild-grown as well as planted trees.

Apiosphaeria (see Dothidella)
Auricularia delicata (Fr.) P. Henn. and other undetermined
    species; associated with woodrots. Scattered.
Cercospora bauhiniae Syd.; leafspot. W. Indies, C. America.
    C. spilosticta Syd.; leafspot. Venezuela.
Chaconia alutacea Juel; rust. British Honduras.
Cladosporium herbarum (Pers.) Lk.; on old leaves. General
    collections. C. urediniphilum Speg. (hyperparasite of the
    rust Uredo cyclotrauma on Pithecalobium in Paraguay).
Clitocybe tabescens Bres. = Armilariella tabescens (Scop.)
    Sing. rootrot. Subtropical USA.
Colletotrichum erythrinae Ell. & Ev. = C. gloeosporioides
    Penz.; anthracnose leafspot, twig lesion, dieback. Wide-
    spread.
Coniothecium oligomerum Speg.; on foliage. Paraguay.
Corticium salmonicolor Berk. & Br.; branch end disease.
    W. Indies.
Coryneum missionum Speg.; leaf infection. Argentina.
Diatrypiopsis guaranitica Speg.; on dead branch. Paraguay.
Diorchidium acanthostephum Syd.; leafrust. Brazil.
Dothidella controversa Speg., synonyms: D. platyasca Speg.
    and Apiosphaeria controversa Arx.; leaf and stem spot
    (subepidermal). Argentina, other countries in southern S.
    America. Probably Phyllachora timbo Rehm is identical,
    see below.
Exosporium sp. (? = Helminthosporium sp.); on leaf. Florida.
Fomes spp.; woodrots. Scattered but widespread.
Hexagona polyramma Mont.; associated with rotted wood.
    (? Locality in S. America.)
Hysterographium pithecolobii Seav.; twig disease. Puerto
    Rico.
Lenzites sp.; woodrot. Scattered.
Loranthaceae (at least three genera observed, flourishing
    growths, no doubt several parasitic species under each
    genus among them, but no collections for absolute identi-
    fications made). C. America, W. Indies.
Marasmius equirinis Moell.; horse-hair blight. S. America.
Maravalia pallida Arth. & Thaxt.; leafrust. Trinidad, C.
    America, S. America.
Meliola pithecolobii Stev. & Tehon; black mildew leafspots.
    Trinidad, Puerto Rico. M. pithecolobiicola Speg.; black
    mildew leafspots. Argentina. M. venezuelana Garces O.;
    black mildew. Colombia, Venezuela.
Meloidogyne sp.; rootknot nematode on seedlings. El Salvador.
Micropeltis pithecolobii Bat.; leaf infections. Brazil.
Microstroma pithecolobii Lam. = M. pithecolobii Speg.; leaf-
    spot. Brazil, W. Indies, Paraguay, Argentina.
Microthyriolum subimperspicuum Speg.; leaf infection. Para-
    guay.
Microthyrium pithecolobii Bat. & Costa; leaf disease. Brazil.
Mycena citricolor (Berk. & Curt.) Sacc. = Omphalia flavida
    Maubl. & Rang.; leafspot on nursery seedlings near diseased
    coffee. El Salvador.

Omphalia liliputiana Speg. ; a saprophyte (one of many) on
  rotting twigs. Paraguay, other countries.
Pestalotia funerea Desm. and/or P. palmarum Cke. ; twig
  lesions, secondary invasion in leafspots caused by primary
  parasites. Neotropics.
Phomopsis sp. ; twig blight. Florida.
Phyllachora timbo Rehm (? perhaps same as both Dothidella
  controversa Speg. and Stigmochora controversa Th. & Syd.);
  tarspot. Widespread in both C. America and S. America.
Phyllosticta pithecolobii Young; leafspotting. Puerto Rico,
  Texas, Dominican Republic. P. pithecolobii Young f. sp.
  monensis Young; leafspotting. Puerto Rico's Mona Island.
Physalospora fusca Stev. (its conidial form is a Diplodia);
  branch disease, die-back. W. Indies, Florida.
Ravenelia spp. (on the whole the host trees are quite tolerant
  to these leafrusts, a few are listed with examples of coun-
  tries where reported): R. amazonica Syd. in Brazil, Para-
  guay. R. bifenestrata Mains in Mexico. R. minuta Syd.
  in Brazil. R. oligotheles Speg. in Costa Rica and Argen-
  tina. R. pileolarioides Syd. in Brazil. R. pithecolobii
  Arth. in Florida, Puerto Rico, Mexico, Cuba and Venezuela.
Septoria pithecolobii Roum. ; leafspot. Paraguay.
Sphaerella pithecolobiicola Speg. ; leaf disease. Paraguay.
Stegastroma theisseni Syd. ; foliage disease. S. America.
Stenella araguata Syd. ; leafspot. Venezuela.
Stigmachora (see Phyllachora)
Stilbum cinnabarinum Mont. ; branch canker. Widespread.
Uredo bomfimensis P. Henn. ; brown leafrust. Brazil. U.
  cyclotrauma Speg. ; leafrust. Paraguay. U. pithecolobii
  P. Henn. ; twig deforming rust. Brazil.
Uromyces albescens Syd. ; leafrust. Brazil.
Virus ?, severe leafmottle in El Salvador.
Xylaria hypoxylon (L.) Grev. f. sp. biceps (Speg.) Th. ; on
  dead wood. ? Paraguay.

PLUMERIA ALBA L. , P. ACUMINATA Ait. (= P. ACUTIFOLIA
Poir.), P. RUBRA L. Popularly grown, these three small
milky-juiced ornamental trees are native to the American tropics
(widespread in the Orient), with specially fragrant flowers, fine
grained wood used for unusual furniture and carving; multiplied
by florists and nurserymen. Of very numerous common names
some are: "alelí ," "azucena," "flor de mayo" Sp. , "Jasmin de
cayenne," "frangipana" Po. , "frangipani" (named for an ancient
perfume), "tree jasmin" En.
  Ascochyta plumeria P. Henn. ; leafspot. Brazil.
  Capnodium sp. and other undetermined genera; sootymolds.
    Neotropics.
  Cassytha filiformis L. ; green dodder. In island areas near
    seashores. W. Indies.
  Coleosporium domingense (Berk.) Arth. = C. plumeriae Pat. ;
    golden leafrust, defoliation. Dominican Republic, Florida,
    Guatemala, Mexico, Bahamas, Cuba, Guadeloupe, Trinidad,
    Panama.

Cuscuta americana L. ; yellow dodder. W. Indies.

Cutting disease, failures to root for various unstudied reasons.

Leaf scorch, undetermined physiological disease. Puerto Rico.

Meliola tabernamontanae Speg. ; black mildew. Neotropics.

Phoradendron flavescens Nutt. ; parasitism of branches by a
woody bush. Florida. (Other Loranthaceae of several
genera attack the trees.)

Phymatotrichum omnivorum (Shear) Dug. ; rootrot. Texas.

(? Puccinia sp. reported in S. America as rust, probably
Coleosporium.)

Rhizoctonia (? solani); cutting infection. Cuba, Puerto Rico.

Septoria plumeriae Sacc. & Syd. ; zonal leafspot. C. America,
Bahamas, W. Indies.

Struthanthus spp. ; parasitism on branches by drooping woody
bushes. Costa Rica. (See Phoradendron.)

White papery leafspot. Physiological. Puerto Rico.

POA spp. Indigenous as well as introduced species of cold and
cool tropical country grasses, grown in lawns, pastures, and for
hay production. Disease studies on them have not been intensive,
and are mostly reports from the cooler parts of Brazil, Argentina, Chile, and Peru.

Amerosporium insulare Speg. ; leaf disease.

Ascochytella lilloana Petr. ; leafspot.

Claviceps purpurea (Fr.) Tul. ; inflorescence disease.

Coniothyrium fuegianum Speg. ; leaf and culm lesions.

Erysiphe graminis DC. (also called Oidium); powdery mildew.

Lophodermium oxyascum Speg. ; leaf disease.

Neospegazzinia pulchra Petr. & Syd. ; leaf damage.

Phoma fuegiana Speg. and P. microspora Pat. ; black leafspots.

Pleospora sphaerelloides Speg. ; ragged leaf, leafspot.

Puccinia spp. (all leafrusts): P. clematidis Lagh., P. exigua
Pat., P. graminis Pers., P. poae sudeticae (Westd.)
Joerst., P. poarum Niels., P. subandina Speg., and other
species.

Septoria macrosperma Speg. ; leafspots.

Stagonospora antarctica Speg. ; killing leaves.

Uredo porophila Speg. ; leafrust.

Uromyces fuegianus Speg. = U. chubutensis Speg. ; rusts.

POINCIANA see DELONIX

POLIANTHES TUBEROSA L. A Mexican unique soft herb (related
to agave), man-selected and man-nurtured so long that it is not
found in the wild. It is of considerable economic return to
small growers, market women and florists for its fragrant,
much esteemed long lasting flowers; "azucena, " "nardo, " "ja-
cinto occidental" Sp., "angélica, " "tuberosa" Po., "tuberose"
En.

Alternaria sp. (variable spores); rare flower injury. Guatemala market.

Botrytis cinerea Fr. ; flower brown-mold. C. America.

Cercospora sp. ; leafspot. El Salvador.
Colletotrichum sp. ; secondary following insect feeding punc-
    tures on stem, and leaf bases. Guatemala.
Erwinia carotovora (L. R. Jones) Holl. ; market softrot. El
    Salvador, Costa Rica, Puerto Rico, Virgin Islands.
Helminthosporium sp. ; leaf and stemspot. Texas.
Leaftip burn, unidentified cause. El Salvador.
Meloidogyne sp. ; root nematode. Florida, W. Indies.
Pythium sp. ; rootrot. Puerto Rico.
Rhizoctonia solani Kuehn; collar rot. El Salvador.
Root failure (diagnosed as from heavy acid soil). Honduras.
(? Rosellinia sp.); black rootrot. Guatemala.
Salt injury, from ocean spray. El Salvador, Puerto Rico.
Wilt, combined insect injury and bacterial infection. El
    Salvador.

POLYMNIA SONCHIFOLIA Poepp. & End. A plant in Andean Ecua-
    dor, Bolivia, Peru, Venezuela and Colombia at medium to high
    altitudes, this thistle-like perennial tuberous rooted important
    crop harvested at irregular times. On "ripening" in storage (it
    is not sold in markets but is popular among mountain people) it
    becomes a sweet tuber, eaten like a fruit; the most usual com-
    mon name is "yacón, " also "jiquimilla, " but on (non-Christian)
    religious festal days is known as "cocashka. " (This crop has
    had only a minimum amount of needed scientific attention, and
    very little plant disease observation. From conversations with
    a traveling priest in Peru, I learned native gardeners watched
    for any disease, removed notably affected leaves and stalks which
    they fed to their llamas, and were told to move their gardens at
    specified intervals decided upon by village headmen.)
    Erysiphe polygoni DC. ; the conidial state is Oidium. Powdery
        mildew.
    Phyllachora perlata Syd. ; tarspot.
    Puccinia polymniae Diet. & Neger; rust.
    Uredo polymniae Jacks. & Holw. ; rust.
    Uromyces polymniae Diet. & Neger; rust.

POPULUS ALBA L. , P. CANADENSIS Moench and P. NIGRA L.
    From the Northern Hemisphere and grown in South American
    areas influenced by Antarctic coolness, these fast growing trees
    produce white wood which is much needed and used for shipping
    boxes and construction, basket work and other purposes; "pobo, "
    "chopo, " "alamo" Sp. , "choupo, " "abele" Po. , "abele, " "poplar"
    En.
    Agaricus and Agrocybe (see Pholiota)
    Agrobacterium tumefaciens (E. F. Sm. & Town.) Conn. ;
        crowngall. Transmission through human manipulations.
        Sporadic occurrence.
    Ascochyta populorum Vogl. ; leafspotting. Argentina.
    "Capnodium"; sootymold. Widespread.
    Ciboria poronioides Speg. ; disease on bud, leaf, petiole.
        Argentina.

Collybia velutipes (Curt.) Sacc.; sapwood rot. Chile.

Cytospora chrysosperma (Pers.) Fr.; bark disease of older
   trees, cutting bench disease in nursery. Chile, Argentina,
   Bolivia.

Diplodia mutila Fr. (= Botryodiplodia sp.); dieback. Chile.

Dothidea sphaerioides Fr.; leaf disease with stromatic surface
   developments. Chile.

Eudarluca australis Speg. is the perfect state of Darluca
   filum (Biv.-Bern.) Cast.; the hyperparasite of Melampsora
   medusae in Colombia.

Flammula argentina Speg.; possibly a weak parasite, mushroom growths adjacent to and on fallen logs. Argentina,
   other parts of S. America.

Fomes applanatus (Pers.) Gill; trunk heartrot. Chile.

Hadrotrichum populi Sacc.; on leaves. Chile.

Hydnum pudorinum Fr.; stump and woodrot. Scattered. H.
   platense Speg. = Auriscalpium villipes (Lloyd) Snell &
   Dick; trunk rot. Argentina.

Loranthaceae, phanerogamic parasites. Scattered.

Melampsora spp. all rusts with usual countries of report;
   M. albertensis Arth. in Argentina. M. allii-populina Kleb.
   in Chile and Argentina (especially devastating in the Delta
   on Populus alba). M. laricipopulina Kleb. in Chile and
   Argentina. M. medusae Thuem. in Argentina and Colombia. M. populina Jacq. in Chile and Uruguay. M. popul-
   nea Karst. = M. aecidioides Schr. in Brazil, Argentina
   and Chile. M. rostrupii Wag. in Argentina.

Mycosphaerella populorum Thomp. apparently the asco-state
   of Septoria; leafspot. Widespread.

Nematodes, mostly rootknot. Argentina.

Pholiota crassivela Speg. with synonyms: Agaricus crassi-
   velus Speg., P. edulis P. Henn., P. algerita Brig.,
   Agrocybe aegerita Sing.; on trunks of diseased trees.
   Argentina, Chile.

Phomopsis sp.; branch disease. Argentina.

Phyllosticta populorum Sacc. & Rehm; leafspotting. Argentina.

Phytophthora citrophthora (Sm. & Sm.) Leon., and P. cryptogea
   Peth. & Laff.; rootrots. Argentina.

Pleospora sclerotioides Speg.; spot and tissue-rot disease on
   leaves. Argentina.

Polyporus adustus Fr.; trunk woodrot. Argentina.

Rosellinia necatrix (Hart.) Berl.; rootrot. Argentina.

Schizoxylon hormosporum Speg.; twig disease. Argentina.

Septoria musiva Pk. = S. populi Desm.; probably the most
   common leaf disease, stem-canker. Argentina, Uruguay.

Sphaceloma populi (Sacc.) Jenk.; leaf and twig spot-anthracno-
   sis. Brazil, Chile, Argentina.

Sphaerella crassa Auers.; leaf infections. Chile.

Taphrina aurea (Pers.) Fr.; catkin deformity. Chile.

Trametes hispida Bagl. and T. trogii Berk.; rots on branch
   and trunk. Argentina.

Typhula candida Fr.; wood infection. Chile.

Valsa ambiens Pers.; branch and twig canker.   Widespread.
Wood stain (? Ceratocystis sp.?) shown to me by wood carver
in El Salvador who had secured some poplar wood from
Argentina.

PORTULACA OLERACEA L. The low-growing, well-known salad
and cooking herb, adapted in the tropics to producing under poor
to good conditions, harvested in innumerable house-yard gardens,
important for dietary reasons, is popular and relished by people
in out-of-the-way areas, but I have never seen it listed or mentioned in cropping programs, or in analyses of monetary values
from it; "verdolaga," "verdolaga comestible" Sp., "beldroega"
Po., "portulaca" En.
   Albugo portulacae (DC.) Kuntze = Cystopus portulacae (DC.)
      Lév.; white rust of leaves and stems.   Neotropical.
   Capnodium spp. (may be other fungus genera); sootymolds
      under shade trees.   W. Indies.
   Cassytha filiformis L.; the green slow-growing dodder.
      Puerto Rico.
   Colletotrichum gloeosporioides Penz. the conidial and common
      state, its perfect stage is Glomerella; anthracnose spot on
      leaf, tip of branch disease, stem lesion.   W. Indies, C.
      America (probably S. America).
   Cuscuta americana L.; the yellow fast-growing dodder.
      Puerto Rico.   (Note: C. umbellata HBK. is reported on
      wild portulaca in the Dominican Republic.)
   Cystopus (see Albugo)
   Erwinia carotovora (L. R. Jones) Holl.; softrot in material
      accidentally stored over a weekend.   Costa Rica.
   Injury, from mechanical bruising, and common on remaining
      lush foliage after harvest.   El Salvador, Guatemala.
   Leafscorch, leaf split of edges, apparently physiological cause.
      Puerto Rico.
   Meloidogyne sp.; nematode rootknot.   Subtropical USA, Puerto
      Rico, Costa Rica, Peru.
   Mineral deficiencies causing chlorosis and other injuries in
      poor soils.   Lack of iron, zinc, boron, calcium.   (Proof
      of these lacks was seen on coffee trees adjacent, showing
      typical known symptoms of minor element deficiency
      troubles.)   Costa Rica.
   Mycena citricolor (Berk. & Curt.) Sacc. (by inoculation);
      leafspot, branch-tip destruction.   Puerto Rico.
   Pestalotia sp.; leafspot.   Costa Rica.
   Phyllosticta sp.; leafspot.   Brazil.
   Phymatotrichum omnivorum (Shear) Dug.; rootrot.   Texas.
   Phytophthora sp.; rootrot.   Puerto Rico.
   Pythium sp.; seedling damping-off.   Puerto Rico.
   Rhizoctonia (? solani), very fine mycelium; damping-off, basal
      rot of stem.   Costa Rica, El Salvador.
   Virus, mosaic mottle (undetermined) in Costa Rica.

POTHOS see SCINDAPSUS

PRUNUS spp. Favorite cultivated trees of the north temperate zone, brought to selected ecologic areas for fruits in the American tropics, these adaptable species are, in some places, successful. (Note: Wild species are indigenous to the American tropics but not covered here. When old known cultivated kinds were introduced they brought with them some (but not all) of their own long-known diseases and in the American tropics these trees are encountering new diseases; there are a few diseases that for the moment are unique to the American tropics. Tropical plant pathologists are not finished with their lists of parasites on the introduced Prunus species, let alone on the wild tropical American species in forest and jungle.) Common names are: (1) P. ARMENIACA L. = ARMENIACA VULGARIS Lam., "damasco," "albaricoque" Sp., "abricotiero" Po., "apricot" En.; (2) P. AVIUM L., "guindo," "cerezo dulce" Sp., "cereja-doce" Po., "sweet cherry" En.; (3) P. CERASUS L., "cerezo," "ciruelo" Sp., "cerezeira" Po., "cherry," "sour cherry" En.; (4) P. COMMUNIS Fritsch = AMYGDALIS COMMUNIS L. = P. AMYG-DALIS Stokes, "almendro" Sp., "amendoeira" Po., "almond" En.; (5) P. DOMESTICA L., "carozo," "ciruelo" Sp., "ameixeira" Po., "plum" En.; (6) P. PERSICA Sieb. & Zucc. = AMYGDALIS PERSICAE L. = P. VULGARIS Mill., "durazno," "melocotón" Sp., "pessegueiro" Po., "peach" En.

    Agrobacterium tumefaciens (E. F. Sm. & Town.) Conn.; crown gall, California (2, 3, 4, 6), Texas (1, 2, 5), Ecuador (4, 6), Peru (4, 5, 6), Colombia (5, 6), Bolivia (6), Uruguay (6), Argentina (4, 5).

    Aphelenchoides parietinus Steiner; root-region nematode. Brazil.

    Aphelenchus avenae Bastian; root-region nematode. Brazil.

    Armillaria mellea Vahl = Armilariella mella (Fr.) Karst.; rootrot. California (1, 2, 4, 5, 6).

    Aspergillus niger v. Tiegh.; fruitrot. California (6).

    Asterinella puiggari (Speg.) Th.; black mildew leafspots. W. Indies (5).

    Asteromella (see Phyllosticta)

    Botryosphaeria dothidea (Moug.) Ces. & deNot.; branch canker, fruit disease. Florida (6), Brazil (6).

    Botrytis cinerea Pers.; fruit graymold. California (1, 2, 3, 5, 6), Chile (6), Bolivia (6), Argengina (6).

    Caryospora putaminum (Schw.) deNot.; seed disease. California and Florida (6), Brazil (6).

    Cephalothecium roseum Cda.; fruit pinkmold. California (6), Ecuador (6).

    Cercospora cerasella Sacc. = C. circumscissa Sacc. its perfect state is Mycosphaerella cerasella Aderh.; leafspot, fruitspot and leaf shothole. Argentina (2, 3, 4, 5), Brazil (1, 2, 3, 5), Subtropical USA (5).

    Cercosporella persicae Sacc.; powdery attack on under side of leaf. Argentina (6).

    Ceriomyces pulchellus Speg.; root disease. Argentina (6).

    Chlorosis, physiological, various causes such as temporary

weather changes, minor element deficiences, other soil problems.

Cladosporium carpophylum Thuem. (synonyms: C. condylonema Pass., C. epiphyllum (Pers.) Mart., also the conidium-bearing fungus Fusicladium cerasi Sacc. = F. amygdali Duc., F. pruni Duc., and Asterula beyerinckii Sacc. with its perfect form Venturia cerasi Aderh.); common and severe fruitscab, freckle, branch and twig canker. California and Texas (1, 4, 5, 6), Honduras (6), Guatemala (6), Colombia (6), Ecuador (1, 4, 6), Peru (6), Brazil (2, 3, 6), Argentina (3, 6), Chile (4). C. herbarum Fr.; fruitrot complex. California (2).

Coccomyces hiemalis Higg. = Cylindrosporium padi Karst.; leafblight, spot and shothole, may attack fruit. Texas and California (1), Chile (2, 3).

Colletotrichum (see Gloeosporium)

Coniothecium persicae Speg.; on decorticated wood. Argentina (6).

Corticium salmonicolor Berk. & Br.; branch pink-disease, leafrot. Brazil (6).

Coryneum carpophilum (Lév.) Jauch or C. beijerinckii Oud.; blight and spot attacks on leaves, buds, stems, fruits. California (1, 3, 4, 5), Mexico (1), Colombia (4, 5, 6), Ecuador (6), Peru (6), Bolivia (2, 5, 6), Uruguay (1, 2, 5, 6), Chile (1, 2, 4, 5, 6), Argentina (1, 2, 3, 4, 6).

Criconemoides morgense Taylor; root nematode. California (6), Brazil (6), Argentina (6). C. xenoplax Raski; root nematode. Chile (2, 6).

Cryptovalsa platensis Speg.; twig disease. Argentina (1).

Cylindrocladium scoparium Morg.; leaf disease. Argentina (5).

Cytospora candida Speg.; bark disease. Argentina (6). C. leucostoma Sacc.; bark and twig disease. Mexico (5), Chile (4, 5, 6), Brazil (6).

Darluca australis Speg. (D. filum (Biv.-Bern.) Sacc.), is the conidial state of Eudarluca australis Speg.; hyperparasite of rusts on all host species. Widespread.

Diatrype macrothecia Speg. = Valsa macrothecia (Speg.) Kuntze; decay of pole. Argentina (3).

Dimerosporium pannosum Speg.; black mildew leafspots. Paraguay (5).

Diplodia or Botryodiplodia. Sometimes given as: D. natalensis P. Evans., D. persicina Grove, D. theobromae (Pat.) Now., the perfect state is Botryosphaeria quercuum (Schw.) Sacc.; dieback. Neotropics (5, 6 and probably on all spp. of host).

Erwinia amylovora (Bur.) Winsl. et al.; fireblight of twig and fruit. Florida and Texas (1, 5).

Eutypa heteracantha Sacc. f. sp. pruni-cerasi Sadeb.; on dead trunk. S. America (3).

Exidia glandulosa (Bull.) Fr. var. scutelliformis Speg.; branch infection. Argentina (3).

Exoascus (see Taphrina)
Fomes applanatus (Pers.) Gill. and F. leucophaeus Mont.;
   root diseases, heartrot of trunk. Chile (6).
Fumago vagans Pers.; sootymold. Ecuador and widespread
   (6).
Fusarium lateritium Nees; twig and branch disease. Argen-
   tina (5, 6); F. roseum Lk.; dieback complex. Chile (1,
   4, 6).
Fusicladium (see Cladosporium)
Gloeosporium armeniacum Speg. = Dothiorella armeniaea
   (Speg.) Arx; a branch anthracnose. Brazil (6), Argentina
   (1). G. amygdalinum Brizi = Colletotrichum gloeosporoides
   Penz. (its perfect state is Glomerella cingulata (Stonem.)
   Sp. & Schr. but this not reported occurring on Prunus spp.
   in American tropics); leafspot, fruitrot, green fruit drop,
   stem disease. Brazil (1, 6). Probably widespread.
Graphium cinerellum Speg.; twig death. Argentina (on Prunus
   spp.).
Gummosis disease, undetermined complex. Uruguay (6).
Helicotylenchus nannus Steiner; nematode at roots. Brazil (6).
Hemicycliophora sp.; nematode at roots. Peru (1, 6).
Heterodera schachtii Schm.; nematode at roots. Brazil (6).
Lenticelosis, disease of lenticels, unkown cause. Colombia
   (6).
Leptothyrium pomi (Mont.) Sacc.; fruitspot, flyspeck. Argen-
   tina (4, 5).
Megalocladosporium carpophilum Vien.-Bour.; fruit crack and
   rot. Uruguay (6).
Melanomma cucurbitarioides Speg.; bark disease. Argentina
   (6).
Meloidogyne spp. (M. exigua Goeldi, M. javanica Chitw. and
   others); rootknot nematodes. Peru (1, 5, 6), Chile (4, 6),
   Brazil (6), Argentina (6).
Monhystera stagnalis Bastian; nematode at roots. Brazil (6).
Monilinia fructicola (Wint.) Honey the asco-state, is sometimes reported as Sclerotinia fructicola (Wint.) Rehm which
   is the conidial fungus; fruit brownrot, blossom blight, twig
   infection. California (6), Guatemala (6), Colombia (6),
   Ecuador (6), Peru (6), Bolivia (6), Venezuela (6), Brazil
   (6), Argentina (3, 5, 6). M. fructigena (Aderh. & Ruhl.)
   Honey the asco-state of the conidial fungus Monilia fructigena Aderh. & Ruhl.; fruit brownrot, twig disease. Chile
   (6), Peru (6). M. laxa (Aderh. & Ruhl.) Honey the asco-
   state has the conidial fungus Monilia laxa Sacc. & Vogl. =
   Sclerotinia cinerea Schr.; fruit brownrot and mummy, twig
   blight. California (6), Honduras (6), Uruguay (6), Brazil
   (6), Argentina (6), Chile (5, 6).
Mucor stolonifer Ehr.; fruitrot. Chile (6).
Mycena citricolor (Berk. & Curt.) Sacc.; leafspot (by inoculation on 6). Costa Rica.
Mycosphaerella (see Cercospora)
Nectria cinnabarina Tode; branch and twig disease. Ecuador
   (6).

Oidium leicoconium Desm. is the conidial white powdery state
   of Sphaerotheca pannosa (Wallr.) Lév.; mildew of foliage
   and fruits. Subtropical USA, Colombia, Ecuador, Peru,
   Bolivia, Chile, Argentina, Brazil (6). In Chile and Argen-
   tina (noted on 1), in Ecuador (4), in Peru (5), in Uruguay
   (5).
Otthiella collabens Speg.; branch-rot disease. Argentina (1).
Paratylenchus sp.; nematode found in rootzone. Chile (2, 6).
Pellicularia koleroga Cke.; branch and leaf rot. Mexico (6).
Penicillium crustaceum Fr.; fruitrot. California (6), Colom-
   bia (6).
Periola cerasicola Speg.; fruit disease. Argentina (2).
Phoma persiciphila Speg.; stem-canker. Argentina (6).
Phomopsis sp.; dieback, stem-canker. Chile (6), Colombia
   (6), Uruguay (6), Argentina (4, 6).
Phorodendron spp.; parasitic phanerogams in tree tops.
   Colombia (3, 5, 6).
Phrygilanthus verticillatus Eichl.; parasitic phanerogam.
   Argentina (6).
Phyllosticta spp.; all cause leaf collapse and leafspots: P.
   cerasella Speg. in Argentina (3). P. cerasicola Speg. =
   Asteromella cerasicola (Speg.) Rup. in Argentina (3). P.
   prunicola Sacc. in S. America (3).
Phymatotrichum omnivorum (Shear) Dug.; root disease.
   Texas (5).
Physalospora spp. (reported as perfect states of Diplodia-
   Botryodiplodia spp.); branch and twig attack. Brazil (6).
   Dieback and gumming. Florida (6).
Podosphaera sp. reported as possible asco-state of Oidium
   in Ecuador (6) and Peru (5), and Chile (1, 6).
Polyporus cinnabarinus Fr., P. pinsitus Fr., P. versicolor
   Fr., and P. velutinus Fr.; all trunk woodrots in S.
   America (6) and other species of host.
Pratylenchus sp.; rhizosphere nematode inhabitant. Chile
   (1, 2, 6).
Protomyces persiciphilus Speg.; causing disease of green
   leaves. Argentina (6).
Psittacanthus cuneifolius Blume; parasitic flowering bush on
   host branch. Argentina (3, 5, 6).
Pythium ultimum Trow.; seedling disease. Argentina (6).
Rhabdospora persiciphila Speg.; a branch disease. Chile (6).
Rhizoctonia solani Kuehn the parasitic but non-sporuliferous
   form of Thanatephorous cucumeris (Frank) Donk; causes
   seedling damping-off. Peru (5). (The same parasite
   thought to be causing twig blight in Argentina (6). How-
   ever, this may be an aerial species type and not solani.)
Rosellinia necatrix (Hartig) Berl.; rootrot. Chile (2, 3),
   Argentina (6). R. pepo Pat.; white rootrot. El Salvador
   (6).
Schizophyllum alneum (L.) Schr.; branch and twig attack.
   Chile (6).
Schizoxylon bagnisianum Speg. var. minus Speg.; on fallen
   twigs. Argentina (5, 6).

Sclerotinia sclerotiorum (Lib.) dBy. = Whetzeliana sclerotiorum (Lib.) Korf & Dumont; occasional flower blight,
  green-fruitrot. California (1, 4, 6), Ecuador (6).
Sclerotium rolfsii (Curzi) Sacc. (perfect state is Pellicularia
  rolfsii (Sacc.) West); seedling bottom-rot disease. Texas
  (6).
Septobasidium spp. all felty fungi associated with insects in
  S. America: S. natalense Couch (on 5, 6). S. pseudopedicellatum Burt (on 6). S. saccardinum Mar. (on 3,
  5, 6).
Sirococcus persicae Speg.; on dead branches. Argentina (6).
Sphaeropsis sp.; branch canker. S. America (6).
Sphaerostilbe sp.; root disease. S. America (5).
Sphaerotheca (see Oidium)
Stereum purpureum (Pers.) Fr.; sporadic rootrot. Uruguay
  (5, 6), Argentina (1, 5).
Taphrina andina Palm with synonyms: T. atkinsonii Ray, T.
  reichel Werd., Exoascus andinus Sacc.; causes leaf curls
  and malformations. Mexico, Colombia, Ecuador (on 1). T.
  cerasi Sadeb.; leaf curl. Chile (3). T. deformans (Fckl.)
  Tul. = Exoascus deformans Fckl.; the classic leafcurl,
  sometimes witches' broom. In all countries on peach (6).
  T. pruni (Fckl.) Tul.; fruit deformation. Argentina (5),
  Uruguay (5).
Thecopsora areolata (Fr.) Magn.; crusty rust sori on leaves.
  Caribbean Islands (6).
Tranzschelia discolor (Fckl.) Tranz. & Litv.; leafrust. Ecuador (4, 6), Chile (5), Guatemala (6). Tranzschelia pruni-
  spinosae (Pers.) Diet.; common rust on leaf, severe defolia-
  tion. Argentina (1, 3, 4, 5, 6), Bolivia (2, 5, 6), Brazil
  (1, 4, 5, 6), Chile (2, 4, 5, 6), Colombia (1, 5, 6), Costa
  Rica (6), Ecuador (4, 6), El Salvador (6), Haiti (6), Honduras (6), Jamaica (6), Mexico (6), Peru (5, 6), Subtropical
  USA (1, 2, 3, 4, 5, 6), Venezuela (6). (Note: In the
  American tropics the common Prunus rust appears to me
  to be Tranzschelia pruni-spinosa. However, T. punctata
  Arth., as well as T. discolor Tranz. & Litv., are like-
  wise given, as well as Uredo persicae Speg., U. pruni
  Cast., and U. prunorum Lk.; Puccinia pruni Pers., P.
  pruni-spinosae Pers. also are used. For purposes of this
  list of diseases I am not going to attempt to out-do
  Tranzschelia specialists, and will call all these T. pruni-
  spinosae. Certainly from my standpoint it is best to use
  this simplification at this time.)
Trichodorus sp.; nematode found in root soil. Chile.
Tylenchorhynchus sp.; nematode collected in root soil. Chile.
Valsa cincta Fr. in Ecuador and V. leucostoma Pers. in
  Brazil; both involved in dieback (6).
Venturia (see under Cladosporium carpophylum)
Viruses: Peach Mosaic in Argentina (2, 4, 6), Peach Wart in
  Argentina (3, 6), Peach Yellows in Argentina (6).
Xanthomonas pruni (E. F. Sm.) Dows.; bacterial leafspot and

branch canker.   Brazil (6), Uruguay (5, 6), Argentina
(5).
Xiphinema americanum Cobb; root nematode.   Brazil (6),
    Chile (1, 2, 4, 5, 6).

PSIDIUM GUAJAVA L. and P. CATTLEIANUM Sab.   The true
    domesticated species selected out of several wild species of
    native trees of tropical America grown in plantations or back
    yards.   (Note: These two species appear to be about equally
    diseased by the same parasites where I have been able to study
    them in mixed plantings.) The commercial products are famed
    jellies, jams, and popular confections made without use of
    sugar, from their strongly fragrant fruits; "guayaba" Sp.,
    "goiabeira" Po., "guava" En.
    Acrostalagmus albus Fr., fungus fruiting on old foliage.
        Dominican Republic.
    Aegerita webberi Fawc.; fungus of whitefly on guava.   Florida,
        C. America.
    Aschersonia cubense Berk. & Curt.; fungus parasite on scale-
        like insect on guava.   Colombia, El Salvador.   A. paraen-
        sis P. Henn.; fungus of guava insect.   Brazil.   A. tur-
        binata Berk.; fungus of a sucking insect on guava (but of
        special note since the fungus also diseases the host).
        Dominican Republic.
    Asterina psidii Ryan; black leafpatches.   Puerto Rico.   A.
        psychotriae Ryan; black patch (but over-grown by several
        other fungi).   Dominican Republic.   ? A. puiggarii Speg.
        (see Opeasterinella).
    Asterolibertia crustacea Hansf.; on leaf.   Dominican Republic.
    Botryodiplodia theobromae Pat. (Diplodia) = Botryosphaeria
        quercuum (Schw.) Sacc.; dieback, twig and branch canker.
        Reports from Honduras, El Salvador but undoubtedly wide-
        spread.
    Botryosphaeria dothidea (Moug.) Ces. & deNot.; canker and
        branch dieback.   Florida.
    Capnodium (see Hypocapnodium)
    Catacauma subcircinans (Speg.) Th. & Syd. is given as
        causing scab-like leafspots in some countries.   (See, in
        this connection, Phyllachora subcircinans.)
    Caudella psidii Ryan = Asterinella puiggari (Speg.) Th.; black
        patches but not scabs on leaf.   Reported in W. Indies, and
        in some lists considered possible synonym of Phyllachora
        subcircinans, see mention of this under that binomial.
    Cephaleuros virescens Kunze; algal scurf-spot of leaf, fruit
        malformation (often primary injury by this parasite is
        followed by secondary fruit rotting organisms).   Common
        in moist areas, from Subtropical USA, to southern S.
        America.
    Cephalosporium lecanii Zimm.; a fungus on a scale insect.
        Widespread.
    Cercospora psidii Rang.; leafspots on upper surface of leaf.
        Brazil, Puerto Rico, Florida.   C. sawadae Yama.; spot
        disease on lower leaf surface.   Brazil.

Clitocybe tabescens (Scop.) Bres. = Armilariella tabescens
    (Scop.) Sing. ; root and crown rot.   California.
Colletotrichum gloeosporioides Penz. (perfect state sometimes
    found is Glomerella cingulata (Stonem.) Sp. & Schr.),
    synonyms: Colletotrichum psidii Putt. , C. psidii Curzi,
    Glomerella rufomaculans Berk. , G. psidii Del. ; cause of
    common anthracnose lesions of leaves, twigs, fruits.
    Neotropics.
Elsinoë pitangae Bitanc. & Jenk. is the asco-state of Spha-
    celoma psidii (see below).
Fomes fructicum Speg. ; heartrot of trunk.   Argentina. (Heart-
    rot of old trunk, with Fomes sp. fruiting bodies seen in
    both Guatemala and Costa Rica.)
Fusarium spp. (two kinds, one roseum-like and other solani-
    like); associated with twig and branch disease.   Costa Rica,
    Honduras, Puerto Rico, probably widespread.
Ganoderma sp. ; stumprot.   Puerto Rico.
Glomerella (see Colletotrichum)
Hypocapnodium guajavae (Bern.) Speg. = Capnodium guajava
    Bern. or Limacinia guajavae (Bern.) Sacc. & Trott. ; sooty-
    molds.   Neotropics.
Linhartia hoehnelii Rehm; rare leafspot.   Brazil.
Lophodermium subtropicale Speg. ; branch disease.   Argentina.
Loranthaceae, numerous genera and species are found on wild
    and cultivated guavas in areas where bird vectors are com-
    mon. (A few of these phanerogamic parasites are noted in
    this list, these and many others are seen as abundant in
    some situations where birds abound near jungle growths.)
Meliola psidii Fr. = M. amphitrica Fr. ; common black mildew
    and leaf patches.   Neotropics (favored by partial shade and
    moist conditions).
Meloidogyne spp. ; rootknot nematode to which guavas seem
    somewhat tolerant.   Widespread.
Mycena citricolor (Berk. & Curt.) Sacc. ; luminescent leafspot
    (natural occurrence in high mountain rainforest).   Puerto
    Rico.
Opeasterinella puiggarii Speg. , synonyms: Asterina puiggarii
    Speg. and Asterinella puiggarii (Speg.) Th. ; black mildew
    on leaves.   Paraguay and nearby countries. (Possibly
    Caudella psidii is a synonym.)
Oryctanthus cordifolius Urb. ; phanerogamic parasite on guava
    branches.   El Salvador, Guatemala.
Pellicularia koleroga Cke. ; branch disease and leafrot.   Col-
    lected in Costa Rica, Guatemala, Panama, and Puerto Rico,
    reported from Florida.
Penicillium spp. ; fruit rots.   Widespread.
Pestalotia glandicola Guba, P. psidii Pat. , and P. funerea
    Desm. ; fruit diseases, leaf and twig lesions.   Reports
    from Cuba, Ecuador, Brazil, Peru, Puerto Rico, Honduras,
    Costa Rica, no doubt more widespread.
Phoma psidii P. Henn. ; black lesion on leaf.   Scattered.
Phoradendron spp. , flowering bushes parasitic on host branches:

<u>P</u>. <u>acinacifolium</u> Mart. & Eichl. in Surinam.   <u>P</u>. <u>gracili-</u>
<u>spicum</u> Trel. in Costa Rica, Colombia, Venezuela.   <u>P</u>.
<u>obtusissimum</u> Eichl. in Surinam.   <u>P</u>. <u>rensoni</u> Trel. in El
Salvador.
Phyllachora <u>cayennensis</u> (Fr.) Cke.; leafspots.   Brazil, Surinam.   <u>P</u>. <u>goyazensis</u> P. Henn. = <u>Catacauma goyazensis</u> (Cke.)
Th. & Syd.; dull black tarspot.   Bolivia, Brazil, Colombia,
Guatemala, Honduras, Costa Rica, Paraguay.   P. subcir-
<u>cinans</u> Speg. = <u>Catacauma subcircinans</u> (Speg.) Th. & Syd.;
a shiny black tarspot, when immature a more diffuse
growth.   Brazil, Argentina, Paraguay, W. Indies.   P.
<u>tropicalis</u> Speg.; leafspots with scab-like stroma on both
upper and lower leaf surfaces.   Brazil, Argentina, Colombia,
Panama, Venezuela.
Phyllosticta <u>aracaicola</u> Bat.; small leafspot.   Brazil.   <u>P</u>.
<u>guajavae</u> Viégas; leafspot.   Brazil.
Phymatotrichum <u>omnivorum</u> (Shear) Dug.; rootrot.   Texas.
Physalospora sp.; on foliage.   Brazil.
Phytophthora <u>cinnamomi</u> Rands; rootrot.   Honduras.
Placosphaeria <u>guajavae</u> Bat.; leaf disease.   Brazil.
Pleurotus <u>magnificus</u> Rick; agaric on decayed wood.   S.
America.
Polyporus <u>versicolor</u> L.; trunk rot.   California.
Porella sp.; leafsmother from small leafy liverwort.   Costa
Rica.
Psittacanthus <u>calyculatus</u> Don; large red-flowered parasitic
bush on main trunk of host.   El Salvador.
Puccinia <u>psidii</u> Wint. = <u>P</u>. <u>sugneurophila</u> Speg.; defoliating
rust of foliage, and attacks fruit (favored by moderate
warm temperatures and high relative humidity).   Believed
to be widespread in distribution, special reports from
Puerto Rico, Cuba, Costa Rica, Jamaica, Trinidad, Ecua-
dor, Colombia, Brazil, Argentina, Uruguay.
Pycnopeltis <u>circinata</u> Bat. & Cayao; spots on leaves.   Brazil.
Rosellinia <u>bunodes</u> (Berk. & Br.) Sacc. and <u>R</u>. <u>pepo</u> Pat.;
root diseases.   C. America.
Scolecopeltis <u>psidii</u> Bat.; leaf infection.   Brazil.
Septobasidium <u>saccardinum</u> March.; felt-disease.   Scattered.
(? Septoria <u>rufomaculans</u> Berk. said to be imperfect state of
the long studied ascus-bearing fungus known as <u>Glomerella</u>
<u>rufomaculans</u> Sp. & Schr. = from later work G. <u>cingulata</u>
(Stonem.) Sp. & Schr.).
Spegazzinia <u>meliolicola</u> P. Henn.; hyperparasite of black mil-
dew disease fungus.   S. America.
Sphaceloma <u>psidii</u> Bitanc. & Jenk.; conidial phase of an
<u>Elsinoë</u>, causing spot anthracnose-scab on foliage and fruit.
Reports from Florida, Brazil, probably widespread.
Strigula <u>complanata</u> Fée; white lichen-spot.   Widespread.
Struthanthus <u>marginatus</u> Blume; weeping vine-like woody branch
parasite.   Paraguay, Brazil.   <u>S</u>. <u>orbicularis</u> Blume; another
vine-like woody branch parasite.   El Salvador, Guatemala.
Syncephalis <u>ubatubensis</u> Viégas; fruitrot.   Brazil.

Tondusia fuscata Bat.; leaf mildew. Brazil.
Trabutia cayennensis Sacc. and T. tropicalis Speg.; leaf
    lesions. S. America.
Trichomerium portoricensis Speg.; leaf black mold disease.
    Puerto Rico. T. psidii Bat.; leaf black mold. Brazil.
Tubercularia leptosperma Speg.; branch bark disease. Para-
    guay.
Valsa guayavae P. Henn.; in dieback complex. C. America,
    S. America.
Zythia psidii Bat.; branch disease. Brazil.

PUERARIA PHASEOLOIDES Benth. A controllable leguminous vine,
    recently introduced to the Americas from the Asiatic tropics for
    experimental ground cover, found adaptable to warm country;
    "pueraria," "kudzú de tropico" Sp. Po., "pueraria," "tropical
    kudzu" En.
    Botryodiplodia sp.; in secondary lesion on stems injured by
        workmen. El Salvador.
    Cassytha filiformis L.; green relatively slow-moving dodder.
        Florida.
    Cercospora pueraricola Yamam.; leafspot. Costa Rica, El
        Salvador.
    Cladosporium herbarum Lk.; recovered from ragged leaf
        edges. Costa Rica.
    Cuscuta americana L.; yellow fast dodder. Puerto Rico.
    Fusarium sp.; in isolations from diseased stems and root-
        knot affected roots. El Salvador.
    Meliola sp.; black mildew on leaves and stem. Puerto Rico.
    Oidium sp.; white powdery mildew. Puerto Rico.
    Phoma sp.; stem lesion. El Salvador.
    Rhizoctonia microsclerotia Matz; web-blight. C. America,
        Ecuador (also seen there on temperate zone kudzu). R.
        solani Kuehn; pre-emergence destruction of planted seed,
        and damping-off. Costa Rica, El Salvador.

PUNICA GRANATUM L. Ancient and Classicly known Old World
    bush-like tree, brought to New World tropics and subtropics,
    grows best in dryer conditions where disease appears les serious,
    bears tough shelled attractive red fruits; "carozo," "granada"
    Sp., "romã" Po., "pomegranate" En.
    Alternaria sp.; fruit rot. California.
    Aspergillus sp.; fruit rot. C. America, W. Indies.
    Botrytis cinerea Pers.; storage graymold rot. Common in
        many markets after long hauls and extended storage
        periods.
    Bursting on trees of fruits, physiological. Puerto Rico.
    Cercospora punicae P. Henn. with synonyms C. punicae Syd.,
        C. lythracearum Heald & Wolf (is commonly found as the
        conidial or imperfect state of Mycosphaerella lythracearum
        Wolf); causes leafspot and blotch, fruitblotch, defoliation.
        Probably general distribution. Special reports from Mexico,
        El Salvador, Guatemala, Honduras, Subtropical dry areas
        in USA, Bermuda, Puerto Rico, Venezuela, Brazil.

Clitocybe tabescens (Scop.) Bres. = Armilariella tabescens
    (Scop.) Sing. ; rootrot. Subtropical USA.
Colletotrichum gloeosporioides Penz. (no perfect state seen);
    anthracnose spots on foliage, fruitrot. Florida, Puerto
    Rico, El Salvador.
Coniothyrium sp. ; stem canker. S. America.
Coryneum carpophilum Jaunch; fruit rot. Argentina.
Cytospora punica Sacc. ; associated with dieback. C. America.
Ganoderma sessile Murr. ; stumprot. Puerto Rico.
Hadrotrichum (see Sphaceloma)
Megalonectria caespitosa Speg. ; trunk and branch canker.
    Brazil, Argentina.
Meloidogyne spp. ; rootknot nematodes. Scattered.
Mycosphaerella (see Cercospora)
Nematospora coryli Pegl. ; fruit dryrot. California (symptoms
    of same trouble seen in Guatemala City market).
Omphalia stuckertii Speg. ; root disease. Argentina.
Pellicularia koleroga Cke. ; thread-web blight of branch and
    leaf. Florida, Guatemala (dried sample sent to clinic in
    Costa Rica, said to come from Venezuela).
Penicillium expansum Lk. ; market bluemold of fruit. Neo-
    tropics.
Phoma sp. ; fruit black spot, black blotch. Brazil.
Phyllosticta granata Rang. ; leafspot. Brazil.
Phymatotrichum omnivorum (Shear) Dug. ; rootrot. Texas, un-
    confirmed report from Mexico.
Phytophthora parasitica Dast. ; canker on low part of stem.
    Argentina, Puerto Rico.
Sphaceloma punicae Bitanc. & Jenk. (also classified at one
    time as Hadrotrichum sp.); spot anthracnose of leaf and
    fruit. Argentina, Brazil, Texas, Dominican Republic,
    Louisiana.
Witches' broom, undetermined cause. Puerto Rico.

PUYA CHILENSIS Mol. and P. COERULEA Miers (classified by
    some botanists in genus PITCAIRNIA). Handsome, tall, dryland
    terrestrial bromeliads of the American tropics with spike-edged
    leaves, showy salable flowers, planted for ornament and hedges,
    brought from wilds by indigenes for valued barter; some common
    names are "eríza, " "piña cortadora, " "chagual, " "cardón, "
    "puya" Sp. (The diseases hereunder are all in Chile and typical
    dryland inhabitants.)
    Anthostomella puyaecola Speg. ; leafdisease with stromatic
        crusts with the similar A. valparadisiaca (Speg.) Sacc. =
        Entosardaria, and A. vestita Speg. ; leafrot.
    Chilemyces valparadisiacus Speg. ; associated with leafrot
        (possibly secondary).
    Circinotrichum maculiforme Nees; a leaf disease.
    Entosordaria (see Anthostomella)
    Jaffuela chilensis Speg. ; leafspot, sunken into mesophyll.
    Metasphaeria puyae Speg. ; leaf infections.
    Microthelia puyae Speg. = Didymospheria puyae (Speg.) Sacc. ;
        partly buried fungus in dead leaf.

Omphalia stuckertii Speg. = Marasmiellus stuckertii (Speg.)
    Sing.; root disease.
Phoma leptospora Speg. and P. puyae Speg.; both are leaf-
    spots.
Pleospora proteosperma Speg. and P. puyae Speg.; both leaf
    disease fungi.
Sirococcus puyae Speg.; apparently on dead leaf.
Sphaerella puyae Speg.; leaf disease.
Stemphyliopsis valparadisiaca Speg. = Stemphyliomma valpara-
    disiaca (Speg.) Sacc. & Trav.; on dead leaves.
Trichopeziza valparadisiaca Speg.; disease of leaf base.
Venturia puyae Speg.; leaf lesion.

PYRUS see MALUS

QUAMOCLIT COCCINEA (L.) Moench = IPOMOEA COCCINEA L.
    and Q. PENNATA Boj. = I. QUAMOCLIT L. Very similar
    native popular ornamental vines in gardens and on wire or
    switch-made fences and used in soil erosion control; grows to
    about three meters, has abundant scarlet flowers, finely divided
    leaves and some common names: "bejucocito rojo," "bejuco
    ciprés" Sp., "cypress vine" En.
        Albugo ipomoeae-panduratae (Schw.) Sw.; white rust. Sub-
            tropical USA, Puerto Rico, Virgin Islands, El Salvador,
            Panama (probably well distributed).
        Cercospora (? ipomoeae Wint.); leafspot. Puerto Rico.
        Coleosporium ipomoeae (Schw.) Burr.; leafrust. W. Indies,
            C. America.
        Meliola sp.; black mildew spots. Costa Rica, Guatemala.
        Meloidogyne sp.; mild rootknot. Florida.
        Mycena citricolor (Berk. & Curt.) Sacc.; luminescent leaf-
            spot. Costa Rica (occurrence both by inoculation and later
            natural occurrence found in field).
        Oidium sp.; white powdery mildew. Costa Rica.
        Phoma sp.; black blast of foliage. Costa Rica.
        Phyllosticta sp.; leafspot. Costa Rica.
        Puccinia crassipes Berk. & Curt. (synonyms: P. macro-
            cephala Speg., P. opulenta Speg.); rust. Argentina, Para-
            guay, Brazil.
        Rhizoctonia solani Kuehn is the very common non-spore
            bearing state of the basidiospore bearing soil fungus Thana-
            tephorous cucumeris (Frank) Donk; damping-off. Puerto
            Rico.

QUERCUS spp. Trees in mountain forests mostly of the northern
    sector of the American tropics, used for charcoal and some for
    minor structures, perhaps not of much importance commercially,
    a few species of small to large trees; "roble," "encina" Sp.,
    "carvalho," "roble" Po., "tropical oak" En.
        Antidaphne viscoidea Poep. & Endl.; phanerogamic parasite.
            Costa Rica.
        Armillaria mellea Vahl = Armilariella mellea (Fr.) Karst.;
            rootrot. California.

Cassytha filiformis L.; green and pungent-smelling dodder on
  young trees, near beach.   Florida.
Cercospora polytricha Cke.; leafspot.   Subtropical USA.
Coprinus fuscescens (Schaeff.) Fr.; agaric associated with
  half buried decayed oak branches.   Costa Rica.
Coryneum foliicolum Fckl.; leaf disease.   C. America.
Cronartium spp. (rusts that are alternate forms of pine
  rusts); C. cerebrum Hedgc. & Long, C. fusiforme Hedgc.
  & Hahn, C. quercuum Miy., C. strobilinum Hedgc. &
  Hahn.
Cuscuta exaltata Eng.; dodder on seedlings.   Texas.
Cylindrosporium kellogii Ell. & Ev.; leafspot.   C. America.
Daedalea ambigua Bres.; woodrot.   Subtropical USA, C.
  America, Mexico (scattered occurrence).
Dendrophthora costaricensis Urb.; phanerogamic parasite on
  branch.   Costa Rica.
Diplodia longispora Cke. & Ell. (Botryodiplodia sp. =
  Botrosphaeria quercuum (Schw.) Sacc.); twig blight.   Sub-
  tropical USA, and no doubt widespread.
Dothidella janus (Berk. & Curt.) Hoehn.; leafspot.   Subtropical
  USA.
Endothia gyrosa (Schw.) Fr.; on dead wood of branch and
  root.   Subtropical USA.
Epiphytes, often exuberant growths of mosses, ferns, lichens,
  algae, liverworts, bromeliads, orchids, ericaceae, and
  other kinds of plants make heavy weights that break off
  branches of old trees in medium cool moist mountain
  forests of Central America.
Fomes spp. (sporocarps producing abundant spores, fleshy
  parts not long lived in tropics in moist areas where insects
  are abundant): F. applanatus (Pers.) Gill, F. australis
  Cke., F. calkinsii (Murr.) Sacc., F. igniarus (L.) Kickx,
  F. marginatus Fr., F. marmoratus (Berk. & Curt.) Cke.,
  and probably other species.   All cause butt and log heartrot.
Fracchiae cucurbitarioides Speg.; wood decay.   Argentina.
  F. heterogenea Sacc.; wood (and bark) decay.   Argentina.
  F. cucurbitarioides Speg. f. sp. querci-sessiliflora Speg.;
  rot disease on branch and trunk.   Argentina.
Ganoderma curtisii (Berk.) Murr. and G. lucidum (Leyss.)
  Karst.; both cause spongy white heartrot.   Subtropical USA,
  C. America.
Godroniopsis quernea (Desm.) Diehl & Cash; dieback.   Sub-
  tropical USA.
Helostroma album (Desm.) Pat.; on foliage.   (? Locality in
  S. America.)
Hymenogaster albellus Mass. & Rod., and H. ruber Harkn.;
  decay complex.   C. America.
Hypoxylon annulatum (Schw.) Mont., and H. marginatum (Schw.)
  Berk.; canker, sapwood rot.   Subtropical USA, S. America.
Lentinus lecomtei Schw.; sapwood rot.   Florida.
Leptosphaeria dryadea (Cke. & Harkn.) Sacc.; leafspot.   Wide-
  spread.

Loranthus spp. ; parasitic large and small mistletoe-like
  bushes. Widespread.
Macrophoma mexicana Sacc; leaf disease. Mexico.
Marasmius crinis-equi Kal. = M. equiricinis Berk. ; the
  horse-hair blight fungus. Scattered.
Microsphaera alni (DC) Wint., M. alphitoides Griff. & Maubl.,
  M. quercina Burr. probably all are perfect state of
  Oidium sp. ; the notable generally distributed white powdery
  mildew.
Microstoma album Sacc. ; a white fungus disease on underside
  of leaf. Argentina.
Monochaetia desmazieri Sacc. ; leafspot. Subtropical USA.
Morenoella quercina (Ell. & Mart.) Th. ; leafblotch. Sub-
  tropical USA, C. America, Mexico.
Mycosphaerella spp. ; leafspots. C. America.
Oidium alphitoides Griff. & Maubl. or O. quercinum Th. ;
  conidial state of Microsphaera (see above).
Orobranche sp. ; root attack by a phanerogam. Mexico.
Paratylenchus sp. ; rhizosphere inhabiting nematode. Chile.
Passalora melioloides Tracy & Earle; black leafspot. Florida.
Phomopsis sp. (? = Phoma sp.); branch and leaf and acorn
  disease. Scattered occurrence.
Phoradendron spp. (parasitic flowering bushes on host
  branches); P. crispum Trel. in Costa Rica, P. flavescens
  Nutt. in Subtropical USA and C. America, P. scaberrimum
  Trel. in Mexico (probably several other species and in
  other countries).
Phyllosticta spp. ; leafspotting diseases. Subtropical USA.
Phytophthora citrophthora (Sm. & Sm.) Leonian; tree base
  ink disease. Argentina.
Pluteus cervinus (Schaef.) Sacc. f. sp. petasatus Fr. ; asso-
  ciated with wood decay. S. America.
Polyporus spp. ; canker, heart, butt, and other kinds of rots
  of living tree. Occurrence widely scattered.
Poria cocos Wolf; white heartrot (also develops as tuckahoe
  sclerotia in some cases). Subtropical USA. P. unita Cke. ;
  spongy heartrot. Subtropical USA.
Pratylenchus sp. ; rhizosphere inhabiting nematode. Chile.
Pseudovalsa longipes (Tul.) Sacc. ; twig canker, dieback.
  Texas, Costa Rica.
Psittacanthus americanus Mart. ; parasitic flowering bush (pro-
  duces "wood rose"). Costa Rica. P. calyculatus Don;
  another "wood rose" disease. El Salvador.
Ptychogaster cubensis Pat. ; white heartrot. Florida, Cuba,
  El Salvador.
Rosellinia spp. ; rootrots. C. America.
Scleroderma spp. ; associated with dying roots. S. America.
Septobasidium spp. ; felt (fungus growth over scale-infested
  branches). In many countries.
Septoria quercicola (Desm.) Sacc. ; leafspot. Florida.
Stagonospora virens Ell. & Mart. ; a leaf disease. Florida.
Stereum spp. ; woodrot, at times possibly saprophytic. Wide-
  spread.

Struthanthus marginatus Blume and S. orbicularis Blume;
    destructive vine-like woody parasites on branches.   C.
    America, Panama, Mexico, Colombia.
Taphrina caerulescens (Mont. & Desm.) Tul.; leafblister
    disease.   Subtropical USA, (one leaf with the blister
    symptom sent to me by mail from Mexico).
Trabutia conzattiana Sacc., T. erythrospora (Berk. & Curt.)
    Th. & Syd., and T. quercina Sacc. & Roum.; all are
    rounded and irregular leafspots (stromatic fungus growth
    under leaf cuticle), tarspot-like.   Subtropical USA, Mexico.
Trametes hydnoides (Sw.) Fr.; a frequently seen woodrot.
    Neotropical.
Trechispora subtrigonosperma Rogers & Jacks.; woodrot with
    Hydnum-like sporocarps.   Scattered.
Trophotylenchulus floridensis Raski; root-attacking nematode.
    Florida.
Xiphinema americanum Cobb; nematode, in root soil.   Florida.

RANDIA spp.   Small spiny trees often growing on poor soils of the
West Indian, Central American, and South American coasts.
Important sources of firewood, protected as land stabilizing
agents, fruit juices employed as ink; "escambrón," "tintero"
Sp., "thorn tree," "ink tree" En.
    Aecidium spp., leaf rusts some reports: A. abscendens Arth.
        in Subtropical USA, Puerto Rico, Costa Rica, A. pulveru-
        lentum Arth. in Mexico, Panama, A. randiae P. Henn. =
        A. basanacanthae P. Henn. in S. America.
    Armillaria sp.; rootrot.   Scattered.
    Cassytha filiformis L.; the slow, enveloping: green dodder.
        Puerto Rico.
    Cephaleuros virescens Kunze; algal leafspot.   W. Indies.
    Chaetophoma chlorospora Speg.; leaf disease.   Paraguay.
    Colletotrichum (? gloeosporioides); anthracnose leafspots.
        Puerto Rico.
    Corticium salmonicolor Berk. & Br.; branch-and-leaf pink-
        disease.   Neotropics.
    Cuscuta americana L.; the fast twining and sap sucking yellow
        dodder.   Puerto Rico (and another smaller, slower growing
        Cuscuta sp. found in Puerto Rico and Costa Rica).
    Dimerosporium pannosum Speg.; black leafblotch.   Paraguay,
        Brazil.
    Elsinoë puertoricensis Jenk. & Bitanc.; scab spots on leaf
        and fruit.   Puerto Rico (? more widespread).
    Fusicladium sp.; leaf disease.   S. America.
    Loranthus spp.; large woody bush parasites on host branches.
        Puerto Rico.
    Marasmius sp.; thread blight (no web produced).   Costa Rica,
        El Salvador.
    Meliola psychotriae Earle; black mildew-spots.   W. Indies,
        probably widespread.
    Phthirusa spp.; low-growing mistletoe-like branch parasites.
        Costa Rica.

Phyllachora randiae Rehm = Trabutia randiae (Rehm) Th. &
    Syd.; tarspot. Trinidad, Puerto Rico, Santo Domingo,
    Bolivia, Virgin Islands, probably widespread.
Schizophyllum commune Fr.; associated with dying branches.
    Puerto Rico.
Strigula spp.; lichen leafspots white to yellow and brown.
    Puerto Rico, Virgin Islands, Costa Rica, Guatemala,
    Panama.
Taphrina randiae Rehm; leaf swellings.   Brazil.
Trabutia (see Phyllachora)

RANUNCULUS spp.   Several ornamentals, cool and moist-land
    flowering herbs, some imported to grow in gardens; "botón de
    oro" Sp., "botão-de-ouro," "ranúnculo" Po., "buttercup,"
    "ranunculus" En.
    Aecidium ranunculacearum DC., A. ranunculi Schw., and
        A. ushuwaiensis Speg.; rusts in S. America (the latter
        described from material collected in Tierra del Fuego).
    Aphelenchoides fragariae Chr.; leaf-infecting nematode.
        Florida.
    Entyloma ameghinoi Speg.; leaf smut.   Argentina (Patagonia),
        Chile.
    Erysiphe polygoni DC. the ascoform, its conidial state is
        Oidium sp.; white powdery mildew.   Widespread.
    Oidium = Erysiphe.
    Peronospora ficariae Tul.; downy mildew.   Scattered.
    Puccinia spp. are typical rusts: P. andina Diet. & Neg.
        in S. America. P. gibberulosa Schr. especially in
        Argentina. P. nubigena Speg. in Argentina.
    Pythium spp.; rots and damping-off.   Widespread.
    Septoria asiatica Speg.; leafspot on imported species of host.
        Argentina.
    Synchytrium andinum Lagerh.; brown leafwarts.   Ecuador.

RAPHANUS SATIVUS L.   Brought in from the north temperate
    zone and found to grow in special ecological conditions in the
    American tropics, it is the European long known salad (and
    cooking) root crop.   Disease information for Neotropical patholo-
    gists on this crop in the American tropics is more or less de-
    pendent on USA and European publications.   There has not been
    the pressure to make extensive and intensive studies of its
    diseases in the American tropics, since the crop, thus far, is
    not grown in the vast quantity (and is still not so popular) as it
    is in gardening areas of northern metropolitan parts of Europe
    and the USA; "rabano" Sp., "rabanete" Po., "radish" En.
    Albugo candida Pers. = Cystopus candidus (Pers.) Lév.;
        white rust of leaves, on stems and pods of seedplants.
        Probably quite common and general, some special reports:
        El Salvador, Subtropical USA, Guatemala, Honduras,
        Venezuela, Peru, Bolivia, Uruguay, Brazil.
    Alternaria herculea (Ell. & Mart.) Elliott; leafspot. Vene-
        zuela, Argentina.

Aphanomyces raphani Kendr. ; blackroot. Florida.

Cercospora cruciferarum Ell. & Ev., C. bloxami Berk. &
   Br., Cercosporella albo-maculans Sacc. ; leafspots. Widespread.

Erwinia carotovora (L. R. Jones) Holl. ; slimy market softrot. Neotropical.

Erysiphe polygoni DC. = Oidium sp. ; white powdery mildew.
   Scattered.

Fusarium oxysporum Schl. f. sp. raphani Kendr. & Sny. ;
   wilt. California.

Meloidogyne sp. ; rootknot. Brazil, Subtropical USA.

Oidium (see Erysiphe)

Peronospora parasitica Pers. with synonyms: P. parasitica
   dBy., P. brassicae Gaum. ; downy mildew. Subtropical
   USA, Brazil, Argentina, Chile.

Phoma lingam (Tode) Desm. ; disease lesions on seed producing plants. Subtropical USA, Guatemala.

Phymatotrichum omnivorum (Shear) Dug. ; rootrot. Texas.

Plasmodiophora brassicae Wr. ; clubroot (not reported on
   radish, but the fungus occurs in Mexico, Peru, Brazil,
   and Argentina on cabbage, and radishes are very susceptible.)

Pythium ultimum Trow. ; damping-off in warmer soils (another undetermined Pythium in coolest, wet areas).   El
   Salvador.

Rhizoctonia microsclerotia Matz is the parasitic hyphal state
   that bears no spores, its perfect state is Pellicularia
   microsclerotia (Matz) Web. ; the cobweb leafrot. Subtropical USA, W. Indies, El Salvador. R. solani Kuehn
   (perfect sporulating state is Thanatephorous cucumeris
   (Frank) Donk); damping-off, stemrot. Guatemalan highlands, El Salvador highlands, Peru, probably widely spread.

Sclerotinia sclerotiorum (Lib.) = Whetzeliana sclerotiorum
   (Lib.) Korf & Dumont; crownrot, market watery rot. C.
   America, Chile.

Viruses: Beet Curly Top in California, Enation Mosaic in
   Guatemala, Mosaic Mottle in Trinidad.

Xanthomonas campestris (Pam.) Dows. ; blackrot. Scattered
   records, but probably common in occurrence.

RAUWOLFIA spp. Some widely grown tropical shrubs and small
   trees containing alkaloids (reserpines) used since prehistory by
   indigenes for tranquilizing purposes, now employed in worldwide
   medical practice. There are few reports of diseases on these
   hosts.

   Aecidium ochraceum Speg. ; rust. S. America.

   Cephaleuros virescens Kunze; algal leafspots. Costa Rica
      (mountain area).

   Cercospora liebenberghii Syd., C. rauwolfiae Chupp & Muller;
      leafspots of various sizes and configurations. Costa Rica,
      Mexico, Venezuela.

   Dothiorella proteiformis Speg. ; on diseased and decayed twigs.
      Argentina.

Eutypella ludibunda Sacc. ; branch decay fungus. S. America.
Leaf edge crumple and necrosis, cause unknown. Costa Rica.
Meliola tabernamontanae Speg. ; black mildew of leaf. Widespread.
Phymatosphaeria argentina Speg. with synonyms: Myriangium argentinum (Speg.) Sacc. & Syd. , M. argentinum (Speg.) P. Henn. , M. duriaei Mont. & Berk. ; cortex and bark disease. Argentina.
Phytophthora sp. and Pythium sp. combined; cutting failures. C. America.
Sootymold (unidentifiable specimen, immature). Mexico.

RAVENALA MADAGASCARENSIS Gmel. Its origin Madagascar, it is planted in many parts of the warmest American tropics, a uniquely ornamental, tall specimen type of tree with on its trunk broad musa-like large leaves, set in two ranks like a fan, the petiole cups always containing water; "palma de viajero" Sp. Po. , "traveler's palm" En.
Botryodiplodia sp. (Diplodia); blackened petiole lesions. El Salvador.
Botrytis sp. ; isolation from diseased leaf edge. El Salvador.
Fomes sp. ; on decayed trunk. Colombia.
Fusarium (saprophytic type); on dead trunk tissue. El Salvador. F. oxysporum Schl. f. sp. cubense (E. F. Sm.) Sny. & Hans. ; typical wilt (possibly race 4) in Costa Rica.
Leaf chlorosis and ultimate decline. Cause physiological. Guatemala.
Meliola sp. ; leaf blotches. Costa Rica.
Nigrospora sp. ; leaf spotting. El Salvador.
Pestalotia sp. ; isolation from leaf. El Salvador.
Strigula spp. ; various colors of lichen spots on old leaves. Puerto Rico.
Trunk dry-rot. Undetermined Basidiomycete. Guatemala.

RHEEDIA EDULIS Pl. & Tr. Native to the American tropics, a large to medium-sized tree with subacid, relished, much used edible fruits; "mamey cimarón, " "chaparrón" Sp. , "limão do matto" Po. , "acid mamey" En.
Algae (masses in tree crotch) probable associate cause of trunk "leak disease, " in rainforest locality. Puerto Rico, Panama, Cuba.
Cephaleuros sp. ; leafspot. Puerto Rico.
Dothiorella pakuri Speg. ; leafspots in Argentina, El Salvador, Puerto Rico.
Loranthaceae, several different kinds, woody bushes parasitizing host tree branches. C. America.
Micropeltis rheediae Rehm; leaf blotch-spot. S. America, C. America.
Rosellinia sp. ; rootrot. Honduras.
Strigula complanata Fée; white lichen leafspot. C. America, Puerto Rico.
Yeast; causing a fruit rot. Honduras.

RHODODENDRON INDICUM Sw. = AZALEA INDICA L. and R.
MUCRONATUM Don = A. LEDIFOLIA Hook. (other species and
hybrids). All introduced from temperate zones, grown in
mountains and cool locations, sold by florists and growers as
exotic flowers and flowering bushes; "azalea" En. Sp. Po. The
disease list below undoubtedly will be much increased as the
hosts are produced and used more and more popularly.

Armillaria mellea Vahl = Armilariella mellea (Fr.) Karst.;
    rootrot. California.
Botrytis cinerea Pers.; disease blight of flower and branch
    tip, as well as of seedlings and cuttings. Scattered in
    several areas.
Cladosporium herbarum Pers.; leaf edge and flower spot.
    Widespread.
Clitocybe tabescens (Scop.) Bres. = Armilariella tabescens
    (Scop.) Sing.; rootrot. Florida, Alabama, Louisiana.
Colletotrichum gloeosporioides Penz. (perfect state not re-
    ported); common anthracnose. Guatemala, Subtropical
    USA.
Cuscuta gronovii Willd.; dodder. Florida.
Exobasidium spp.; leaf and shoot galls: E. azaleae Peck in
    Argentina. E. discoideum Ell. in Brazil. E. rhododendri
    Cra. in Chile. E. vaccinii (Fckl.) Wor. in Subtropical
    USA, Jamaica.
Mycosphaerella polyspora Johans. f. sp. octospora Rang.;
    large leafspot. Argentina, Brazil.
Ovulinia azalea Weiss; flower disease, limb blight. Sub-
    tropical USA.
Pellicularia koleroga Cke.; web-thread blight. Louisiana,
    Florida, Guatemala.
Pestalotia rhododendri (Sacc.) Guba; flower and leaf lesions.
    Noted especially in Chile, probably general in occurrence.
Phomopsis sp.; dieback. Subtropical USA, Panama.
Phyllosticta rhododendri Brun.; leafspot. Subtropical USA,
    Chile, probably more widespread.
Phytophthora cinnamomi Rands; rootrot, little leaf. Wide-
    spread.
Pucciniastrum myrtillii (Schum.) Arth.; rust. Subtropical USA.
Pycnostysanus azaleae Mason = P. resinae Lind.; leafscorch.
    Panama.
Rhizoctonia solani Kuehn = Thanatephorous cucumeris (Frank)
    Donk; damping-off of seedlings, cutting disease, stemrot.
    Widespread in warm areas.
Septoria azaleae Vogl., S. azaleae-indicae Maubl.; leaf
    disease. Brazil, Argentina.
Strigula complanata Fée; white lichen leafspot. Brazil.

RICINUS COMMUNIS L. Originally from tropical Africa, it is
grown widely in the American tropics; its brilliantly marked
seeds are extracted for medicinally important character, and
seed oil is marketed as lubricant in fine machinery. Often
planted as an ornament it is a small tree to tall shrub; "higue-
rillo" Sp., "mamoneira" Po., "castor bean" En.

Alternaria sp.; leafspot.   Florida.
Botryodiplodia theobromae Pat., B. natalensis P. Evans, also
　　called Diplodia, is the conidial state of the asco-form
　　which is Botryosphaeria quercuum (Schw.) Sacc. = Physa-
　　lospora rhodina Berk. & Curt.; an often severe disease of
　　stem and pods, often dieback.   Neotropical and of special
　　note in Subtropical USA, Ecuador, C. America, and Brazil.
Botryosphaeria dothidea (Moug.) Ces. & deNot.; severe stem-
　　canker.   C. America, Florida.
Botryotinia ricinis (Godf.) Whet. (the perfect state of Botrytis
　　sp.); it causes graymold severe inflorescence blight.
　　Specially seen in C. America, Cuba, St. Thomas, Puerto
　　Rico, and reported in Brazil and Ecuador but is much
　　more widespread.   (Sometimes called Sclerotinia.)
Capnodium sp.; sootymold (on old, small trees).   Puerto Rico.
Ceratocystis sp. = Ceratostomella sp.; branch death, wood-
　　stain.   C. America.
Cercospora canescens Ell. & Mart.; the smaller leafspots.
　　W. Indies, C. America.   C. coffeicola Berk. & Cke. =
　　Cercosporina coffeicola (Berk. & Curt.) Speg.; medium
　　large leafspots.   Costa Rica.   C. crotonophila Speg.; leaf-
　　spot.   Brazil (may be more widespread).   C. ricinella
　　Sacc. & Berl. (synonyms: Cercosporina ricinella (Sacc. &
　　Berl.) Speg., Cercospora ricini Speg., C. albido-maculans
　　Wint.); the largest spot, the so-called "white" leafspot.
　　Subtropical USA, Puerto Rico, Barbados, Trinidad, Nica-
　　ragua, Guatemala, El Salvador, Honduras, Peru, Vene-
　　zuela, Argentina, Brazil, Ecuador.
Chromocytospora ricinella Speg. (Phomopsis-like); black twig
　　disease.   Argentina.
Cladosporium sp.; mold on fruits.   Brazil.
Clitocybe tabescens Bres. = Armileriella tabescens (Scop.)
　　Sing.; rootrot.   Florida.
Colletotrichum gloeosporioides Penz. (is the conidial state of
　　the asco-fungus Glomerella cingulata (Stonem.) Sp. & Schr.,
　　the latter state seldom if ever noted on castorbean plants);
　　common anthracnose often present but usually mild in ef-
　　fect.   Neotropics.
Diaporthe ricini Speg.; branch disease.   Argentina.
Didymella ricini Ell. & Ev.; disease on stem of plants grown
　　as annual.   Subtropical USA.
Eutypa spp.; all are trunk cancers, branch lesions: E. lata
　　(Pers.) Tul., E. ludibunda Sacc. and E. ludibunda Sacc. f.
　　sp. ricini-communis Speg.   These are apparently (but not
　　certainly) confined to southern S. America.
Fusarium spp. spores often seen, sometimes these fungi are
　　isolated from diseased tissues, are generally semi-parasitic
　　and saprophytic types accompanying other disease or injury
　　from other primary cause.   Neotropical.
Irpex cartilagineus Speg. (Hydnum-like sporocarp.); trunk rot.
　　Argentina.
Laestadia guarapiensis Speg.; branch disease.   Paraguay.

Macrophomina phaseoli (Taub.) Ashby; stem tissue disease
with multitude of small buried black specks. Peru.

Meloidogyne sp.; rootknot nematode. Subtropical USA, Peru,
Honduras. (Perhaps widespread as its mild effects did not
kill plants where seen by me and thus occurrences can be
undiscovered.)

Mycena citricolor (Berk. & Curt.) Sacc.; leafspot. Costa
Rica.

Peroneutypa heteracantha Sacc. & Berl. f. sp. ricini-communis Speg.; branch, stem, and trunk lesions. Argentina.

Phoma ricinicola Speg.; branch disease. Argentina.

Phyllosticta ricini Rostr.; leafspot. Brazil.

Phymatotrichum omnivorum (Shear) Dug.; dryland rootrot.
Texas.

Physalospora abditis Stev. and P. rhodina Cke.; stem tip-
collapse, branch lesion, dieback, capsule disease. W.
Indies, Honduras (moist area). (See also under Botryodiplodia.)

Phytophthora palmivora Butl.; stemrot near ground line, rootrot. Costa Rica. P. parasitica Dast. var. nicotianae (de
Haan) Tuck.; stem canker, rootrot. Puerto Rico.

Pleospora sp.; leafspot with gray center and marked stippling.
C. America.

Pseudomonas solanacearum E. F. Sm.; bacterial wilt. Scattered.

Rhizoctonia microsclerotia Matz = Pellicularia microsclerotia
(Matz) Web.; is spiderweb blight of foliage. C. America,
S. America. Typical R. solani Kuehn = Thanatephorous
cucumeris (Frank) Donk which latter state bears basidia
and basidiaspores; damping-off, shoot canker, base-strangle.
El Salvador, Guatemala, Honduras, Peru, Florida, Ecuador.

Schizophyllum commune Fr.; on dead stems. C. America,
Florida.

Sclerotinia (see Botryotinia)

Sclerotium rolfsii (Curzi) Sacc.; non-sporulating state seen of
a highly parasitic fungus that is studded with small sclerotia,
causing ground-line rot. Florida, Trinidad.

Sphaerella ricinicola Speg.; in branch disease complex. Argentina.

Strigula complanata Fée; lichen white spotting on older trees.
C. America.

Thyridium argentinense Speg.; lesions killing branches.
Argentina.

Tryblidiella rufula (Spreng.) Sacc. = Rhytidhysterium rufulum
(Spreng.) Petr.; on dead branches. El Salvador, Costa
Rica.

Virus causing chlorosis and malformed leaves (apparently
spread by white fly) in Honduras.

Xanthomonas ricinicola Dows.; bacterial leafspot. Brazil,
Nicaragua.

ROLLINIA DELICIOSA Saff.   Medium-sized fruit tree of northern

Brazil (West Indies types somewhat inferior), produces abundantly a much liked fruit; most common name is "biribá" Po.
Sp. En.   This tree deserves considerable attention respecting
disease as where seen in the West Indies and Honduras many
small trees appeared weak and the foliage severely damaged.
The following few diseases have been reported:

Colletotrichum sp. ; anthracnose.   Brazil.

Glomerella cingulata (Stonem.) Sp. & Schr. (? the perfect
    state of the Colletotrichum given above) causes dieback
    and fruitrot in Florida.

Rhizopus nigricans Ehr. ; market and storage fruitrot.   Brazil.

Uredo cherimoliae Lagerh. = Phakopsora cherimoliae Cumm. ;
    rust.   Brazil, Cuba, Subtropical USA, Trinidad.

ROSA sp.   These are woody trailing and bush plants, originally
from the Old World but that grow with care under a wide range
of tropical conditions and are of great value to florists and
growers as they produce the almost universally salable flowers
and, moreover, are often in pots or tubs in the tropics.   Flowering plants, with such common names as; "rosa, " "rosal" Sp. ,
"rosa, " "roseira" Po. , "rose" En.   (Note: A crop so specialized
as rose, when brought to the tropics, requires much more attention as to disease there than has been given it up to the present.
There are comparatively few tropical pathologists and these are
under great pressure to work on tropical foods and fibers crops.
In Costa Rica in my home garden were 22 rose bushes and in one
day I counted on them 36 diseases, of which four were to me unidentified leafspots, but there were two unexpected root diseases
I did determine and these I believed had never been reported
before on this host.)

Actinonema (see Diplocarpon)

Agrobacterium tumefaciens (E. F. Sm. & Town.) Conn;
    crowgall.   Texas, Jamaica, Argentina.

Alternaria sp. (small and irregular spores, possibly A. al-
    ternata (Fr.) Keiss. or A. circinans (Berk. & Curt.) Bolle);
    a few leafspots.   Scattered in subtropical USA, Chile,
    Costa Rica, W. Indies.

Armillaria mellea Vahl. = Armilariella mellea (Fr.) Karst. ;
    rootrot.   Subtropical USA.

Aspergillus spp. ; dieback complex and on old market flowers.
    Costa Rica, Honduras, Cuba, Guatemala, El Salvador.

Basidiomycete hyphae of undetermined fungus, basal rot of
    old bush.   Puerto Rico.

Botryodiplodia theobromae Pat. , with synonyms such as: B.
    rosae Berk. & Curt. = Microdiplodia rosae Tassi (often
    the genus is given as Diplodia) is the conidial form of the
    asco-state Physalospora rhodina Cke. = Botryosphaeria
    quercuum (Schw.) Sacc. ; dieback, twig and cane canker.
    Subtropical USA, Guatemala, Costa Rica, Puerto Rico, El
    Salvador, probably widespread.

Botryosphaeria ribis Gross. & Dug. = B. dothidea (Moug.)
    Ces. & deNot. ; canker, stem collapse.   Scattered in occur-
    rence.

Botrytis cinerea Pers. ; bud-blight, flowering branch tip
    disease, fruit dryrot.   Brazil, Argentina, Chile, Venezuela,
    Uruguay, Puerto Rico, Subtropical USA.
Capnodium coffeae Pat. and C. citri Berk. & Desm. ; sooty-
    molds (the former near coffee harboring sucking insects,
    the latter near insect-infested citrus).   Puerto Rico, Costa
    Rican semi-dry area.
Cassytha filiformis L. ; slender dodder (infection by inocula-
    tion).   Puerto Rico.
Cephaleuros virescens Kunze; algal leaf scurf.   Costa Rica,
    Puerto Rico, Virgin Islands.
Cercospora spp. (possibly some synonymy): C. hyalina Muller
    & Chupp; leafspot.   Colombia, Brazil.   C. puderi Davis;
    small leafspot, budspot, petalspot.   El Salvador, Mexico,
    Subtropical USA, Costa Rica.   C. rosicola Pass. (its per-
    fect state is Mycosphaerella rosicola Davis), the most com-
    mon leafspot, it is a large leafspot.   Widespread.   C.
    rosigena Bat. & Sil. ; leafspot.   Brazil.   There is also a
    leafspot, C. rosigena Thorp, reported from Texas.
Cladosporium sp. ; on old dying leaf.   W. Indies, Chile.
Clitocybe tabescens (Scop.) Bres. = Armilariella tabescens
    (Scop.) Sing. ; rootrot.   Florida.
Colletotrichum gloeosporioides Penz, the common parasitic
    conidial state of Glomerella cingulata (Stonem.) Sp. &
    Schr., both reported and seen.   Common anthracnose
    results in dieback, perfect state occurs on canes.   Puerto
    Rico, Mexico, Costa Rica, probably widespread perhaps
    general in occurrence.
Coniothyrium fuckelii Sacc. = C. rosarum Cke. & Harkn. , the
    perfect ascostate is Leptosphaeria coniothyrium Sacc. ; the
    common canker, leafspot, dieback.   Widespread but mostly
    reported in Subtropical USA, Argentina, Uruguay, Brazil.
Corticium salmonicolor Berk. & Br. ; pink disease.   Costa
    Rica by inoculation.
Cryptosporella umbrina Jenk. & Wehm. = Diaporthe umbrina
    Jenk. ; crowncanker, bud-and-flower disease.   Florida,
    California, Brazil.
Cuscuta americana L. ; fast growing yellow colored dodder.
    W. Indies, C. America.   C. indecora Choisy; dodder.
    Florida, Texas, El Salvador,   C. paradoxa Raf. ; dodder.
    Florida, Texas.
Cutting bed disease, rooting failure, roots when grown are
    weak, sparse and slender.   Cause is combined excessive
    bacterial and nematode population.   Puerto Rico, Costa
    Rica.
Cyathus montagnei Tul. f. sp. brasiliensis Speg. ; bird's-nest
    fungus on decaying bush wood.   Brazil.   C. vernicosus
    (Butl.) DC. f. sp. chilensis Tul. ; bird's-nest fungus on
    decaying rose bush wood.   Chile.
Cylindrocladium scoparium Morg. ; canker, seedling rootrot.
    Subtropical USA, Costa Rica, Argentina.
Cytospora sp. ; recovered as an isolate from collapsing cane.
    Puerto Rico.

Diaporthe eres Nits.; cane, twig and bud disease.   Subtropical
    USA.   (For D. umbrina see Cryptosporella.)
Diplocarpon rosae Wolf = Marssonina rosae (Lib.) Lind. (its
    imperfect state is Actinonema rosae Fr., and this is often
    reported, a few times determined as Marsonia rosae Br.
    & Cav.); the severe and common blackspot, defoliation,
    occasional cane canker.   Most common report is of the
    imperfect fungus, the Diplocarpon state, is more often
    seen in cool edges of tropics and in such a country as
    Guatemala or as Costa Rica in affected plants in high hill
    gardens.   Studies in Costa Rica, Bolivia, Subtropical USA,
    Brazil, Peru.
Diplodia (see Botryodiplodia)
Erwinia carotovora (L. R. Jones) Holl; bacterial softrot of
    flower stems in flower markets.   Neotropical.
Eutypa heteracantha Sacc.; foliage disease.   S. America.
Fabraea rosae Maubl.; leaf disease.   Uruguay.
Fusarium oxysporum Schl.; rootdisease but not a wilt of the
    plant (fungus culture typical).   Puerto Rico, reported also
    from Argentina.   Also Fusarium sp., undetermined species
    isolated many times from dieback and leafspot complexes.
    Costa Rica.
Glomerella (see Colletotrichum)
Leaf edge necrosis, physiological.   Puerto Rico.
Leaf scorch from salt spray.   Puerto Rico.
Leptosphaeria (see Coniothyrium)
Marsonia and Marsonina (see Diplocarpon)
Meloidogyne sp.; rootknot nematode.   Subtropical USA, Puerto
    Rico.   (Complaints from several other countries of inter-
    cepting diseased shipments from temperate zone and neigh-
    boring tropical areas.)
Microdiplodia (see Botryodiplodia)
Monochaetia sp.; leaf infection.   Mexico.
Mycena citricolor (Berk. & Curt.) Sacc.; luminescent leafspot.
    Costa Rica.
Mycosphaerella (see Cercospora)
Nectria sp.; sporocarps on dying cane in moist chamber.
    Costa Rica.
Nematodes, numerous and unidentified, inhabiting rhizosphere
    soil.   Chile.
Nigrospora sp.; in branch dieback complex.   Costa Rica,
    Puerto Rico.
Oidium leuconicum Desm. (its perfect state is Sphaerotheca
    pannosa (Wall.) Lév. var. rosae Wr., according to many
    in the American tropics but it is also given as S. humuli
    (DC.) Burr. by others); the common white powdery mildew
    on leaves, shoot flush, flower buds.   Almost always where
    rainy seasons are not too long and rains not too heavy.
Pellicularia koleroga Cke.; thread-web blight of foliage.
    Florida and Louisiana in Subtropical USA, Guatemala high-
    land area, Peru.
Peronospora sparsa Berk.; downy mildew on cuttings and new
    flush.   Florida, Guatemala, Brazil.

Pestalotia spp. ; cause both primary and secondary disease
developments: P. compacta Berk. & Curt. ; on leaf, bud,
and stem. Florida. P. funerea Desm. ; associated with
other fungi in dieback, and causes leaf lesion on main vein
and small cankers on stem. Puerto Rico, Costa Rica. P.
suffocata Ell. & Ev. ; leafspot. Jamaica.

Pezizella oenotherae (Cke. & Ell.) Sacc. ; spots on leaf and
cane. Subtropical USA.

Phragmidium disciflorum (Tode) James (synonyms, P. mucronatum (Pers.) Schl., P. subcoticium Lk.); common leaf
and stem rust that may almost disappear during rainy-
seasons. Mexico, Guatemala, Colombia, Bolivia, Ecuador,
Venezuela, Peru, Uruguay, Paraguay, Argentina, Brazil,
Puerto Rico.

Phyllosticta rosae Desm. ; leafspotting. Brazil, Dominican
Republic.

Phymatotrichum omnivorum (Shear) Dug. ; dry-land rootrot.
Texas.

Physalospora (see Botryodiplodia)

Pratylenchus sp. ; nematode inhabitant of rhizosphere soil.
Peru.

Pseudomonas syringae Van Hall; leaf and twig bacterial spot.
Scattered.

Rhizoctonia solani Kuehn is the parasitizing hyphae that are
rarely collected on this host, the sporulating state is
Thanatephorous cucumeris (Frank) Donk (I found it once
in a mountain garden in Panama); causes occasional
disease at base of cane, cutting bed disease, cutting
failure after transplanting. From isolations found in:
Puerto Rico, Guatemala, Panama, Costa Rica, it is
probably widespread.

Rootrots of several kinds, still to be studied, resulting from
what appears to be Basidiomycetes and a complex of bacteria and semiparasitic fungi.

Rosellinia bunodes (Berk. & Br.) Sacc. ; rootrot following infection by other fungi and insect feeding. Costa Rica. R.
pepo Pat. ; rootrot, primary infection near diseased cacao
tree. Costa Rica. R. nicatrix (Hartig) Berl. ; rootrot in
Argentina.

Sclerotium rolfsii Sacc. (sporulating state not found by me);
soil-line rot of stem and other parts that touch the fungus.
Florida, Texas, Puerto Rico.

Septoria rosae Desm. ; small dark leafspots, defoliation.
Peru, Subtropical USA, Puerto Rico. S. rosae-aruensis
Sacc. ; small leafspots. Dominican Republic. S. rosarum
West. ; leafspot. Brazil. S. rosarum West. var. lepto-
sperma Speg. ; leaf disease, cane-canker. Brazil.

Sphaceloma rosarum Jenk. ; spot anthracnose on leaves and
stem. Brazil, Argentina, Uruguay, Guiana, Chile, Venezuela, Colombia, Peru, Subtropical USA, W. Indies, C.
America.

Stagonospora rosarum Bat. & Silva. ; bud disease, mummy.
Brazil, Argentina.

Stilbum aurantium-cinnabarinum Speg., and S. cinnabarina
    Wor.; cankers often of red color on matured cane, infec-
    tion may follow flower cut.   Probably these parasites are
    widespread, special note taken in Brazil, Trinidad, Argen-
    tina, Paraguay, El Salvador, Guatemala, Puerto Rico.
Stomatochroon lagerheimii Palm; leaf infection by systemic
    algal parasite.   Guatemala.
Strigula spp.; various colors and textures of lichen spots on
    leaves and stems.   Puerto Rico (hill gardens).
Trichoderma viride Pers.; repeatedly isolated in dieback
    branches.   Costa Rica.
Tylenchorhynchus sp.; nematode inhabitant of rose soil.   Peru.
Valsa ambiens Pers.; on dead cane.   Costa Rica.
Viruses undetermined, but symptoms include mosaic, severe
    chlorosis, leaf and stem dwarf, necrotic leaf spots, tip
    necrosis.   Scattered (seldom, if ever, in rainforest areas).
Xiphinema sp.; nematode in rose soil.   Chile.
Xylaria sp.; specimen sent from Colombia to temporary clinic
    in Panama.

RUBUS spp.   A number of woody and prickly, multiple-stemmed
    bushes grown in relatively small fields, roadside plots and
    jungle clearing edges, where their stems or canes bear much
    relished and marketed fruits; "frambuesa," "zarzamora" Sp.,
    "amora preta," "framboesa" Po., "bramble berry," "blackberry"
    En.
    Anthostomella appendiculosa (Berk. & Br.) Sacc., and A.
        rubicola (Speg.) Sacc. & Trott.; on dead and dying canes
        both in S. America.
    Aulographium valdivianum Speg.; on dead branch.   Chile.
    Botryodiplodia valdivianum Speg.; cane dieback.   C. America,
        W. Indies, Chile (probably widespread).
    Botrytis cinerea Pers.; fruit-rot in market.   Guatemala.
    Calonectria pachytrix Rehm; stem disease.   Chile (but also
        a Calonectria seen in El Salvador).
    Cephaleuros virescens Kunze; scurfy algal leafspot not es-
        pecially spectacular.   Guatemala, Costa Rica, Florida,
        Puerto Rico (probably more widespread).
    Cercospora (see Mycosphaerella)
    Coniothyrium fuckelii Sacc., and C. phacidioides Speg.; both
        cane cankers.   Argentina.
    Cordella rubicola Speg.; leaf disease.   Chile.
    Curvularia cymbisperma Ell.; leaf disease.   Ecuador.
    Cuscuta spp.; dodders.   C. America, W. Indies.
    Cyphella albo-violascens (Alb. & Schw.) Karst.; on dead cane,
        probably saprophytic.   Southern S. America.
    Didymosphaeria araucana (Speg.) Sacc. = Microthelia araucana
        Speg. in Chile.   D. gregaria Speg. = M. gregaria (Speg.)
        Kuntze in Argentina, both lichen problems on canes and
        twigs.
    Elsinoë veneta (Burk.) Jenk.; perfect state of Sphaceloma
        necator Jenk. & Shear; spot anthracnose.   Occurrence
        scattered but in many countries.

Fusarium roseum Lk. and F. semitectum Berk. & Rav. ; both
     in attack on fruit peduncle, leaf disease. Argentina, (the
     former also in Costa Rica).
Gymnoconia intersticialis Lagh. ; rust (orange color). Hon-
     duras.
Heterodera sp. ; nematode root-zone soil inhabitant. Chile.
Hyalotheles dimerosperma Speg. ; mildew. Brazil.
Hypoderma virgultorum DC. ; twig infection. Chile.
Hypospila rubicola Speg. ; on dead shoot (? semi-saprophyte).
     Chile.
Kuehneola spp. , are the yellow stem rusts, sometimes on
     leaves: K. andicola Diet. (also given as K. andinicola
     Diet.) in Chile, Falkland Isl. , Tierra del Fuego. K.
     guatemalensis Cumm. in Guatemala. K. loesneriana
     (Arth.) Jacks. & Holw. in C. America, Ecuador, Bolivia,
     Brazil. K. uleana Syd. in Brazil. K. uredinia Arth.
     (also given as K. uridinis Arth.) in Subtropical USA, Ar-
     gentina.
Leptothyrium rubicola Speg. in Chile, and L. ushurvaiensis
     Speg. in Cape Horn, both flyspeck-like leaf spotting disease.
Mainsia spp. , are the common yellow leaf rusts: M. clara
     Jacks. & Holw. in S. America. M. columbiensis Kern,
     Thur. , & Whet. = Spirechina columbiensis Kern & Whet.
     in Brazil. M. cundinamarcensia (Mayor) Jacks. = Uromy-
     ces cundinamarcensis Mayor or Spirechina cundinamarcensis
     Diet. , well spread in S. America. M. holwayii Jacks. =
     M. imperialis (Speg.) Lindq. or Uredo imperialis Speg. ,
     widespread in many countries from Mexico to Argentina.
     M. lagerheimii (Magn.) Jacks. & Holw. (synonyms are
     Uromyces lagerheimii Magn. , U. andinus Lagerh. ,
     Spirechina lagerheimii Kern & Whet.) in S. America. M.
     peruviana Jacks. in Peru. M. pitteriana Jacks. (synonyms
     are Spirechina pitteriana Arth. , Uredo ochraceo-flava P.
     Henn. , Uromyces pitterianus (P. Henn.) Jacks.) in Costa
     Rica and Brazil. M. quitensis (Lagh.) Jacks. & Holw. =
     M. mayorii Jacks. or Uromyces quitensis Lagh. , wide-
     spread in S. America. M. rubi Jacks. = Spirechina rubi
     (Diet. & Holw.) Arth. or Uromyces rubi Diet. & Holw. in
     Mexico and C. America. M. rubi-urticifolia (Mayor) Jacks.
     = Uromyces rubi-urticifolii Mayor in Brazil, Colombia.
     M. standleyii Cumm. in Guatemala. M. tenella Jacks. &
     Holw. in Ecuador and Haiti. M. variabilis (Mayor) Jacks.
     & Holw. = Spirechina variabilis Diet. or Uromyces varia-
     bilis Mayor in Colombia, Ecuador, Venezuela.
Melanopsamma europea Speg. ; cane and twig disease. Argen-
     tina.
Meliola puiggarii Speg. (= Appendiculella calostroma Hoehn. ,
     Irene calostroma (Desm.) Hoehn. , I. manca (Ell. &. Mart.)
     Stev. , I. puiggarii (Speg.) Doidge) and Meliola calostroma
     Hoehn. ; the causes of black mildew on foliage. Widespread.
Meloidogyne sp. ; rootknot nematode. Brazil.
Microthelia (see Didymosphaeria)

Mollisia rubicola Pat.; a severe cane-canker. Widespread.

Mycosphaerella confusa Wolf = Cercospora rubi Sacc. which
    is its imperfect or conidial state; leaf blotch, large leaf-
    spot. Subtropical USA, Guatemala.

Oidium erysiphoides Fr. is the conidial organism and seasonal
    in occurrence; white powdery mildew. Widespread.

Pellicularia koleroga Cke.; web-thread blight. Louisiana,
    Costa Rica.

Phragmidium rubi-idaci (DC.) Karst.; rust (dark colored).
    Guatemala.

Phyllosticta rubi P. Henn., P. ribesicida Speg.; both cause
    round leafspot. Brazil.

Physalospora idaei Sacc.; part of dieback complex. Chile.

Platysphaeria argentina Speg.; twig and cane disease. Argen-
    tina.

Rhizopus nigricans Ehr.; mold on market fruit. Guatemala.

Septoria spp., leafspot and defoliation: S. campoi Speg. in
    Chile, S. ribis Desm., is widespread, and S. rubi West.,
    widespread.

Sphaceloma (see Elsinoë)

Spirechina (see Mainsia)

Stereum complicatum Fr.; associated with cane decay. Chile.

Uredo (see Mainsia)

Uromyces (see Mainsia)

SACCHARUM OFFICINARUM L. Grows best in the hot tropics and
    is the giant cultivated grass which stores sugar in its fat culms
    (called barrels) and is the result of natural crosses between wild
    canes, and then subsequent folk selections, and although its origin
    is from the South Pacific Islands and tropical Asia, in the last
    two centuries it has been subjected, in 37 countries in the Neo-
    tropics, to disease study and breeding by one of the largest
    bodies of specialists of any crop in the Western Hemisphere
    tropics; "caña de azúcar" Sp., "caña-de-açúcar" Po., "sugar-
    cane" En. (Note: Some organisms were reported that are now
    thought of as saprophytes, a number of weak parasites. Names
    of most surface-inhabiting myxomycetes, bacteria and algae are
    not in the present list, but a number of fungi, some described
    long ago by Spegazzini and others, are included not only for
    historical reasons but also as no one can tell when some newly
    bred and selected cane may show susceptibility to one of these
    organisms that at present is possibly not considered even re-
    motely dangerous.)

    Acremoniella atra (Cda.) Sacc.; on leaves. Scattered in S.
        America.

    Acrostalagmus albus Preuss f. sp. dichotoma Speg. (= Ver-
        ticillium sp.); an insectivorous fungus. Colombia, Argen-
        tina, W. Indies.

    Alectra brasiliensis Benth.; phanerogamic root parasite.
        Brazil, Panama, Trinidad.

    Algae ("green scum") associated with poor growth on wet
        soil. Costa Rica.

<u>Alternaria</u> <u>alternata</u> (Fr.) Keiss = <u>A</u>. <u>tenuis</u> Nees; small
    spored leaf and cane mold. At most a very weak parasite.
    Occasional in W. Indies, Subtropical USA, Brazil.
Amoebosporus (see <u>Plasmodiophora</u>)
<u>Antennaria</u> (see <u>Fumago</u>)
<u>Anthostomella</u> (see <u>Rosellinia</u>)
<u>Aphelenchoides</u> sp.; nematode collected in rhizosphere.
    Puerto Rico.
<u>Apiospora</u> <u>camtospora</u> Penz. & Sacc.; fungus acting as
    secondary invasion following other parasites on old leaves.
    Puerto Rico.
<u>Apiosporium</u> <u>bergii</u> Speg.; infection of dying leaf. Argentina.
<u>Arthrobotrys</u> <u>superba</u> Cda.; on aging leaves. Puerto Rico.
<u>Ascochyta</u> <u>sacchari</u> Bat.; leaf spotting. Brazil.
<u>Aspergillus</u> spp., saprophytic and epiphytic molds (a few
    questionably associated with leaf infections): <u>A</u>. <u>argentinus</u>
    Speg. in Argentina. <u>A</u>. <u>flavus</u> Lk. is widespread. <u>A</u>.
    <u>fumigatus</u> Fres. is common. <u>A</u>. <u>glaucus</u> (L.) Lk. var.
    <u>tonophilus</u> Speg. said to cause distinct leaf infection in
    Argentina. <u>A</u>. herbariorum (Wigg.) Fisch. is widespread.
    <u>A</u>. <u>nidularis</u> (Eid.) Wint. is widespread. <u>A</u>. <u>niger</u> v. Tiegh.
    often encountered. <u>A</u>. <u>parasiticus</u> Speare on cane mealy-
    bugs in Puerto Rico. <u>A</u>. <u>penicilloides</u> Speg. in Argentina
    and other countries. <u>A</u>. <u>repens</u> (Cda.) dBy. & Wr. is
    widespread. <u>A</u>. <u>sydowii</u> (Bainier & Sartory) Thom and
    Church is scattered but widespread. <u>A</u>. <u>terreus</u> Thom is
    scattered.
<u>Asterostroma</u> <u>cervicolor</u> (Berk. & Curt.) Mass.; on basal
    leafsheaths. Puerto Rico.
<u>Atylenchus</u> <u>decalineatus</u> Cobb; rootzone soil nematode. Florida.
<u>Bacillus</u> <u>flavidus</u> Fawc. = <u>B</u>. <u>sacchari</u> Speg.; toprot. Argen-
    tina. (See also Xanthomonas.)
Basisporium (see <u>Nigrospora</u>)
<u>Belondira</u> <u>cylindrica</u> Thorne and B. <u>tenuidens</u> Thorne; Rhizo-
    sphere nematodes in Puerto Rico.
<u>Botryodiplodia</u> <u>theobromae</u> (Pat.) Now. (B. <u>natalensis</u> P.
    Evans), also given by some as <u>Diplodia</u>, has the asco-state
    Botryosphaeria quercuum (Schw.) Sacc. = <u>Physalospora</u>
    <u>rhodina</u> Berk. & Curt.; dryrot of old canes, occasional
    stemrot of mature cane, sett-cane rot. Probably in all
    countries but special reports from Barbados, Brazil,
    British Guiana, Colombia, Cuba, El Salvador, Dominican
    Republic, Ecuador (seen on eating cane in market), Florida,
    Jamaica, Mexico, Puerto Rico, St. Thomas, Trinidad.
<u>Calonectria</u> <u>gigaspora</u> Mass.; on leaf sheaths following insect
    damage. Trinidad.
Capnodium (see <u>Fumago</u>)
<u>Cassytha</u> <u>filiformis</u> L.; the olive-green slow acting dodder.
    Puerto Rico, El Salvador.
<u>Cephalosporium</u> <u>sacchari</u> Butl.; wilt, stunting, hollow stem.
    (Occurrence expected on high dry hills in Central America
    and West Indies). Argentina, Barbados, Brazil, Colombia,

Cuba, El Salvador, Guadeloupe, Guatemala, Mexico, Subtropical USA but infrequent, Trinidad.

Ceratocystis paradoxa (Dade) Mor. (synonyms: Endoconidiophora paradoxa (Dade) Davids., Ceratocystis paradoxa
Mor., C. paradoxa (de Sey.) Mor., Thielaviopsis paradoxa
Hoehn., Sphaeronema adiposum Butl., Ceratostomella adiposum Sart., C. paradoxa Dade); causes black rot of setts,
black rot in storage piles (occurrence more common in
moist land areas). Reports from all important cane areas.

Cercospora atrofiliformis Yen, Lo, & Chi.; black stripe in
Brazil (some time ago known only in Orient). C. kopkei
Krueger = C. longipes Butl.; oval yellow to brown leaf-
spots in all C. America, in Subtropical Alabama, Florida,
Louisiana, Texas in USA, in W. Indies islands especially
Barbados, Cuba, Puerto Rico, in Mexico and Panama, and
in S. America especially noted in the Guianas, Colombia,
Ecuador, Peru, Venezuela, Argentina, Brazil. C. sacchari
deHaan; leafspot in all C. America, Mexico and Panama,
and perhaps more widespread. C. vaginae Krueger causes
redspot on leaf and large redspot of leafsheath, most in
Argentina, Barbados, Brazil, British Guiana, British Hon-
duras, Costa Rica, Colombia, Cuba, Dominican Republic,
Florida, Haiti, Honduras, Jamaica, Louisiana, Mexico,
Nicaragua, Panama, Peru, Puerto Rico, Texas, Trinidad.

Chaetomella sacchari Delacr.; sheath disease. Peru.

(? Chaetophoma maydis Speg. = Lasiophoma maydis Speg.?)
an occasional report of occurrence on dying leaves. Argentina, Paraguay.

Chaetostroma sacchari Speg. = C. saccharicolum Sacc. & Syd.;
leaf and sheath attack. Described in Argentina, probably
in much of southern S. America.

Chlorosis from physiological causes such as unusual cold,
excess of fertilizer, excess salt or lime, manganese deficiency, iron deficiency, nitrogen deficiency. Scattered.

Cladoderris dendritica (Pers.) Berk.; a Basidiomycete involved in stalk death, sometimes in newly cleared land.
S. America.

Cladosporium herbarum (Pers.) Lk.; gray-green mold, leaf-
edge disease. Mexico, Panama, C. America, and probably
widespread (dryseason disease).

Clathrus (see Ithyphallus)

Claviceps sp.; ergot, an inflorescence disease. Rather rare,
noted in Colombia.

Cochliobolus (see Helminthosporium)

Colletotrichum falcatum Went (with synonyms C. graminicola
(Ces.) Wils. and C. lineola Cda.) has as its perfect state
Physalospora tucumanensis Speg. its synonym is Glomerella
tucumanensis v. Arx. & Müll. but the fungus most commonly recovered is the conidial form; the severe redrot
disease seen in all sugarcane countries and islands of the
Neotropics. (See also Sphaeria.)

Coniosporium (see Papularia)

Coniothyrium melanosporium (Berk.) Sacc. = C. sacchari
  (Mass.) Prill.; infections of leafsheath. Mexico, Guate-
  mala, Argentina.
Corethropsis elegans Speg.; on leaves and sheaths. Argentina.
Corticium sasakii (Shirai) Matsu = Rhizoctonia ferruginea
  Matz. (? = C. arachnoideum Berk.); sclerotial disease at
  plant base and when bent to ground, root rot. Louisiana,
  Puerto Rico, Panama, probably in other countries.
Creonectria (see Nectria)
Criconemoides sp.; rhizosphere inhabiting nematode. Puerto
  Rico.
Curvularia lunata (Wakk.) Boed.; leaf-spot and leaf-edge
  disease, secondary in rootrot. Florida, Louisiana, Puerto
  Rico, British Guiana, Surinam.
Cuscuta americana L.; yellow fast-growing dodder. Puerto
  Rico.
Cuscuta spp.; other dodders in Puerto Rico, Guatemala, El
  Salvador, Panama, Peru.
Cytospora sacchari Butl.; leafsheath rot, blade-spot. Argen-
  tina, Bolivia, Brazil, British Guiana, British Honduras,
  Costa Rica, Cuba, Dominican Republic, Florida, Mexico,
  Nicaragua, Peru, Puerto Rico.
Dematium laevisporum Speg.; culm disease. Argentina.
Dendrophoma saccharicola Av.-Sacc.; on decaying leaves.
  Argentina, Brazil.
Diaporthe austro-americana Speg. (with synonyms D. sacchari
  Speg. and D. zeina Speg.); culm decay. Argentina.
Dimeriella sacchari (de Haan) Went. when combined with
  Eriosphaeria sacchari Went.; red and/or purple leafspot.
  Cuba, Trinidad.
Dinemasporium sacchari P. Henn.; brown leafspots. Peru.
Diplodia (see Botryodiplodia)
Diplogaster sp.; a nematode in rhizosphere soil. Puerto Rico.
Ditylenchus sp.; a nematode in rhizosphere soil. Puerto Rico.
Elsinoë sacchari (Lo) Jenk. & Bitanc.; is the asco-state of
  Sphaceloma sacchari Lo the cause of white-scab and streak.
  Brazil, Haiti, Florida, Puerto Rico.
Endoconidiophora (see Ceratocystis)
Epicoccum neglectum Desm.; leaf blackening. Mexico.
Eriosphaeria (see Dimeriella)
Eucephalobolus sp.; a root-feeding nematode. Puerto Rico.
Eurotium sacchari Speg. and E. argentinum Speg., both are
  asco-states of Aspergillus described in Argentina in 1898
  and 1899.
Euryachora sacchari Av.-Sacc.; stroma on leaf. Brazil.
(Eutypella an incorrect name for Eurotium.)
Fomes spp. (rather dwarf-sized sporocarps) associated with
  at least three-year-old cane clumps partly dead. Puerto
  Rico, Guatemala, Panama.
Fumago sacchari Speg. (with synonyms F. pannosa Berk.,
  F. pannosa Speg., Triposporium pannosum (Berk.) Speg.,
  Polychaeton pannosum (Berk.) Speg., Antennaria subantarc-
  tica Speg., Capnodium sp.); dryseason sootymold, fumagina,

black fungus that flourishes on insect honeydew.   In all
sugarcane growing areas.

Fusarium spp. are readily isolated from many diseased sugarcane samples, but it seems only five of the organisms are
strictly pathogens: 1. F. moniliforme Sheld. is the conidial
state of the ascomycetous Gibberella fujikuroi (Saw.) Ito =
G. moniliforme (Sheld.) Winel.; the common cause of wilt,
twisted top or escalera, toprot or pokkah boeng, collar-rot,
and sett-rot.   Where more susceptible clones or varieties
are grown the diseases caused appear scattered but to
occur when conditions are suitably warm (which is when
sugarcane flourishes best), and in any area.   2. F. oxysporum Schl. causing systemic invasion in Uruguay.   3.
F. roseum Lk. which has the asco-state Gibberella roseum
(Lk.) Sny. & Hans. is a weak but often present parasite
that increases by secondary action the severity of cane rots.
Seems widely distributed.   4. F. solani (Mart.) Sacc. has
been isolated from diseased culms.   El Salvador, Guatemala.   5. Fusarium sp. isolated from rootrot in Peru,
Colombia, Panama, El Salvador.

Galls on culm, of physiological cause?, may be accompanied
by multiple stem development.   Probably widespread and
is considered a curiosity, partial list of countries where
found is Barbados, Cuba, Jamaica, Louisiana in USA,
Puerto Rico, Nicaragua, Peru.

Glenospora (see Nigrospora)

Gleocercospora sorghi Bain & Edg.; a zonate leafspot.
Louisiana (symptoms seen but no spores found in Guatemala).

Glomerella (see Colletotrichum)

Gnomonia iliau Lyon; leafsheath and cane infections.   Brazil,
Louisiana.

Gomosis, gumming and often leafstreak, cause not determined,
possibly bacterial but not well proved.   Scattered occurrence.

Graphium pistillarioides Speg.; leafmold in Argentina.   G.
sacchari Speg. = Graphiopsis sacchari Speg. = Graphiopsis
sacchari (Speg.) Goid.; leafmold in Puerto Rico.

Gymnosporium vinosum Berk. & Curt.; on dead leaves.   Cuba.

Haplographium sacchari Speg.; on leafblade and leafsheath.
Argentina.

Haplolaimus sp.; nematode found near roots.   Puerto Rico.

Helicotylenchus sp.; nematode found in rhizosphere.   Puerto
Rico, Bolivia.

Helminthosporium spp.; leaf and leafsheath attacking fungus.
(In plantations in or close to the heat equator, fields are
most severely attacked on hillsides somewhat above the
floodplain.)   1. H. rostratum Drechs.; leafspot sometimes
oval and purplish.   W. Indies.   2. H. sacchari (de Haan)
Butl. = H. ocellum Faris; common eyespot, ring spot,
occasionally brownstripe.   In all sugarcane growing countries
and islands.   3. When H. sacchari (ocellum) is combined

with Leptosphaeria sacchari it causes severe and charac-
teristic ringspot.   Special reports or observations from
Alabama in USA, Argentina, Barbados, Brazil, British
West Indies, Colombia, Cuba, Dominican Republic, Florida
in USA, Panama, Peru, Puerto Rico.   4. H. purpurascens
Bourne; target spot.   Cuba, Florida, Costa Rica, El
Salvador.   5. H. stenospilum Drechs., its perfect state
is Ophiobolus stenospilus Carp. = Cochliobolus stenospilus
(Carp.) Mat. & Yam.; causes brownstripe in Florida,
Georgia, Louisiana, Mexico, Puerto Rico, Venezuela,
Brazil, Jamaica.

Hendersonia sacchari Speg.; on ripening plants, causes scorch
  symptoms.   Widespread.   (Compare with Stagonospora.)

Hendersonina sacchari (deHaan) Butl.; collar rot.   Argentina,
  Brazil, Costa Rica.

Himantia guttulifera Speg.; wilting of leaves with fungus on
  inner surfaces.   Argentina.   H. sacchari Speg.; developed
  as growth on diseased culm surface in Argentina.   H.
  stellifera Johnst. its perfect state apparently is Odontia
  saccharicola Burt., a Hydnum-like fungus, but the Himantia
  state was parasitic causing leafsheath adhesion and sheathrot disease.   Reported from Puerto Rico, Trinidad, Cuba,
  Jamaica, British Guiana, Colombia.   (Note: This genus is
  often discarded by mycologists but in my observations had
  a useful purpose as it delimited a velvety and web-like
  growth, with no pellicle or sporuliferous spreading hymenium; I believed it not Rhizoctonia, Corticium, Pellicularia,
  or Thanatephorous.)

Hormiactella sacchari Johnst.; part of the withertip complex.
  Florida, Jamaica, Puerto Rico.

Hypochnus sacchari Speg.; (aerial Rhizoctonia-like fungus
  attack) web blight on apical leaves.   Argentina, W. Indies.

Ithyphallus rubricundus (Bosc) Fisch. = Clathrus columnatus
  Bosc; stinkhorn root damage, also a saprophyte on cane
  debris.   Florida, Puerto Rico, in Costa Rica it developed
  in wet season.

Knyaria (see Tubercularia)

Lepiota lycoperdinea Speg.; sugarcane field toadstool (? root
  injury on cane from toxic fungus exudation?).   Argentina.
  Lepiota sp.; fairy ring in sugarcane field in Panama.

Leptosphaeria michotti (West.) Sacc.; leafspot.   Brazil.   The
  more common L. sacchari deHaan (apparently several
  synonyms: L. saccharicola P. Henn., L. spegazzinii Sacc.
  & Syd., L. spegazzinii Sacc. & Syd. var. minor Speg., L.
  tucumanensis Speg.); cause of circular spot and ringspot.
  Distributed in all countries and islands.

Ligniera (see Plasmodiophora)

Linospora sacchari Av. -Sacc.; leafspot.   Brazil.

Lisea australis Speg. and L. australis Speg. var. sacchari
  Speg. (? = Gibberella) on dead stems in Argentina.

Longidorus sp.; rhizosphere nematode.   Puerto Rico.

Lophodermium sacchari Lyon; leaf disease.   Puerto Rico,
  probably more widespread.

Marasmius sacchari Wakk. = M. plicatus Wakk. ; base rot,
   root disease.   Subtropical USA, W. Indies, C. America,
   S. America. (Note: This synonym is tentative and it
   needs modern scrutiny in mycology and pathology, including
   intensive laboratory cultural research and host ranges.
   Fungi of the genus Marasmius, in addition to being root
   and plant-base-rots like M. sacchari, also have characteristics like M. pulcher (Berk. & Br.) Petch that parasitizes by use of aerial thread-like cords; other types
   similar, M. oreades (Bolt.) Fr. or M. sarmentosus Berk.
   that may be mostly saprophytes, and Crinipellis (Marasmius)
   perniciosus (Stahel) Singer, which infests meristematic tissues of expanding buds and enters into a singularly subtle
   systemic existence in tissues of its host.)
Melanconis sacchari Mass. was given close attention and
   classic description by Spegazzini in 1895, was later named
   Pleocyta sacchari (Mass.) Petr. & Syd., as well as
   Strumella sacchari Cke., Trichospheria sacchari Mass.
   and Trullula sacchari Ell. & Ev. and its much used name
   Phaeocytostroma sacchari (Ell. & Ev.) Sutton; the severe
   rind disease.   With improved growing methods and tolerant
   varieties it has practically disappeared, but at one time it
   was the most important disease of the crop and found in all
   established areas.
Meloidogyne spp. ; rootknot nematodes.   Widespread with some
   special observations in Puerto Rico, Bolivia, and Brazil.
Metasphaeria saccharicola Speg. ; associated with culmrot in
   Argentina.   M. usteri Speg. ; a leaf disease in Brazil.
Microdiplodia melaspora (Berk.) Griff. & Maubl. ; culm attack
   and sometimes in a complex associated with rind disease.
   Puerto Rico, S. America.
Microtypha saccharicola Speg. ; on decaying cane plants.
   Argentina.
Monilia sitophila (Mont.) Sacc. its perfect state is Neurospora
   sitophila Shear & Dodge; on burned leaf trash, weak leaf
   parasite.   Special note in El Salvador, Puerto Rico, Guate-
   mala, Costa Rica, but Neotropical in distribution.
Mycosphaerella sacchari Speg. = M. sacchari (Speg.) Seav.
   & Char. ; leaf disease, association at times with withertip.
   Cuba, Puerto Rico, Argentina.   M. striatiformans Cobb;
   leafsplit. Jamaica.
Myriogenospora aciculispora Vizz. ; leaf-binding, leaf-crush.
   Louisiana, Argentina, Brazil.   M. paspali Atk. ; leaf-crush.
   Brazil.
Myrothecium verrucaria (Alb. & Schw.) Ditm. ; a weak-leaf-
   parasite.   W. Indies, S. America.
Necrosis of leaf margins from high soil salinity.   Peru.
Nectria flavociliata Seav. ; ascocarps on dying stalks.   Puerto
   Rico, Guatemala, El Salvador, probably in more countries.
   N. laurentiana Marchal = Creonectria laurentiana (March.)
   Seav. & Chard. ; on dead and dying cane stalks.   W. Indies
   and C. America. N. saccharicola Speg. ; stalk disease.
   Argentina.

Nigrospora sacchari Speg. = N. oryzae (Berk. & Br.) Petch (synonyms: N. maydis (Garov.) Jech., Glenospora sacchari Speg., Basisporium gallorum Moll.); cause of black mold, increases leaf spot severity. Argentina, Ecuador, Puerto Rico, Subtropical USA, and is probably in all cane growing countries.

Odontis (see Himantia)

Oedocephalum bergii Speg.; leaf and leafsheath infections. S. America.

Olpidium sacchari Cook; invasion of weak roots. Puerto Rico.

Oospora tomentella Speg.; moldyness on old leaves. Argentina.

Ophiobolus (see Helminthosporium)

Ophiognomia sacchari Speg.; weak parasite, leaf attack. Argentina.

Ozonium sacchari Speg.; non-sporulating injurious hyphae on lower leafsheaths. Argentina.

Panagrolaimus sp.; nematode in rootzone soil. Puerto Rico.

Papularia arundinis Fr. (synonyms: Coniosporium arundinis (Cda.) Sacc., C. sacchari Speg.), P. sphaerosperma Hoehn., P. vinosa (Berk. & Curt.) Mason all cause black smudge and occasional leaf midrib disease. Scattered reports from Puerto Rico, Louisiana, Colombia, Argentina.

Pellicularia sasakii Ito; banded sclerotial disease. Subtropical USA, W. Indies.

Penicillium platense Speg.; infection on inside leaves. Argentina.

Peniophora flavido-alba Cke. and P. sacchari Burt are gemma developing fungi, found on plant bases, possibly saprophytes, in S. America.

Periconia atra Cda. and P. sacchari Johnst.; disease of aging leaves. Puerto Rico, Louisiana, Guatemala.

Pestalotia fuscens Wakk. f. sp. sacchari Wakk.; leafspot. Cuba.

Phoma heterospora Speg.; black leafblotch. Argentina.

Phyllachora sacchari P. Henn.; shiny black leafspot. Brazil, Argentina.

Phyllosticta sacchari Speg. (synonyms: P. saccharicola Speg., P. saccharicola P. Henn., P. sacchari-majoris Bat. & Vit.); leafspotting often linear with purple edges. Argentina, Brazil, Bolivia, Florida, Guatemala, Puerto Rico, Virgin Islands. P. sorghina Sacc.; secondary infection with other leafspots. Subtropical USA, Puerto Rico.·

Physalospora (see Botryodiplodia)

Physalospora tucumanensis (see Colletotrichum falcatum)

Phytophthora erythroseptica Pethy., and P. megasperma Drechs.; cutting or sett rots. Subtropical USA.

Plasmodiophora vasculorum Matz (synonyms: Amoebosporus vascularum Cke., Sorosphaeria vascularum (Matz) Schr., Ligniera vascularum (Matz) Cook); stunting vascular disease, dry top collapse. Puerto Rico, Cuba, Barbados, Colombia, Venezuela. (Note: This original name drew controversy and other designations have been given as is seen in

synonyms. The causal organism seems unique in plant
pathology and some workers have questioned its occurrence.
I confirmed presence of this sugarcane disease in two old
fields growing the "suave" variety of cane in widely sepa-
rated locations in Puerto Rico. Microscopic examinations
of both fresh and stained host tissues displayed in diseased
vasculars the amoeba-like structure of the parasite. I
did not find any development that was a spore, but in some
invaded cells small eliptical bodies were seen. I made no
attempt to culture the organism or to make inoculations
with it.)

Plectospira gemmifera Drechs. ; rootlet rot. Louisiana.
Pleospora infectoria Fckl. var. sacchari Speg. ; on leaf
    lesions. Argentina.
Polyporus spp. ; rather rare but spectacular occurrence of
    dwarf fungus fruiting bodies on old culms and on dead
    clumps of the host. W. Indies, C. America.
Polyscytalum sacchari Speg. ; rare leaf infection. Argentina.
Polystictus sanguineus Sacc. ; probably a saprophyte. Scat-
    tered.
Poria ambigua Bres. ; on dead culms. Florida, Guatemala,
    Peru, Bolivia.
Pratylenchus pratensis Filip. ; root feeding nematode. Costa
    Rica, Puerto Rico, may be widespread.
Puccinia kuehnii (Krueg.) Butl. (= Uredo kuehnii (Krueg.)
    Wakk. , Uromyces kuehnii Krueg. , Uredo ravennae Maire);
    rust. Scattered in W. Indies and C. America, Mexico.
Pyrenochaeta sacchari Bitanc. ; gray leafspot. Brazil.
Pyricularia grisea (Cke.) Sacc. = P. oryzae Cav. , with
    perfect state Tridiothecium grisea Cke. ; minor leafspot.
    Costa Rica, Guayana. P. sacchari Av. - Sacc. (is possibly
    synonymous with P. grisea); leafspot. Brazil.
Pythium spp. ; about 17 different species reported on decayed
    and diseased sugarcane roots. Some species often re-
    corded are: P. arrhenomanes Drechs. , P. apanidermatum
    Fitz. , P. graminicolum Subr. , P. ultimum Trow, P. ir-
    regulaire Buis. Widespread.
Radopholus similis Thorne; root nematode. Costa Rica.
Rhabditis sp. ; nematode found in soil near plants. Puerto
    Rico.
Rhinocladium sacchari Speg. ; fungus weakly parasitic on leaf-
    sheaths. Argentina.
Rhizoctonia spp. : R. ferruginea Matz (see Corticium), R.
    grisea (Stev.) Matz (see Sclerotium), R. pallida Matz;
    rootrots of weak plants. Barbados, Florida, Puerto Rico.
    R. solani Kuehn (the spore producing state is Thanate-
    phorous cucumeris (Frank) Donk); causes root and basal
    leafsheath rot. Widespread but noted specially in Louisiana,
    El Salvador, Guatemala, Puerto Rico, St. Thomas, Costa
    Rica, Peru, Colombia.
Rolling of heart leaves, believed to be physiological. Peru.
Rosellinia spp. : R. bunodes (Berk. & Br.) Sacc. ; black

infections of root and plant base with sporocarps on dead
cane surfaces. Costa Rica. R. necatrix (Hartig) Prill. ;
root infection. Bolivia. R. paraguayensis Speg. and its
conidial state Anthostomella paraguayensis Speg. ; root and
plant-base rot. Paraguay, Argentina, Guatemala, Costa
Rica, Honduras, Puerto Rico. R. pulveracea (Ehr.) Fckl. ;
on dead canes, possible saprophyte. Barbados, Cuba,
Puerto Rico.

Rotylenchus similis (Cobb) Filip. ; nematode root attack.
Louisiana.

Schizophyllum commune Fr. ; cane disease following wounds
made by cultivation practices. Florida, Louisiana, Mexico,
Argentina, Cuba, Brazil, Dominican Republic, Haiti, Martinique, Mexico, Puerto Rico, El Salvador, Costa Rica, Peru,
Brazil, probably in all cane countries.

Scirrhia lophodermioides Ell. & Ev. ; subepidermal disease of
cane. Puerto Rico, El Salvador.

Sclerospora macrospora Sacc. (synonyms S. sacchari Miy.,
Sclerophthora macrospora Thir., Shaw & Maras.); downy
mildew. Louisiana, Ecuador, Bolivia, Peru, becoming
widespread.

Sclerotium griseum Stev. = Rhizoctonia grisea (Stev.) Matz;
grayrot of setts. Puerto Rico, Barbados, Brazil, Colombia. S. rolfsii (Curzi) Sacc. its perfect state is Pellicularia rolfsii (Sacc.) West; redrot at groundline. Louisiana,
Texas, Florida, Puerto Rico, Mexico, Cuba, Jamaica,
British Honduras, Trinidad, Dominican Republic, Colombia,
Peru, Brazil.

Sorosphaera (see Plasmodiophora)

Spegazzinia tessartha (Berk. & Curt.) Sacc. = S. ornata Sacc.
in Puerto Rico and S. tucumanensis Speg. in Argentina
found on dying and dead leaves and canes, probably Neotropical. (Note: A genus of fungi, with ornamented spores,
considered weak to medium parasites and saprophytes, but
deserve attention as they have possibility of developing
severe disease mutants. One species alone, S. tessartha
(often seen as a saprophyte) is reported on at least the
main cultivated species of fourteen important crop genera.)

Sphaceloma (see Elsinoë)

(Sphaerella is sometimes given for Mycosphaerella.)

Sphaeria dematrium Pers., its imperfect state is Colletotrichum dematium (Pers.) Grove; leaf and stalk disease.
Puerto Rico.

Sphaerulina sacchari Speg. = S. sacchari P. Henn. ; a common
circular to oblong leafspot. Widespread.

Sporocybe sacchari Speg. ; lesions on leafsheaths and culms.
Argentina.

Stachybotrys pulchra Speg. ; is at most weakly parasitic on
leaves. Argentina.

Stagonospora sacchari Lo & Ling; leafscorch. Panama,
Argentina.

Stigmella sacchari Speg. = Sporidesmium bakeri Hughes var.

sacchari (Speg.) Hughes or Sporidesmium sacchari Speg.;
weak parasite of sugarcane. Argentina.
Stigmina pulchella Speg. = S. sacchari Speg.; on leaves,
sheaths, culms. Argentina.
Striga euphrasioides; parasitic flowering plant attacking roots.
Dominican Republic, British West Indies.
Strigula complanata Fée; white lichenspot on leaf. Puerto
Rico, Costa Rica.
Tangletop, mechanical. Barbados, Panama, Florida, Jamaica,
Cuba, Puerto Rico, Peru, Brazil.
Thielaviopsis (see Ceratocystis)
Treleasia sacchari Speg., its conidial state is Treleasiella
sacchari Speg.; a leaf disease. Argentina.
Trichoderma koningi Oud., in Peru and T. lignorum Tode =
T. viride Pers., widespread, both are subterranean para-
sites causing root and seed-cane (or sett) rot, green mold-
rot.
Tubercularia saccharicola Speg. = Knyaria saccharicola (Speg.)
Cif.; on dead canes. Puerto Rico, Dominican Republic,
Argentina.
Tylenchorhynchus sp.; nematode near roots. Peru.
Tylenchus similis Cobb; nematode attack of roots. Jamaica.
Uromyces (see Puccinia)
Ustilago scitminea Syd. = U. sacchari Rabenh. or U. ravennae
Rabenh.; black whip and tip smut. One of the three most
important sugarcane diseases. In Argentina (in 1940),
British Guiana (1949), in Argentina, Brazil, Paraguay and
Bolivia (1965), also reported in Uruguay, Colombia, British
Honduras, Trinidad. U. scitminea Syd. var. saccharum-
officinarum Mundk. & Hirsch.; in Argentina, Bolivia,
Brazil, Paraguay (reported from Trinidad, Guyana, Domini-
can Republic but by some believed questionable).
Venturia sterilis Speg.; uncertainly parasitic on leaves. Argentina.
Viruses: 1. Sugarcane Mosaic (particle is a long flexuous
rod), in old susceptible varieties one of most severe
diseases in Subtropical USA, W. Indies, C. America, S.
America. (Note: Considered at one time ninth of the ten
greatest crop diseases in the world. In areas near the
heat equator it is most common and severe in plantings at
low to medium low altitudes, less at 1000 meters altitude
or above, and does not spread well in wet rainforest conditions and is rare in high Andean plantings. Control is by
use of tolerant and resistant clones that usually are given
varietal number designations.) 2. Chlorotic Streak is widespread from Louisiana to southern S. America. 3. Ratoon
Stunt (the particle character is a small polyhedral and not
a mycoplasm) it may be local in its field occurrence, is
in at least 22 countries. 4. Streak symptom reported
from Brazil, may be different from Chlorotic Streak.
White Spots or bands, from cold injury. Florida, Louisiana,
El Salvador, Costa Rica, Argentina, Brazil, Guatemala,
Colombia, Caribbean Islands.

Witches' brooming or multiple bud, cause undetermined.
Fairly rare but is spectacular and readily eliminated by
roguing at harvest.

Xanthomonas spp., bacterial diseases in some reports given
generic names of: Bacillus, Bacterium, Phytomonas, and
Pseudomonas. X. albilineans (Ashby) Dows.; bacterial scald
and wilt. Scattered but in practically every cane country.
X. axonoperis Starr & Garces; bacterial gumming on both
the grass Imperata brasiliensis and sugarcane. Colombia.
X. rubrilineans (Lee et al.) Starr & Burk.; redstripe, top-
rot. Barbados, Guianas, Jamaica, Guadeloupe, Haiti,
Dominican Republic, Honduras, Puerto Rico, Martinique,
Mexico, Nicaragua, Trinidad, Argentina, Colombia. X.
rubrisubalbicans (Chr. & Edg.) Savul.; mottle stripe, rare
but reports from: Barbados, Colombia, Guadeloupe,
Jamaica, Martinique, Nicaragua, Peru, Puerto Rico, Venezuela, Subtropical USA. X. vasculorum (Cobb) Dows.;
gomosis or gumming. Caribbean Islands, C. America,
Mexico, Panama, Brazil, Guianas, Colombia.

Xiphinema americanum Cobb; root ectonema. Puerto Rico,
Brazil.

Xylaria apiculata Cke.; sporocarps on stubble but possibly a
weak parasite on senescent culms. Puerto Rico, Ecuador,
Costa Rica, El Salvador.

SAINTPAULIA IOANTHA Wendl. Its origin the Usumbara hills of
northeast Africa, this is a tender, small but very attractive and
greatly variable fleshy-leaved flowering perennial, of fairly recent
introduction to the American tropics but used widely by commercial florists, and increasingly popular among homeowners whether
rich or of very modest means; although far from being a true
violet plant common names are "violeta africana" Sp. Po.,
"African violet," "usumbara violet" En.

Aphelenchoides fragariae (Ritz.-Bos) Christie; leaf lesion
disease from nematode infection. Florida.

Basal rot (many unidentified fungi). Puerto Rico, Costa Rica.
(Note: Diagnoses for growers in both these countries disclosed a complex of about a dozen apparent causal fungi;
these organisms were not isolated individually, identified,
and tested for proof of pathogenicity through standard Koch's
Postulate procedures.)

Botrytis cinerea Pers.; moldy leafrot. W. Indies, C. America,
Florida.

Colletotrichum sp.; leaf and stem lesion. Costa Rica.

Fusarium sp.; root browning and collapse. Guatemala.

Meloidogyne sp.; rootknot nematode (in untreated potting soil).
C. America.

Penicillium spp.; secondary at edges of leaf scorch lesions.
Puerto Rico, Costa Rica.

Pestalotia sp.; isolated from browned stem. Costa Rica.

Phytophthora sp. (possibly P. cinnamomi); root- and crownrot. Costa Rica.

Pythium sp.; root and cutting rots (symptom is transluscent,

watery tissue rot).   Puerto Rico, Costa Rica, reported
    symptoms from Panama and Guatemala.
Rhizoctonia solani Kuehn (Thanatephorous cucumeris (Frank)
    Donk is the spore-bearing state); plant collapse.   Puerto
    Rico, Costa Rica, probably everywhere.
Ringspot.   (Said to be physiological, but I could not repeat
    results giving ringspot one year in one household planting,
    when plants from the same source were "treated in identical
    manner" in another household.)   Puerto Rico, Costa Rica,
    widespread but scattered.
Sclerotium rolfsii (Curzi) Sacc. (the perfect state is Pelli-
    cularia rolfsii (Sacc.) West); ground rot when inoculated in
    both Florida and Guatemala.
Scorch of leaf lamina, cause unidentified.   Puerto Rico.
Wither-tip of leaves and flower petals.   Cause is unproved
    but appears in certain cases as seasonal, resulting from
    undue exposure to sun.   Costa Rica.

SALIX spp.   A few species are economically important in the
    American tropics (one is mostly an ornament), as fast-growing
    soil-conserving windbreak and pole trees, producing light durable
    wood for structures, used in fashioning musical instruments,
    making of "carbón" (cooking charcoal), and the source of withies
    for basket-making and wickerwork.   Three species that seem to
    have had some plant disease attention, with common names indi-
    cated, are: (1) S. BABYLONICA L., an ornamental; "sauce
    llorón" Sp., "salgueiro ornamentál" Po., "weeping willow" En.;
    (2) S. HUMBOLDTIANA Willd. = S. CHILENSE Mol., an im-
    portant pole and windbreak tree; "sauce amargo" Sp., "salgueiro
    de chile" Po., "chile willow" En.; (3) S. VIMINALIS L., noted
    for its slender and lithe branches; "sauce mímbre" Sp., "vimeiro"
    Po., "basket willow" En.   (A disease if reported on a definite
    species named, will be thus indicated by the appropriate number
    within parentheses; when reports are on miscellaneous species no
    number will be indicated.)
    Agaricus spp., all are reports from Argentina and on decaying
        logs: A. disseminatus Pers., A. fascicularis Speg. =
        Hypholoma megapotamicum Speg. (2), A. portegnus Speg. =
        Pleurotus portegnus Speg. (2), A. tigrensis Speg. = Crepido-
        tus tigrensis Speg. (2) and A. variabilis Pers.
    Anthostomella limitata Sacc.; branch infection.   Argentina (2).
    Botryodiplodia sp.; dieback, stem and branch lesions.   Peru.
    Botryosphaeria ribis Gross. & Dug. = B. dothidea (Moug.)
        Ces. & de Not.; branch canker.   Brazil (3).
    Botrytis flavo-cinerea Speg.; on tree trunk.   Argentina (2).
    Calloria rubella (Pers.) Karst.; branch woodrot.   Argentina
        (2).
    Capnodium salicinum Mont.; foliage sootymold.   Uruguay.
    Cercospora salicina Ell. & Ev.; leafspot.   Peru, Argentina,
        Brazil.
    Ceriomyces schnyderianus Speg., and C. spongia Speg., both
        from trunk rot.   Argentina.

Clavaria dubiosa Speg.; bark disease.   Argentina (2).
Clitocybe leccata (Scop.) Quél.; semi-parasitic at tree base.
    S. America (2).
Colletotrichum sp.; anthracnose on leaves and twigs.   Argen-
    tina.
Coprinus fuscescens (Schaeff.) Fr.; probable saprophyte.   S.
    America.
Corticium lacteum Fr.; ? epiphytic.   S. America.
Crepidotus mollus Schaeff.; wood decay (? saprophyte).   S.
    America.
Cryptostictus platensis Speg.; twig disease.   Argentina (2).
Cryptovalsa platensis Speg.; branch disease.   Argentina (2).
Cuscuta americana L.; dodder.   Florida.
Cyphella villosa (Pers.) Karst.; wood saprophyte.   Argentina
    (2).
Cytospora chrysosperma (Pers.) Fr.; branch disease.   Argen-
    tina.   C. salicis (Cda.) Rab.; branch infection.   Argentina.
Daedalea trametes Speg. = Trametes daedalia Speg.; woodrot
    of trunk.   Argentina (2).
Delitschia congregata Speg.; on decaying logs.   Argentina.
Diaporthe catamarensis Speg., D. eres Nits., D. humboldtiana
    Speg. (possibly synonyms) all causing twig dieback.   S.
    America (2, 3).
Diatrypella exigua Wint., D. nigro-annulata (Grev.) Nits.;
    branch infections.   S. America.
Diplodia salicina Lév.; dieback.   Brazil (3).
Discula microsperma (Berk. & Br.) Sacc.; on wythes.
    Brazil (3).
Eutypella bonariensis Speg. = Valsa bonariensis Speg.; branch
    disease.   S. America (2).   (See Valsa below.)
Fomes applanatus (Pers.) Gill.; root and heart rot.   Chile
    (1).   F. australis Fr.; root and heart rot.   S. America.
    F. xylocreon Speg.; root and heart rot.   Argentina (2).
Fusarium spp.; secondary invasions following primary para-
    sites.   C. America, S. America.
Geopyxis marasmioides Speg.; on rotted wood.   ? Locality in
    S. America.
Graphium fissum Pruess.; on wood.   Widespread.   G. chloro-
    cephalum (Speg.) Sacc.; trunk disease and woodstain.
    Argentina (2).
Helminthosporium cerinum Speg. subsp. cerinellum Speg. is
    the conidial form of Dasyscypha cerina (Pers.) Fckl.; leaf
    disease, canker and woodrot.   Argentina.
Helotiella velutina Speg.; on decaying log.   Argentina.
Hymenochaete bonaerensis Speg.; a bark disease.   Argentina.
Hypholoma appendiculatum Bull.; possibly saprophytic agaric
    on decayed log.   Argentina.
Hypocrea pezizaeformis Speg., branch rot.   Paraguay.
Hypoxylon serpens (Pers.) Fr.; branch dryrot.   Widespread.
Hysterium bonarense Speg.; bark infection.   Argentina (2).
Hysterographium bonaerense Speg.; on old bark and wood.
    Argentina (2).

Irpex cartilagineus Speg. = Xylodon cartilagineus (Speg.)
  Kuntze; trunk infection. Argentina (2).
Isaria ceratioides Speg.; wood decay. Argentina (2).
Lasiosphaeria racodium (Pers.) Ces. & deN.; woodstain.
  Widespread.
Lentinus eximius Speg. = L. spegazzinii Sacc. & Cubani and
  L. schnyderi Speg. = L. nigripes Fr. = Panus schnyderi
  (Speg.) Sing.; rots of logs and trunk base in Argentina (2).
Lisea australis Speg.; rot of dead branch. Argentina.
Lophiosphaera bonaerensis Speg.; dieback. Argentina (2).
Lophiostoma vicinissimum (Speg.) Cke.; disease lesions on
  cortex of stem and trunk. Argentina (2).
Loranthaceae, mistletoe-like parasites, various genera and
  species, no collections made; branch and trunk diseases.
  C. America.
Marasmius sphaerodermus Speg.; on fallen twigs. Argentina
  (2). M. sarmentosus Berk. (sometimes given as M.
  equirinus Muell., in W. Indies.); causing "horsehair blight"
  on stems and leaves. Panama.
Marssonina nigricans Ell. & Ev.; stem and foliage lesions,
  anthracnose. Argentina, other countries in S. America
  (1, 3).
Melampsora humboldtiana Speg. with synonyms: M. abieti-
  capraearum Tub., M. americana Arth., M. americana
  Jorst, M. epitea Th.; leafrust. Widespread. M. repentis
  Plowr.; leafrust. S. America.
Monilia platensis Speg.; wood discoloration after sawing and
  planing. Argentina.
Naemaspora citrinula Speg.; branch canker. Argentina (2).
Nectria hematochroma Speg. = Cucurbitaria haematochroma
  (Speg.) Kuntze; bark disease. Argentina (2). N. platense
  Speg. = Cucurbitaria plantense (Speg.) Kuntze; bark disease.
  Brazil (2). N. vulgaris Speg. = the conidial state is
  Verticillium tubercularioides Speg.; dieback and branch
  disease. Argentina (2).
Odontia argentina Speg.; trunk and tree-base rot. Argentina
  (2).
Panus laciniato-crenatus Speg.; disease of living tree trunk.
  Argentina (1, 2).
Pellicularia isabellina (Fr.) Rogers; seedling tree diseases.
  S. America.
Peziza hyalino-alba Speg. = Humaria hyalino-alba Speg.;
  trunk and branch infections. Argentina (2).
Pholiota pseudofascicularis Speg.; tree-base disease. Argen-
  tina.
Phoradendron macrophyllum Cock. (Also see above Lorantha-
  ceae.) In C. America and subtropical USA, mistletoe-like
  parasites.
Polyporus spp.; tree base-and-root rot. S. America (1, 2).
Poria vitrea Pers.; wood and branch rot. S. America (2).
Psittacanthus coccineus Patsch.; mistletoe-like parasite. Peru.
Ramularia rosea (Fckl.) Sacc.; leafspots. Argentina (2).

Rosellinia spp. ; (probably on all species of host) root rots.
    C. America.
Rutstroemia firma (Pers.) Karst. ; on foliage. Argentina (2).
Septobasidium saccardinum Speg. ; felt disease, associated with
    scale insects. Brazil (3).
Sphaceloma murrayae Gods. & Jenk. ; scab anthracnose. Sub-
    tropical USA, Argentina (1, 3).
Sporotrichum murinum (Lk.) Bon. ; leaf and twig disease. S.
    America.
Stereum spp. heart-wood and trunk rot both on living and
    felled trees: S. argentinum Speg. in Argentina. S. atro-
    zonatum Speg. in Argentina (1). S. ochraceo-flavum
    (Schw.) Sacc. in S. America. S. purpureum Pers. in
    Chile (3).
Stilbum minutissimum Speg. ; disease of tree cortex. Argentina.
Tremella lutescens Pers. ; on dead wood. S. America (2).
Typhula candidad Fr. ; wood infection. S. America (2).
Valsa humboldtiana Starb. ; branch and other wood attack.
    Brazil (2).
Verticillium tubercularioides Speg. ; its asco-state is Nectria
    vulgaris Speg. (synonyms: N. ochroleuca (Schw.) Berk.
    and Creonectria ochroleuca (Schw.) Seav.); branch and
    twig disease. Argentina (2).
Xerotus conicus Speg. ; on decayed trunk. Argentina (2).
Xylaria cristata Speg. ; rootrot. Argentina (2). X. fasiculata
    Speg. ; rootrots. Argentina (2).

SAMANEA SAMAN (Jacq.) Merr. = S. SAMAN (Willd.) Merr. =
    PITHECOLOBIUM SAMAN (Jacq.) Benth. Native to the West
    Indies region they are large flat-topped important shade trees
    along roadways, in pastures, parks and market squares (plazas),
    its wood used in semipermanent structures and for kitchen fuel;
    "samán" Sp. Po., "rain tree, " "saman" En. (Note: Sucking insects
    flourish on this tree and under certain conditions may excrete
    drops of sap in such quantity that resting in the shade can be an
    unpleasant experience.)
    Aphanopeltis aequetoriensis Syd. ; on old leaves. Panama and
        scattered in C. America.
    Botryodiplodia sp. ; dieback. C. America.
    Capnodium spp. ; sootymolds. Widespread.
    Cercospora samaneae Chupp & Muller; leafspot. Common on
        older leaves. Widespread.
    Cissus sp. ; so-called "liana" climbing vines crawling over
        branches breaking trees and seen in warm wet areas. C.
        America.
    Cladosporium herbarum Fr. ; on senescent leaves. W. Indies,
        C. America.
    Colletotrichum (? gloeosporioides); anthracnose. Widespread.
    Cuscuta americana L. ; yellow dodder on foliage of large trees.
        W. Indies.
    Epiphytes such as: bromeliads, orchids, ferns, mosses,
        climbing bamboo, ericaceous plants that develop abundantly

on scaffold branches and result in tearing away of limbs.
Rainforest areas.

Fusarium oxysporum Schl. f. sp. samanea Wellman; root infection and systemic disease causing wilt and death.
Puerto Rico, St. Croix. F. sp. (roseum type); secondary
following primary infections by other fungi and insect injury.
Associated with dieback. C. America.

Heartrot or "caries" in old pasture trees, starts at base apparently and extends inward and upward. Various fungi can
be isolated. Scattered.

Loranthus sp. ; phanerogamic (mistletoe-like) large parasitic
bush on branches. El Salvador.

Marasmius spp. ; cord blights. Panama.

Meliola pithecolobiicola Speg. ; mold spots on leaf. W. Indies,
Panama.

Meloidogyne sp. ; rootknot. Subtropical USA.

Microstroma pithecolobii Lamk. ; leaf-mold. Puerto Rico.

Mycena citricolor (Berk. & Curt.) Sacc. ; leafspot on seedlings.
Puerto Rico.

Odontia sp. ; branch rot. Costa Rica.

Omphalia sp. ; rootrot. Peru.

Pestalotia sp. ; part of dieback complex. Costa Rica.

Phoma-like leafspot. W. Indies.

Polyporus sp. ; on roots of uprooted tree. Guatemala.

Poria sp. ; rot of branch torn away from trunk by load of
epiphytes. Guatemala.

Ravenelia sp. ; a mild leafrust. Widespread.

Strigula (? complanata); lichen leafspot of raised type and
yellowish color. Puerto Rico.

Struthanthus spp. ; phanerogamic (mistletoe-like) recumbent
bushes attacking branches. El Salvador, Costa Rica.

Trametes hydnoides (Sw.) Fr. ; heartrot. S. America.

Woodstains, causes undetermined. C. America.

SAMBUCUS spp. Perennial bushes grown in mountains and other
cool parts of the tropics for windbreaks and field edge markers.
Fruit is eaten and short lengths of canes are buried for fertility
around hills of maize, potatoes, cucurbits; "saúco" Sp. , "sabugo, "
"sabugueiro" Po. , "tropical elder" En.

Aecidium sambucinum Arth. & Cumm. ; rust. S. America.

Bertia australis Speg. ; old cane decay. Argentina.

Botryodiplodia theobromae Pat. ; branch disease, cane-tip rot.
Guatemala.

Botryosphaeria dothidea (Moug.) Ces. & de Not. ; canker, cane
dieback. Mexico, Guatemala.

Cercospora sp. ; leafspot. Guatemala, El Salvador, Mexico.

Diaporthe circumscripta Otth. ; stem canker. Scattered in S.
America.

Eutypa tuyutensis Speg. ; on fallen branch. Argentina.

Fusarium sp. ; cane disease. El Salvador.

Melanomma fuscidilum Sacc. ; stem disease. Argentina.

Nectria sp. ; cane decay. Guatemala.

Oidium sp.; powdery white mildew.   El Salvador, Guatemala.
Rhizoctonia (coarse mycelium and not R. microsclerotia);
    causing typical web blight of leaves.   Guatemala.
Sphaeropsis aguirre (Speg.) Sacc.; canker.   Argentina.
Valsa tuyutensis Speg.; stem infection.   Argentina.
Xylaria sp.; rootrot.   C. America.

SANSEVIERA ZEYLANICA WILLD. (this binomial is usually given
but sometimes questioned).   A flowering perennial whose long
stiff leaves contain fine strong fiber used for various purposes,
some times grown as an ornament and a visitor deterrent;
"espada del diablo," "planta del culebra" Sp., "espino de
diablo," "sanseviéria" Po., "bowstring hemp," "sanseviera" En.
    Aspergillus niger v. Tiegh.; basal rot especially under moist
        conditions.   Puerto Rico, Costa Rica, El Salvador, Colom-
        bia, Venezuela.
    Botryodiplodia theobromae Pat.; dieback.   Puerto Rico, El
        Salvador.
    Cassytha filiformis L.; green dodder, love vine.   Puerto Rico.
    Cephaleuros virescens Kunze; algal leafspot.   Florida, Puerto
        Rico.
    Chaetophoma sansevieriae Tassi; leaf lesions.   Widespread.
    Colletotrichum gloeosporioides Penz. = Gloeosporium sanse-
        vieriae Verw. & duPl.; anthracnose.   Florida, Puerto Rico,
        probably widespread.
    Erwinia carotovora (L. R. Jones) Holl. is apparently primary,
        followed by E. aroideae (Towns.) Holl. where seen by me;
        soft rots.   Subtropical USA, W. Indies, El Salvador, Costa
        Rica, probably in S. America.
    Fusarium moniliforme Sheld.; sunken leafspot.   Florida,
        Costa Rica.
    Haplographium chlorocephalum (Fres.) Grove; leaf degenera-
        tion.   Argentina.
    Leafspots various types (sterile).   Puerto Rico.
    Meloidogyne sp.; rootknot nematode.   Florida.
    Menoidea sp.; infection of aging leaf.   Argentina.
    Trichoderma sp.; isolation from diseased root.   Puerto Rico.

SAPOTA ACHRAS see ACHRAS SAPOTA

SCHINUS MOLLE L. and S. TEREBINTHIFOLIA Radii.   Trees indigenous to South America, these are the most common species,
grown as ornamentals for their graceful foliage and panicles of
red berries; "pimento del Brasil" Sp., "arvore pimento-orna-
mental" Po., "pepper-tree" En.
    Amphisphaeria majuscula Speg.; branch wood infections.
        Argentina.
    Armillaria mellea Vahl = Armilariella mellea (Fr.) Karst.;
        rootrot.   California.
    Capnodium sp.; sootymold.   Puerto Rico, Panama, probably
        widespread.
    Cercospora schini Syd.; leafspot.   Argentina.

Clitocybe tabescens (Scop.) Bres. = Armilariella tabescens
  (Scop.) Sing.; rootrot. Subtropical USA.
Colletotrichum gloeosporioides Penz. = Glomerella cingulata
  (Stonem.) Sp. & Schr.; anthracnose. Florida and W. Indies.
Ganoderma orbeforme Fr.; root and tree-base decay. S.
  America.
Irenopsis coronata (Speg.) Stev.; leaf spot-mold. Paraguay
  and other nearby countries.
Meliola amphitricha Fr., M. brasiliensis Speg., M. ludibunda
  Speg.; black mildews in Brazil. M. malacotricha Speg.;
  black mildew in Puerto Rico.
Myxosporella schini Carranza; twig and foliage disease.
  Argentina.
Oidium erysiphoides Fr.; powdery mildew. S. America,
  Guatemala.
Phoradendron sp.; mistletoe-like upright bush, a branch para-
  site. Brazil.
Phrygilanthus repens Patsch.; recumbent parasitic shrub at-
  tacking branches. Peru.
Phymatotrichum omnivorum (Shear) Dug.; rootrot. Texas and
  Arizona in Subtropical USA.
Phytophthora cactorum (Leb. & Cohn) Schr. var. applanata
  Chest. and P. parasitica Dast.; stem and seedling diseases.
  S. America.
Psittacanthus coccineus Patsch.; upright shrub on swollen part
  of host branch. Peru.
Pythium spp.; seedling diseases. Widespread.
Rosellinia pepo Pat.; a white rootrot. Panama.
Strigula complanata Fée; lichen leafspot. Puerto Rico.

SCINDAPSUS AUREUS Engler = POTHOS AUREUS Lind., with yellow
  spotting, and S. PICTUS Hassk. = POTHOS sp. with its white
  and gray spotting. Large climbing vines and the most hardy and
  common species of the genus, both popular ornamentals introduced from the East Indies; "trepapalo armarillo," "trepapalo
  común" Sp., "arão trepador" Po., "climbing arum," "ivy-arum,"
  "pothos" En.
    Capnodium spp.; sootymolds (that may not always be Capnodium)
      under many kinds of shade trees. Widespread.
    Cephaleuros virescens Kunze; algal leafspot. Puerto Rico.
    Cladosporium sp.; leaf edge infections. Guatemala, Puerto
      Rico, Panama.
    Colletotrichum gloeosporioides Penz. (asco-state occurring on
      aging diseased stems is Glomerella cingulata (Stonem.)
      Sp. & Schr.), an old synonym is Colletotrichum pothi
      Koord. and perhaps the most convenient usage might well
      be C. gloeosporioides Penz. f. sp. pothi for certainly this
      type of Colletotrichum is adapted to the conditions of para-
      sitism on Scindapsis in its epiphytic life habit. It causes
      anthracnose disease of leaf and stem and can subsist in-
      ternally with no outward signs. Puerto Rico, El Salvador,
      Panama.

Leafscorch, apparently physiological cause. Puerto Rico.
Lichen (green) causes blemished leaf appearance. W. Indies.
Liverwort (miniature leafy types) epiphytic on leaves and
   petioles and climbing roots. Rainforest area in Puerto
   Rico.
Phyllosticta aricola Weedon; large shiny leafspot. El Salvador.
   P. pothicola Weedon; small leafspots. Guatemala, El Sal-
   vador, Costa Rica, Panama.
Pythium splendens Braun; rootrot. Florida, W. Indies.
Rhizoctonia solani Kuehn (without sporulating state), cutting
   disease and stem rot in nursery. Puerto Rico.
Rosellinia bunodes (Berk. & Br.) Sacc.; black veins in roots
   ending in death, roots in soil near coffee trees with this
   disease. Costa Rica.
Strigula complanata Fée; white lichen leafspots. Puerto Rico.

SCIRPUS spp. Much valued rushes that are important in certain
   localities where they are used in weaving mats, wall dividers,
   baskets and the like. There has been minor attention to
   diseases on them in the warm tropics where these plants grow
   wild and in pond and stream edges; "weavers rush," "sweet
   rush" En. Literature is of a scanty nature but from it and a
   few observations, indications are that certain groups of diseases
   are as follows:
      Culm injury from Endothella sp., Pestalotia sp., Balansia sp.
      Leaf black mildews from Meliola spp.
      Leafspots from Phyllachora spp., Anthostomella sp., Cerco-
         sporella sp.
      Rusts from Uromyces spp. and Puccinia spp.
      Smuts from Cintractia spp.

SCORZONERA HISPANICA L. From Europe a fairly recent root-
   crop-introduction for hillside gardens and cool tropics. Thus
   far in the Neotropics it has only a few diseases reported on it;
   "salsifí negro" Sp., "cercefi" Po., "black salsify" En.
      Albugo tragopogonis Pers.; leaf disease called white-rust.
         Guatemala, ? Widespread.
      Cuscuta sp. (slender type); dodder. El Salvador.
      Erwinia carotovora (L. R. Jones) Holl.; soft rot. El Salva-
         dor, Guatemala.
      Fusarium sp.; isolated from root lesion. Guatemala.
      Rhizoctonia sp.; mycelium on plant base, causing constriction.
         Guatemala market.
      Watery leaf lesions. Cause undetermined (neither bacteria
         nor fungi isolated). El Salvador in volcano garden.

SECALE CEREALE L. An ancient north European crop, planted
   in the American tropics in cool, low fertility soils, uses are for
   bread, livestock grain food, pasturage, cover crops, and the
   resistant straw is made into thatch and mats. (It is not a popular crop and in the American tropics has not had intensive
   disease study but is a leading product in Argentina in some cold
   areas); "centeno" Sp., "centeio" Po., "rye" En.

Boerlagella argentinensis Speg.; culm infections. Argentina.
Cladosporium herbarum (Pers.) Lk.; spores from unhulled
    seed. Mexico.
Claviceps purpurea (Fr.) Tul., the imperfect state is Aphacelia
    segetum Lév.; ergot. Mexico, Ecuador, Uruguay, Bolivia,
    Brazil, Argentina, Colombia, Chile.
Colletotrichum cereale Manns; anthracnose. Brazil. C.
    graminicola (Ces.) Wils.; leaf and culm disease. Mexico,
    Texas, Brazil, probably widespread.
Erysiphe graminis DC. var. secalis Em. March.; powdery
    mildew. Widespread.
Fusarium oxysporum Schl. (reported as F. orthoceras App. &
    Wr.); footrot of seedlings. Argentina.
Gibberella cyanogena (Desm.) Sacc. (reported as G. saubinettii
    (Mont.) Sacc.), and G. zeae (Schw.) Petch; both cause scab,
    footrots, blights. Argentina, Texas.
Helminthosporium sativum Pam., King & Bakke; leaf-blotch
    and rootrot. Subtropical USA, Argentina. H. tritici-
    repentis Died. and its asco-state which is Pyrenophora
    tritici-repentis (Died.) Drechs.; leaf diseases. Brazil.
Leptosphaeria sp. (possibly herpotrichoides); culm infections.
    Argentina.
Ophiobolus graminis Sacc., O. cariceti (Berk. & Br.) Sacc.;
    take-all, footrot. Mexico, Andean S. America.
Penicillium spp.; moldy grain. Mexico.
Pseudomonas coronafaciens (Elliott) Stev.; halo blight from
    inoculations. Argentina.
Puccinia glumarum (Schw.) Eriks. & P. Henn.; leaf yellow-
    rust. Colombia, Chile, Uruguay, Argentina. P. graminis
    secalis Eriks. & E. Henn.; stem rust. Peru, Uruguay,
    Colombia. P. rubigo-vera secalis (Eriks.) Carl. (this is
    also given by some as P. dispersa Eriks. & E. Henn.);
    leaf redrust. Colombia, Uruguay, Chile, Argentina.
Pyricularia grisea (Cke.) Sacc.; culm and leaf disease in areas
    near old hill (dry-land) rice plantations. Scattered.
Rhynchosporium secalis (Oud.) J. J. Davis (synonyms are R.
    graminicola Heins and Marssonina secalis Oud.); leafscorch.
    Mexico, Argentina.
Sclerotium rolfsii (Curzi) Sacc., its basidial-state, though
    seldom seen, is Pellicularia rolfsii (Sacc.) West; ground-
    level rot. Argentina.
Scolecotrichum graminis Fckl.; leaf-stripe. Brazil, Uruguay.
Septoria secalis Prill. & Del.; spotting injury on leaves
    and leaf sheaths. Subtropical USA, Mexico.
Spegazzinia tessartha (Berk. & Curt.) Sacc.; leaf disease.
    Argentina.
Sphacelia (see Claviceps)
Tilletia caries (DC.) Tul. = T. secalis (Cda.) Kuehn (has
    rough spores); bunt. Argentina, California, probably wide-
    spread. T. contraversa Kuehn; dwarf bunt. Argentina,
    Uruguay. T. foetida (Wallr.) Liro = T. Levis Kuehn (has
    smooth spores); bunt. Argentina, may be the most wide-
    spread.

Urocystis occulta (Wallr.) Rabh.; flagsmut.   Subtropical USA,
    Argentina, Brazil.
Ustilago hordei (Pers.) Lagh.; covered smut.   Uruguay,
    Argentina.   U. nuda (Jens.) Rostr.; loose smut.   Uruguay,
    Argentina.   U. spegazzinii Hirsch.; stem smut.   Argentina.
Virus: Hoja Blanca disease of rice by inoculation in Colombia.
Xanthomonas translucens L. R. Jones, Johns. & Reddy f. sp.
    secalis (Reddy, Godkins, & Johnst.) Hagb.; bacterial blight.
    California, possibly in Mexico.

SECHIUM EDULE (Jacq.) Sw.   A robust and variable semi-herba-
    ceous perennial vine indigenous to tropical America, very im-
    portant as a vegetable grown in moderately warm, not hot, lands,
    producing edible fruits, edible large tendrils, and edible swollen
    starchy roots; "huisquil," "chayote" Sp., "chayote," "chuchu,"
    "chuchuiero" Po., "chayote," "christophine" En., and by market
    women: "cho cho," "tallote," "huisayote," "quisquíl" Ind.
    Ascochyta sp.; leafspot.   S. America.
    Asterina sp.; leaf blotchy-spot.   Guatemala.
    Botryodiplodia theobromae Pat. is the conidial but disease-
        causing state of the ascomycete Botryosphaeria quercuum
        (Schw.) Sacc. = Physalospora rhodina Berk. & Curt.; large
        dark fruit-lesion, pedicel-rot, vine-canker.   Puerto Rico,
        El Salvador, Guatemala, Dominican Republic.
    Botrytis sp.; flower disease, market fruitrot.   C. America in
        mountains.
    Cassytha filiformis L.; slow green dodder.   Puerto Rico.
    Capnodium spp.; sootymolds.   C. America, S. America.
    Cephaleuros virescens Kunze; algal spot on leaf, vine, and
        sunken fruitspot.   C. America, W. Indies, Panama, Vene-
        zuela (probably widespread).
    Cercospora spp. all causing leafspots of varying severity:
        C. brachypoda Speg. in W. Indies, S. America, C. citrul-
        lina Cke. is widespread, C. cucurbitae Ell. & Ev. in Sub-
        tropical USA, C. hibisci Tracy & Earle in W. Indies, C.
        sechii Stev. (is common and some others listed here may
        be synonyms of it or the first named) in Subtropical USA,
        Cuba, Panama, W. Indies, Venezuela, Brazil, C. malayen-
        sis Stev. in Puerto Rico.
    Colletotrichum gloeosporioides Penz. (I did not find its asco-
        state); mild-type anthracnose, fruit spot.   Subtropical USA,
        C. America and C. orbiculare (Berk. & Mont.) v. Arx =
        C. lagenarium (Pass.) Ell. & Halst.; diffuse anthracnose
        leaf and fruit lesions with more tendency to expand over
        more tissues.   Florida, El Salvador, Guatemala, Panama.
    Cuscuta spp. (several unidentified but different species ap-
        parent); dodders.   El Salvador, Guatemala.
    Erwinia spp.; softrot bacteria in markets on tendrils, roots,
        and fruits displayed for sale.   C. America, Ecuador, and
        probably widespread.
    Erysiphe cichoracearum DC = Oidium sp.; powdery-white mil-
        dew, dryseason disease especially on new foliage.   Guate-
        mala, Nicaragua, Panama, Cuba.

Fusarium sp.; isolated from root lesions.   Costa Rica.
? Ganoderma sp.; rootrot of chayote adjacent to a sporophore
    on a dying rubber tree.   Peru.
Helminthosporium sechicola J. A. Stev.; leaf disease.   Puerto
    Rico, C. America, S. America.
Leandria momordieae Rang.; on aging foliage.   Brazil.
Meloidogyne sp.; rootknot nematode.   El Salvador.
Mycena citricolor (Berk. & Curt.) Sacc. = Omphalia flavida
    (Cke.) Maubl. & Rangel; leafspot from inoculation.   Costa
    Rica.
Mycosphaerella citrullina (C. O. Sm.) Gross; fruit blackrot,
    leafspot.   Puerto Rico.
Phyllosticta sechii Young; leafspot.   Puerto Rico, El Salvador,
    Guatemala.
Pseudomonas solanacearum E. F. Sm.; bacterial wilt.   Guate-
    mala, Puerto Rico, Cuba.
Pseudoperonoplasmopara cubensis (Berk. & Curt.) Rost.;
    downy mildew.   W. Indies, C. America.
Red lesions (unknown cause) on stem and leaf.   Guatemala.
Rhagadolobium cucurbitacearum (Rehm) Th. & Syd.; leaf tar-
    spot.   Widespread.
Rhizoctonia sp. (aerial parasitic hyphae); attacking edible
    tendrils and rest of foliage.   Costa Rica.   R. solani Kuehn
    its basidial state is Thanatephorous cucumeris (Frank) Donk;
    damping-off of seedlings, fat-root disease, stem lesions,
    fruit injured if lying on infested soil.   El Salvador, Costa
    Rica.
Rhizopus nigricans Ehr.; rot of fruits and edible tendrils.   In
    markets all over Guatemala.
Rootrot (undetermined cause); black decay causing bitter flavor.
    Guatemala, Nicaragua.
Rosellinia sp.; rootrot, no effect on flavor.   Guatemala and
    Honduras highlands.
Scab spots on fruits (undetermined cause).   Scattered.
Sclerotium rolfsii (Curzi) Sacc.; rot at ground line.   Texas.
Septoria cucurbitacearum Sacc.; leaf spotting.   Mexico, Costa
    Rica.
Verticillium sp.; rare wilting.   S. America.

SERJANIA spp.   Ornamental vines, and some important kinds are
    used for the juices that stun fish.   The list of disease parasites
    which follows (without annotations) is from miscellaneous tropical
    American sources.   Two common names of the fish-stunning vines
    are "barbasco" and "pedilla" Sp. Ind.
    Aecidium serjaniae P. Henn.
    Botryodiplodia sp.
    Calonectria guarapiensis Speg., and 3 or 4 others.
    Catacauma serjaniae (Speg.) Chard.
    Cephaleuros virescens Kunze.
    Dasystica sapindophila Speg.
    Dexteria pulchella Stev.
    Diplacella pauliniae (Gonz. Frag. & Cif.) Syd.

Fusarium sp.
Meliola spp. (several).
Nectria spp.
Phyllachora spp., over a dozen named.
Puccinia arechavaletae Speg.
Sphaerella confarta Speg.
Strigula spp.
Xanthospora melanostoma Speg.

SESAMUM INDICUM L. = S. ORIENTALE L.   Originally from the
   Orient, under tests in several parts of the Western tropics it
   has shown to bear well and is a crop whose seed is the source of
   a fine edible oil, cake from the extraction is of special value in
   animal feed, and seeds themselves are eaten in breads, confec-
   tions, and other human foods; "ajonjoli," "sésamo" Sp., "ajon-
   jolí," "gergalim" Po., "sesame" En.
      Alternaria sesamicola Kam (? synonyms: A. sesami (Kowa.)
         Moh. & Beh. and A. solani (Ell. & Mart.) Sor.); blast,
         leafspotting, stem cankering.   Florida, C. America, Vene-
         zuela.
      Botryodiplodia theobromae Pat.; isolation from top-diseased
         plants.   Nicaragua.   (Often reported as Diplodia.)
      Canker of large stalks but no proof of fungus or bacterial
         cause.   El Salvador.
      Cercospora sesami Zimm.; small brown and circular spots on
         leaves, petioles, stems, and capsules.   Common in moist
         areas.   Widespread.
      Colletotrichum sp.; isolations from old stalks.   El Salvador.
      Cylindrosporium sesami Hansf.; angular leafspot.   W. Indies,
         Subtropical USA, Venezuela.
      Fusarium oxysporum Schl. in warm slightly acid soils (it
         deserves being designated as f. sp. sesami); systemic
         wilt.   Nicaragua, Dominican Republic, Colombia, Ecuador,
         Venezuela, Peru.   F. solani (Mart.) Sacc., if the asco-
         state is found it will be Nectria haematococca Berk. &
         Br.; stem lesions and rot of plant base.   Nicaragua.
      Helminthosporium sesami Miyaki; leaf and capsule blotch,
         stem canker.   Nicaragua, Ecuador, Colombia.
      Macrophomina phaseoli (Maubl.) Ashby; in warm areas, in
         dryseasons causes charcoal rootrot, wilt.   Subtropical
         USA (dry areas), Jamaica, Ecuador.
      Microxyphium aciculiforme Cif., Bat., & Nas.; sootymold.
         W. Indies.   (Podoxyphium sp. has also been reported as
         a sootymold on this crop in the West Indies and this may
         be a synonym.)
      Nematodes (not determined as to genus or species), feeding
         on rotted ends of Fusarium infected roots.   Nicaragua.
      Oidium sp.; white powdery mildew.   Scattered in S. America.
      Phytophthora cactorum (Leb. & Cohn) Schroet.; collar rot,
         rootrot.   El Salvador, Santo Domingo, Peru.   P. para-
         sitica Dast.; stem canker, collar canker.   Venezuela,
         Argentina.   (Phytophthora-Pythium, isolated in El Salvador

and in Costa Rica from seedling damping-off and rootrot,
and mycological characters were not distinctly either genus.)
Pseudomonas sesami Malk. ; bacterial leafspot (seed carried).
  Florida, Texas. P. solanacearum E. F. Sm. = Xantho-
  monas solanacearum (E. F. Sm.) Dows. ; bacterial wilt.
  El Salvador, Arizona.
Pythium ultimum Trow. ; root disease. Florida, Nicaragua.
Rhizoctonia crocorum DC. is the imperfect state of Helicobasi-
  dium purpureum Pat. ; the violet colored collar-and-root
  rots. El Salvador, scattered in Guatemala. R. solani
  Kuehn (imperfect but parasitizing state of the basidial fungus
  Thanatephorous cucumerus (Frank) Donk); common severe
  rootrot, damping-off in newly emerging seedlings. Sub-
  tropical USA, Santo Domingo, Guatemala, El Salvador,
  Venezuela, Colombia, Peru.
Sclerotium rolfsii (Curzi) Sacc. mycelial state of Pellicularia
  rolfsii (Sacc.) West; rot at soil line. Florida, El Salvador,
  Nicaragua, Guatemala, Colombia.
Viruses: Mosaic of mottle-type in Venezuela. Philody-type
  also in Venezuela.
Xanthomonas (see Pseudomonas)

SETARIA spp. Over 20 species of grasses both introduced and
  indigenous to the American tropics, important in pastures,
  poultry feed, and ground cover plantings; "yerba [hierba] de
  pasto" Sp. , "pastinho de lavoura" Po. , "pasture grass" En.
  Diseases of these crops are in the process of being more fully
  studied.
  Acrothecium falcatum Tehon; leaf disease. Puerto Rico.
  Angiospora cameliae (Mayor) Mains; rust. W. Indies, C.
    America.
  Balansia claviceps Speg. (imperfect state is Ephelis mexi-
    cana Fr.); spike stromatic disease. Florida, Cuba,
    Mexico, Costa Rica, Bolivia, Argentina. B. trinitensis
    Cke. & Mass. (= Ephelis trinitensis Cke. & Mass.);
    spike disease. Trinidad, Costa Rica.
  Beniowskia penniseti Wakef. and B. sphaeroidea (Kalch. &
    Cke.) Mason; leaf spot diseases. W. Indies, S. America.
  Cerebella (see Tolyposporum)
  Cladosporium herbarum Lk. ; on old heads and other senescent
    growth. Widespread.
  Claviceps purpurea (Fr.) Tul. , C. ranunculoides Moell. , C.
    setariae Herter are all ergots. In medium dry country.
    Scattered.
  Curvularia falcata (Tehon) Boed. ; leafmold, and C. geniculata
    (Tracy & Earle) Boed. ; rootrot. Puerto Rico.
  Dinemosporium graminium Lév. ; on leaf, weak parasite.
    Widespread.
  Ephelis (see Balansia)
  Eudarluca australis Speg. ; hyperparasite of grass rusts.
    Widespread.
  Helminthosporium spp. ; leaf spots and blast. Subtropical USA,
    Mexico, Costa Rica, Panama, Ecuador.

Melasmia setariae Atk. ; leafspot.   Subtropical USA.
Meliola chaetochloae Stev. ; black mildew.   In moist areas,
   widespread but not spectacular.
Mycena citricolor (Berk. & Curt.) Sacc. = Omphalia flavida
   (Cke.) Maubl. & Rang. ; leafspot from infection bodies
   splashed in rain drops from diseased coffee.   Costa Rica.
Phyllachora spp. all causing tarspots: P. chaetochloae Stev.
   in Trinidad, C. cornispora Atk. in Puerto Rico, Panama,
   Barbados and probably widespread, P. setariaecola Speg.
   in Brazil, Ecuador, Paraguay, Argentina and P. striatella
   Maubl. in Brazil.
Puccinia spp. are all rusts: P. atra Diet. & Holw. in
   Mexico, P. cameliae (Mayor) Arth. in Puerto Rico, Colom-
   bia, P. catevaria Cumm. is scattered, P. chaetochloae
   Arth. in Florida, W. Indies, P. levis (Sacc. & Bizz.)
   Magn. scattered, P. setariae Diet. & Holw. in Chile,
   Mexico, Subtropical USA, P. substriata [alternate hosts of
   this rust are species of Solanaceae, including tobacco] Ell.
   & Barth. in C. America, Bolivia, Brazil, Argentina.
Pyricularia grisea (Cke.) Sacc. ; leafspot.   Subtropical USA.
Sclerospora graminicola (Sacc.) Schroet. ; downy mildew.
   Subtropical USA, Mexico.
Sphacelia grisea Speg. ; culm disease.   Argentina.
Sphacelotheca pamparum (Speg.) Cling. = S. magnusiana
   (Waldh.) Cif., = Ustilago pamparum Speg. ; spike and seed
   smut. Widespread.   (Note: For convenience it is to be
   noted that Ustilago produces sori on almost any part of the
   host; Tilletia is similar to it but in host ovaries only, often
   the spore masses are more dusty and the spores are larger.
   The Ustilaginoidea, false smuts, are more primitive fungi
   that develop in diseased flower spikes the remarkable
   swollen structures that are dark colored gemmae.)
Sporotrichum peribebuyense Speg. = Beniowskia, see above;
   leafspot. S. America.
Staurophoma panici Hoehn. ; culm disease.   S. America.
Tilletia zundellii Hirsch. ; smut or bunt.   Argentina.
Tolyposporium pampeanum Speg. = Cerebella annae Speg. ;
   spike disease. Argentina and nearby countries.
Uredo setariae Speg. = Uromyces leptodermus Syd. ; leafrust.
   Florida, Panama, Costa Rica, Puerto Rico, Argentina,
   probably widespread. U. puttemansii Rangel; rust.   Panama,
   W. Indies, Brazil, Argentina.
Ustilaginoidea ochracea P. Henn. and U. setariae Bref. ; false
   smuts in grass spikes or panicles. Some special reports
   from Brazil, W. Indies, Subtropical USA, probably scattered widely.
Ustilago crameri Koern. ; headsmut.   Widespread.   (Sometimes U. neglecta Niessl. and U. setaria Rabenh. are also
   reported.)
Viruses: Panicum Mosaic Virus in Texas.   Mottle and leaf
   malformations seen in W. Indies but no virus proved.
   Sugarcane Mosaic Virus in Peru, Florida, and other locations.

SIDA spp.   Indigenous woody bushes, important to householders who
use the strong fiber and the cured stems to make brooms and
brushes.   Some species are grown in pastures for natural control of ectoparasites of animals; "escobilla, " "escoba fuerte" Sp. ,
"gauxuma, " "vassouras" Po. , "sida, " "broom bush" En.   (Note:
The plants are common, tough, self seeding.   Since they are of
greatest service to peasantry and indigenous peoples there has not
been, thus far, enough economic pressure to study the diseases,
except in the case of the plants being wild sources of virus that
affect nearby cultivated crops.   What is known by me about the
diseases is from almost accidental observation and personal collection.)

> Asterina diplocarpa Cke. and A. peraffinis Speg. ; both cause
> black leaf mildew in at least Brazil and nearby countries,
> and the former is common in Puerto Rico, probably wide-
> spread.
>
> Cassytha filiformis L. ; a slow growing green dodder.   W.
> Indies.
>
> Cercospora densissima Speg. ; a common large leafspot.   W.
> Indies, southern S. America.   C. hyalospora Muller &
> Chupp; leafspot.   Brazil.   C. micranthae Muller & Chupp;
> leafspot.   Brazil.
>
> Cercosporina sphaeralceicola Speg. = Cercospora sphaeral-
> ceicola (Speg.) Chupp; leafspot.   S. America.
>
> Cuscuta americana L. ; yellow dodder.   W. Indies and C.
> boldinghii Urb. (and other unidentified species); fast
> growing dodders.   W. Indies.
>
> Dietelia verruciformis P. Henn. ; leafrust.   Argentina.
>
> Dothidea collecta (Schw.) Ell. & Ev. ; stem canker disease.
> Subtropical USA.
>
> Entyloma sidae-rhombifoliae Cif. ; leafsmut.   W. Indies.
>
> Globulina trichocarpa Syd. ; stem weakening infection.   Scat-
> tered occurrence.
>
> Irenopsis molleriana (Wint.) Stev. ; black mildew.   Puerto
> Rico.
>
> Meliola molleriana Wint. (? possibly same as the Irenopsis);
> Widespread.
>
> Meloidogyne sp. ; rootknot nematode.   Brazil, no doubt else-
> where.
>
> Oidium erysiphoides Fr. ; powdery white mildew.   Virgin
> Islands, El Salvador, Venezuela.
>
> Phymatotrichum omnivorum (Shear) Dug. ; dryland rootrot.
> Texas.
>
> Puccinia spp. , all leafrusts: P. heterospora Berk. & Curt.
> in almost all tropical America, P. lobata Diet. in Chile,
> P. malvacearum Bert. in S. America and P. platyspora
> (Speg.) Jacks & Holw. = Uromyces platyspora Speg. in
> S. America.
>
> Sclerotium rolfsii (Curzi) Sacc. ; is the non-sporuliferous para-
> sitic state of a rare plant-base rot in moist areas.   Florida,
> Costa Rica.
>
> Septoria spp. , causes leaf spotting and defoliation: S. guaximae

Viégas in Brazil, S. heterochroa Desm. in Costa Rica and
S. sidicidae Speg. in Argentina.
Uromyces (see Puccinia)
Virus, an infectious brilliantly yellow chlorosis is widespread
and deserves further study. Infection studies with white
fly show it causes tobacco leafcurl, and in association with
other crops causes tobacco streak, yellow stunt of tomato,
dwarf bean mottle, and such reactions. In Brazil a virus
of Sida has been transmitted mechanically.

SOJA   see GLYCINE

SOLANUM MELONGENA L. or S. MELONGENA L. var. ESCULEN-
TUM Nees [see also S. MURICATUM; S. QUITOENSE; and S.
TUBEROSUM, following three entries]. An important vegetable
(introduced from Europe but originally from southern Asia) that
is, in the American tropics, an erect perennial sometimes over
1 meter high; "berenjena" Sp., "berinjela" Po., "eggplant" En.
   Alternaria solani (Ell. & Mart.) Sor.; rare blast in seedbed,
      in large plants causes a concentric zoned leafspot. Sub-
      tropical USA, Puerto Rico, Brazil. (Apparently seedborne
      but not reported as severe in warm tropics on this host.)
   Ascochyta hortorum (Speg.) Smith in Uruguay. A. lycopersici
      Brun. in Venezuela. A. phaseolorum Sacc. in Brazil.
      All cause leafspots and occasionally infect fruits especially
      the latter species.
   Botryodiplodia spp. = Diplodia spp. (the Botryosphaeria state
      not seen); fruit spot, fruit stem-end disease, branch
      disease. Scattered in dryer areas of W. Indies and C.
      America.
   Botrytis cinerea Pers.; flower disease, fruitrot. Brazil,
      Venezuela.
   Cassytha filiformis L.; green dodder (by inoculation). Puerto
      Rico.
   Cephaleuros virescens Penz.; algal leafspot. Panama, Guate-
      mala.
   Cercospora melongenae Welles = C. capsici Heald & Wolf;
      leafspot. California, Haiti, Dominican Republic, Brazil.
      C. solani Thuem. (= C. solani Feuill., C. nigrescens
      Wint., C. feuilleuboisii Sacc.); leafblast, sometimes lesions
      with indistinct edges. Brazil. C. solani-torvi Gonz. Frag.
      & Cif.; leafspot. Puerto Rico, Cuba, Dominican Republic,
      Trinidad, Jamaica, El Salvador, Guatemala, Guadeloupe,
      Martinique, Haiti, Colombia, Ecuador. C. trichophila
      Stev.; leafspot. Puerto Rico, S. America.
   Cladosporium fulvum Cke.; velvet leafmold. Puerto Rico,
      Dominican Guatemala. C. herbarum (Pers.) Lk.; black
      moldy blotches on leaves and stems. W. Indies, C.
      America, probably S. America.
   Colletotrichum atramentarium (Berk. & Br.) Taub. = C.
      phomoides Chester. (The one on eggplant may be con-
      sidered as a race of C. atramentarium); fruitrot and small

sunken spotting. W. Indies. C. gloeosporioides (Penz.)
Sacc. (has many synonyms, some commonly noted on this
host are C. piperatum Ell. & Ev., C. cradwickii Bancr.,
Gloeosporium melongenae Ell. & Hadst.); it causes damping-
off, common anthracnose, is seedborne. Neotropical.
Cuscuta spp.; dodders. C. America.
Diaporthe (see Phomopsis)
Erwinia carotovora (L. R. Jones) Holl.; bacterial softrot.
Neotropical.
Erysiphe cichoracearum DC.; white powdery mildew. Known
as "Oidium." Widespread, often most notable in dry season.
Fusarium spp.; in Brazil an unidentified species causes wilt
and vascular disease, another increases the damping-off
complex, and another is a main cause of a fruit peduncle
disease. F. solani (Mart.) Sacc. with asco-state Nectria
haematococca Berk. & Br. = Hypomyces ipomoea (Halst.)
Wr., on stem cankers. Guatemala, Costa Rica.
Galls, on stem and branch (no microorganism isolated). El
Salvador.
Helicotylenchus nannus Stein.; root-feeding nematode. Brazil.
Meliola solani Stev.; black mildew. Puerto Rico, El Salvador,
Venezuela.
Meloidogyne spp.; rootknot nematodes. Subtropical USA,
Jamaica, Cuba, Brazil.
Mycena citricolor (Berk. & Curt.) Sacc. (by inoculation)
caused large leafspot. Puerto Rico.
Oidium (see Erysiphe)
Peronospora tabacina Adam; downymildew in seedbeds. Sub-
tropical USA.
Pestalotia sp.; isolated from leaf and stem lesions. El
Salvador.
Phomopsis vexans (Sacc. & Syd.) Harter (= in culture Diaporthe
vexans Gratz); fruit spot and mummy, soil rot, damping-off,
branch canker. Seedborne and in practically every country.
Phyllosticta solani Ell. & Mart.; small leafspots. Subtropical
USA., W. Indies.
Phymatotrichum omnivorum (Shear) Dug.; dry country rootrot.
Dry Subtropical USA.
Phytophthora spp. cause cottony fruit rots: P. cactorum (Leb.
& Cohn) Schr. in El Salvador, P. capsici Leon, in Vene-
zuela, P. infestans (Mont.) dBy. also causing foliage blight
in Dominican Republic, Chile, P. parasitica Dast. causing
stem lesions, in Neotropics, and P. parasitica Dast. var.
nicotianae (de Hann) Tucker causing foliage blight in Puerto
Rico.
Pratylenchus sp.; nematode found in root zone. Brazil, Peru.
Pseudomonas solanacearum E. F. Sm. = Xanthomonas solana-
cearum (E. F. Sm.) Dows.; bacterial wilt. Subtropical USA,
Cuba, El Salvador, Puerto Rico, Trinidad, Panama, Surinam, Venezuela, Brazil.
Puccinia solanita (Schw.) Arth.; leafrust. Trinidad, Subtropical
USA.

Pythium spp. causing cottony leakrot of fruit in countries reported: P. aphanidermatum (Edson) Fitzp. in California,
Florida, Venezuela, P. butleri Subr. in Puerto Rico, P.
debaryanum Hesse in Puerto Rico, Subtropical USA, C.
America, P. miriatylum Drechs. in Florida, and P. vexans
dBy. in Brazil.

Rhizoctonia solani Kuehn = its spore state, Thanatephorous
cucumeris (Frank) Donk; damping-off (very often combined
with species of Fusarium and Colletotrichum), stemrot,
occasional fruitrot. Widespread. R. microsclerotia Matz =
Pellicularia microsclerotia (Matz) Weber; stem disease, web
blight. Puerto Rico, Costa Rica.

Rhizopus stolonifer (Ehr.) Lind.; fruitrot. Subtropical USA,
Jamaica.

Rosellinia bunodes (Berk. & Br.) Sacc.; rootrot. Panama.

Sclerotinia sclerotiorum (Lib.) dBy. = S. whetzeliana (Lib.)
Korf & Dumont; foliage disease, stalk rot. Subtropical
USA, W. Indies, El Salvador, Brazil.

Sclerotium rolfsii (Curzi) Sacc. (basidiospore state is Pellicularia rolfsii (Sacc.) West); soilrot. Subtropical USA,
Puerto Rico, Surinam.

Septoria lycopersici Speg.; small and numerous leafspots.
Costa Rica, Peru.

Sporidesmium melongenae Thuem.; leafdisease. Brazil.

Strigula complanata Fée; lichen white spots on branches of
old plant. Panama.

Trentepohlia aurea Mart.; tomentose algal epiphyte. Panama.

Verticillium albo-atrum Reinke & Barth. and V. dahliae Kleb.;
chlorosis and vascular wilt. Brazil (cool planting areas).

Viruses: "Calice gigante" (bigbud), Tomato vira-cabeca,
Potato leafroll, Potato Y, Tobacco mosaic (TMV), Tomato
"broto crespo" all five in Brazil. Tobacco mosaic in Subtropical USA, W. Indies. "Peste negra" or "Corcova" in
Argentina. Cucumber Mosaic virus in Florida.

Xylaria spp.; isolation from canker on old stem. Panama.

SOLANUM MURICATUM Oit. (= S. VARIEGATUM Ruiz & Pavón,
S. GUATEMALENSE Rous) [see also S. MELONGENA, preceding
entry, and S. QUITOENSE and S. TUBEROSUM, following two
entries]. A native herbaceous fruit plant important especially to
peasant gardeners and indigenes distant from cities; "katchan,"
"xachum," "pepino dulce" Sp. Ind. (Diseases have been very
sketchily studied. The following list is from diseases reported
by gardners or seen personally.)

    1. Black rot of fruit. 2. Black mildew (? Meliola). 3.
    Branch and stem canker (? Fusarium). 4. Chlorosis
    (physiological). 5. Damping-off (? Pythium, ? Rhizoctonia).
    6. Fruitrot (fungus but undetermined). 7. Leaf collapse
    (probably Phytophthora). 8. Leaf spot, small (? Phyllos-
    ticta). 9. Leafspot large and expanding (? Cercospora).
    10. Rootrot (? Phytophthora). 11. Tip disease of young
    plant (? Sclerotinia).

SOLANUM QUITOENSE Lam. [see also S. MELONGENA and S.
MURICATUM, preceding two entries, and S. TUBEROSUM,
following entry]. This fruit plant which is semi-woody, perennial, above waist high and comes from South America, has been
grown for countless generations for the fruits which are much
valued in native village markets and in some cities. The plant
is tough and generally disease tolerant; "naranjillo" Sp., "lulo,"
"lula-lulo" Ind.

 Algae (both green filamentous and single-celled) on old plants
  at ground level hastening host plant senescence. Costa
  Rica.
 Blossomend rot of fruit. Cause not studied, Guatemala.
 Botryodiplodia spp. (spores of various sizes, no isolations
  or determinations attempted); dieback and fruit stem end
  rot. C. America, Panama, probably widespread.
 Botrytis sp.; infection of petals of old flowers. (Did not
  appear to interfere with fruit set.) Costa Rica. Botrytis
  sp., gray growth (possibly same as immediately above),
  causing additional decay of fruit after primary infection of
  Phytophthora. Costa Rica.
 (? Capnodium sp.); sootymold. Seems especially severe on
  this host's pubescent leaves. Widespread.
 Cephaleuros virescens Kunze; algal leafspot. Panama, and
  probably widespread.
 (? Cladosporium?), spores typical but in numerous trials im-
  possible to isolate living fungus; severe leaf injury. Costa
  Rica.
 Gnomonia-like fruiting bodies of fungus on diseased leaf
  petioles that had fallen. Panama, Ecuador.
 Liverwort (miniature foliose type); epiphyte on leaves and on
  old branches. Ecuador.
 Meliola sp.; black mildew. Costa Rica.
 Mycena citricolor (Berk. & Curt.) Sacc. = Omphalia flavida
  (Cke.) Maubl. & Rang.; leafspot from inoculation. Costa
  Rica.
 Penicillium sp.; on old leaf specimen. Costa Rica.
 Peronospora tabacina Adam; downy mildew of seedlings.
  Seen in Guatemala and Panama.
 Phytophthora sp. (not P. capsici) from collapsed stalk.
  Ecuador. P. capsici Dast.; branch canker. Guatemala.
 Pseudomonas solanacearum E. F. Sm.; mild bacterial wilting.
  El Salvador, Guatemala, Mexico.
 Puccinia araucana Diet. & Neger; rust. Chile, Ecuador,
  Bolivia, Argentina.
 Pythium (? butleri); isolation from stalk lesion. Costa Rica.
 Rhizoctonia solani Kuehn; hyphae on base of plant without in-
  vasion. Panama.
 Rhizopus sp.; black growth on discarded fruit shells. Costa
  Rica.
 Rootrot. Black root tips, discolored vasculars. No other
  plant symptoms. Panama.
 Scab-spots on leaf and branch. Cause undetermined. Ecuador.

Strigula complanata Fée; white lichen-spot.   Panama.
Trichoderma (? viride); isolated from discolored root.
    Panama.
Virus, Mosaic Mottle in Nicaragua, Guatemala.
Yeast, wild type, in fruit discarded by gardener before taking
    it to market.   (Brought to clinic.)   Costa Rica.

SOLANUM TUBEROSUM-ANDIGENUM = S. TUBEROSUM L. and/or
S. ANDIGENUM Juz. & Buk., sometimes given thus: S. TUBERO-
SUM L. subsp. ANDIGENA (Juz. & Buk.) Hawkes [see also S.
MELONGENA; S. MURICATUM; and S. QUITOENSE, preceding
three entries]. A supremely important good flavored and nutritious tuberous vegetable, native to the Americas, widely grown
under moist cool country conditions and the famous food of millions in the Neotropics; "papa," "patata" Sp., "batata," "bata-
tinha" Po., "potato," "white potato" En., "poyuy," "pa-apa",
two Indian names of many, and the skinless, desiccated tuber
after processing in the Andes is often called "chuño." (Note:
It seems that the type called S. andigenum (one of a long list of
described tuberous species) is a short-day plant with medium-
sized tubers and relatively thin leaf lamina and may be the
original species from which came S. tuberosum the long-day
segregate with larger tubers and thicker leaves, grown by
Indians on the island of Chiloe under long-day conditions there;
and from this was taken the first tubers to be cultivated in the
temperate zone, and was the type Linnaeus described. In Peru,
Colombia, Chile, Ecuador, Venezuela, Central America, and
Mexico the ancient agriculturists used for original and later
plantings the more variable S. andigenum. In Argentina, Uruguay, Brazil, and Paraguay there have been later introductions
of S. tuberosum after much reselection in North America and
Europe. At the present time potatoes of the American tropics
are mostly from much intercrossing. There seem to be little
or no great differences in parasite lists noted on Andigenum or
on Tuberosum potatoes.) [In the disease list following, I have
indicated within brackets short notes of suggestions of ecological
effects on certain diseases.]
    Actinomyces (see Streptomyces)
    Aecidium cantense Arth.; rust of leaf, fruit and stem; some-
        times called the Peruvian rust and the mountain rust.
        Peru, Honduras. (There is possible transmission to potato,
        from wild relatives, of other rusts such as: A. hispaniolae
        Kern, Cif., & Thurs. in the Dominican Republic, A. ju-
        juyensis Lindq. in Argentina, A. mundulum Jacks. & Holw.
        in Chile, and A. papudense Jacks. & Holw. in Chile.)
    Alternaria solani (Ell. & Mart.) Sor.; leaf bladespot, rarely
        tuber rot. [A medium high highland disease.] Scattered.
    Armillaria mellea Vahl = Armilariella mellea (Fr.) Karst.;
        mushroom rootrot. [Not in coolest areas.] Chile, Florida.
    Aspergillus niger v. Tiegh.; part of tuber rot complex.
        Sparingly in markets of C. America, Colombia, Ecuador.
    Asterina spp.; black mildew on dried leaf from Brazil. (There

are about ten species of this fungus reported on wild
Solanums in the American tropics, but only one leaflet
with immature fungus on it was seen or reported on cultivated potato.)

Bacillus amylobacter v. Tiegh. ; tuber decay. Guatemala. B.
mesentericus (Flugge) Migula = B. mesentericus Trev. ; the
slimy decay of seed pieces.   Florida, W. Indies, ? Colombia (seen but no isolation).

Bacteria that are epiphytic on tubers, I found they invade
fungus hyphae, but there is no proof of parasitism on host.
El Salvador, Costa Rica.

Basidiomycete (unidentified); dry-rot and rhizomorph production.
Peru.

Black heart, physiological from poor storage.   Early stage is
tuber brownspot. [Lack of oxygen in storage.] In markets
of C. America, Jamaica, Argentina, Florida, Peru.

Boron deficiency; leaf chlorosis, tip blight. C. America.

Botrytis cinerea Pers. ; stem grayrot tuber infection. [Seen
most during warm drizzly seasons.] Peru, Colombia,
Guatemala.

Caconema radicicola (Greef) Cobb; root nematode. Ecuador.
(? Capnodium sp. = ? Netrocyzbe sp.); sootymold. Argentina,
Brazil, Mexico.

Cercospora solani Thuem. (synonyms: C. solani Feuill., C.
solanicola Atk., C. nigrescens Wint., C. fevilleuboisii
Sacc.); leafblotch, brown leafspot. Subtropical USA, Venezuela, Guatemala, Brazil. [Disease becoming severe
towards close of wet season.]

Chrysophlyctis (see Synchytrium)

Cladosporium fulvum Cke. ; dark velvet spots on leaf. [Rare
on potato, a wet season disease noted especially when dry
season is delayed in arrival.] W. Indies, Subtropical USA,
Colombia, Peru. C. herbarum (Pers.) Lk. ; on stem of
old vine, on senescent leaf. C. America, S. America.

Colletotrichum atramentarium (Berk. & Br.) Taub. = C.
solanicola O'Gara; anthracnose dot lesions of leaf, stem,
stolon. Noted especially in Ecuador, Brazil, Peru, and
isolated in Costa Rica, but is probably widespread. (An
old fungus name sometimes seen is Vermicularia varians
Duc.)

Corynebacterium sepedonicum (Spieck. & Kotth.) Skapt. &
Burkh. ; bacterial ringrot of tuber. [More often in warm
moist planted soils, lower plantings in Andean countries
and low mountain valleys of Panama, Mexico and Guatemala.] Subtropical USA, Panama, Mexico, Guatemala,
Peru, Ecuador, Colombia, Venezuela, Brazil, Bolivia.

(Crossopsora opposita Syd. ; rust on wild potato relatives in
Ecuador is possible future danger on cultivated potato.)

Cuscuta spp. ; dodders. Scattered.

Degeneration, physiological stunting. [From high temperature,
long abnormal dryspell, dry wind exposure.] C. America,
Puerto Rico, Virgin Islands, Argentina.

Dendryphion obstipum Poll.; mold on leaves. Mexico.
(Didymopsora spp.; leafrusts on wild potato relatives in Brazil
    are possible future danger to potato.)
Ditylenchus destructor Thorne; root and stem nematode.
    Brazil.
Erwinia aroideae (Town.) Holl. and combined with E. caro-
    tovora (L. R. Jones) Holl.; common typical tuber softrots.
    [Under unsanitary and moist conditions of market and
    storage, most severe in warm markets.] Widespread and
    probably in all countries. E. phytophthora (App.) Holl.
    (synonyms reported: E. atroseptica (v. Hall) Jenn., E.
    carotovorum (L. R. Jones) Dows., Pectobacterium caro-
    tovorum (L. R. Jones) Waldee); insect distributed typical
    blackleg ("pierna negra," "canela preta") lenticel infection,
    some seedpiece rot. Subtropical USA, C. America, W.
    Indies, and in S. America with possible exception of Argen-
    tina.
Erysiphe (see Oidium)
Fasciation, hereditary abnormality results in stem flattening.
    Guatemala.
Fusarium spp., there are about five fusarial diseases on
    potato in tropical America as follows: 1. F. oxysporum
    Schl. (with such synonyms as F. euoxysporum Wr., F.
    oxysporum Schl. forma 1 Wr., F. oxysporum Schl. f. sp.
    batatas (Wr.) Sny. & Hans.) causing plant wilt. 2. F.
    oxysporum combined with F. solani (Mart.) Sacc. causing
    wilt and stem cankers. 3. F. roseum Lk. causing vascular
    discoloration but no wilt, crop failure. 4. F. roseum Lk.
    f. sp. sambucinum (Fckl.) Sny. & Hans. causing vascular
    infection but not death. 5. F. solani (Mart.) Sacc. and/or
    F. solani (Mart.) Sacc. f. sp. eumartii (Carp.) Sny. &
    Hans. cause of the troublesome stolon disease, both the
    dry and jelly rots of tubers. All these diseases are wide-
    spread in occurrence. [Most common in slightly acid,
    friable soils in warm areas or in warm seasons.]
Gliocladium sp.; tuber rot. Subtropical USA.
Helicobasidium (see Rhizoctonia crocorum)
Helicotylenchus nannus Stein.; nematode of root and tuber.
    Brazil.
Helminthosporium (? solaninum Sacc. & Syd.); leaf spotting.
    Chile, Brazil.
Heterodera rostochiensis Wr.; the troublesome "golden nema-
    tode." [Probably came from the Peruvian Sierra; is a cool
    area disease and has forced field rotation on potato growers
    since pre-Incan times.] Peru, Ecuador, Chile, California,
    Argentina.
Hollow heart, physiological. Especially noted in Argentina,
    Peru, and Subtropical USA, probably widespread.
Hypomyces ipomoeae (Halst.) Wr. = H. solani Reinke &
    Barth., reported numerous times, is the perfect state of
    Fusarium solani. Such Ascomycetes sometimes not
    reported as disease collections are usually of the imperfect

Fusarium state, found long before the ascocarps have had
a chance to develop. Widespread.

Iron deficiency, whitish leaf chlorosis, poor growth. Guatemala, El Salvador.

(Isaria sp. the famed "Muscadine" fungus secured from dead
immature stage of a Lepidopterous insect collected in
Colombia in a fold of dead potato leaf. On moist chamber
treatment the fungus grew through rotted leaf lamina tissue.)

Macrophoma sp.; leafblight and stem lesions. Peru, Guatemala.

Macrophomina phaseoli (Maubl.) Ashby; gray stemrot, charcoal
rot. Subtropical USA, Peru. [Not common, apparently
requires warm-dry condition.]

Magnesium deficiency results in leaf chlorosis, stem streaks.
C. America.

Meloidogyne spp.; nematode root and tuber infections, and
some countries where reported: M. arenaria (Neal) Chitw.
in Jamaica, M. exigua Goeldi in Brazil, M. hapla Chitw.
in Subtropical USA, and Venezuela, M. incognita (Kof. &
White) Chitw. in Brazil, Guyana, Jamaica, Venezuela,
Peru, Argentina, and Subtropical USA, M. javanica (Treub.)
Chitw. in Florida, Panama, Venezuela, Brazil. [These are
all rootknot nematodes, sometimes distributed through being
in seed tubers, not generally disturbing in cool mountain
locations, apparently most common in lowland fields.]
They can be expected to appear in almost any country for
in addition to countries specifically mentioned above the
disease is reported from all countries of C. America, from
Colombia, Bolivia, and I saw pictures of "knotted roots"
from Mexico.

Mycelium radicis Auct.; a mycorrhizal fungus that is occasionally the cause of root necrosis. California; in El
Salvador reported on potato plants grown between coffee.

Mycena citricolor (Berk. & Curt.) Sacc. = Omphalia flavida
(Cke.) Maubl. & Rang.; leaf spot from laboratory inoculation. Costa Rica.

Mycosphaerella sp.; leafspot. Venezuela.

Nectria spp.; sunken stem lesions studded with asco-carps.
Ecuador.

(? Neocosmospara vasinfecta E. F Sm.; some mycologists
do not recognize this name; however, in El Salvador a
semi-parasitic fungus with perithecia and other characters
close to those described for this in literature was seen on
old potato vine stem and tuber. No isolations were secured.
It is still an accepted disease organism reported in California.)

Nitrogen deficiency, weak growth, stunt, light green foliage,
no tubers. Costa Rica.

Oidium solani Vañha (= Erysiphe cichoracearum DC. is the
asco-state); the conidial form is powdery mildew or ashy-
dust mildew but is not often serious on the host but common and widespread. [The Oidium state follows occurrence

of seasons of heavy dews while strong rains wash away and damage this mildew. The asco-state is found at high altitudes or in lower areas where seasons are cold and reasonably dry.]

Ojo-de-pollo of unknown cause, sunken spots on tuber. Peru.

Paratylenchus sp.; nematode causing apparent root injury. Chile.

Penicillium glaucum Lk. and other unidentified species; cause infections of injured market tubers. C. America, Mexico, Chile.

Phleospora sp.; lesions on stems. Bolivia.

Phoma solanicola Prill. & Delacr.; stemcanker. Brazil, Peru, Guatemala, Costa Rica. [Apparently in highland and coolest plantings.]

Phycomycete, undetermined but the cause of tuber decay in Peru.

Phyllachora fluminensis Th. = Physalospora fluminensis Th.; rare black shiny leaf spots. Argentina, Brazil.

Phyllosticta (species uncertain, but P. ciferrica Gonz.-Frag. and P. hortorum Speg. have been described on wild solanums, the former in Brazil and the Dominican Republic, and the latter in the Orient); leafspot. Peru.

Phymatotrichum omnivorum (Shear) Dug.; soil rot. Dry Subtropical USA, said to be in Mexico but no confirmation of report.

Phytopthora spp., apparently four diseases: 1. P. capsici Leonian; leaf and stem infection in Puerto Rico. 2. P. erythroseptica Pethy.; pinkrot. [In cool country.] Peru, Argentina, Bolivia. 3. P. infestans (Mont.) dBy.; blast blight. (In the north temperate zone it is given the common name "late blight" but in parts of the tropics this name is very confusing and among peasants not used except in parroting outside specialists. It is the most serious disease of potato.) [Often called rainyseason disease, in warm countries of C. America and northern S. America it is most severe in the mountains. In the coolest mountains (Andes) it is mostly a tuber disease but in lower valleys is a leaf blight. It is a cosmopolitan disease that is well known as a cause of famines.] 4. P. parasitica Dast. var. nicotianae (deHaan) Tucker; tuber rot disease. Puerto Rico and W. Indies.

Polysaccopsis hieronymi (Schr.) P. Henn. = Urocystis hieronymi Schroet.; smut-galls of stems, branches, leaves. Not only on cultivated potato but on several wild tuber producing relatives. Mexico, Brazil, Bolivia, Peru, Argentina.

Potassium deficiency causes leaf bronzing, marginal necrosis, poor growth, small to no tubers. Guatemala.

Pratylenchus pratensis (de Man) Filipj.; tuber nematode in Subtropical USA and Brazil, probably more widespread. P. steineri Lorde.; another tuber nematode in Brazil and Chile.

Pringsheimia sp. ; leafspot. S. America.

Pseudomonas garcae Amar. , Teix. , & Pinh. ; bacterial leafspot. [May occur near coffee trees diseased with this
bacterium.] Brazil. P. solanacearum E. F. Sm. =
Xanthomonas solanacearum (E. F. Sm.) Dows. ; bacterial
wilt and brown slimerot. Attacks many tuberous but wild
close relatives of potato although some show evidence of
tolerance. [A large number of solanums are susceptible
and may serve as holdover hosts. Reported from the relatively warm areas.] Occurrence is in all potato growing
countries in the American tropics.

Puccinia pittieriana P. Henn. ; a common leafrust (underside
of leaves). (Note: Over 20 species of this genus have
been reported by potato men, the first in 1898, to be the
cause of rusts of wild Solanaceae, widespread in the American tropics. Most are typical leaf rusts, some are yellow
types, others black to brown; some deform leaf, branch,
and stem. There is no means of predicting if some of the
species may or may not develop races (mutants) that will
cause disease in some of the constantly newer potatoes
produced by modern breeding methods.) Reported from
Mexico, Costa Rica, Colombia, Peru, Bolivia, Ecuador
and probably more widespread. P. substriata Ell. &
Barth. (synonyms: P. tubulosa (Pat. & Gaill.) Arth.,
Aecidium solanophilum Speg.). Colombia, Argentina. This
species attacks both potato and other wild and cultivated
solanums and has grasses as alternate host. It is possible
the P. substriata of potato in southern S. America is a
race separate from the P. substriata on several solanums
in the tropics of C. America, W. Indies, and Subtropical
USA.

Pythium spp. ; e.g. P. catenulatum Mat. , P. debaryanum
Hesse, P. spinosum Saw. , P. ultimum Trow. ; watery-leak
tuber decay in storage and market. Reports are infrequent
but trouble is probably widespread. Special work on it has
been done in Panama, W. Indies, Subtropical USA, Brazil,
Peru.

Rhizoctonia crocorum (Pers.) DC. with its asco-state Helicobasidium purpureum Pat. ; violet rootrot. Argentina, C.
America. [In El Salvador it is in potatoes grown at a
medium cool elevation planted between coffee trees and it
became diseased with this parasite spread from the coffee.]
R. solani Kuehn is the non-sporuliferous, disease-causing,
hyphal state most usually found and isolated and cultured;
the spore-and-basidium state is given as Thanatephorous
cucumeris (Frank) Donk (the R. solani is perhaps the best
known plant disease fungus in the world); it causes stem
and root rot, stem and sprout canker, tuber black-scurf.
No doubt in all tropical American potato countries. [Not
found in coolest potato areas.] R. microsclerotia Matz. =
Pellicularia microsclerotia (Matz) Weber; web disease on
foliage. Costa Rica. [Warm moist-season disease.]

Rhizopus nigricans Ehr. = R. stolonifer Lind; market tuber
and storage rot.   Nearly ubiquitous, some notable reports
from Subtropical USA, Mexico, W. Indies.
Rosellinia bunodes (Berk. & Br.) Sacc.; root rot.   Special
reports in Colombia and Jamaica. R. pepo Pat.; root rot.
Notably in Jamaica, Colombia, Costa Rica, Guatemala, but
in other countries. [Both cause severe rootrots, stolon and
tuber infections that are more common in soils with ample
humus, tending towards being warm, some most serious
occurrences in newly cleared lands.]
Sclerotinia minor Jagger; stem disease in Subtropical USA.
S. sclerotiorum (Lib.) dBy. = Whetzeliana sclerotiorum
(Lib.) Korf & Dumont found in Subtropical USA, Guatemala,
Bolivia, Peru.
Sclerotium rolfsii (Curzi) Sacc. with its basidial state, Pelli-
cularia rolfsii (Curzi) West; footrot, seedpiece decay, wilt.
[It is usually scattered in old replanted lands where the
season may be warm and moist.]   Trinidad, St. Lucia,
Cuba, Dominican Republic, Haiti, Puerto Rico, Florida,
Panama, Colombia, Peru, Uruguay, Argentina.
Septoria lycopersici Speg. "strain A"; speckle and annular
ringspot of leaf. [Is a highland disease.] Peru, Dominican
Republic, Argentina, Venezuela. (This species is perhaps
most severe on tomato and other wild solanums but note
should be taken of S. pseudo-quinae Pat. and S. solanio-
phila Speg., which are severe leafspots of solanums in S.
America and may be of future concern if they also spread
to potato.)
Spondylocladium atrovirens Harz = S. atrovirens Thaxt.;
silver scab. [A wet cool "mountain disease."] Mexico,
Ecuador, Colombia, Peru, Brazil, Argentina.
Spongospora subterranea (Wallr.) Lagh.; stem canker and
tuber powdery-scab. [In the Andean region it is common
in coolest growing areas; it is known in C. America as
a "high altitude disease" and it appears to have originated
in Peruvian wild potatoes.] Subtropical USA, Mexico, C.
America including Panama, S. America.
Stemphylium consortiale (Thuem.) Groves & Skolko in Peru,
and S. solani Weber in Florida; both minor leafspots at
present.
Stilbum cagnii Av.-Sac.; stem canker.   Brazil.
Stomatochroon lagerheimii Palm; systemic leaf injury by alga.
Guatemala.
Streptomyces scabies (Thaxt.) Waks. & Henr. = Actinomyces
scabies (Thaxt.) Gussow; common scab on tuber, stolon,
root. [Generally in more neutral to slightly alkaline soils.]
In all countries.
Stysanus stemonitis (Pers.) Cola; tuber spot.   S. America.
Sunburn, physiological cause from over exposure of tubers to
sun. Occasional.
Synchitrium endobioticum (Schilb.) Pers. = Chrysophlyctis
endobioticum Schilb.; wart. Mexico, Bolivia, Uruguay,

Peru, Ecuador, Chile, Brazil, Argentina. [Found in the central Andes and higher parts of Mexico, questionably reported in one or two fields in Guatemala, is spoken of as a "mountain disease."]

Thecaphora solani Barrus; tuber smut, gangrene, "buba."
[Occurs over a wide range of ecological conditions.]
Mexico, Panama, Colombia, Venezuela, Peru, Ecuador, Bolivia.

Tuber cracks, from growth during much moisture fluctuation. Florida, C. America, probably other countries.

Tylenchorhynchus sp.; rhizosphere nematode. Peru, Chile.

Verticillium albo-atrum Reinke & Berth. = V. lateritium Berk.; vascular wilt, tuber dryrot. Subtropical USA, Chile, Ecuador, Peru, Colombia. [Not in rain forest country.] V. cinnabarinum (Cda.) Reinke & Berth. (the conidial state of Nectria inventa Pethy.); tuber systemica disease. Scattered. [In warm-soil plantings.]

Viruses (a few of the 29 items are no doubt tentative, and more viruses may be present; a regional list of confirmed viruses in the American tropics is much needed; countries of occurrence are only those from which definitive reports have come to me): 1. Abutilon Mosaic in Brazil. 2. Andean (latent) mosaic found in Andes of Peru, Colombia. 3. Aster Yellows often causes purpletop in Mexico, Guatemala, Costa Rica, Panama, Colombia, Peru. 4. Aucuba Mosaic (Solanum virus 9) in Mexico, Costa Rica, Colombia, Ecuador, Brazil. 5. Common mosaic (see Virus X). 6. Cork (see Tobacco Rattle). 7. Crinkle (see Paracrinkle, Potato Crinkle, and Virus A + X). 8. Eggplant Mosaic (by inoculation) in Trinidad. 9. Leaf Drop Streak (strain of Y) in Peru. 10. Leaf Roll (Solanum virus 14) or degeneration disease in Florida, Guatemala, Costa Rica, Dominican Republic, Nicaragua, Colombia, Venezuela, Ecuador, Peru, Brazil. 11. Paracrinkle (potato virus M) in Venezuela. 12. Potato Crinkle (A + X) in Florida, C. America, Argentina. 13. Potato Virus S (possibly a strain of Paracrinkle) in Venezuela. 14. Rugose (X + Y) in Florida, Nicaragua, Peru. 15. Spindle Tuber (known in Peru since prehistoric times) in Florida, Peru, Brazil. 16. Streak, also see Leafdrop, in Venezuela, Argentina. 17. Tobacco Rattle (Corky Ring Spot, Tobacco virus PMTV, corcho) in Peru, Florida. 18. Tobacco White Streak in Brazil. 19. Tomato Spotted Wilt (Solanum virus 1) top necrosis and wilt, bronzing, streak-like symptoms, called "vira-cabeça," in Nicaragua, Argentina, Brazil, El Salvador. 20. Top Necrosis (possible strain of vira-cabeça), necros do topo in Brazil also found in Venezuela. 21. Vein-Yellowing in Ecuador, Colombia. 22. (Virus, disease not sap transmissible, Colombia.) 23. Virus A in Costa Rica, Ecuador, Colombia, Venezuela, Peru. 24. Virus A + X, crinkle in Venezuela, Colombia, Ecuador. 25. Virus X (also known as Potato Virus X + Y, Rugose Mosaic, Latent

virus X, Solanum Virus 1, Common mosaic, Calico mosaic)
causes mottle, occasionally apical wilt, at times streak and
rugosity in Florida, Costa Rica, Nicaragua, Jamaica, Guatemala, Haiti, Dominican Republic, Guyana, Venezuela, Ecuador, Colombia, Bolivia, Brazil, Peru, Argentina. [Under
a wide range of ecological conditions.] 26. Virus X +
Cucumber mosaic virus, found in Brazil. 27. Virus Y
(Solanum virus 2) is possibly the well known Rugose Mosaic
causing, in addition to rugosity, top necrosis or acronec-
rosis and veinal mosaic. Costa Rica, Guatemala, Nicaragua,
Venezuela, Uruguay, Colombia, Bolivia, Ecuador, Brazil.
[Often in high mountains.] 28. Witches' broom, apparently
disease caused is scattered in Argentina. 29. Yellow mottle
from virus, causes deformations in the higher grown fields.
Nicaragua, Ecuador, Colombia, Peru.

Xiphinema americanum Cobb in Brazil and Chile and X. brasiliense Lorde., in Brazil. Both root ectonemas.

Xylaria apiculata Cke.; tuber rot. Florida.

SORGHUM spp. (the genus is given by some as HOLCUS, by some
included under ANDROPOGON). Increasingly valuable Old World
introductions for dry season growing and semi-desert areas,
these large grass crops are used as both annuals and perennials
in the American tropics: (1) S. HALEPENSE (L.) Pers. =
HOLCUS HALEPENSE L. and is sometimes designated as S.
ARUNDINACEARUM Stapf., a forage and cutting-grass crop;
"zacate jonson" Sp., "grama de sorgo" Po., "Johnson grass"
En. (2) S. VULGARE Pers. = ANDROPOGON ARUNDINACEUS
Scop., the important grain crop having such varieties as durra,
kafir, milo, feterita; "sorgo," "maicillo," "millo" Sp. Po.,
"massambará" Po., "grain sorghum" En. (3) S. VULGARE var.
SUDANENSE (Piper) Hitch., used in cut and grazing forage, said
to be better than local grasses in some areas; "hierba sudan"
Sp., "grama sudan" Po., "sudan grass" En. (4) S. VULGARE
Pers. var. TECNICUM (Koern.) Jav., limited use for brooms;
"maicillo de escoba" Sp. Po., "broom corn" En.

   Alternaria alternata (Fr.) Keiss. = A. tenuis Nees, small
      spored. Costa Rica (found once with other spores on
      blackhead of species (2)).
   Ascochyta sorghi Sacc.; rough leafspot. Mexico, Subtropical
      USA, Panama, Costa Rica. A. sorghina Sacc., leafspot
      similar to above. Subtropical USA (1, 2).
   Aspergillus spp. (cultures of various colors); head mold,
      seedrot in storage. Widespread (1, 2).
   Balansia claviceps Speg.; on inflorescence. Report from
      Cuba (2).
   Botryodiplodia (? theobromae) also reported as Diplodia sp.;
      stalk infections. Scattered in C. America, Brazil (2).
   Cassytha filiformis L.; green dodder. Puerto Rico (3).
   Cercospora graminicola (Ces.) Wils.; leafspot (1). C. sorghi
      Ell. & Ev.; gray leafspot (1, 2). Where these hosts are
      grown.

Chlorosis, physiological causes, from manganese deficiency
and excess lime (2). Occasional in W. Indies and C.
America. Probably in S. America.

Cintractia (see Sphacelotheca)

Cladosporium spp. ; glume disease, often on aging leaves,
leafmold (1, 2, 3, 4). Where hosts are grown.

Cochliobolus carbonus Nelson = Helminthosporium carbonum
Ullstr. ; leaf disease. Mexico (1, 2).

Colletotrichum falcatum Went that some observers think may
be a special form; redspot, stalkrot, seedling blight. Subtropical USA, Panama, El Salvador, Nicaragua, Dominican
Republic (2). C. dematium (Pers.) Grove; irregular and
elongate leafspots. Widespread (1, 2, 3, 4).

Corethropsis sp. ; part of headmold complex. Costa Rica (2).

Curvularia lunata (Wakk.) Boed. ; head panicle infection,
senescent leaf edge disease. W. Indies, C. America,
Brazil (2, 4).

Cuscuta americana L. ; yellow dodder. Puerto Rico, El Salvador (2, 3).

Diplodia (see Botryodiplodia)

Ellisiella sp. ; by some considered a Colletotrichum. However, in S. America on these hosts believed different.
Leaf lesions. On all four hosts.

Epicoccum sp. ; weak parasite on stem and leaf. S. America
(2).

Fusarium spp. , and numerous cultivars of them, are on
record in reports from most of the American tropics.
These parasites cause widespread diseases on all host
species, and are in two great groups as follows: First,
the F. roseum Lk. group (macroconidia are from medium
broad to slender with "foot" and having pointed end with
microconidia borne separately), with reports of such cultivars as avenaceum, culmarum, equiseti, gramineum,
semitectum, its perfect state is Gibberella pulcaris (Fr.)
Sacc. var. minor Wr. = G. roseum (Lk.) Sny. & Hans.
f. cerealis (Cke.) Sny. & Hans. (By some, G. zeae (Schw.)
Petch is thought separate but others consider it synonymous
with G. roseum.) These are most associated with rootrots,
localized stalk rots, stalk canker-like lesions, and leaf-
molds. Second, the F. moniliforme Sheld. group (macroconidia somewhat slender having a "foot" and end pointed
with microconidia borne in chains that may collapse into
heads) has had a cultivar reported as lactis. Perfect state
is Gibberella fujikuroi (Saw.) Wr. = G. fujikuroi (Saw.)
Wr. var. subglutinans Edw., Wr., & Rg.: synonym is G.
moniliforme (Sheld.) Winel. These commonly cause systemic leaf and stalk infections and crumpling. Both groups
are involved in moldyness of the more compactly structured
heads of grain.

Gibberella (see Fusarium)

Gloeocercospora sorghi Bain & Edg. ; zonate leafspot. Subtropical USA, Mexico, El Salvador, Nicaragua, Argentina
(1, 2, 3).

Guignardia sp.; leaf disease.   El Salvador (2).

Helminthosporium sorghicola Lefeb. & Sherw.; targetspot.
    Subtropical USA, Mexico, W. Indies, C. America (1, 2, 3).
    H. sudanensis Gonz. Frag., & Cif. (this may be a spe-
    cialized form of H. turcicum); leaf blight.   Dominican Re-
    public, and nearby islands (3).   H. turcicum Pass.; leaf
    blight and blast, burn.   Widespread (1, 2, 3, 4).

Leptosphaeria sacchari deHaan; ringspot.   El Salvador, Puerto
    Rico.   Panama (2).

Macrophoma sorghicola Speg.; large black spots on leaf and
    stalk.   Argentina (2).

Macrophomina phaseoli (Maubl.) Ashby; stem dryrot, dryseason
    disease.   Mexico, Dominican Republic, El Salvador, Guate-
    mala, Panama (2).

Meloidogyne sp.; rootknot nematode.   Subtropical USA, scat-
    tered in C. America (2, 3, 4).

Mycena citricolor (Berk. & Curt.) Sacc. = Omphalia flavida
    (Cke.) Maubl. & Rangel; leaf spot by inoculation.   Costa
    Rica (2).

Nematodes, spiral group in rhizosphere.   Collected in Chile
    (2).

Nigrospora oryzae (Berk. & Br.) Petch; head blackening.   C.
    America (1, 2, 4).

Pellicularia filamentosa (Pat.) Rog. (aerial type); leaf banding,
    target spot.   C. America (3).

Penicillium spp.; moist season seed-mold, grain-head disease.
    Panama, Costa Rica, Guatemala, probably widespread (2).

Periconia circinata (Mang.) Sacc.; root disease.   Subtropical
    USA in dry areas (2).

Phoma insidiosa Tassi; stalk disease.   Argentina (2).

Phyllosticta sorghina Sacc.; leafspotting.   Subtropical USA,
    Argentina, Puerto Rico, Brazil (1, 2, 3, 4).

Physopella zeae (Mains) Cumm. & Ram.; rust.   Brazil (2).

Pratylenchus sp.; nematode collected in rhizosphere.   Chile (2).

Pseudomonas andropogoni (E. F. Sm.) Stapp; bacterial leaf-
    stripe.   Subtropical USA, W. Indies, C. America (often in
    mountain plantings), Argentina, Brazil (1, 2, 3, 4).   P.
    syringae v. Hall = P. holci Kendr.; bacterial eyespot.
    Honduras, Argentina, Subtropical USA (2, 3).

Puccinia purpurea Cke. = P. sorghi-halpense Speg.; leaf
    rust, "red" rust.   Widespread (1, 2, 3, 4).

Pyricularia grisea (Cke.) Sacc. (perfect state is Trichothecium
    griseum Cke.); leaf disease near rice plantings.   Costa
    Rica (3).

Pythium ultimum Trow. and other species; seedling blights.
    Subtropical USA, Mexico, Puerto Rico, C. America, probably
    widespread (1, 2, 3, 4).

Radopholus similis (Cobb) Thorne; nematode feeding on roots,
    stunt, plant failure.   C. America (2, 3).

Ramulispora sorghi (Ell. & Ev.) Olive & Lefeb.; leaf sooty-
    stripe, leaf spot.   Subtropical USA in moist areas, Argentina
    (1, 2, 3, 4).

Spathodea companulata   408

Rhizoctonia solani Kuehn the mycelial and pathogenic stage of
   Thanatephorous cucumeris (Frank) Donk; root and plant base
   attack, seedling damping off. C. America, Peru, probably
   widespread (1, 2, 3, 4).
Sclerospora graminicola (Sacc.) Schroet., ? and/or S. sorghi
   (Kulk.) Weston & Uph.; downy mildew, witches' broom,
   stunt, not common. Texas, Mexico, Costa Rica, Nicaragua,
   Panama, Argentina, Brazil (1, 2, 3).
Sclerotium rolfsii (Curzi) Sacc. is the sclerotial and pathogenic state of Pellicularia rolfsii (Sacc.) West; red sheathrot. Subtropical USA, W. Indies (2).
Selenophoma bromigena (Sacc.) Sprague; stalk disease. Costa
   Rica (2).
Sphacelotheca cruenta (Kuehn) Potter = S. holci Jacks.; black
   loose-smut (1, 2, 3, 4) is often most spectacular in dry
   season. Subtropical USA, Puerto Rico, Mexico, Jamaica,
   Nicaragua, Haiti, El Salvador, Brazil, Argentina (1, 2, 3,
   4). S. reilana (Kuehn) Clint. (synonyms: Sorosporium
   reilanum (Kuehn) McAlp., Ustilago abortifera Speg.); head-
   smut but seldom abundant. Jamaica, Brazil, Argentina,
   Chile, C. America, W. Indies, Mexico, Subtropical USA
   (2). S. sorghi (Lk.) Clint., synonyms: Cintractia sorghi
   (Lk.) Hirsch., S. sorghi (Pass.) Speg., Ustilago sorghi
   Pass., U. sorghicola Speg.; covered smut is the most
   common of all these smuts. Widespread (1, 2, 3, 4).
Sphaerophragmium sorghi Batista & Bez.; rust. Brazil (2).
Uredo sorghi Pass.; rust. Barbados, St. Vincent (2, 3).
Urocystis agropyri (Preuss.) Schroet.; flagsmut. Costa Rica
   (2).
Ustilago (see Sphacelotheca)
Viruses, all hosts 1, 2, 3, 4 seem susceptible to these five
   viruses: Cucumber Mosaic strain Commelina-Celery in
   Florida, El Salvador, Cuba. Maize Dwarf in Subtropical
   USA, Brazil, El Salvador. Sugarcane Mosaic in all Neo-
   tropics. Sugarcane Mosaic Rugose strain in Brazil.
   Sugarcane Mosaic Streak strain in Nicaragua. (Sugarcane
   Ratoon Stunting apparently unreported on Sorghum from
   American tropics but sorghum proved susceptible in Queens-
   land.)
Xanthomonas holcicola (Elliott) Starr. & Burkh.; bacterial
   streak. Subtropical USA, scattered in C. America and W.
   Indies, probably in S. America (2, 3, 4).

SPATHODEA COMPANULATA Beauv., large, warm country tree
   from African tropics, used as an ornamental for its handsome
   habit, foliage, and its showy flowers. Only a minimum amount
   of work has been done on the pathology of this much planted
   ornamental tree; "tulipán Africano" Sp. Po., "African tulip" En.
   Botryodiplodia = Diplodia, probably two or three species; die-
      back. Costa Rica.
   Cephaleuros virescens Kunze; small algal leafspot. Honduras.
   Colletotrichum sp.; anthracnose of leaf and twig. Costa Rica.

Corticium salmonicolor Berk. & Br.; pink disease on branch. Costa Rica, El Salvador.

Felt growths (unidentified) on young branches. Panama.

Fumago sp. and Capnodium sp.; sootymolds. Panama, Costa Rica, Honduras.

Loranthaceae, various seen and unidentified. On branches. Panama.

Nematodes feeding on base of diseased cuttings. El Salvador, Guatemala.

Oidium sp.; white powdery mildew. Colombia, Venezuela, widespread.

Penicillium sp.; on flower stalk. Guatemala, Honduras.

Phyllosticta moscosoi Cif. & Gonz. Frag.; leafspot. W. Indies, Caribbean Islands.

Polyporus lignosus Kl.; wound parasite, log decay. Puerto Rico.

Rosellinia bunodes (Berk. & Br.) Sacc.; root disease. El Salvador.

Strigula spp.; lichen leafspots. Widespread.

Woodstain, unidentified. Panama.

SPINACIA OLERACEA L. Introduced from the temperate zone, in Central America it is usually produced in mountain gardens, a cool temperature garden herb grown for greens; "espinaca" Sp., "espinafre" Po., "spinach" En.

Albugo occidentalis Wils.; white rust. Subtropical USA, Guatemala.

Alternaria spinaciae Allesch. & Noack; leafspot. Brazil.

Cercospora beticola Sacc.; leafspot. Mexico, Subtropical USA, Venezuela.

Cladosporium macrocarpum Pruess; leaf disease. Chile.

Colletotrichum spinaciae Ell. & Halst.; anthracnose. Texas, El Salvador.

Erwinia carotovora (L. R. Jones) Holl.; bacterial soft rot in market and shipment. Widespread.

Meloidogyne sp.; rootknot nematode. Brazil.

Peronospora effusa (Grev.) Ces., and P. spinaciae Laub.; downy mildews. Argentina, Chile.

Phyllosticta chenopodii Sacc.; leafspot. Argentina.

Phymatotrichum omnivorum (Shear) Dug.; rootrot. Texas.

Pythium spp.; rootrot. El Salvador, Guatemala.

Rhizoctonia solani Kuehn (pathogenic mycelial state only reported); damping-off, stem rot. Peru, El Salvador.

Virus, distinct mottling but not transferred by inoculations. El Salvador.

SPONDIAS spp. All are important trees tolerant to a wide range of tropical conditions and producing abundant plum-sized fruits. First species introduced from Polynesia, others from American tropics: S. CYTHEREA Sonn. = S. DULCIS Forst.; called "cajá-manga," "ambarella," "pomine cythera," "cajá mombín estranjera." S. LUTEA L.; "hog-plum," "mombín," "ceruela."

S. MOMBIN L. = S. PURPUREA L., "jocote," "ceruela roja,"
"mobín rouge," "cajá-vermelha," is most widely occurring and
used not only for its fruit, but as growing posts to resist insects and rot, and as shade. S. TUBEROSA Arruda, "imbu,"
"mombín de Brasil," is grown for both fruit and edible swollen
roots. (Although important, because all these trees grow
readily, often are self seeded, and seem generally disease tolerant, little is studied of their pathology. All appear to have about
the same parasitic attacks.)

Botryodiplodia sp.; dieback. Widespread.
Cephaleuros virescens Kunze; algal leafspot. Brazil.
Cercospora mombin Pet. & Cif.; leafspot. Venezuela.
Cerotelium alienum (Syd. & Butl.) Arth.; rust. Report is
   from Puerto Rico but believed to be in other countries.
Chaetopeltopsis tenuissima (Petch) Theiss.; fruit flyspeck.
   Puerto Rico, C. America, probably widespread.
Colletotrichum gloeosporioides Penz. (its perfect state is
   Glomerella cingulata (Stonem.) Sp. & Schr.; foliage
   anthracnose, fruit rot. Neotropical.
Daedalea elegans Spreng.; woodrot. Puerto Rico.
Endothia havanensis Bruner; branch and twig disease. Cuba.
Fomes lamaoensis Murr.; honeycomb woodrot. W. Indies,
   C. America.
Fruit decays from unidentified fungi. Widespread.
Ganoderma zonatum Murr.; on stumps and fallen logs. Puerto
   Rico, C. America.
Irenopsis portoricensis Stev.; leaf black spot and smudge.
   W. Indies.
Loranthus sp.; mistletoe-like parasite on branch. El Salvador.
Meliola comocladiae Stev.; black mildew. Puerto Rico, El
   Salvador, Costa Rica.
Meloidogyne sp.; rootknot nematode. Florida.
Phthirusa sp.; scandent parasitic bush on branch. El Salvador.
Polyporus occidentalis Kl.; woodrot. Widespread.
Psittacanthus sp.; stiff large upright bush parasitic on branch.
   El Salvador.
Schizophyllum commune Fr.; branch dryrot. Widespread.
Sphaceloma spondiadis Bitanc. & Jenk.; spot anthracnose,
   scab. Widespread.
Stomiopeltis sp.; foliage disease. Brazil.
Tillandsia sp.; epiphytic smothering plant. W. Indies.
Yeast (wild types, unidentified) fruit decay. Widespread.

STACHYTARPHETA spp. = VALERIANOIDES spp. Herbs becoming
   woody, notable for their abundant flowers, native to tropical
   America and used as specimen plants or in hedges, in Middle
   America and in South America; vigorous, colorful, bush-like;
   "rabo-de-gato," "canastillo" Sp., "balunakuta" Po., "bretonica,"
   "bush verbena" En.

Aecidium (see Endophyllum)
Asteridiella callista (Rehm) Hansf. = Meliola cookeana Speg.,
   A. valerianae Hansf.; black mildew, leaf smudge. Puerto

Rico, Guiana, Venezuela, Ecuador, Trinidad, and other countries.

Botryodiplodia sp. = Diplodia sp. (its asco-state has not been recovered); leaf base disease, stem canker.   Costa Rica.

Botrytis sp.; flower disease.   Costa Rica.

Capnodium sp.; sootymold (under trees with sucking insects). Widespread.

Cercospora papillosa Atk.; leafspot.   Puerto Rico.   C. stachytarphetae Ell. & Ev.; leafspot.   Barbados, Grenada,   Puerto Rico, Bahamas, but probably widespread.

Cuscuta americana L.; yellow fast-growing dodder.   Puerto Rico.

Dieback of old flower spikes, a complex of numerous unidentified weak parasites.   C. America.

Endophyllum stachytarphetae (P. Henn.) Whet. & Olive = Aecidium stachytarphetae P. Henn.; leafrust.   Widespread.

Gemmiferous fungus (not Mycena citricolor), tawny infection bodies, leaf spot, decay back to stem.   Costa Rica (in yard of thatched shack in rain-forest area).

Helminthosporium capense Thuem.; stem and flower spike disease.   C. America, S. America.

Irene glabroides (Stev.) Toro = Irenina glabroides Stev., synonym Meliola glabroides Stev.; black mildew.   Widespread.

Meliola (see Irene)

Mycena citricolor (Berk. & Curt.) Sacc. = Omphalia flavida (Cke.) Maubl. & Rangel; round leafspot with central yellow kernel.   Puerto Rico.

Puccinia urbaniana P. Henn.; leaf rust.   Florida, Grenada, St. Vincent, Puerto Rico, Cuba, Jamaica, Trinidad, Venezuela, Panama, Colombia.

Rhizoctonia solani Kuehn (its basidiospore state seen in Costa Rica, Thanatephorous cucumeris (Frank) Donk); basal rot. Costa Rica, Guatemala.

Root disease (not Fusarium, Rosellinia, or Rhizoctonia) undetermined.   Costa Rica.

Trichothyrium dubiosum (Bomm. & Rouss.) Th. = T. reptans (Berk. & Curt.) Hughes; a black mildew.   Scattered in W. Indies and South America.

Virus, mottle and slight malformation of leaf, no determination. Florida.

STENOTAPHRUM SECONDATUM (Walt.) Kuntze, a grass grown for pastures and lawns in warm areas, is characteristically creeping and tenacious; "gramineo de San Agustin," "cinta," "herbaje del prado" Sp., "grama de jardím," "pasto" Po., "St. Augustine grass" En.

Algae, green, in wet lawns.   Both filamentous and non-filamentous types.   Smothering effect.   Puerto Rico and probably elsewhere.

Cassytha filiformis L.; slow-growing brownish-green dodder. Puerto Rico.

Cercospora setariae Atk.; gray leafspot.   Widespread.
Cuscuta spp.; fast-growing dodders.   W. Indies, C. America.
Echinosphaeria rhizocphila Speg.; (? parasite on Rhizoctonia?).
    S. America.
Helminthosporium sp.; footrot.   Subtropical USA.   Helmin-
    thosporium spp. unusually large spores with 14-15 cells
    and some unusually small spores with 3-5 cells; footrot,
    leafspots, base deteriorations.   Puerto Rico.
Himantia stellifera J. R. Johnst.; mycelial growths on stolons.
    Puerto Rico.
Ithyphallus murrillii West; wet season fungus.   Florida,
    Puerto Rico.
Meliola stenotaphri Stev.; black mildew.   Widespread.
Mycena citricolor (Berk. & Curt.) Sacc. (perfect state not
    found); on grass adjacent to coffee tree harboring the fungus.
    Puerto Rico.
Myxomycetes, various, in moist lawns.   W. Indies, Sub-
    tropical USA, Costa Rica.
Nigrospora sphaerica (Sacc.) Mason; found in association with
    differing leafspots and injuries.   Florida, W. Indies, El
    Salvador.
Phyllachora macorisensis Chard.; leafspot.   Dominican Re-
    public.
Physiological problems: chlorosis, tip dying, stunting from
    poor soils.   Scattered.
Pleospora sclerotioides Speg. var. vulgatissima Speg. (? =
    P. vulgatissima Speg.), = P. herbarum (Fr.) Rab.; leaf-
    spot.   Argentina.
Puccinia stenotaphri Cumm.; leafrust.   W. Indies.
Pyricularia grisea (Cke.) Sacc. (perfect state is Trichothecium
    griseum Cke., but not found by me).   Subtropical USA,
    Nicaragua, Cuba, Puerto Rico, Brazil.
Rhizoctonia solani Kuehn (perfect state not seen); basal attack,
    leaf banding disease, lawn brownpatch.   Subtropical USA,
    Puerto Rico, Costa Rica, El Salvador, probably widespread.
Sclerotinia sp.; dollar-spot in lawn.   Florida.
Uromyces ignobilis (Syd.) Arth.; leafrust.   Puerto Rico.
Ustilago affinis Ell. & Ev. in Brazil and Puerto Rico and
    U. affinis Ell. & Ev. var. hilariae (P. Henn.) Fisch. &
    Hirsch. in S. America; both cause inflorescence smut.
Virus: Sugarcane Mosaic Virus in Florida, no doubt in many
    countries and the Panicum Mosaic Virus in Texas.

STIZOLOBIUM see MUCUNA

SWIETENIA MACROPHYLLA King, and S. MAHOGANI (L.) Jacq.
    The two species most sought commercially; the disease list
    below is of reports and observations on them.   These trees are
    native to tropical America, best specimens grow scattered among
    other kinds of trees (reports are that often mahogany fails from
    disease if planted in pure stands).   A log is of great value as
    it is considered the foremost cabinet and carving wood of the

world, if given a chance all grow to be large trees; "mahogán,"
"caoba," "caobana" Sp., "mógono," "acaju" Po., "mahogany,"
"baywood," "Honduras mahogany" En., "maghogán" Arawaki Ind.

Agaric, unidentified (yellow color, tough stipe was 3 cm.
tall), on fallen decayed branch. Puerto Rico.

Axonchium arcuatum Thorne; nematode rhizosphere inhabitant.
Puerto Rico.

Botryodiplodia (? theobromae); dieback. Widespread.

Botrytis sp.; on dying leaf edges (moist chamber cultures).
Puerto Rico.

Capnodium sp.; sootymold. Widespread.

Cassytha filiformis L.; weak growing dodder on young tree.
Puerto Rico.

Cephaleuros (minimis ?); twig infection. Costa Rica.

Cercospora subsessilis H. & P. Sydow = Cercosporina domingensis Cif.; leafspots, premature defoliation. Reports
specially from W. Indies and Venezuela, probably widespread.

Cladosporium sp.; on ragged leaf. (Specimen sent anony-
mously from Panama to El Salvador.)

Colletotrichum gloeosporioides (Penz.) Sacc. (the asco-state
is Glomerella cingulata (Stonem.) Sp. & Schr.; anthracnose.
Widespread.

Cord disease (? Marasmius sp.); causing weakened foliage on
branch. Panama.

Corticium salmonicolor Berk. & Br.; pink disease of branch.
C. America, S. America (scattered occurrence).

Crinipellis rubida Pat. & Heim.; on branch wood. Venezuela.

Curvularia sp.; leaftip burn. Puerto Rico.

Defoliation, unknown cause, very severe in moist mountain
area in Honduras.

Dolichodorus sp.; nematode rhizosphere inhabitant. Peru.

Epiphytic fungi (no spores found, apparently non-pathogenic
hyphae), on twigs and leaf. Puerto Rico (rain forest
ecology).

Fomes sp.; white colored pocket rot of trunk. In some cases
at base of tree in presence of termites. Panama, scattered
in W. Indies, Honduras, Costa Rica.

Fusarium roseum-type, associated with small branch lesions.
Scattered.

Meliola swietenia Cif. & Frag.; black mildew spots on leaf
and twig. Dominican Republic, Puerto Rico.

Oidium sp.; thin white powdery mildew. Costa Rica.

Pellicularia koleroga Cke.; leafrot. Guatemala, other parts
of C. America.

Pestalotia funerea Desm.; leaf and twig lesions on struggling
young tree over-topped by palms. Costa Rica. P. swieteniae Gonz. Frag. & Cif.; leafspot, petiole and young
twig lesions. Dominican Republic, Jamaica, Puerto Rico,
Florida, Honduras (probably rather widespread).

Phoma sp.; black lesions with entire edges on leaf and stem.
El Salvador.

Phoradendron haitiense Urb., in Santo Domingo and Haiti and
P. robustissimum Eich., in Honduras and El Salvador,
both are large bushes that are mistletoe-like parasites on
host tree branches.

Phyllachora balansae Speg.; tarspot, black to reddish-black
leafspots. Dominican Republic, Honduras, Jamaica. P.
swieteniae Petr. & Cif.; small tarspot. Santo Domingo,
Florida, Cuba.

Phyllosticta swieteniae Alvarez-G.; seedling blight and leafspot. Puerto Rico, Honduras, British Honduras, Guatemala, Puerto Rico.

Phytophthora sp.; isolation from young dying seedling in
city park. Costa Rica.

Pleospora sp.; leafspot. Puerto Rico, Honduras.

Psittacanthus calyculatus Don; large leaved bush parasitizing
tree branches (kills host branch from swollen point of
attack out to tip). El Salvador.

Rhizoctonia sp. (? solani); brown and shallow cortex lesions
on bases of seedling trees. El Salvador, Puerto Rico.

Rosellinia spp. (R. aquila (Fr.) deNot., R. bresadolae Theiss.,
R. bunodes (Berk. & Br.) Sacc., R. goliath Speg.) have
all been considered cause of mahogany rootrot problems in
the West Indies and parts of S. America.

Seedling decline, complex of weak parasitic fungi invading
along with ant and other insect work at plantlet base in
seedbed. Costa Rica, Puerto Rico.

Septobasidium spp.; felty sheaths of various characteristics on
stem and twigs. Widespread but scattered.

Stem (small branch) cankers, that heal over. Venezuela.

Struthanthus sp.; short bushy mistletoe-like branch parasites.
Honduras.

Tillandsia sp.; smothering so-called long moss, in moist and
warm locations. Scattered.

Trametes corrugata (Pers.) Bres.; dead fallen branch woodrot.
Widespread.

Trophurus logimarginatus Román; rhizosphere nematode.
Puerto Rico.

Tympanopsis acanthostroma (Mont.) Muller & Arx; wood decay.
Puerto Rico.

Witches' broom at branch tip. Unknown cause, rare. Guatemala.

Xylaria sp.; associated with dieback of twigs. Puerto Rico.

TABEBUIA PENTAPHYLLA Hemsl. = T. PALLIDA (Lindl.) Miers
(the genus is sometimes given as TECOMA). Origin in Central
America but well distributed in the American tropics, this is a
medium tall, fine tree protected for natural regeneration in
forest and jungle, noted for its long trunk of light colored but
solid commercially valuable wood, and when planted is scattered
in woodlands or for ornament, as well as for shade in cacao and
coffee; "poa, " "pouri, " "carubo de campo, " "roble blanco, "
"macquelisua" Sp. Ind.

Aecidium simplicius Arth. & Johns.; leafrust.   Cuba, and
    scattered in W. Indies.
Amauroderma brittonii Murr.; stump decay.   W. Indies.
Apiosphaeria guaranitica (Speg.) Hoehn. = Munkiella
    guaraintica Speg.; black and fuzzy leafspot.   W. Indies,
    Paraguay, Brazil, Panama.
Arthrobotryum tecomae P. Henn.; leaf disease.   S. America.
(? Capnodium sp.); sootymold.   Puerto Rico, probably wide-
    spread.
Cephaleuros virescens Kunze; algal leafspot.   W. Indies, C.
    America, Panama, probably widespread.
Cercospora spp.; all cause leafspots and defoliation: C. jahnii
    Syd. in Trinidad, C. leprosa Speg. in Brazil, Paraguay,
    C. tecomae Chupp & Viégas in Brazil, and C. stenolobiicola
    (Speg.) Chupp = Cercosporina stenolobiicola Speg. in Brazil.
Cylindrosporium aureum Speg. = Libertiella aurea Speg.;
    twig and stem rot disease.   Paraguay, Argentina, Brazil.
Elsinoë tecomae Viégas; spot anthracnose.   Brazil.
Helicobasidium sp.; branch disease.   El Salvador.
Hypospila ospinae Chard.; leafspot.   W. Indies.
Irenina arachnoidea (Speg.) Stev. = Meliola arachnoidea Speg.
    or Asteridiella arachnoidea (Speg.) Hansf.; scanty to
    medium thick black mildew.   S. America, W. Indies.
Libertiella (see Cylindrosporium)
Loranthus sp.; tall parasitic bush on host branch.   El Salvador.
Meliola arachnoidea see Irenina, M. bidentata Cke. and M.
    tecomae Stev.; black mildews in Puerto Rico.
Meloidogyne incognita acrita Chitw.; rootknot.   El Salvador.
Munkiella (see Apiosphaeria)
Mycena citricolor (Berk. & Curt.) Sacc. = Omphalia flavida
    (Cke.) Maubl. & Rang.; leafspot under rainforest conditions.
    Puerto Rico.
Mycosphaerella tecomae Viégas; translucent leafspot.   Brazil.
Oidium sp.; white powdery mildew at onset of dryseason.
    W. Indies, C. America.
Phthirusa sp.; low-growing parasitic bushes on host branch.
    El Salvador.
Phyllachora sordida Speg.; stromatic leafspots.   Brazil.   P.
    tabebuiae (Rehm) Th. & Syd.; shiny tarspot.   Brazil.
Pluteus albo-rubellus (Mont.) Pat.; agaric decay of fallen
    branch.   Puerto Rico, Nicaragua.
Prospodium appendiculatum (Wint.) Arth. = Uredo lilloi Speg.;
    rust in many areas.   P. couraliae Syd.; rust in British
    Honduras, Costa Rica.   P. plagiopus (Mont.) Arth.; rust in
    Florida, Puerto Rico, Cuba.   P. tabebuiae Kern; rust in
    Dominican Republic, Puerto Rico.   P. tecomicola (Speg.)
    Jacks. & Holw.; rust in Honduras.   P. venezuelanum
    Kern; rust in Venezuela.
Psittacanthus calyculatus Don; large bush with distinct swelling
    at point of attack on host branch.   El Salvador.
Root failure of seedlings when transplanted, undetermined
    complex of numerous fungi and possibly nematodes.   El
    Salvador.

Septoria cucutana Kern & Toro; leafspot.   Scattered.
Strigula complanata Fée; lichen leafspot.   W. Indies, Panama,
   C. America, and probably widespread.
Uredo (see Prospodium)
Virus, mottle of leaf in Venezuela.   ? Virus, witches' broom.
   Puerto Rico.   Similar symptoms seen in Colombia in trees
   used as cacao shade, later with fungus-caused witches'
   broom but the shade tree disease apparently not caused by
   cacao fungus.
Weakspot of leaves, sunken small round spots yielding no
   fungus on isolation.   El Salvador, Costa Rica.

TABERNAEMONTANA CORONARIA Willd.   Well adapted to a wide
range of American tropical conditions, a popular ornamental
bush introduced from warmer parts of the temperate zone but
originated in the Indian-Malaya region, it has white flowers and
handsome foliage; "canestillo, " "jasmino blanco" Sp. Po.,
"crape-jasmine" En.
Aecidium ochraceum Speg. ; rust of leaf of Tabernaemontana
   sp. in Paraguay.
Botryodiplodia theobromae Pat. asco-state is Botryosphaeria
   quercuum (Schw.) Sacc. = Physalospora rhodina Berk. &
   Curt. (Conidial state often reported as Diplodia); leaf
   disease, stem-canker, lesion at flower-cutting-stub.   Widespread.
Cephaleuros virescens Kunze; algal leafspot.   Costa Rica,
   Puerto Rico, Guatemala.
Cladosporium (? herbarum); large leafspot.   Florida, Costa
   Rica.
Clitocybe tabescens = Armilariella tabescens (Scop.) Sing. ;
   mushroom rootrot.   Florida.
Colletotrichum gloeosporioides (Penz.) Sacc. = Gloeosporium
   tabernaemontanae Speg., the occasionally found asco-state
   is Glomerella cingulata (Stonem.) Sp. & Schr. ; mild an-
   thracnose.   Stem dieback following flower cutting readily
   develops producing the Glomerella state in wet season.
   Widespread but often not spectacular in disease effect.
Cuscuta americana L. ; dodder (by inoculation).   Puerto Rico.
Diplodia (see Botryodiplodia)
Hemileia jurensis Syd. ; yellow leafrust.   Brazil.
Hypospila ospinae (Chard.) Chard. & Toro; leaf spot.   Puerto
   Rico.
Meliola tabernaemontanae Speg. ; black mildew.   Widespread.
Mycena citricolor (Berk. & curt.) Sacc. = Omphalia flavida
   (Cke.) Maubl. & Rang. ; leafspot, natural infestation (and
   by inoculation).   Costa Rica.
Nectria sp. ; on old diseased stem in mountain garden.   Costa
   Rica.
Rosellinia bunodes (Berk. & Br.) Sacc. ; black rootrot.   Costa
   Rica.
Sootymolds (unidentified).   Scattered.
Strigula sp. ; yellow lichen leafspots.   W. Indies, C. America,
   Panama.

Tryblidiella rufula (Spreng.) Sacc. = Rhytidhysterium rufulum
 (Spreng.) Petr.; associated with dead stem wood.   Puerto
 Rico, Panama.
Xylaria sp.; branch tip disease.   Costa Rica.

TAGETES ERECTA L. and T. PATULA L. (there are ten or more
 wild species).   Herbaceous natives of the middle American
 tropics, early taken to Europe where they were incorrectly said
 to be from Africa, and seeds returned to New World tropics,
 becoming handsome popular low-growing ornamentals and free-
 flowering garden favorites sometimes used in funerals; "clavel
 de muerto," "clavelón" Sp., "cravo-de-defunto," "cravó" Po.,
 "African marigold" En.   (Very little work has been done on the
 pathology of these two flower producers in the American tropics.
 Native florists save their own seeds and habitually change their
 bed locations frequently.)
   Acrasia sp.; slime mold in moist areas.   Puerto Rico.
   Botrytis cinerea Pers.; flower blight.   C. America.
   Cercospora tageticola Ell. & Ev.; leafspot.   Florida, W.
     Indies, C. America, probably widespread.
   Diaporthe tageteos Speg. = D. arctii (Lasch) Nits.; stem
     canker, leaf infection, diseased flower head.   Widespread.
   Erwinia sp.; flower and cut-stem rot in market.   Guatemala.
   Fusarium spp.; probably about four species but unidentified:
     a wilt, a leaf disease, stem necrotic area, flower invasion.
     Guatemala, El Salvador.
   Meliola sp.; black mildew.   W. Indies (occasional).
   Oidium sp. = Erysiphe lamprocarpa (Wallr.) Lév.; white
     powdery mildew.   Widespread.
   Phyllachora ambrosiae (Berk. & Curt.) Sacc.; shiny leaf-
     spot.   S. America.
   Phytophthora sp.; stem base-rot.   Widespread in wet tropics.
   Pseudomonas solanacearum E. F. Sm.; bacterial wilt.
     Puerto Rico.
   Puccinia indecorata Jacks. & Holw.; rust.   Bolivia.   P.
     tageticola Diet. & Holw.; common rust.   Widespread.
   Rhizoctonia solani Kuehn is the mycelial pathogenic state of
     Thanatephorous cucumeris (Frank) Donk (the latter is
     epiphytic and was on dry plant tissues); causes damping-
     off of seedlings and base-rot of old plants.   Mexico, W.
     Indies, C. America.
   Sclerotinia ? (unidentified, sclerotial formations in mixed
     host-fungus structures); plant rot in wet season.   El
     Salvador, Guatemala.
   Tip blights of flower and old leaves, from combined poor
     growing conditions and attack by such weak fungi as
     Cladosporium, Penicillium, a black-spored possible Curvu-
     laria, and on Aspergillus.   C. America.
   Viruses: Aster Yellows Virus in Mexico, Florida, Guatemala,
     Common Cucumber Mosaic Virus in Florida and the es-
     pecially virulent Cucumber Mosaic Virus variety Celery-
     Commelina in Florida, Cuba.

TAMARINDUS INDICA L.   From the Orient; an introduction to many
parts of the Western tropics and increasingly valued for its
abundant crops of pods full of sweet pulp used in food, drinks
and medicine, and the logs that yield a handsome red wood, are
these tall and symmetrical ornamental fruit and shade trees;
"tamarindo" Sp. Po., "tamarind" En.   (Study of their diseases
in the Americas is very sketchy.)

   Aspergillus fumigatus Fres.; on poorly stored pods.   Honduras.

   Capnodium sp. (= Fumago); sooty mold on lower branches.
      W. Indies, C. America, probably in S. America.

   Cephaleuros virescens Kunze; algal leafspot.   Widespread.

   Colletotrichum gloeosporioides Penz. (= Gloeosporium tama-
      rindii P. Henn.); anthracnose disease.   Puerto Rico, Vene-
      zuela, Panama, Guatemala, Honduras, El Salvador, no
      doubt in S. America.

   Leaf spots, unidentified but with following symptoms: a
      speckle; small spots sunken and white; a large type trans-
      lucent with dark brown edge, spot scabby and brown, a
      shiny red spot.

   Loranthaceae, phanerogamic woody bushes parasitizing host
      branches were notable in C. America and no collections
      for exact identifications made but diagnoses from the ground
      indicate genera are: Loranthus, Oryctanthus, Phoradendron,
      and Phthirusa (see below).

   Meliola amphitricha Fr.; black mildew.   C. America.

   Meloidogyne sp.; rootknot.   Florida.

   Phthirusa spp., in Surinam: P. pirifolia (HBK.) Eichl., P.
      seitzii Krug. & Urb., P. squamulosa Eichl.

   Root diseases, possibly both Rosellinia and Pythium (only a
      few isolations tried but without successful results).   El
      Salvador, Honduras.

   Stomatochroon lagerheimii Palm; systemic algal leaf infection.
      Puerto Rico.

   Strigula spp.; leafspots of white and gray and green-blue
      colors.   Widespread.

   Yeast, wild type, causing unpleasing pulp fermentation.   Guate-
      mala.

TECOMA   see   TABEBUIA

TECTONA GRANDIS L.   Originally from the Indian-Malaya tropics,
this large tree has been famed for thousands of years for its
valuable resistant structural and ornamental wood, and seeds of
it were introduced over a century ago to the American tropics
where it is proving to be large, fast-growing, and adaptable;
"teca" Sp. Po., "teak" En.   (Its diseases in the Americas are
being slowly determined.)

   Cephaleuros virescens Kunze; algal leafspot.   Puerto Rico.

   Cercospora tectoniae Stev.; leafspot.   Trinidad.

   Chlorosis, possibly excess of lime (although in the Orient the
      tree is said to produce its best wood on calcareous soils).
      Costa Rica.

Cyathus sp.; birds-nest fungus associated with wood decay.
___Scattered in C. America.
Daedalea elegans Spreng.; white agaric on rotted fallen branch.
___Reported from Puerto Rico.
Fusarium (solani-type); rootlet rot.   Honduras.
Leaf distortion, physiological.   Costa Rica.
Limber-trunk of young trees, physiological.   Honduras, Guate-
___mala.
Loranthus sp.; phanerogamic branch parasite.   Panama.
Marasmius sp.; long round and shiny cord causing foliage
___disease.   Costa Rica.
Polyporus zonalis Berk.; old plank decayed, and on rotted log.
___Puerto Rico, Honduras.
(? Rhizoctonia solani?); damping-off of seedlings.   Panama,
___Honduras.
(? Rosellina sp.   ?); rootrot of exposed root on fallen tree.
___Costa Rica.
Virus, unidentified, leafmottle and malformation in C. America.

TELANTHERA see ALTERNANTHERA

TEPHROSIA sp. = CRACCA spp.   About seven species of perennial
___low to high bushes used for temporary agricultural shade and
___soil cover; its crushed leaves which contain rotenone have been
___applied in water solutions to crops as an insecticide in the
___Americas since prehistoric time; "barbasco, " "chilapate. "
___(Disease information listed here is all from scattered reports
___in middle America.)
___Corticium salmonicolor Berk. & Br.; pink disease.   C.
_____America.
___Macrophomina phaseoli (Maubl.) Ashby; charcoal rot.   W.
_____Indies.
___Parodiella perisporioides (Berk. & Curt.) Speg.; common
_____black mildew.   Subtropical USA, W. Indies.
___Pellicularia koleroga Cke. (aerial type); web-thread blight of
_____branches.   W. Indies, C. America.
___Phymatotrichum omnivorum (Shear) Dug.; root rot.   Texas.
___Phytophthora parasitica Dast. var. nicotianae (de Haan)
_____Tucker; stem rot, leaf decay.   W. Indies, C. America.
___Pythium aphanidermatum (Edson) Fitz.; damping-off of
_____seedlings, decay of cuttings.   W. Indies.
___Ravenelia epiphylla (Schw.) Diet.; leafrust.   Subtropical USA.
_____R. caulicola Arth.; leaf and stem rust.   Puerto Rico,
_____Bahamas.   R. irregularis Arth.; leafrust.   Mexico.   R.
_____talpa (Long) Arth.; leafrust.   Mexico, Nicaragua.
___Rhizoctonia dimorpha Matz; leaf, stem, and pod rot.   Puerto
_____Rico.   R. solani Kuehn (perfect state not seen or reported);
_____damping-off, rootrot, stem lesion.   W. Indies, C. America.
___Rosellinia sp.; root disease of old plants.   Costa Rica.
___Sclerotium rolfsii (Curzi) Sacc. (spores borne on Pellicularia
_____rolfsii (Sacc.) West); wilt, rot at plant base.   Trinidad,
_____Puerto Rico, ? Nicaragua.
___Virus, an unidentified leaf mottle in Nicaragua.

TERMINALIA CATAPPA L. A large tree brought from southeast
Pacific islands, planted widely in the American tropics and now
is also self sown along streams and ocean shores, known for its
edible seeds ("almonds") and producing domestic wood for posts
and cooking fires; "amendoéira tropica" Po., "almendro" Sp.,
"tropical almond," "myrobalan" En.

Botryodiplodia theobromae Pat. also reported as Diplodia sp.
(its asco-state is Botryosphaeria quercuum (Schw.) Sacc. =
Physalospora rhodina Sacc.); associated with dieback. Ecuador, Puerto Rico.

(? Capnodium sp.); sootymold. Neotropics.

Cassytha filiformis L.; olive-green dodder. Puerto Rico, St.
Thomas, can be expected on young trees in seaside locations of any country or island in the Caribbean region.

Cephaleuros virescens Kunze; algal leafspot. Studied in Puerto
Rico, is widespread.

Cercospora catappae P. Henn. is widespread from W. Indies
to south Brazil. C. geraisensis Chupp is in Brazil. Both
cause large leafspots.

Colletotrichum gloeosporioides Penz., with its perfect state
Glomerella cingulata (Stonem.) Sp. & Schr.; common anthracnose. Neotropics.

Cuscuta americana L.; fast growing, golden-yellow dodder.
St. Croix.

Diplodia (see Botryodiplodia)

Fomes lamaoensis Murr.; honeycomb trunkrot. Costa Rica.

Fusicoccum microspermum Har. & Karst.; on dead twigs,
part of dieback complex. Puerto Rico. F. microspermum
Har. & Karst. f. sp. catapae Gonz. Frag. & Cif., special
type of above. W. Indies.

Loranthaceae, bushy parasites on host branches. Puerto Rico,
Virgin Islands, C. America.

Meliola sp.; black mildew. Puerto Rico.

Nigrospora sp.; found in spore scrapings from diseased leaf.
El Salvador.

Pestalotia disseminata Thuem.; leafspot. Florida, Costa
Rica.

Phoma terminaliae P. Henn.; black lesions on leaf and twig.
W. Indies, S. America.

Phomopsis sp.; leafspot. Subtropical USA, W. Indies.

Phyllosticta terminaliae P. Henn.; leafspot. Brazil and
southern S. America, Florida.

Polyporus (see Trametes)

Schizophyllum commune Fr.; dieback, rifted bark of dead
branch. W. Indies, C. America.

Sphaceloma terminaliae Bitanc.; spot anthracnose. Brazil.

Stomatochroon lagerheimii Palm; algal leaf lesion, systemic
invasion of leaf petiole. Puerto Rico.

Strigula complanata Fée; white lichen leafspots. Widespread.

Trametes corrugata (Pers.) Bres. = Polyporus corrugatus
Pers.; wood decay. Scattered but widespread.

Tryblidiella rufula (Spreng.) Sacc. = Rhytidhysterium rufulum
(Spreng.) Petr.; branch and twig dieback. Widespread.

Uredo terminaliae P. Henn.; leafrust.   Brazil.

TETRAGONIA EXPANSA Murr.   Brought from East Asia, a low-
growing lush leaved herb used for greens; "espinaca" Sp. Po.,
"new-zealand-spinach," "tropical spinach" En.
   Cercospora tetragoniae (Speg.) Chupp = C. tetragoniae (Speg.)
      Jacz.; leafspot.   El Salvador, Puerto Rico, Brazil, Argen-
      tina.
   Erwinia carotovora (L. R. Jones) Holl.; market softrot.   C.
      America.
   Helminthosporium sp.; leafspot.   Texas.
   Meloidogyne sp.; rootknot nematode.   Subtropical USA.
   Rootrot, unidentified dryrot disease, in sandy locations,
      Puerto Rico, El Salvador.
   Virus: Beet Curly Top Virus in California.   Cucumber
      Mosaic Virus var. Celery-Commelina in Florida.

THEA SINENSIS L. = CAMELLIA SINENSIS (L.) Kuntze.   Grown in
   China tropics for internal use since prehistoric times, later becoming important in international trade whereupon it was planted
   in other cool/tropical American, Oriental and African areas, it
   is a small tree selected for stimulant-containing leaves which
   after special curing and drying processes are used to prepare
   the most universally enjoyed pleasure drink in all the world (the
   crop is being studied and increased in the American tropics
   where it grows commercially in three countries and in experimental gardens in about 11 more); "té" Sp., "chá" Po., "tea"
   En. (Note: Disease research on tea has not been reported
   fully in tropical America.   Wherever the crop is being established
   growers depend a great deal on advice about pathology from
   Oriental and European sources.   Perhaps the best studies in our
   hemisphere (still in progress) are in Peru; much is going on in
   Argentina, Brazil, and C. America also.)
   Armillaria mellea Vahl = Armilariella mellea (Fr.) Karst.;
      shoestring mushroom rootrot.   C. America.
   Botryodiplodia theobromae Pat. = Diplodia cacaoicola P. Henn.
      (the asco-state is Botryosphaeria quercuum (Schw.) Sacc. =
      Physalospora rhodina Berk. & Curt.); dieback following leaf
      pluck, cutting disease causing rootlet failure.   Peru, Costa
      Rica, Subtropical USA.
   Capnodium theae Boed. = Antennariella sp.; sootymold.   Peru,
      Brazil.
   Cephaleuros virescens Kunze; algal leafspot, red-rust.   Sub-
      tropical USA, Costa Rica, Peru.
   Cercoseptoria theae (Cav.) Curzi; leafspot.   Peru.
   Cercospora theae (Cav.) deHaan; eye leafspot.   Costa Rica,
      Peru.
   Cercosporella theae Petch; small eyespot, leaf malformation.
      Peru.
   Cladosporium herbarum Lk.; brown leafpatches, also along
      edge of senescent and ragged old leaf.   El Salvador,
      Guatemala.

Colletotrichum gloeosporioides Penz. with synonyms: C. camelliae Mass., Gloeosporium thea Zimm. has the perfect state of Glomerella cingulata (Stonem.) Sp. & Schr.; anthracnose leaf and twig lesions, red-brown leafspot, cutting disease, systemic invasion of plant remaining symptomless until subjected to unusual physiological stresses when one branch after another is killed. Costa Rica, Peru, El Salvador, Jamaica, Guatemala, Subtropical USA, Bolivia, Argentina, Peru, Brazil.

Corticium salmonicolor Berk. & Br.; leafrot, branch pinkrot. Scattered in Costa Rica, El Salvador, Peru and no doubt other locations.

Cuscuta americana L.; the coarse yellow dodder. Panama.

Diplodia (see Botryodiplodia)

Elsinoë leucospila Jenk. & Bitanc.; scab anthracnose in Brazil, and for a Sphaceloma on tea see below, while another scab, Elsinoë theae Bitanc. & Jenk., is reported on Thea sp. in Brazil.

(Exobasidium vexans Mass., the most dread disease of tea in the Orient, is at present not known in the Americas, but there is good chance for its spread to this hemisphere as the crop is increased.)

Fomes lamaoensis Murr. = F. noxius Corner; brown honeycomb rootrot of old dying bush in El Salvador. F. lignosus Kl.; crumbly rootrot. Guatemala.

Fusarium sp.; cutting-bed disease. Peru.

Ganoderma lucidum (Leys.) Karst.; rot and hard sporocarps at base of old dead bush. Panama.

Guignardia camelliae (Cke.) Butl.; leaf disease. California.

Hendersonia sp.; in dieback complex. Scattered in S. America.

Marasmius (? equirinus); black-cord blight. Peru.

Meloidogyne sp.; root nematode. Subtropical USA.

Mycena citricolor (Berk. & Curt.) Sacc. = Omphalia flavida (Cke.) Maubl. & Rang.; leafspot, severe defoliation in Peru. (Natural infection in my garden at edge of Costa Rican coffee plantation.)

Pellicularia koleroga Cke. (sometimes given as Corticium koleroga (Cke.) Hoehn.); leaf-rot and branch disease. Peru.

Pestalotia guepini Desm.; leafspot and twigblight. Subtropical USA. P. theae Saw.; zonal leafspot, grayspot, disease of tips on flush growth. Peru.

Phyllosticta erratica Ell. & Ev. = Pleospora tetrasepta Alv.-Garc.; ringspot. Subtropical USA, Puerto Rico.

Physalospora (see Botryodiplodia)

Pleospora (see Phyllosticta)

Rhizoctonia solani Kuehn (soil inhabiting parasitic fungus); infection of cuttings in bed. Costa Rica.

Rosellinia bunodes (Berk. & Br.) Sacc., R. necatrix (Hartig.) Berl. = Dematophora necatrix Hartig., R. pepo Pat. and R. radiciperda Mass.; all root diseases. Peru, Costa Rica.

423   Theobroma cacao

Sclerotinia sclerotiorum (Lib.) dBy. = Whetzeliana sclerotiorum
    (Lib.) Korf & Dumont; sclerotial disease of flush growth.
    Chile.
Sphaceloma sp.; leaf anthracnose.   Subtropical USA.
Sphaerostilbe repens Berk. & Br.; rootstain disease.   W.
    Indies.
Stilbella sp.; stem canker.   Peru.
Strigula (probably more than one species, various colors and
    sizes and textures); lichen leaf and twig spots.   Peru, C.
    America.

THEOBROMA CACAO L. (there are some 20 related species, a few
    are large jungle trees).   A relatively small tree with multiple
    trunks bearing pods closely held to their bark, which contain
    sweet pulp and seed.   On extraction and processing the commer-
    cially important seeds become a flavorsome dry staple product,
    the "beans" of commerce.   Aside from many local names the
    most common in business and modern growing are "cacao,"
    "cho-co-lát," Sp., "cacau," "cacauiero" Po., "cacao," "cocoa,"
    "chocolate tree" En.
    Acanthosyris paulo-alvimii Barr.; medium tall jungle tree
        which by root parasitism kills cacao.   Brazil.
    Actinomyces (see Streptomyces)
    Alternaria tenuis Nees = A. alternata (Fr.) Keiss.; small
        spored type causing blackened smudge on dried "beans."
        Dominican Republic.
    Anthostomella bahiensis (Hemp.) Speg. (its perfect state is a
        Calonectria); stem disease.   Brazil.
    Aphelenchus coccophilus Cobb; nematode at roots.   Venezuela.
    Armillaria mellea Vahl = Armilariella mellea (Fr.) Karst.;
        white rootrot, collar crack.   Mexico, Ecuador, Guatemala,
        Brazil, Colombia.
    Aspergillus delacroixii Sacc. & Syd. and A. fumigatus Fres.
        associated with pod rot and mold, A. flavus Lk. and A.
        niger v. Tiegh. associated with mustiness of dried "beans"
        and on discarded pod shells.   In various countries.
    Asterostomella paraguayensis Speg.; black mildew-like leaf
        disease.   S. America, Panama.
    Atradiella sp.; in podrot complex.   Brazil.
    Bacteria, unidentified, found in injured cushiongalls in Nica-
        ragua, and found as part of cause of failure of cuttings to
        root, in Ecuador and Colombia and Costa Rica (probably
        widespread).
    Botryodiplodia theobromae Pat. (synonyms are Diplodia theo-
        bromae (Pat.) Now., D. cacaoicola P. Henn., D. natalen-
        sis P. Evans, Lasiodiplodia theobromae (Pat.) Griff. &
        Maubl.), its perfect state is often given as Physalospora
        rhodina (Berk. & Curt.) Cke. = Botryosphaeria quercuum
        (Schw.) Sacc.; causes pod charcoal-rot and mummy, chupon
        wilt, dieback, cutting root failure.   In all countries.
    Botryosphaeria ribis Gros. & Dug. = B. dothidea (Moug.)
        Ces. & deNot.; stem-canker, occasional pod-spot.   Brazil,
        Trinidad.

Branch dieback on escaped cacao tree in relatively untouched
jungle growth, and from isolations the following fungi were
recovered: species of Fusarium, Colletotrichum, Botryo-
diplodia, bacteria, a brown Basidiomycete.   Costa Rica.
Calonectria bahiensis (see Anthostomella)
Calonectria rigidiuscula (see Fusarium rigidiuscula)
Calostilbe striispora (Ell. & Ev.) Seav. = Sphaerostilbe
     musarum Ashby; stem-canker, rootrot.   Trinidad and
     Tobago, Jamaica, Surinam, Nicaragua, Colombia, Vene-
     zuela.
Calostilbella calostilbe Hoehn.; stem-canker.   Trinidad.
Capnodium spp. (and other fungus genera); sootymolds sub-
     sisting on insect honeydew deposits on foliage and pods
     (often in dryer areas and in some parts called dryseason
     disease).   Neotropics.
Cephaleuros virescens Kunze = C. mycoidea Kars.; algal leaf-
     spot, redrust, occasionally a twig canker, increased in
     rainyseason.   W. Indies (including all islands in Caribbean),
     C. America, Guyana, Brazil, Ecuador.   Probably in all
     cacao growing moist-warm areas.   C. minimis, twig
     disease.   Puerto Rico.
Cephalothecium roseum Cda.; infection of trunk bark some-
     times affecting flower cushions.   Colombia, Ecuador.
Ceratocystis fimbriata Ell. & Halst. = Ceratostomella fim-
     briata (Ell. & Halst) Elliott, said to be always on trees
     infested with the small bark borer Xyleborus, in cooler
     geographic areas; trunk vascular discoloration and branch
     death, fruit rot entering often through stem or on pod side
     that is tight held onto trunk.   Scattered but fairly wide-
     spread, in some countries certain areas have more severe
     disease than others as in Venezuela most disease is near
     the coast, in Ecuador common in low main district but al-
     most unknown in eastern part, in W. Indies islands it is
     more severe in foothill areas.   (? C. moniliformis (Hedgc.)
     Moreau; woodstain.   Costa Rica.)
Cherelle wilt, pods shrivel while young and small, they then
     dry while still hanging onto fruiting cushion, cause is
     mainly physiological.   Isolations have yielded fungus species
     of Botryodiplodia, Fusarium, Phytophthora, Ceratocystis,
     Pestalotia.   Widespread.
Choanephora cucurbitarum (Berk. & Curt.) Thaxt.; blossom
     blight and drop, and infection of harvested "beans."   Scat-
     tered.
Circinella spinosa v. Tiegh. & LeMan; a mucor-like mold on
     curing "beans."   Colombia.
Cladosporium herbarum (Pers.) Lk. = C. theobromicola Av.-
     Sac. (spores plump perhaps it should be considered a
     variety of C. herbarum); injury of mature leaves and on
     dried "beans."   Costa Rica, Ecuador.
Colletotrichum gloeosporioides Penz. (asco-state is Glomerella
     cingulata (Stonem.) Sp. & Schr.), this one anthracnose
     fungus known for over a century is in all cacao countries

(has many synonyms, a few: C. luxificum v. Hall &
Drost. [which latter was isolated from witches' broom
tissue in Surinam in 1898, again in Demarara in 1906, at
one time believed possible cause of witches' broom], C.
theobromae App. & Str., C. theobromicola Del., C. fructi-
theobromae Av.-Sac.); large leafspot, may be systemic and
host show no symptoms, in some cases chupons are weakened finally die, podrot, parasite follows into twigs through
apertures left when leaf drops.   Neotropics.

Corticium salmonicolor Berk. & Br. (sometimes reported as
    C. stevensii Burt); leafrot, pink-disease of branch, rainy-
    season disease, most severe in warm-moist ecology.
    Puerto Rico, Dominican Republic, Trinidad and Tobago,
    Guatemala, Surinam, El Salvador, Ecuador, Colombia,
    Brazil.

Crinipellis (see Marasmius)

Cunninghamella sp. (? = Choanephora sp.); mold of harvested
    pods.   Colombia.

Cuscuta spp. (numerous species, 17 reported in Brazil);
    dodders not common on cacao but occasional.   Bolivia,
    Brazil.

Cuscuta americana L.; dodder in nursery.   Nicaragua.

Cutting disease, symptoms are on cut ends in bedding soil,
    tissue discoloration and roots weak or fail to grow.   In
    Ecuador a complex of organisms present: bacteria in ex-
    cess numbers, many saprophytic nematodes, Fusarium spp.,
    Botryodiplodia spp.

Diplodia (see Botryodiplodia and see Macrophoma)

Epiphytes, phanerogams of many kinds: large liverworts,
    micro-liverworts, mosses, algae, ferns, orchids, brome-
    liads, etc., cause smothering and poor fruit set.   In old
    plantations, often near jungle in rainforest ecology.

Epiphytic fungi on old leaves, hyphae tightly attached to host
    epidermis but without spores or other such structures as
    gemmae or chlamydospores.   Puerto Rico, Costa Rica,
    Ecuador.

Eutypa erumpens Mass.; bark-and-wood rot.   W. Indies,
    Trinidad.

Fomes lamaoensis Murr. = F. noxius Cor.; brown rootrot,
    trunkrot, stumprot.   Ecuador.   F. lignosus Kl.; old-trunk
    rot.   C. America, Dominican Republic, Trinidad & Tobago.

Fusarium spp.: 1. F. moniliforme Sheld.; branch disease.
    Scattered in W. Indies and C. America, Colombia. 2. F.
    roseum Lk. from many isolations of such diseases as non-
    productive flower-cushion, leaf lesion, rootlet rot, dieback
    of small chupon, pedicel-rot, pod-rot.   Costa Rica, Nica-
    ragua, Panama, Ecuador, Peru. 3. F. theobromicolum
    Av.-Sac.; pod disease.   Brazil. 4. Fusarium spp.; asso-
    ciated with red tissue inside pod, seed sprouted and then
    died, leaf sector lesion with red border, pod peduncle rot.
    5. F. rigidiuscula Berk. & Br., synonyms: F. decemcellu-
    lare Brick, F. flavida Mass., F. cremea Zimm., F.

theobromae Lutz, F. spicariae-colorantes Sacc. & Trott.
(perfect state is Calonectria rigidiuscula (Berk. & Br.)
Sacc.); infection of flower cushion results in both green
and woody and multiple-bud galls and reduced fruit set.
W. Indies including Caribbean islands, Mexico, C. America,
S. America.

Ganoderma applanatum (Fr.) Pat., and G. lucidum (Leys.)
Karst.; rootrot, stumprot and logrot. Surinam, Trinidad
and Tobago, Honduras, Jamaica, Ecuador.

Geotrichum candidum Lk. = Oidium lactis Fres.; cause of
musty flavor in "beans." Widespread.

Helicotylenchus sp.; root zone nematode. Surinam.

Helminthosporium theobromicola Cif. & Gonz. Frag.; leaf
disease. Dominican Republic.

Hymenochaete cacao Berk.; twig and branch canker, wood
disease. Venezuela.

Hypoxylon glomeratum Cke., H. rubigenosum (Pers.) Fr.,
H. truncatum (Schw.) Mill; all in Brazil and the second
also in Haiti, attack on trunk bark and disease of wood
under affected bark.

Hysteropsis brasiliensis Speg.; twig and leaf disease. Brazil.

Irenopsis guianensis (Stev. & Dow.) Stev. = Meliola guianensis
Stev. & Dow.; black mildew leafpatches. Specially noted in
Trinidad and British Guiana, but well spread (seen in Costa
Rica). (Note: This organism parasitized by Spiropes
capensis (Thuem.) Ell. in Trinidad.)

Labrella theobromicola Av.-Sac.; on foliage. Brazil.

Lasionectria cacaolicola Av.-Sac.; branch canker. Brazil.

Leaf scorch of margin, from deficiency of potash. Special
study in Trinidad, found in several other countries.

Leaf twist and curvature, from deficiency of copper. Costa
Rica.

Leptosphaeria theobromicola Cif. & Gonz. Frag.; foliage
disease. Dominican Republic, Haiti, Cuba, Jamaica.

Letendraea bahiensis Speg.; chupon disease. Brazil.

Lophophytum mirabile Schott & Endl.; a low-growing, jungle
phanerogam parasitizing host roots. Brazil.

Macbridella striispora (Ell. & Ev.) Seav. = Nectria striispora
Ell. & Ev.; bark disease. W. Indies, C. America.

Macrophoma vestita Prill. & Del. (possibly a synonym is the
old name Diplodia cacaoicola); attack on both foliage and
root in young plants. Mexico, W. Indies.

Marasmius perniciosus stanel, is the most commonly used
binomial; however, Crinipellis perniciosa (Stahel) Sing. is
also a correct name; cause of witches' broom, a swelling
of chupons and proliferation of swollen unhealthy sideshoots
from the diseased chupons, and killing trees and branches
in a most spectacular manner. (Note: Confirmed studies
show it attacks 12 or more wild cacaos, as well as some
of their close relatives, STERCULIA spp. and HERRANIA
spp. It is also notable that a cacao witches' broom is
known in Africa, caused by Exoascus bussei v. Fab., where

it has persisted for years in the so-called Cameroun
region. See also under Colletotrichum that in earliest
studies witches' broom isolates yielded anthracnose fungi,
making results confusing for a long time.) Occurence
noted in Guiana 1895-1904, described 1915 in Surinam.
Since then in Brazil, Peru, Surinam, British Guiana,
Colombia, Trinidad and Tobago, Bolivia, Ecuador, Venezuela, Panama (? report in El Salvador), still spreading.
Marasmius spp. of miscellaneous kinds causing surface
attacks on foliage: (a) M. equirinus Muell. a black somewhat dull surfaced cord, in Brazil, Trinidad, W. Indies;
(b) M. sarmentosus Berk. the black brightly shining cord
in W. Indies, Panama, scattered in Costa Rica, Brazil;
(c) M. scalpturates Berk. & Curt. a brown-black grooved
cord in Cuba; (d) M. semiustus Berk. & Curt. that spreads
back and forth between banana and cacao plantings, color
from white when young to black, cord-threads in Mexico,
Costa Rica, Nicaragua, Jamaica; (e) M. stenophyllus Mont.
found as a brown cord in several countries; and (f) M.
trichorizus Speg., a black and common cord in almost
every country, specially noted in Mexico, Nicaragua, Costa
Rica, Ecuador, Jamaica.
Melanomma henriquesianum Bres. & Roum.; stem disease.
   W. Indies.
Meliola (see Irenopsis)
Menispora acicola Ell. & Ev.; part of podrot complex.
   Mexico.
Micropeltis theobromae Batista; black mildew on leaves.
   Brazil.
Monilia roreri Cif.; watery podrot, first observed in Ecuador
   1914. Panama, Trinidad, Ecuador, Colombia, Peru, still
   spreading.
Morte subita, sudden death with red exudate. Believed to be
   a complex of organisms (one of which is a seeming special
   form of Botryodiplodia theobromae). Costa Rica, Mexico,
   Brazil, Ecuador, Nicaragua, W. Indies islands.
Mucor butingii Lend., M. circinelloidea v. Tiegh., and M.
   mucedo L.; in processing bins and boxes, all are acci-
   dental spoilage infections of "beans." Widespread.
Mycena citricolor (Berk. & Curt.) Sacc.; an uncommon leafspot, spread from a nearby diseased coffee tree. Costa
   Rica. The less studied Mycena sp. a false witches' broom.
   Haiti, Dominican Republic, Mexico.
Mycogone sp.; mold on processed "beans." Mexico.
Nectria bainii Mass. = Creonectria bainii (Mass.) Seav.; pod
   disease. W. Indies, British Guiana. N. cinnabarina (Tode)
   Fr.; branch canker. Jamaica. N. theobromae Mass.
   (which has an unnamed Fusarium state); branch and leaf
   petiole disease. Trinidad, Mexico, Jamaica, Dominican
   Republic. Costa Rica. Nectria (unidentified species); pod
   rot. Brazil. Nectria (unidentified species); stem and trunk
   cankers, also developed on dried seed. Mexico, Nicaragua,
   Costa Rica, Jamaica.

Oidium (see Geotrichum)
Oospora spp. ; fungi largely saprophytic in nature noted for
    chain-structured bulbils, causing altered flavor in processed
    "beans." Widespread.
Oryctanthus cordifolius (Presl.) Urb., O. ocidentalis (L.)
    Eichl., and O. spicatus (Jacq.) Eichl.; all are short and
    drooping parasitic bushes with external roots that suck sap
    from host tree branches. Fairly widespread from north of
    the Guatemala-Mexican border south into northern S.
    America.
Pellicularia koleroga Cke.; branch thread-web blight and leaf-
    rot. (In moist-warm ecology, and it sporulates mostly in
    rainyseason.) Specially noted in areas of Mexico, W.
    Indies including Trinidad, the Guianas, Nicaragua, Panama,
    Ecuador, Brazil. (Other Pellicularias see under Rhizoctonia,
    and Sclerotium.)
Penicillium candidum Lk., P. roseum Lk., and other species;
    mustyness of "beans" and mold on decaying pod husks.
    Widespread.
Periconia pycnospora Fres. = P. byssoides Pers.; a fungus
    associated with leafscorch. Scattered in C. America.
Pestalotia spp.; in pod disease complex. Reports from Brazil,
    Peru, Costa Rica, Venezuela, no doubt widespread.
Phaeopeltosphaeria sp.; pod shell lesion. Brazil.
Phomopsis sp.; dieback. Peru.
Phoradendron crassifolium (Pohl.) Eichl.; host branch infec-
    tion by a mistletoe-like bush. Surinam. P. latifolium
    Eichl.; a large mistletoe-like bush causing swelling of host
    branch. Brazil. P. piperoides (HBK.) Trel., P. rensoni
    Trel., and P. robustissimum Eichl. are all three mistletoe-
    like parasites in C. America.
Phthirusa pyrifolia (HBK.) Eichl.; abundant drooping and small-
    leaved parasitic bushes with aerial roots that suck sap from
    cacao branch and trunk. Mexico, Nicaragua, El Salvador,
    Guatemala, Costa Rica, Panama. P. theobromae Eichl.;
    a more vigorous type of the genus, but a seemingly specially
    adapted species. Surinam, Guyana, Venezuela, Brazil,
    Colombia.
Phyllosticta theobromae Alm. & Cam. and P. theobromae Alm.
    & Cam. f. sp. dominicana Cif. & Gonz. Frag.; leafspotting.
    Dominican Republic. P. theobromicola Vinc.; leafspot.
    Brazil.
Physalospora (see Botryodiplodia)
Physarum spp.; myxomycetes on pods piled after harvest.
    Dominican Republic.
Phytophthora cactorum (Leb. & Cohn) Schr.; branch canker.
    W. Indies. P. palmivora Butl. (synonyms: P. faberi
    Maubl., P. theobromae Colem., P. omnivora dBy.);
    chupon wilt, podrot, canker of branch, bud wilt. Neo-
    tropics.
Podosporium theobromicola Av.-Sac.; foliage disease. S.
    America.

Polyporus lignosus Kl., P. maximus (Mont.) Overh., P. pavonius (Hook.) Fr.; woodrots of trunk-and-tree bases. Colombia, no doubt other countries.

Polystictus cinnabarinus Fr. and P. persooni Fr.; wood-infecting fungi of old trees. S. America, C. America.

Porella spp.; miniature leafy liverworts on old cacao foliage. Puerto Rico, Mexico, Costa Rica, Guatemala, Panama, Venezuela, ? Northern Brazil.

Poria subserpens Murr.; wood rot. Bolivia.

Pseudosaccharomyces apiculatus Kloec.; organism found in curing bins while "beans" are being processed. Scattered reports.

Psittacanthus calyculatus (DC.) Don; tall and vigorous parasitic bush that forms a large knot where it attacks on the cacao host branch. Guatemala, Costa Rica, El Salvador.

Pterulia sp.; stem decay (die-back complex). Colombia.

Ramularia necator Mass.; seedling cotyledon white-mold, damping-off. W. Indies.

Rhizoctonia solani Kuehn (spore-state not reported but is Thanatephorous cucumeris (Frank) Donk); seedbed and cutting-bench disease. Can be expected anywhere under general conditions for growing the host. A Rhizoctonia sp. is a soil inhabiting parasite but different from above (coarser hyphae, very dark brown in culture); rootrot, black collar of young trees. Haiti, Guatemala, Dominican Republic. Another Rhizoctonia sp. of aerial character (abundant spore state tentatively designated as Pellicularia filamentosa (Pat.) Rog.); causing discrete necrotic leafspot. Costa Rica.

Rosellinia bunodes (Berk. & Br.) Sacc. is the black rootrot, much most common, widespread, R. hartii Mass. rootrot, occasional, R. pepo Pat. is the white-strand rootrot,' widespread, R. paraguayensis Starb. is occasionally reported from Mexico, Guatemala, W. Indies islands, Colombia.

Rotylenchulus reniformis Lindf. & Oliv.; nematode at roots. Surinam, Peru.

Saccharomyces spp.; yeasts involved that are valuable in fermentation processing. Widespread.

Schyzophyllum alneum (L.) Schr.; dryrot of branch. Ecuador.

Sclerotium rolfsii (Curzi) Sacc. (is the parasitic mycelial state of Pellicularia rolfsii (Sac.) West); the soil-line rot of nursery plants. Dominican Republic.

Septobasidium spongia (Berk. & Curt.) Pat.; bark disease. Cuba.

Septoria theobromicola Cif. & Gonz. Frag.; leaf spots. Dominican Republic.

Sphaeronema sp.; stem lesion, podspot. Ecuador.

Sphaerostilbe (see Calostilbe)

Spicaria colorans DeJonge; a branch infection. Puerto Rico, C. America, S. America. S. lateritia Cif. (much of the time in association with Calonectria rigidiuscula) a canker-like branch infection. Dominican Republic.

Spiropes (see Irenopsis Note)

Stachylidium theobromae Turc.; dieback.   Costa Rica.

Stilbella sp.; on dead branches.   C. America.

Stilbospora cacao Mass.; (? saprophyte) on dead leaves.
Scattered.

Stomatochroon lagerheimii Palm; systemic habit of alga-
caused leafspot.   Puerto Rico (a rare find there), Costa
Rica.

Streptomyces sp. = Actinomyces sp.; occasional tightly appressed fungus growths on pods and seeds in contact with
soil. W. Indies.

Strigula complanata Fée; white lichen-spot.   Widespread (I
have noted in some cases abundant lichen but no alga, although the latter is an inseparable part of the Strigula).

Strionemadiplodia sp.; in podrot complex.   Brazil.

Struthanthus aduncus (Meyer) Don; scandent parasitic shrub
with many exogenous roots that sink suckers into host
branches. Widespread. S. marginatus (Desr.) Bl.; parasite about as immediately above.   Costa Rica. S. orbicularis (HBK.) Bl. also in Costa Rica. S. radicans Bl.
is a parasitic chrub noted for its scandent character and
most abundant exogenous roots with suckers. Brazil.

Thielaviopsis paradoxa (De Seyn.) Hoehn.; podrot. Haiti,
Cuba, Ecuador, Costa Rica.

Thyridaria tarda Bancr.; defoliation and dieback, staghead.
Colombia.

Torula herbarum Lk.; woodstain disease.   Scattered.

Trametes cubensis (Mont.) Sacc., and T. serpens Fr.; both
wood rots of old trees.   W. Indies, northern S. America.

Trentapohlia sp.; algal hair-like tufts growing epiphytically.
Puerto Rico, probably many other countries.

Triblidiella rufula (Spreng.) Sacc. = Rhytidhysterium rufulum
(Spreng.) Petr.; secondary on dead twigs. Widespread.

Trichoderma sp.; recovered in isolates made from cushion
gall.   Nicaragua.

Tripospermum fructigenum (Rabh.) Hughes; sootymold.
Bolivia.

Ustulina zonata (Lév.) Sacc.; woodrot of fallen material and
in stumps.   Widespread but not in large quantity.

Verticillium ochro-rubrum Desm.; in dieback complex.
Colombia.

Viruses: West Indian Mosaic first seen in Trinidad in 1943
has strain A called Cacao Red Mottle and Strain B called
Cacao Vein Clearing in Trinidad.   Cacao Dented Leaf in
Colombia, Dominican Republic.   Cacao Mosaic in Costa
Rica and Venezuela.

Wilt, physiological.   Sudden and spectacular, collapse so fast
that leaves do not become chlorotic.   Scattered.

Xylaria spp.; branch disease.   C. America, a few instances
in Puerto Rico, Trinidad.

TRAGOPOGON PORRIFOLIUS L.   An introduction from Mediterranean countries, in Western tropics, especially in Central

America and the large islands of the West Indies, produces best
in high hill gardens. It has had slowly increasing popularity
(but little disease study), is a biennial herb with thick roots
that are flavorsome when cooked; "salsifí, " "planta de ostra"
Sp. Po., "salsify, " "vegetable oyster" En.

Albugo tragopogonis Pers. ; whiterust. El Salvador, Guate-
    mala, Costa Rica, Cuba, Puerto Rico, Panama, Venezuela,
    Chile, Argentina.
Cercospora tragopogonis Ell. & Ev. ; leafspot. El Salvador,
    Honduras, Costa Rica, probably many other countries.
Colletotrichum sp. (large spores); anthracnose lesions on
    stem, leafspot, El Salvador.
Erwinia carotovora (L. R. Jones) Holl. ; market root softrots.
    Subtropical USA, El Salvador, Nicaragua, Guatemala,
    probably in any market.
Erysiphe cichoracearum DC. ; powdery mildew. El Salvador,
    Guatemala, possibly widespread.
Fusarium sp. ; (isolated from dry rot of root). El Salvador.
Rosellinia bunodes (Berk. & Br.) Sacc. (near coffee tree
    roots diseased with the fungus); black rootrot. El Salvador.
Sclerotinia sclerotiorum (Lib.) dBy. = Whetzeliana sclerotio-
    rum (Lib.) Korf & Dumont; basal rot. Guatemala.
Ustilago trapogonis-pratensis (Pers.) Rouss. ; leafrust. (Men-
    tioned in a letter from Argentina.)

TRICHOLAENA ROSEA Nees = T. REPENS (Willd.) Hitch., a grass
    from south and west Africa, introduced for green feeding, pasture
    and ground cover (and for its ornamental appearance, the wine
    colored plume-like inflorescence), it has proved to be easily
    naturalized; "yerba de natál, " "arrocillo colorado, " "ilusión"
    Sp., "olahoomgoombi" (West African name), "wine grass, " "natal
    grass" En.

Cerebella andropogonis Ces. (conidial state said to be Spha-
    celia sp.); headmold associated with ergot. Widespread.
Claviceps purpurea (Fr.) Tul. ; ergot. Scattering occurrence
    but widespread.
Curvularia sp. ; leafspot, culm infection. W. Indies, C.
    America.
Meloidogyne sp. ; rootknot. Florida.
Nigrospora sp. ; spores from diseased hay. Puerto Rico.
Puccinia levis (Sacc. & Bizz.) Magn. ; rust. Widespread.
    P. levis (Sacc. & Bizz.) Magn. var. tricholaenae (H. &
    P. Syd.) Rama. & Cumm. ; rust. Puerto Rico.
Rhizoctonia (? solani), mycelium cells unusually thickened;
    cause of a basal rot. Puerto Rico.
Saprophytic bacteria, fungi, nematodes in grass cut green for
    feeding, no attempts at identity of these very numerous
    organisms, most of which disappeared on drying for hay
    purpose. Costa Rica.
Sorosporium (see Ustilago)
Sphacelia (see Cerebella)
Ustilago argentina Speg. = Sorosphorium argentinum Speg. ; in-
    florescence smut. Argentina, other countries in S. America.

TRIFOLIUM spp. (T. INCARNATUM L., T. PRATENSE L., T.
REPENS L.). Introductions of clovers from the temperate zone,
so long well known there, and found usable in some cool to
coolest parts of the American tropics, the famous leguminous
soil-building, cover, and pasture herbs; "trebols" Sp., "trevos"
Po., "clovers" En.
 Aecidium (see Uromyces)
 Alternaria sp. ; uncommon foliage disease. Colombia, Peru,
   Venezuela.
 Aphelenchoides parietinus (Bast.) Stein. ; root-zone nematode.
   Brazil.
 Black stem, undetermined cause (probably physiological). El
   Salvador.
 Botryodiplodia sp. = Diplodia sp. ; isolated from darkened and
   shrunken stem lesions. Occasional in dry hay samples.
   Scattered.
 Botrytis sp. ; moldy blight. Subtropical USA, C. America,
   Colombia.
 Cassytha filiformis L. ; green-bronze dodder. Florida coastal
   dunes.
 Cercospora stolziana Magn. ; leafspot. S. America. C.
   zebrina Pass. ; most common leafspot. Nicaragua, Guate-
   mala, Subtropical USA, Mexico, Uruguay, Brazil.
 Colletotrichum trifolii Bain & Essary (this is an unusual
   species, seed borne); anthracnose. Subtropical USA,
   Guatemala, El Salvador, S. America.
 Curvularia trifolii (Kauff.) Stev. ; stem and leaf disease.
   Found in herbarium specimens. Costa Rica and Guate-
   mala, probably in S. America.
 Cuscuta spp. ; dodders. Scattered in Neotropics (slowed
   growth in cool locations).
 Cymadothea trifolii (Pers.) Wolf (its conidial state is Poly-
   thrincium trifolii Kunze); black sooty leafspot. Subtropical
   USA, Chile, Colombia, Uruguay.
 Ditylenchus dipsaci (Kuehn.) Filip. ; root & stem nematode.
   Chile.
 Dothidella trifolii (Bayl.) Ell. & Stansf. ; leaf disease.
   Uruguay.
 Erysiphe polygoni DC (conidial state is Oidium sp.); white
   powdery mildew. Peru, Guatemala, occurs rarely in Sub-
   tropical USA.
 Fusarium sp. ; from diseased root. Colombia.
 Helicotylenchus nannus Stein. : root nematode. Brazil.
 Hemicycliophora sp. ; rhizosphere infesting nematode. Chile.
 Heterodera schachtii Schm. var. trifolii Goffort; root in-
   vading nematode. Subtropical USA, Chile.
 Meloidogyne sp. ; rootknot nematode. Chile, Brazil.
 Monhysteria stagnalis Stein. ; root nematode. Brazil.
 Oidium (see Erysiphe)
 Olpidium (? trifolii); leafcurl. Guatemala.
 Ovularia trifolii Speg. = Pseudovularia trifolii Speg. ; leaf
   disease. Argentina.

Paratylenchus sp.; a nematode found near clover roots. Chile.
Peronospora trifoliorum dBy.; downy mildew. Scattered in
  El Salvador, also in Chile, and in Argentina.
Phanerogamic, low-growing uncollected herbaceous parasites,
  on many legumes in tropical America, some (but unde-
  termined) occur on clovers in C. America.
Phymatotrichum omnivorum (Shear) Dug.; dry-land rootrot.
  Subtropical USA.
Polythrincium (see Cymadothea)
Pratylenchus pratensis (deMan) Filip.; root nematode. Brazil,
  Chile.
Pseudopeziza medicaginis (Lib.) Sacc.; black leafspot. Argen-
  tina. P. trifolii (Biv.-Bern.) Fckl.; leafspot. S. America.
Pseudoplea trifolii (Rostr.) Petr.; leafscorch and wither.
  Subtropical USA, El Salvador, Guatemala, Colombia.
Pyrenopeziza sp.; leafspotting and deformation. Colombia.
Pythium spp.; rootrot. Scattered occurrence in plantings in
  warm locations. C. America.
Rhizobium sp.; the needed root nodule bacterium. If not
  present must be introduced.
Rhizoctonia microsclerotia Matz. is the hyphal and actively
  parasitizing aerial state (Pellicularia microsclerotia (Matz)
  Weber, is the spore state); web-blight. Subtropical USA.
  R. solani Kuehn the parasitic state of Thanatephorous cu-
  cumeris (Frank) Donk; causes rootrot. Widespread but
  not severe in coolest climates.
Sclerotinia minor Jagger; wilt, special scloerotial disease.
  Chile. S. sclerotiorum (Lib.) dBy. = Whetzeliana
  sclerotiorum (Lib.) Korf & Dumont; common sclerotial
  wilt disease. Chile, Colombia, Argentina (occasionally S.
  trifoliorum Eriks. is given as the causal organism).
Sclerotium rolfsii (Curzi) Sacc. = Pellicularia rolfsii (Sacc.)
  West; soil-rot, wilt. Subtropical USA, Puerto Rico in
  warm area.
Septoria sp. (and see immediately below); leafspotting and
  defoliation. C. America.
Stagonospora compta (Sacc.) Diet. = Septoria compta Sacc.;
  leaf spotting and drop. Chile. S. trifolii Ell. & Ev.;
  leaf blast. Colombia.
Stemphylium sarcinaeforme (Cav.) Wilts.; large leafspot.
  ' Brazil.
Trichodorus sp.; nematode found in rootzone. Chile.
Tylenchorhynchus sp.; nematode found in rootzone. Chile.
Uromyces elegans (Berk.) Lagh.; rare rust in subtropical
  USA. U. flectens Lagh.; common brown rust on leaf and
  petiole in Andean areas of Chile, Bolivia, Ecuador, Peru,
  Venezuela. T. nerviphilus (Grog.) Hotson; rust. Colombia,
  Brazil, Argentina. U. trifolii (Hedw. f.) Lév.; in cooler
  parts of Guatemala, Colombia, Chile, Uruguay, Bolivia,
  Venezuela. U. trifolii-megalanthi (Diet. & Neger) Jacks.
  & Holw. (prior name was Aecidium trifolii-megalanthi
  Diet. & Holw. and seen specially on TRIFOLIUM

PERUVIANUM in Brazil, Peru, Chile). U. trifolii-repentis
(Liro) Arth. ; rust in Argentina, and in Jamaica on plants
growing at very high altitudes.
Viruses: Mosaic viruses scattered in S. America.
Xanthomonas phaseoli (E. F. Sm.) Dows. ; bacterial blight.
Guatemala.

TRIPSACUM LAXUM Nash and T. LATIFOLIUM Hitch. Important
and persistent large native grasses propagated in moist-warm
farms entirely by vegetative means, a heavy-producing green-cut
forage; "pasto de Guatemala, " "zacate gama, " "gama verde" Sp. ,
"gama grass, " "Guatemala grass" En. (Note: The common
management of this valued crop is to cut the green, heavy growth
at least every six months, sometimes a little more often. It is
only in abandoned fields that plants have a chance to mature;
there is little possibility for many serious diseases to multiply
on good, well-handled plantings of gama.
    Angiospora pallescens (Arth.) Mains. = Physopella zeae
        (Mains) Cumm. & Rama. (synonyms: Dicaeoma pallescens
        Arth. & Fromme, Puccinia pallescens Arth. , Uredo pallida
        Diet. & Holw.) small more-or-less cuticle-covered rust
        lesions. Reports are mostly from the northern part of the
        Neotropics, more especially subtropical USA, Mexico,
        Guatemala, Nicaragua, El Salvador, Puerto Rico, Honduras.
    Bacterial softrot (? Erwinia sp. ?) of planting setts left over
        a weekend in close bundles. Honduras.
    Cladosporium sp. (spores from old leaf used for riding mule
        bedding). Honduras.
    Claviceps purpurea (Fr.) Tul. ; inflorescence disease, ergot
        (rare find). Honduras.
    Colletotrichum graminicola (Ces.) Wils. ; leaf anthracnose-
        spot, sett-rot, sett-failure. C. America.
    Coniosporium tripsaci Gonz. Frag. & Cif. ; stem disease,
        plant base-rot, weakened growth. W. Indies.
    Erwinia (see Bacteria)
    Fusarium moniliforme Sheld. (its asco-state is Gibberella
        fujikuroi (Saw.) Wr. = G. moniliforme (Sheld.) Winel.);
        wilt and leaf shredding. Scattered, seen in abandoned
        plants in Guatemala. Fusarium sp. (spores recovered
        from culm lesions in moist chamber studies); culm canker.
        Costa Rica.
    Helminthosporium sp. ; large leaf lesion. Guatemala.
    Leafspot (undetermined, no spores secured, no isolations ob-
        tainable), lesions large, spreading with red-brown edges,
        and cattle left a bundle of severely spotted leaf blades.
        El Salvador.
    Phyllachora tripsacina Petr. & Cif. ; shiny leafspots, no en-
        largement. Guatemala, El Salvador, Honduras.
    Physopella (see Angiospora)
    Puccinia polysora Underw. (= Dicaeoma polysorum (Underw.)
        Arth.); orange colored dusty leafrust. Widespread.
    Pythium spp. ; sett-failure in warm wet land. Honduras.

Viruses: A mild mottle in Guatemala, a streak and dwarfing in Guatemala.

TRITICUM AESTIVUM L. (Note: One species is named here, although in the genetic make-up of wheat grown in tropical America, other species genes are included. The purpose of this list is to make available to tropical workers some information on diseases reported in their region; some fungi that are obviously saprophytes are not included. Disease alone is noted, and not differences on symptoms from host breeding and selection refinements, nor are results given of race studies on some of the fungi.) The famous long known wheat, the important cereal crop from the north temperate zone where winters are cold, growing seasons cool as well as semi-dry, and soils good. Europeans settling in the New World brought wheat to the American tropics, made test plantings and found it failed in many typical warm and rainy tropical areas but was adaptable in somewhat restricted parts where growing seasons are cool, somewhat dry, and soils not excessively subjected to severe deterioration by sun and rain. It is a permanent and highly valued contributor to human food in the American tropics, grown in its proper environment, so long as pathologists are trained and monitor it for diseases and can be called upon to aid in controls; "trigo" Sp. Po., "wheat" En.
  Alternaria alternata (Fr.) Keiss. = A. tenuis Nees; leaf-tip-and-edge necrosis, grain spot. Subtropical USA, Mexico, Guatemala, Peru. A. peglioni Curzi; seedling blight, black grain, straw injury. Mexico, Peru, Uruguay, Argentina, Colombia, Chile.
  Anguina tritici (Steinb.) Filip.; head nematode, cockle, false fungus-blast. Subtropical USA, Brazil, Argentina, other countries.
  Arjona patagonica Hombr.· & Jacq. = A. tuberosa Cav. and A. tuberosa Cav. var. patagonica Literature; root attack by low-growing persistent phanerogamic parasite with small root tubers. Argentina.
  Aspergillus spp.; cause of moldyness of stored grain, also straw mold. Occurrence widespread (spores air borne in profusion): A. candidus Lk., A. flavus Lk., A. glaucus Lk., A. niger v. Tiegh., A. ochraceus Wilh.
  Boron deficiency, poor growth, often arrested head in the boot. Colombia, Venezuela.
  Calcium deficiency, chlorotic and fragile growth. Guatemala.
  Cephalothecium sp.; weak leaf-parasite on dead tissues. Argentina.
  Chlorosis, physiological from extremely wet, wet season. Argentina.
  Chlorosis, physiological from extremely dry, dry season. Mexico, S. America.
  Cladosporium herbarum (Pers.) Lk. (asco-state secured only in culture is Mycosphaerella tulasnei (Janez.) Lind.); glume disease, senescent leaf disease, leaf spot, black tissue. Nicaragua, El Salvador, Brazil, Chile, Colombia, Argentina.

Claviceps purpurea (Fr.) Tul. (has been reported as Cordy-
ceps); ergot, poison grain. Chile, Colombia.
Cold injury, extremely low temperature on plants at boot- or
spiga-time. Argentina, Guatemala (one personal report).
Colletotrichum graminicola (Ces.) Wils. = C. pamparum
Speg., it has a perfect state, Glomerella tucumanensis
(Speg.) v. Arx & Mull.; anthracnose leafspot, culm lesion.
Subtropical USA, Brazil, Mexico, Argentina, Guatemala.
Curvularia Sp.; culm disease. Scattered.
Deficiency symptoms (see names of chemicals)
Dicaeoma (see Puccinia striiformis)
Diplodina verruculosa Speg. (on Triticum secundum Kunth); a
leafspot disease. Tierra del Fuego.
Epicoccum nigrum Lk.; weakly parasitic, leafsmudge, occurred
on dead and dying leaf. Argentina.
Erysiphe is an asco-state seldom recorded in Neotropics (see
Oidium).
Fumago vagans Pers.; sootymold develops along with sucking
insects. Scattered.
1. Fusarium culmorum (W. G. Sm.) Sacc.; seedling rootrot,
culm disease of older plants. Peru, Argentina. 2. F.
culmorum (W. G. Sm.) Sacc. f. sp. cereale (Cke.) Wr.;
severe type culm disease. Argentina. 3. F. graminearum
Schwabe and its perfect state Gibberella zeae (Schw.) Petch;
rootrot, grain-scab, headmold. Argentina, Chile, Colombia,
Brazil, Uruguay, Guatemala, Mexico, Honduras, Nicaragua,
Venezuela. 4. F. moniliforme Sheld. its perfect state is
G. fujikuroi (Saw.) Wr.; rootrot, plant base rot above
ground (in C. America seen in fields growing at low edge
of cool elevations on moist side of hills). Guatemala,
Costa Rica where it seems to have caused many crop
failures in certain parts, Mexico, Uruguay, Argentina.
5. F. poae (Peck) Wr.; glume disease, grain infection.
Argentina. 6. F. roseum Lk. = F. concolor Reink. with
its perfect state G. roseum (Lk.) Sny. & Hans; rootrot,
headmold. Widespread. 7. Fusarium spp. unidentified;
root and culm base rots in Peru, Guatemala with leaf
collapse and browning in Argentina and nearby countries,
some grain lesions but not scab in Colombia, Peru.
Gaeumannomyces graminis (Sacc.) Arx & Olive; a rootrot.
Chile.
Gibberella (see Fusarium)
Heat damage to heading and grain after hot winds during young
head and grain (boot) development. Whitish chlorosis,
grain shrivel. Mexico, Argentina.
Helminthosporium sativum Pam., King., & Bakke (Synonyms:
H. gramineum (Rob.) Ericks., H. tetramera McK., H.
teres Sacc., H. tritici-repentis Died., H. tritici-vulgaris
Nisik.); foot and node rot, spot-blotch of leaf, seedling
death, discolored grain and poor germination. Widespread.
Heterosporium sp.; leaf-and-boot mold. Chile.
Iron deficiency, marked chlorosis, shrivelled head. Scattered.

Leptosphaeria spp. ; culm infections impeding normal grain
    development in Brazil. L. existialis (Mor.) Web. culm
    disease in Colombia. L. herpotrichoides deNot. and L.
    nodorum Mull. culm lesions in Brazil.
Lodging, storm damage (some varieties more susceptible than
    others), can occur in any area.
Meloidogyne sp. ; rootknot. Brazil.
Micrococcus tritici Prill. ; shrivel of grain. Brazil, Argen-
    tina.
Mucor sp. ; on stored grain. Colombia.
Mycena citricolor (Berk. & Curt.) Sacc. ; using only the in-
    fection bodies caused leafspot from artificial inoculation in
    laboratory in Costa Rica.
Nematodes, various unnamed kinds in rhizosphere. Chile,
    Peru.
Nigrospora sphaerica (Sacc.) Mason, notably large spores;
    leaf blackening and decay in senescent condition. Brazil,
    found in straw from Mexico.
Nitrogen deficiency, general chlorosis, weak growth. Scat-
    tered.
Oidium monilioides Lk. , perfect state is Erysiphe graminis
    DC. f. sp. tritici Mar. ; the white powdery mildew which
    is sometimes scanty and does not develop every year.
    Mexico, Guatemala, El Salvador, Nicaragua, Ecuador, Peru,
    Chile, Brazil, Argentina.
Ophiobolus graminis Sacc. (? = O. cariceti (Berk. & Br.)
    Sacc.); footrot, white tip, "espiga blanca, " sometimes in
    slightly alkaline soils in cool areas. Mexico, Uruguay,
    Bolivia, Peru, Chile, Colombia, Argentina.
Penicillium spp. ; bluemold in poorly stored grain. Occasional.
Phoma sp. ; culm and leafsheath disease. Mexico.
Phosphorous deficiency, stunt and bluish top. Brazil, and
    scattered.
Pleospora herbarum (Pers.) Rabh. ; seedling leaf collapse,
    glume infections. Guatemala.
Potash deficiency, weak growth, plants more susceptible to
    fungus infections. Scattered, seen in Mexico.
Pratylenchus sp. ; root nematode. Peru.
Puccinia spp. (Note: Species of this genus have caused many
    serious Old World famines as well as extreme losses in
    North America, and special attention has been given to
    species, varieties, and races of these rusts. There are
    numerous synonyms, some not given in this list, and the
    race problem is still being studied; at present there are
    over 300 numbered race entities in only one species.)
    1. P. graminis Pers. f. sp. tritici Eriks. & E. Henn.
    (synonyms used in American tropics: P. megalopotamica
    Speg. and P. graminis Pers.); typical stem rust (Spegaz-
    zini reported it abundantly sporulating on leaf and leaf-
    sheath). Widespread in all wheat countries of the Neo-
    tropics. This rust has eliminated the crop where grown
    without pathological attention in one after another area in

somewhat warm-moist ecology.   2. P. striiformis West
(synonyms used in American tropics: Dicaeoma glumarum
(Eriks. & E. Henn.) Arth. & Fromme, P. glumarum
Eriks. & E. Henn., P. glumarum Eriks. & E. Henn. f.
sp. tritici Eriks. & E. Henn.); stripe rust, sometimes
severe attack of glumes. Widespread. 3. P. recondita
Rob. (synonyms used in American tropics: P. rubigo-vera
(DC.) Wint., P. rubigo-vera (DC.) Wint. f. sp. tritici
Eriks., P. rubigo-vera (DC.) Wint. tritici (Eriks. & E.
Henn.) Carl., P. clematidis (DC.) Lagh., P. brachypus
Speg.); yellow leafrust, under some conditions brownish to
orange-red leafrust, Spegazzini reported it developing at
times in distinctly linear leaf pustules.   Guatemala, Nicaragua, Chile, Ecuador, Peru, Venezuela, Colombia, Bolivia, Brazil, Argentina, El Salvador, Mexico.
Pyrenochaeta terrestris (Hans.) Gor., Walker, & Lars.; root
   disease.   Subtropical USA.
Pyrenophora teres (Died.) Drechs. (conidial state is Helmin-
   thsoporium teres Sacc.) and P. trichostoma (Fr.) Fckl.;
   stalk diseases, straw injury. Bolivia, Puerto Rico.
Pythium arrhenomanes Drechs., P. oligandrum Drechs., and
   P. ostracodes Drechs. are all root rots.   Subtropical USA,
   determined by symptoms in Ecuador and Guatemala but no
   isolations secured.
Rhizopus sp.; moldy stored grain.   Scattered reports in C.
   America.
Rhynchosporium graminicola Hein.; node disease.   Peru.
Sclerospora macrospora Sacc.; downy mildew.   Mexico, Sub-
   tropical USA.
Sclerotium clavus DC.; plant baserot.   Chile.   The well known
   S. rolfsii (Curzi) Sacc., perfect state is Pellicularia rolfsii
   (Sacc.) West; leafsheath and culm disease, soil-line rot.
   Subtropical USA, Mexico, Puerto Rico.
Septoria nodorum Berk. = Leptosphaeria nodorum (Berk.)
   Mull.; wetseason bootrot, malspiga, glumespot.   Subtropical
   USA, Mexico, Ecuador, Venezuela, Bolivia, Colombia,
   Uruguay, Brazil, Argentina.   S. tritici Rob. = S. tritici
   Desm.; leafblotch, leafspot, stemspot.   In Central America
   only at high elevations.   Guatemala, El Salvador, Ecuador,
   Chile, Uruguay, Bolivia, Brazil, Argentina.
Stemphylium sp.; grain spot.   Subtropical USA, Colombia.
Tilletia caries (DC.) Tul. = T. tritici (Bjer.) Wint. (rough
   spore character); bunt.   Mexico, El Salvador, Guatemala,
   Colombia, Chile, Ecuador, Bolivia, Peru, Venezuela,
   Brazil, Argentina.   The T. contraversa Kuehn; European
   bunt.   (In 1915 collected in Argentina, years later, in
   1937 to 1940, found in both Argentina and Uruguay.)   T.
   foetida (Wallr.) Liro = T. levis Kuehn or T. foetens Berk.
   & Trel.; bunt that has smooth spores.   Chile, Peru,
   Colombia, Ecuador, Bolivia, Venezuela, Brazil, Argentina,
   Subtropical USA, C. America.
Tylenchorhynchus sp.; nematode collected near roots.   Peru.

Uredo lejoderma Speg.; leafrust.   Paraguay.
Urocystis tritici Koern. = U. agropyri (Pr.) Schr.; flagsmut.
    Mexico, Chile, Brazil, and in other countries.
Ustilago tritici (Pers.) Rostr. = U. tritici Jens. or U. carbo
    DC.; spikelet and loose smut.   In all wheat countries.
Viruses: Cucumber Mosaic Virus strain Celery-Commelina in
    Florida.   Dwarfing in Colombia.   Hoja Blanca of Rice in
    Colombia.   Mosaic mottle in Cuba.   Unidentified mottle in
    Rio Grande do Sul in Brazil.
Xanthomonas transluscens (L. R. Jones, Johnson, & Reddy)
    Dows.; bacterial brown necrosis.   Reported but infrequent
    in Peru, occasional but rare now in a few other countries.
"Yellow berry" is English name, "panza blanca" is Spanish,
    cause is physiological from poor or unbalanced nitrogen
    nutrition.   Argentina, Mexico.
Zinc deficiency causes notable stunt, poor seed set.   Scat-
    tered.

TROPAEOLUM TUBEROSUM Ruiz & Pavón.   A native vegetable of
    the Andes closely related to the ornamental nasturtium.   Its
    tubers are brought to local markets often far from where grown.
    It produces in high mountains where most vegetables fail, a nu-
    tritious and flavorsome root crop; with such common names as
    "nasturto" Sp., "aboboreira" Po., "mashua," "isaño" Ind.   No
    scientific disease studies have been either requested by growers
    or made on this crop.
        Blackening of tuber at root connection.   Market material in
        Colombia.
        Cercospora tropaeoli Atk.; leafspot.   Brazil.
        Meloidogyne sp.; rootknot nematode.   Brazil.
        ? Rust (possibly Uromyces sp.); verbal description from a
        non-technical traveler recently in Ecuador.
        Spongospora subterranea (Wellr.) Lagh.; tuber powdery scab.
        Ecuador, Colombia, Peru.
        Tuber dryrot, in markets in Colombia.
        Tuber wet-rot apparently bacterial but no isolations made.
        In market in Ecuador.
        Tuber with sunken spots, not insect work, in markets in
        Colombia, Ecuador.
        Virus, mottle and malformation of roots.   Market in Ecuador.
        "Weather injury," plants stunted, tubers elongated and tough
        (traded at low price).   Market in Ecuador.

ULLUCUS TUBEROSUS Loz. = U. TUBEROSUS Caldas.   A native
    Andean tuberous crop, in some areas a very important food,
    most common in highland-cold tropical country where it is much
    eaten; "timbo," "malloco," "ulluco," "oca-quira," "chiqua,"
    "papa-lisa" (local Indian names).
        Aecidium cantense Arth.; rust.   Special reports on it from
        Peru and Bolivia, probably in the other timbo growing ad-
        jacent high Andes countries.
        (? Asterina sp., leaf smudge-spot seen without microscope.
        Peru.)

Colletotrichum sp. ; spores from old tuber.   Peru.
Fusarium sp. ; spores from old tuber having dry rot lesion.
    Peru.
Heterodora rostochiensis Wr. ; root-and-tuber invasion.   Found
    in plant material far from area where the nematode could
    have been introduced by foreign human intervention.   Peru.
Nematodes, various, from soil near tuber.   Peru.
Oidium sp. ; white powdery mildew.   Peru, probably wherever
    crop is grown.
Rhizoctonia solani Kuehn (perfect state is Thanatephorous
    cucumeris (Frank) Donk); base-of-stem rot, tuber disease.
    Peru.

URENA LOBATA L.   A central American plant grown by peasantry
    for the excellent fiber in its fiber-containing bark that is stripped
    off its stems.   A malvaceous bush-like somewhat woody shrub;
    "malvita, " "cadillo" Sp. , "aramina, " "carrapido" Po. , "urena"
    En.   (It is a disease-tolerant plant with many leafspots and
    other injuries that have never been given much pathological at-
    tention.)
Bacteria (wild types, unidentified) active in experimental retting
    of bark strips to secure fiber.   El Salvador.
Cankers on stem and branches, fungus cause but organism not
    identified.   Fiber discoloration and weakening effect.   C.
    America.
Capnodium (see Morfea)
Cephaleuros virescens Kunze; algal leafspot and leaf-scab.
    Brazil, Costa Rica.
Cercospora malachrae Heald & Wolf; leafspot, some defoliation
    on old stalks.   C. America.   C. urenae Viégas & Chupp;
    leafspot.   Brazil.
Fusarium moniliforme Sheld. , typical conidia reported on seed
    (tropical source unspecified) sent to temperate zone.
Hormisciella rubi Bat. ; sootymold.   Puerto Rico, Costa Rica,
    probably in all C. America and W. Indies.
Leaf-edge disease (unidentified small-spored fungus).   Guate-
    mala.
Meliola-like black mildew of leaf.   Costa Rica, El Salvador.
Morfea moniliforme (Fraser) Cif. & Bat. = Capnodium monili-
    forme Fraser; sootymold.   Subtropical USA, Puerto Rico.
Myriangium duriaei Mont. & Berk. = M. floridanum Hoeh. ;
    parasite of scale insect and invades leaf epidermis of host.
    El Salvador.
Oidium sp. ; white powdery mildew.   Widespread.
Podonectria coccicola (Ell. & Ev.) Petch; parasite of insect,,
    giving confusing appearance of a leaf disease.   W. Indies,
    C. America.
Rootrot, unidentified fungus (study incomplete) on specimens
    lost from military revolt action.   El Salvador.
Virus, a Mild Mosaic mottle in El Salvador.

VACCINIUM spp.   Berry bushes, some from the tropics, some from

the temperate zone, used for test growing, to determine value
for commercialization.   (The disease list is from tropical
sources and is the result of answering consultation requests.)
   Botryodiplodia sp. = Diplodia sp. (see Physalospora); twig
      dieback, disease in some branches from harvest injury
      that becomes infected.   El Salvador and widespread.
   Catacauma paramoense Chard. ; black lesion underside leafspot.
      Brazil, Venezuela.
   Colletotrichum sp. = Gloeosporium sp. ; anthracnose.   C.
      America, Florida.
   Cylindrosporium sp. ; leafspot.   Texas, Guatemala.
   Dendropthora hexasticha v. Tiegh. ; a small parasitic phanero-
      gamic shrub.   Peru, (a similar shrub not identified, on
      host species at jungle edge seen in Costa Rica and in El
      Salvador).
   Exobasidium vaccinii Wor. ; leaf gall.   Subtropical USA, W.
      Indies.
   Meliola nidulans (Schw.) Cke. ; black mildew.   Subtropical
      USA, C. America.
   Microsphaeria alni DC. its conidial state is Oidium; powdery
      mildew.   Widespread.
   Mycosphaerella vacinii (Cke.) Schr. ; leafspot.   Subtropical
      USA, W. Indies.
   Oidium (see Microsphaeria)
   Ophiodothella vaccinii Boyd; flyspeck leafspot.   Subtropical
      USA.
   Pestalotia vacinicola Guba & Zell. ; leafspot.   W. Indies,
      Florida.
   Phoradendron crispum Trel. ; woody bushlike phanerogram para-
      site on bush branch.   Costa Rica.
   Phyllosticta vaccinii Earle; leafspot.   Subtropical USA, Mexico.
   Physalospora obtusa (Schw.) Cke. = Botryosphaeria quercuum
      (Schw.) Sacc. which is the asco-state of Botryodiplodia; die-
      back, leafspot.   Widespread.
   Pucciniastrum myrtilli (Schum.) Arth. ; leafrust.   Subtropical
      USA.
   Ramularia vaccinii Pk. ; leafspot.   Florida.
   Rhytisma vaccinii Schw. ; tarspot.   Subtropical USA, C.
      America.
   Septobasidium sinuatum Couch; brown felt on branch.   Florida,
      C. America.
   Septoria albopunctata Cke. ; leafspot.   Subtropical USA. S.
      lagerheimii Pat. ; leafspot.   Brazil, Ecuador.
   Sphaerodothis circumscripta (Berk.) Th. & Syd. ; leafspot with
      shiny stroma.   Colombia, Peru, Brazil.

VALERIANOIDES see STACHYTARPHETA

VANILLA PLANIFOLA Andr. = V. FRAGRANS (Salisb.) Ames.
   Native in Mexico-Central America, it is a climbing orchid, and
   produces a notably commercially important crop in several
   countries, requiring special horticulture for producing the fleshy

pods which on harvest and processing yield the highly valued
world-renowned vanilla flavor material; "vainilla, " "vaynilla"
Sp. , "baunilha" Po. , "vanilla" En. (several unrecorded Indian
names).

? Amerosporium (see Sphaeropsis)

Botryodiplodia theobromae Pat. = Diplodia cacaoicola P. Henn. ;
  stem canker, pod peduncle disease, leaf-tip disease.
  Puerto Rico, Honduras, Mexico.

Botryosphaeria vanillae (Stonem.) Petch & Ragun. ; leafspot,
  pod canker-spot, vine canker.   Puerto Rico, Honduras,
  Guatemala, Mexico.

Calospora vanillae Mass. ; leaf lesion, leafscorch.   W. Indies,
  Colombia.

Capnodium sp. ; sootymold.   Puerto Rico.

Cassytha filiformis L. ; laboratory infection by the green
  dodder.   Puerto Rico.

Cephaleuros virescens Kunze; algal leafspot and scurf.   Puerto
  Rico, Honduras, Brazil (probably almost anywhere crop
  grown under moist conditions if the organism exists in
  vicinity).

Cladosporium orchidacearum Cke. & Mass. ; spores recovered
  from dried pods that were not fermented.   Honduras.

Colletotrichum gloeosporioides Penz. (some synonyms: C.
  orchidearum Allesch. , C. vanillae Scalia, C. vanillae
  Verpl. & Cl. , Gloeosporium vanillae Cke. , G. cattleyae
  Sacc. , G. rufo-maculans (Berk.) Thuem. , Hypodermium
  orchidearum Cke. & Mass. , Trullula vanillae P. Henn.)
  is the often encountered imperfect state of Glomerella
  cingulata (Stonem.) Sp. & Schr. = Gnomoniopsis vanillae
  Stonem. ; the common and important anthracnose spot of
  leaf, stem, both aerial and submerged root, pod, flower.
  Puerto Rico, Honduras, Mexico, Dominican Republic,
  Panama, Costa Rica, Cuba, Brazil.

Cuscuta americana L. ; yellow dodder laboratory infection.
  Puerto Rico.

Diplodia (see Botryodiplodia)

Fusarium batatas Wr. var. vanillae Tuck. = F. oxysporum
  Schl. f. sp. vanillae (Tuck.) Gord. ; root infection, systemic
  vascular wilt.   Puerto Rico, El Salvador, Honduras, Costa
  Rica, Mexico, (also in islands of Oriental tropics).   Common F. solani (Mart.) App. & Wr. with perfect state
  Nectria haematococci Berk. & Br. = Hypomyces solani
  Reinke & Berth. ; severe rootrot in soil or under mulch,
  no infection by study of wounded inoculated vine.   Puerto
  Rico, Dominican Republic, Panama.   F. solani (Mart.) App.
  & Wr. (apparently a new f. sp.   I call planifoliae as it infected tips of aerial roots, infected wounded and inoculated
  vines, was recovered from vine canker, pure culture red
  in color and did not infect living root or tuber of white
  potato or sweet potato).   Puerto Rico. A Fusarium sp. ;
  leaf-base disease.   Brazil.

Gloeosporium (see Colletotrichum)

Glomerella (see perfect state of Colletotrichum)
Gnomoniopsis (see perfect state of Colletrotrichum)
Hypodermium (see Colletotrichum)
Knyaria avernae-saccas Cif. & Gonz. Frag.; increased leaf
    injury after initial anthracnose. Puerto Rico, Dominican
    Republic.
Lembosia rolfsii Horne; black mildew. Florida.
Leptostroma sp.; small spots on senescent leaf and yellowed
    pod. Honduras.
Leucothyridium vanillae Av.-Sac.; crust on foliage. Brazil.
Lichens, excess smothering growth of foliose and appressed
    types of green, gray, brown, and bronze coloration. Wild
    growth in rainforest ecology. Costa Rica.
Liverworts and mosses. Epiphytic, grown associated with
    foliose lichens (see directly above) in Costa Rica (also very
    small miniature foliose liverworts in Puerto Rico).
Macrophoma vanillae Av.-Sac.; ashy leafspot. Brazil.
Mycena citricolor (Berk. & Curt.) Sacc. = Omphalia flavida
    (Cke.) Maubl. & Rang.; leafspot. Costa Rica, Puerto Rico.
Myiocopron vanillae Av.-Sac.; leaf disease. Brazil.
Nectria peristomata Zimm.; leaf and stem disease. Dominican
    Republic, Puerto Rico, Cuba, Haiti. (For another Nectria
    see Fusarium solani.)
Pellicularia koleroga Cke.; leafrot, thread blight. Costa Rica,
    Guatemala.
Pestalotia vanillae Av.-Sac.; leafspot. Brazil, W. Indies,
    Guatemala, Honduras, Mexico.
Phytophthora cactorum (Leb. & Cohn) Schr.; rootrot. Florida.
    A Phytophthora sp. (said to be not the above); part of the
    complex with Fusarium causing rootrot. Puerto Rico, Hon-
    duras.
Rhizoctonia solani Kuehn (basidial state not recovered); root
    rot, transplant rot, sett failure. Also it is a predisposing
    infection for disease entrance of Fusarium oxysporum f.
    sp. vanillae. Puerto Rico, Dominican Republic.
Rhynchodiplodia citri Bri. & Farn.; scurf. Argentina.
Sclerotium rolfsii (Curzi) Sacc. = Pellicularia rolfsii (Sacc.)
    West; young plant collapse. Puerto Rico.
(? Sphaeropsis sp., possibly Amerosporium orchidacearum
    Speg.); leaf lesions with round black pycnidia. Ecuador.
Strigula complanata Fée; white-lichen leafspot. W. Indies,
    Honduras, probably widespread.
Trentopohlia sp.; tomentose epiphytic algal growth. Scattered
    in C. America, W. Indies.
Trullula (see Colletotrichum)
Tubercularia sp.; stem disease. Brazil.
Uredo scabies Cke.; a rust. Subtropical USA, Mexico, Nica-
    ragua, Panama, Costa Rica, Colombia.

VICIA FABA L. A crop of antiquity originating in the Mesopotamia-
North Africa region and in the American tropics readily adapted
to relatively cool parts where it is grown for the long proved

good protein food in beans and forage for man and beast; "haba"
Sp., "fava" Po., "broad bean" En.

Aecidium (see Uromyces)

Alternaria sp.; leafspot and blast under moist conditions.
Peru, Ecuador.

Aphanomyces sp.; rootrot. Peru.

Arthrobotrys chilensis Allesch. & P. Henn. (? on senescent
foliage) in Chile.

Ascochyta fabae Speg. (? = A. pisi Lib.); leafspot. Argentina,
Brazil.

Botrytis cinerea Pers.; pod brown-spot, stem rot, graymold
leaf disease. Subtropical USA, Argentina, Peru. B. fabae
Sard.; chocolate spot. Chile, Bolivia, Uruguay, Argentina,
Brazil.

Cercospora zonata Wint. (synonyms: C. viciae Ell. & Holw.,
C. fabae Fautr.); ringspot. Ecuador, Colombia, Peru,
Chile, Venezuela, Brazil.

Chaetoseptoria wellmanii J. A. Stev.; round gray-spot.
Mexico, Costa Rica, Panama.

Colletotrichum lindemuthianum (Sacc. & Magn.) Bri. & Cav.;
spots on leaf, pod, on twigs slender elongate lesions.
Chile, Brazil.

Dimerium grammodes (Kunze) Gar.; (synonyms: Parodiella
grammodes (Kunze) Cke., P. perisporioides (Berk. &
Curt.) Speg.); leaf disease. W. Indies, C. America.

Diplosporium album Bon.; on old foliage. Chile.

Erysiphe (see Oidium)

Footrot from complex of fungi, some weakly parasitic: Fusarium spp.; Verticillium spp., Rhizoctonia sp., etc.
Peru, other countries.

Fusarium oxysporum Schl. f. sp. medicaginis (Weim.) Sny.
& Hans., wilt. Subtropical USA, Ecuador. F. roseum
Lk.; base-of-plant rot. Peru, Chile. Fusarium spp.;
root infections. Colombia, Bolivia, Peru, Ecuador.

Heterodera rostochiensis Wr.; root nematode. Peru.

Meloidogyne sp.; rootknot nematode. Subtropical USA.

Mucor sp.; recovered from wilted plants. Peru.

Mycosphaerella sp.; leafspot. Subtropical USA, Ecuador.

Nematospora coryli Pegl. = N. phaseoli Wing.; seed yeast-
spot. Subtropical USA, Mexico, Puerto Rico, Brazil.

Oidium sp. is the commonly seen white powdery mildew
which is the conidial state of Erysiphe polygoni DC. Widespread in occurrence.

Peronospora viciae dBy. synonyms: P. viciae (Berk.) Casp.,
or P. viciaesativae Gaum.; downy mildew. Scattered in
C. America, noted in Argentina.

Phakopsora vignae (Bres.) Arth.; rust. Puerto Rico.

Phoma sp.; leaf, stem, and pod disease. Colombia.

Phyllosticta sp.; small spots on leaf. Disease scattered in
S. America.

Phymatotrichum omnivorum (Shear) Dug.; rootrot. Texas.

Pythium spp.; damping-off. El Salvador, Ecuador, California.

Ramularia sp. ; mealy spreading-leafspot. Mexico.

Rhizobium leguminosarum Frank = Pseudomonas radicola
(Beij.) Moore; desired nitrogen-fixing rootnodule bacterium.
Widespread.

Rhizoctonia solani Kuehn (the basidial state seldom collected
is Thantephorous cucumeris (Frank) Donk); damping-off,
stemrot, rootrot, wilt. W. Indies, C. America, Peru,
Argentina, Ecuador.

Sclerotinia sclerotiorum (Lib.) dBy. = Whetzeliana sclerotiorum (Lib.) Korf & Dumont; stemrot, market podrot.
C. America, California.

Sclerotium rolfsii (Curzi) Sacc. its perfect state is Pellicularia rolfsii (Curzi) West; groundline rot. Puerto Rico,
Florida.

Tylenchorhynchus sp. ; rhizosphere nematode. Peru.

Uromyces fabae (Pers.) dBy. ; leaf-stem-pod rust. Widespread. (Note: This rust species confirmed on authoritative grounds in Subtropical USA, in C. America, and reported as seen on observational basis not only by me but
by others in Uruguay, Ecuador, Chile, Bolivia, Peru,
Colombia, Venezuela. In Brazil and Argentina it is also
given as U. viciae-fabae (Pers.) Schr. Other rusts have
been reported in S. America on other Vicia spp., as
follows: Aecidium porosum Pk. var. anodonata Speg.,
Uromyces clavatus Diet. [synonyms: A. lathyrinum Speg.,
U. lathyrinus Speg.], U. corrugatus Speg., U. johowii
Diet. & Neg., U. nordenskjoeldii Diet.)

Verticillium sp. ; associated with other parasites causing footrot. Scattered in S. America.

Viruses. At least ten viruses are encountered on faba beans
in the American tropics, listed with countries in which
they were noted as follows: Bean Yellow Mosaic in Cuba,
Beet Curly Top in California, Cucumber Mosaic in Subtropical USA, Cucumber Mosaic strain Celery-Commelina
in Florida, Puerto Rico, Pea Enation Mosaic in Argentina,
Common Pea Mosaic in Argentina, Sunflower (Argentine)
mosaic in Argentina, Sweet Pea Streak in Argentina, Tobacco Streak in Brazil, and Vicia Faba Mosaic in Venezuela.

Xanthomonas phaseoli (E. F. Sm.) Dows. ; bacterial blight,
black bacterial rot. Subtropical USA, (now practically
absent), scattered reports in Middle America.

VIGNA SINENSIS (Torner) Savi. An increasingly used annual leguminous crop apparently native to the Africa-Asia tropics grown in
the American tropics for soil cover-rehabilitation, stock feed,
and its round, nutritious and edible seeds; "cáupi," "chicharo de
vaca," "frijol de campo" Sp., "feijo miudo," "feijo branco" Po.,
"black-eyed pea," "black-eyed bean," "cowpea" En. (Note: In
the genus are several other species but this one is the most
planted in the American tropics. It is not only of local value
but it is shipped out as well and is a potentially important export

product. Cured seed is a staple, good for long term storage
and shipment. In addition to its use against hunger and its em-
ployment in over-seas famines, the seed is a special ingredient
in certain chemical manufacturing processes.)

Aristostoma oeconomicum (Ell. & Tracy) Tehon = Amerospo-
rium oeconomicum Ell. & Tracy; cause of rather uncommon
white leafspot. Subtropical USA, Honduras, Haiti, Domini-
can Republic, Nicaragua.
Ascochyta cruenta Sacc.; leafspot, podspot. Honduras, Guate-
mala (dry area), El Salvador, Costa Rica (Pacific slopes).
A. pisi Lib.; leaf and pod spot. Venezuela, Subtropical
USA.
Botrytis cinerea Pers.; podrot, the rainyseason mold on old
leaves at base of plants. Scattered in wet areas.
Bud drop, physiological, possibly minor-element deficiency.
W. Indies, C. America.
Cephaleuros virescens Kunze; algal leafspot. Brazil.
Cercospora caracallae (Speg.) Chupp; gray leafspot. S.
America. The C. cruenta Sacc., with synonyms C. dolichi
Ell. & Ev., C. vignae Ell. & Ev., C. canescens (Sacc.)
Ell. & Mart. (its asco-state is Mycosphaerella cruenta
Lath.); the severe and common large brown leafspot, on
pod it causes redspot. Subtropical USA, West Indian large
and small islands, C. America, Honduras, Nicaragua, El
Salvador, Venezuela, Brazil.
Chaetoseptoria vignae Tehon; small pycnidia on medium sized
irregular leafspot. Guatemala, El Salvador. C. wellmanii
J. A. Stev.; large pycnidia on large and circular gray leaf-
spot. El Salvador, Costa Rica, Honduras, Mexico, Guate-
mala, Panama, Colombia, Venezuela.
Chlorosis, physiological from deficiency in soils of calcium or
iron or manganese. Scattered occurrences.
Cladosporium pisi Cug. & Macc., in Dominican Republic.
C. vignae Gard. in Subtropical USA, C. America. Both
are cause of pod and leaf spot.
Colletotrichum lindemuthianum (Sacc. & Magn.) Bri. & Cav.;
typical common bean anthracnose. Widespread.
Cornyespora cassiicola (Berk. & Curt.) Wei; target spot.
Subtropical USA, W. Indies.
Elsinoë phaseoli Jenk.; pod and leaf scab-spot anthracnose.
El Salvador.
Entyloma sp.; leaf-smut pustules. Brazil.
Erysiphe (see Oidium)
1. Fusarium oxysporum Schl.; wilt. W. Indies, northern
part of S. America. 2. F. oxysporum Schl. f. sp.
tracheiphilum (E. F. Sm.) Sny. & Hans. (? different type
of wilt from above). Subtropical USA. 3. F. roseum Lk.;
seed disease. El Salvador. 4. F. semitectum Berk. &
Rav.; isolated from blackened seed, and from dead branch.
Guatemala, Costa Rica. 5. F. solani (Mart.) App. & Wr.;
root disease. California, Puerto Rico, C. America.
Glomerella spp.; various colored, small and large asco-carps

on old stem lesions, the conidial states not determined. Scattered occurrences.

Helminthosporium vignae Olive; leaf target-spot. Subtropical USA, W. Indies, El Salvador.

(? Irenina or Meliola; black leaf disease fungus, no spores found. El Salvador.)

Isariopsis griseola Sacc.; angular leafspot. El Salvador, Brazil.

Lepiota sp.; on decaying hay. Costa Rica.

Macrophoma subconica Ell. & Ev.; stem canker. Subtropical USA.

Macrophomina phaseoli (Maubl.) Ashby; ashy stem-blight, charcoal disease. El Salvador, Guatemala, Texas.

Meloidogyne exigua Goeldi; rootknot nematode. Subtropical USA, Jamaica, El Salvador.

Mycosphaerella pinodes (Berk. & Blox.) Vest.; wetseason leafspot. Honduras.

Nematospora coryli Pegl., rare findings of seed yeast-spot. Occasional in shipments from tropical America.

Oidium balsamii Mont., the asco-state which is seldom collected is Erysiphe polygoni DC. that is also sometimes given as Leveillula taurica (Lév.) Arn. or Microsphaera alni DC.; the common white powdery mildew. Widespread.

Pellicularia koleroga Cke.; the thread-web blight and leafrot. Costa Rica, Subtropical USA on the east coast from southern North Carolina into Florida.

Phyllosticta phaseolina Sacc. and P. phaseolorum Sacc. & Speg. (sometimes considered synonymous but may be reported separately by workers), both cause defoliating bright colored leafspots. Brazil, Argentina, Costa Rica. P. vignae Speg.; dull appearing leafspot. Argentina.

Phymatotrichum omnivorum (Shear) Dug.; rootrot. Texas and possibly Arizona in Subtropical USA, unconfirmed report from adjacent Mexico.

Physarum cinereum (Batsch.) Pers.; epiphyte giving appearance of leaf disease. Puerto Rico.

Phytophthora spp.; podrots, stem cankers, rootrots. Scattered occurrence.

Pseudomonas syringae Van Hall; red bacterial-leafspot, podspot. Widespread. P. tabaci (Wolf & Foster) Stev.; wildfire (small and coalescing bacterial leafspots). Florida.

Pythium sp.; podrot. Panama.

Rhizobium spp.; bacteria that develop the desired nitrogen-fixing root nodules. Widespread (abundant in best soils).

Rhizoctonia microsclerotia Matz (not uncommonly seen spore-bearing growth is Pellicularia microsclerotia (Matz) Web.); aerial web-blight. Subtropical USA, W. Indies, Nicaragua, El Salvador, Costa Rica. Common R. solani Kuehn; subterranean damping-off, plant base-rot. Widespread. A Rhizoctonia sp., somewhat atypical in culture, profuse spores on stroma in test tubes; zonal leafspot. Mexico, Peru, Panama, El Salvador.

Sclerotinia sclerotiorum (Lib.) dBy. = Whetzeliana sclerotio-
    rum (Lib.) Korf & Dumont; stemrot, white mold.   Texas,
    Panama.
Sclerotium rolfsii (Curzi) Sacc. (this is the mycelial state
    that is usually seen, the sporuliferous state is Pellicularia
    rolfsii (Sacc.) West); rootrot, collar rot, footrot.   Sub-
    tropical USA, Puerto Rico, El Salvador, Nicaragua, Panama,
    Surinam.
Thielavia basicola Zopf.; black-root.   Subtropical USA, Cuba
    (said to develop in fields grown following long standing
    tobacco culture).
Uromyces phaseoli (Reb.) Wint. synonyms: U. vignae Barcl.,
    U. phaseoli (Reb.) Wint. var. vignae (Barcl.) Arth., U.
    appendiculata (Pers.) Fr.; common severe rust.   Widespread.
Viruses, eleven viruses reported, some coming originally from
    other hosts: 1. Beet Curly-Top in California. 2. Cowpea
    (Vigna) Mottle, the typical aphid-borne common systemic
    mottle mosaic is widespread.   3. Cowpea (Vigna) Yellow
    Mosaic is a severe seed-borne chlorosis and malformation
    in California, El Salvador, Costa Rica, Nicaragua, Vene-
    zuela.   4. Cucumber Mosaic is a relatively mild mottle in
    Florida.   5. Cucumber Mosaic strain Celery-Commelina,
    is a severe yellow mottle with malformation and necrosis
    in Florida, Cuba.   6. Eggplant Mosaic through inoculation
    studies in Trinidad.   7. Peach Yellow-Bud Mosaic =
    Tomato Ring-spot in California.   8. Sunflower Virus through
    inoculation studies in Argentina.   9. Tobacco Necrosis by
    inoculation studies in Brazil.   10. Tobacco Streak by inocu-
    lation studies in Brazil.   11. Tomato Spotted-Wilt by inocu-
    lation studies in California.
Xanthomonas vignicola Burkh.; bacterial blight.   Texas, El
    Salvador, probably in several other countries as symptoms
    can be easily confused with other problems.

VINCA ROSEA L. = CATHARANTHUS ROSEUS (L.) Don.   Native to
    both Eastern and Western tropics, the much grown delightful and
    somewhat variable, practically ever-blooming short evergreen
    shrubs; "chuladita," "chula-chula" Sp., "bõa noite" Po., "peri-
    winkle" En., "vinca" Sp. Po. En.
    Camarosporulum santiaguinum Speg.; leaf disease.   Chile.
    Cassytha filiformis L.; slow-growing green dodder.   Puerto
        Rico.
    Cephaleuros virescens Kunze; algal leafscurf.   Puerto Rico
        (in rainforest area), Brazil.
    Chlorosis with stunt and flower bud death.   Apparent deficien-
        cies of such elements as iron and boron.   Cuba, Costa Rica.
    Coleosporium apocynaceaum Cke., leafrust.   Puerto Rico,
        Subtropical USA.
    Colletotrichum gloeosporioides Penz. (synonyms: C. vincae
        Speg., C. vincae Politis), asco-state is Glomerella cingu-
        lata (Stonem.) Sp. & Schr.; anthracnose leafspot and stem
        lesion.   Puerto Rico, Florida, Argentina, El Salvador,
        Costa Rica.

Cuscuta americana L. ; a fat, fast-growing golden-yellow
   dodder. Puerto Rico. C. indecora Choisy; slender dod-
   der. Texas.
Diplodia vincae Sacc. & Wint. ; stem disease, occasional leaf
   infection, dieback. Puerto Rico, El Salvador, Costa Rica,
   Guatemala, Florida.
Meloidogyne sp. ; rootknot. Subtropical USA, Puerto Rico.
Oidium sp. ; mild white powdery mildew. Widespread.
Phytophthora citrophthora (Sm. and Sm.) Leon. ; root disease.
   S. America and P. nicotiane de Haan var. parasitica
   Dast. Waterh. ; stem canker, wilt. Subtropical USA, W.
   Indies, C. America.
Rhizoctonia microsclerotia Matz; web blight of foliage, at
   times discrete leafspots. El Salvador. R. solani Kuehn
   (the common hyphal parasite); basal-rot of whole plant.
   Scattered in W. Indies, and C. America.
Viruses: Aster Yellows in Texas, Common Cucumber Mosaic
   in Florida, El Salvador, Puerto Rico, Virulent Cucumber
   Mosaic strain Celery-Commelina in Florida, Cuba, Puerto
   Rico, Honduras, El Salvador.

VITIS VINIFERA L. Famed Old World perennial woody vine often
   pruned to a short strong trunk and stout "side arms, " early
   introduced to the American Tropics and grown most successfully
   in dry-cool-climates far from rainforest ecology, producing commercially in southern S. America fine wines and dessert fruits,
   in some other locations grown only as exotic fruit; "uva, " "la
   vid" Sp. , "uva, " "vite, " "videira" Po. , "grape, " "grapevine"
   En. (Note: There are three other grape species that produce,
   under difficulties, some occasional fruit in the American tropics,
   but the most important species for fruit and wine is the one
   given. Diseases in this list are mostly from national reports.
   No attempt has been made to secure and search for disease
   records on grapes grown in private vineyards of the tropical
   Americas.)
      Agrobacterium tumefaciens (E. F. Sm. & Town.) Conn; crown-
         gall. California, Argentina, Ecuador, Bolivia, Peru.
      Alternaria vitis Cav. , leafspot, fruitspot. Subtropical USA,
         Chile, El Salvador (on one pot-grown vine).
      Angiospora (see Physopella)
      Apiosporium brasiliensis Noack; a black mold. Brazil.
      Armillaria mellea Vahl = Armilariella mellea (Fr.) Karst. ;
         rootrot. California, Peru.
      Ascochyta chlorospora Speg. ; leafspot. Argentina.
      Aspergillus niger v. Tiegh. ; storage and transit fruit black-
         mold. Special studies in California and Colombia, probably
         widespread.
      Bacteria (Rikketsia-like) associated with Pierces disease. See
         under Virus.
      Bacterium (unidentified); plant rot. Chile.
      Belonolaimus longicaudatus Rav. ; rhizosphere nematode.
         Florida.

Boron deficiency, leaf and fruit spot. Argentina.
Botryodiplodia sp. = Diplodia sp.; dieback. Widespread.
Botryotinia fuckeliana (dBy.) Whet.; foliage disease. S.
   America.
Botrytis cinerea Pers. (synonyms: B. cinerea Fckl., B.
   novaesii Noack); shoot blight, fruit graymold rot. Argen-
   tina, Chile, Brazil, Bolivia, Uruguay, Peru, Ecuador,
   California.
Briosia ampelophaga Cav.; on foliage. Brazil.
Capnodium sp.; sootymold. Colombia.
Cephalosporium zonatum. Saw.; zonate and large leafspot.
   Colombia, El Salvador. (Note: Abundant on old grapevine
   in 1944 setting fruit in patio, other vines close by but in
   the open had no zonate leafspot and no fruit set. Six
   years later the same organism was found on coffee leaves
   in a nearby plantation, causing identical symptoms.)
Cercospora roesleri (Catt.) Sacc.; mild leafspotting. Argen-
   tina. C. vitis (Lév.) Sacc. (with synonyms: C. vitis
   (Lév.) Sacc. var. ruprestris Cif., Psudocercospora vitis
   (Lév.) Speg., C. viticola (Lév.) Speg., C. viticola (Ces.)
   Sacc.); common large brown round to irregular leafspot.
   S. America, W. Indies.
Cicinnobolus cesati dBy.; hyperparasite on white mildew.
   Scattered collections.
Cladosporium viticolum Ces. = C. herbarum (Pers.) Lk.,
   (has been reported as Heterosporium sp.); green-gray
   moldyrot of fruit. California, Puerto Rico, Chile, Brazil.
Colletotrichum gloeosporioides Penz. (synonyms: C. vitis
   Maire., C. amelinum Cav., C. ampelophagum (Pass.)
   Sacc., Gloeosporium amphelophagum (Pass.) Sacc.) the
   perfect state is Glomerella cingulata (Stonem.) Sp. &
   Sch. sometimes given as G. cingulata (Atk.) Sp. & Sch.;
   fruit riperot, leaf and stem black anthracnose lesions.
   Widespread, special studies in Florida, Texas, Chile,
   Uruguay, Costa Rica, Brazil, Jamaica, Ecuador, Haiti.
Coniothyrium diplodiella (Speg.) Sacc.; stem-canker, fruit
   cluster whiterot. Uruguay, Argentina.
Coryneum sp. (probably conidial state of a Valsa); bark
   disease. Argentina.
Criconemoides xenpolax Raski; root region nematode. Chile,
   California.
Crossopsora caucensis (Mayor) Kern, Thur., & Whet. =
   Uredo caucensis Mayor; leaf rust. Southern S. America.
Cryptosporella viticola (Red.) Shear; dead-arm, tendril
   disease. Brazil, Uruguay, California.
Cyphella monacha Speg. and C. uricola Speg. the parasitic
   state of Lachnella uvicola (Speg.) Cke.; twig infections,
   young fruit infections. Argentina.
Didymosphaeria sarmenti (Cke. & Harkn.) Berl. & Vogl.;
   twig disease. California.
Diplodia (see Botryodiplodia)
Elsinoë ampelina Shear = Sphaceloma ampelinum dBy., which

is the imperfect state; spot anthracnose on young fruits and new shoots, berry-scab. Subtropical USA, Puerto Rico, Dominican Republic, Haiti, Colombia, Ecuador, Venezuela, Chile, Uruguay, Brazil, Argentina.

Eutypa heteracantha Sacc. var. vitis-vinifera Speg. imperfect state of Valsa heteracantha Sacc. ; woodrot of stem. Argentina.

Failure to set fruit, physiological, seen in moist-warm areas where there is no growth-break in seasons, often patio-grown vines where roots are crowded and masonry induces considerable heat, vines will set fruit and nearby outside plants will not produce.

Fumago vagans Pers. ; a sootymold. S. America.

1. Fusarium culmorum (W. G. Sm.) Sacc. ; isolated from grape raceme. Chile. 2. F. oxysporum Schl. ; wilt. Brazil. 3. F. roseum Lk. ; secondary invasion in new shoot diseased by anthracnose and rust. Costa Rica.

Graphium cinerellum Speg. ; on dying leaves. Argentina.

Guignardia baccae (Cava.) Taez. = G. bidwellii (Ell.) Via. & Rav. ; blackrot of fruit, spot on branches and leaves. Brazil, Florida, Alabama, El Salvador, Puerto Rico, Jamaica, Venezuela, Uruguay, Chile, Haiti. (Requires moist-country conditions.)

Hansenula anomala (Hans.) H. & P. Syd. ; yeast from grape flowers. Brazil.

Helminthosporium vitis Ces. ; leaf and twig disease. Colombia.

Hemicycliophora sp. ; rhizosphere nematode resident. Chile.

Heterosporium sp. (see Cladosporium)

Hot wind causes severe scorch and shrivel, sometimes in S. America called apoplexy (apoplejía). California, Chile, Argentina.

Isariopsis clavispora (Berk. & Curt.) Sacc. ; irregular leaf-spot, leafedge disease, blight. Florida, Costa Rica, Chile, Peru, Bolivia, Brazil.

Lachnella (see Cyphella)

Macrophoma acinicola (Speg.) Sacc. & Trott. , and M. flacida Vial. & Rav. = Phoma acinicola Speg. ; fruit disease the former in Argentina, latter in Colombia.

Marasmius viticola Berk. & Curt. ; on wild grape vine. Subtropical USA.

Melanconium fuligenum (Scrib. & Viala) Cav. ; fruit bitter-rot, a common warm season disease. Subtropical USA, Brazil, nearby countries.

Meloidogyne sp. ; rootknot. Peru, Chile, Subtropical USA.

Metasphaeria peruviana (Cke.) Sacc. ; on branch. Peru.

Monochaetia ampelophila Speg. ; disease of new shoots. Argentina.

Mycena citricolor (Berk. & Br.) Sacc. ; rare large leafspot from both natural and laboratory infections. Costa Rica.

Mycosphaerella personata Higg. ; brown leafspots. El Salvador, Subtropical USA.

Nectria viticola Berk. & Curt. ; branch disease. Subtropical USA, El Salvador, Guatemala.

Oidium tuckeri Berk. is in the American tropics the almost
   exclusively collected conidial state, the asco-state is
   Uncinula necator (Schw.) Burr.; cause of common white
   powdery mildew.   Widespread, in all countries where
   grapes are grown.   (O. syringae Blum. and U. americana
   Howe are believed to be synonyms for the above.)
Ophiodothis paraguariensis Speg.; on leaf.   Paraguay.
Pellicularia koleroga Cke.; leafrot, web-thread blight.   Peru.
Penicillium glaucum Lk.; blue mold on fruits.   Scattered in
   Chile, Subtropical USA, probably other countries.
Periconia pycnospora Fres.; possibly a saprophyte on dead
   leaves and other plant parts.   S. America.
Pestalotia uvicola Speg. (specially determined in Colombia
   and Peru); anthracnose-like stem lesions, ragged-leaf,
   and associated with fruit deterioration.   Florida, El Salva-
   dor, Costa Rica, Colombia, Peru.
Peziza ascoboloides Bert.; on stems.   Chile.
Phoma uvicola Berk. & Curt.; black leafspot.   Florida,
   Colombia, Chile, Argentina.   (Also see Macrophoma.)
Phyllachora taruma Speg.; shiny leafspot.   Paraguay.
Phyllosticta viticola Sacc. & Speg.; leafspot.   Costa Rica,
   El Salvador, Guatemala.
Phymatotrichum omnivorum (Shear) Dug.; rootrot.   Texas.
Physopella ampelopsidis (Diet. & Syd.) Cumm. & Rama.
   (Synonyms: Angiospora ampelopsidis (Diet. & Syd.)
   Thirum. & Kern, P. vitis (Thuem.) Arth., Uredo vitis
   Thuem., Phakopsora vitis (Thuem.) Syd., P. ampelopsidis
   Diet. & Syd.); common severe rust of leaf.   Widespread.
Plasmopara viticola (Berk. & Curt.) Berl. & DeToni; downy
   mildew of leaf, inflorescence, fruit, shoot.   Widespread.
Pleospora vitis Catt.; leafspot.   Chile.
Pratylenchus sp.; root nematode.   Peru, Chile, California.
Pseudocercospora (see Cercospora)
Rhizoctonia sp.; disease of plant base.   Chile.
Rosellinia necatrix (Hartig) Berl.; root-rot.   Argentina,
   Brazil, Chile.
Saccharomyces apiculatus Reess, S. pasteurianus Reess, S.
   ellipsoideus Hans.; fruit fermentation and market spoilages.
   (See also Hansenula and Trichosporon.)   Widespread.
Sclerotinia fuckeliana (dBy.) Fckl.; tipblight of new growth.
   Uruguay, Chile.   S. sclerotiorum (Lib.) dBy. = Whetze-
   liana sclerotiorum (Lib.) Korf & Dumont; causing tip-
   blight in California.
Septoria ampelina Berk. & Curt. and S. melanopsis Pat.,
   leafspotting fungi.   Brazil.
Shot berry, physiological drying and hard mummy of imma-
   ture fruit.   Scattered occurrence.
Sphaceloma (see Elsinoë)
Sphaeronaema sp.; mild leafspot.   Colombia.
Stereum sp.; trunk disease and death.   Bolivia, Argentina.
Stilbum aurantio-cinnebarinum Speg.; branch and vine canker.
   Argentina, Brazil.

Trichodorus sp.; rhizosphere nematode. Chile.
Triochosporon pullulans (Lind.) Didd. & Lodd.; yeast injury
   of flowers. Brazil.
Trullula atro-fulginea Speg.; shoot death. Argentina.
Tylenchorhynchus sp.; rhizosphere nematode. Chile.
Tylenchulus semipenetrans Cobb; root feeding nematode.
   California.
Uncinula (see Oidium) [U. spiralis Berk. & Curt. reported
   in Barbados believed by some to be U. necator.]
Ustulina vulgaris Tul. = U. zonata (Lév.) Sacc.; trunk and
   root decay of old vines. S. America.
Valsa (see Eutypa)
Valsaria ampelina Av.-Sac.; vine-wood decay. Brazil.
Viruses: Grape Fanleaf in Venezuela. An infectious degenera-
   tion in Argentina, Brazil. Typical mosaic mottle in Puerto
   Rico. Pierces disease = Anaheim disease (included under
   virus as it has been considered for years as virus causation
   but shown in Florida to be caused by Rikettsia-like bac-
   teria) in Florida, California, Argentina, Brazil, Ecuador.
   A rugose disease that may be a strain of leafroll in Argen-
   tina.
Water berry, physiological. Rare excessively soft and juicy
   fruits. Seen in Guatemala market.
Xiphinema sp.; nematode resident in rhizosphere. Chile.
Zinc deficiency causes little-leaf and stunt of new growth.
   California, Ecuador.

WARSZEWICZIA COCCINEA Kl. Native tree in the West Indies, a
   strikingly beautiful ornamental being slowly increased and planted
   in several hot tropical American countries and as far away as
   Ceylon. A rather small tree with many branches terminating in
   spectacular poinsettia-like, but more handsome, long crowded
   racemes of brilliantly red flowers; "chaconia, " "tree poinsettia. "
   (The few diseases I have seen listed here are only from the
   West Indies.)
   Blight, undetermined, of flower sepals.
   Dieback of branches (almost certainly Botryodiplodia sp.).
   Leafspot, undetermined, large with gray center.
   Leafspot, undetermined, small with sunken center.
   Lembosia warszewicziae P. Henn.; black mildew.
   Pellicularia koleroga Cke.; branch disease and leafrot.
   Root disease (no information on cause) probably Rosellinia.
   Scorch of leaf edge, undetermined cause.
   Trichothyrium dubiosum (Boram. & Rouss.) Th.; apparent
      hyperparasite of fungus hyphae on leaf surface.

WEINMANNIA spp. Trees indigenous to South America, which although seldom planted are protected and guarded in the wild,
   yielding ornamental leaves that are often gathered and since prehistoric times used in funerary and religious processions; "palo
   santo, " "teníu, " "trisel. "
   Antennaria scoriadea Berk.; type of sootymold. Argentina,
      Brazil, Chile.

Asterina weinmanniae Syd. ; black smudge growth on leaf
   surface. S. America.
Atichia spp. all moldy growths associated with scale insects:
   A. antarctica (Speg.) Mar. , A. chilensis Cot. , A. cordo-
   bensis (Speg.) Mar. and A. millardeti Rac. ; all in S.
   America.
Brefeldiopycnis membranacea Petr. & Cif. ; a leaf disease.
   W. Indies.
Calonectria stromaticola P. Henn. ; gives appearance of a
   disease but is a parasite of insects.   Scattered in occur-
   rence in Neotropics.
Diatrype weinmanniae Rehm; wood infection.   Brazil.
Fistulina hepatica Fr. ; woodrot.   Scattered.
Meliola mucronata (Mont.) Sacc. ; black mildew.   S. America,
   C. America.
Seynesia colliculosa Rehm; leaf blemish fungus.   Brazil.

XANTHOSOMA SAGITTIFOLIUM Schott, and X. VIOLACEUM Schott,
   perennial strong lush herbs with large wide leaves held hip high
   on long petioles.   At the plant base are numerous good sized
   nutritious subterranean tubers that are a common article of food
   for millions in tropical America.   These crops, known from
   early prehistory, have innumerable common names; a few:
   "bádu, " "yautía, " "coco, " "tanier, " "malanga, " "mangarito, "
   "ocumo, " "tisquisque, " "raafafa" Ind. Sp.
   Acrostalagmus aphidum Oud. ; simulates a fungus attack on
      foliage parts but is in actuality a parasite of aphids.   Wide-
      spread.
   Aecidium aroideum Cke. , in Bolivia and Venezuela, and A.
      dubitabile Lindq. , in Argentina, both leafrusts and probably
      much more widely spread.
   Antennaria sp. ; dryseasons sootymcld (along with aphid infes-
      tation).   Widespread.
   Aphelenchoides parietinus (Bast.) Stein. ; root nematode.
      Brazil.
   Aphelenchus avenae Bast. ; root nematode.   Brazil.
   Armillaria mellea Vahl = Armilariella mellea (Fr.) Karst. ;
      rootrot.   Subtropical USA.
   Axonchium amplicolla Cobb; root nematode.   Brazil.
   Botryosphaeria quercuum (Schw.) Sacc. (= Botryodiplodia
      theobromae Pat. , Diplodia natalensis P. Evans); leafspot,
      stem-canker, tuber-canker.   W. Indies, C. America.
   Cassytha filiformis L. ; green slow dodder.   Puerto Rico.
   Cercospora verruculosa Stev. & Salh. (= C. chevalieri
      Sacc.) and C. xanthosomae Gonz. Frag. & Cif. ; leafspots.
      Former reported in Venezuela, latter in Dominican Re-
      public but are much more widespread.
   Colletotrichum dematium (Pers.) Grove = C. xanthosomatis
      Sacc. that is the conidial organism with its asco-state
      Glomerella xanthosomae (Cif. & Gonz. Frag.) Petr. ; leaf-
      spot, petiole lesion, tuber infection.   W. Indies, Costa
      Rica, S. America.

Cuscuta americana L. (infection by inoculation); fast yellow
   dodder. Puerto Rico.
Erwinia spp. ; market tuber-rots. Widespread.
Eudarluca venezuelana Syd. ; hyperparasite on rust. S.
   America.
Failure, gradual (from undetermined cause or causes) requiring
   replanting of fields, and occasional movement of entire vil-
   lages. W. Indies, C. America.
Fusarium oxysporum Schl. ; wilt, systemic discoloration in
   tuber. Dominican Republic. F. solani (Mart.) Sacc. ;
   tuber dry grayrot. Florida, Puerto Rico, probably C.
   America.
Guignardia xanthosomae Cif. & Gonz. Frag. ; leaf and stem
   disease. Dominican Republic.
Helminthosporium caladii Stev. ; large tan leafspot. W.
   Indies. H. xanthosomae Gonz. Frag. & Cif. ; large leafspot. Dominican Republic.
Meloidogyne exigua Goeldi; rootknot. Brazil.
Mycena citricolor (Berk. & Curt.) Sacc. (Synonyms: Omphalia
   flavida (Cke.) Maubl. & Rang., Stilbum flavidum Cke.);
   leafspot. Costa Rica, Colombia, W. Indies.
Periconia byssoides Pers. and P. pycnospora Fr. ; both are
   mostly saprophytic on dead and dying leaves. W. Indies.
Pestalotia spp. ; small lesions and associated with other
   disease damage. W. Indies.
Phyllosticta araceae Cif. & Gonz. Frag., P. colocasiae
   Hoehn., P. colocasicola Hoehn., P. xanthosomatis Sacc.,
   P. xanthosomatis Petr. & Cif. (uncertain about species of
   five different fungi); all seem to cause the same type of
   leafspot. Mexico, W. Indies, C. America.
Phytophthora parasitica Dast. ; rootrot, wilt. Puerto Rico,
   Virgin Islands, Cuba.
Punctillina salteroi Toro; on leaf. Venezuela.
Pythium ultimum Trow; fungal softrot of leaves, petioles,
   roots. Widespread in moist-warm localities.
Rhizoctonia sp. ; leaf and plant base disease. Dominican Re-
   public.
Rosellinia bunodes (Berk. & Br.) Sacc. ; rootrot. Widespread.
Scleroconium venezuelanum Syd. ; plant stem infection. Vene-
   zuela.
Sclerotium rolfsii (Curzi) Sacc. its perfect state is Pellicu-
   laria rolfsii (Sacc.) West; plant basal-rot and then wilt.
   Florida, Puerto Rico.
Vasculomyces xanthosomae Ashby; tuber dryrot. W. Indies,
   Venezuela.
Viruses: Dasheen mosaic in W. Indies and leaf mottle in
   Puerto Rico.

XYLOPIA spp. Include about four species of small fruit and spice
   trees native to tropical countries that are being slowly increased
   in South and Central America plantings and have considerable
   future potential although as yet having had only causal scientific

Yam 456

horticultural or pathological study in the American tropics; "especia, " "arbol de sazon" Sp. , "condimento" Po. , "spice tree" En.

Aecidium anonae P. Henn. and A. xylopiae P. Henn. ; rusts in Argentina, Uruguay, Brazil.
Calonectria ferruginea Rehm; hyperparasite on black mildews in Brazil and other countries in S. America.
Cercospora xylopiae Viégas & Chupp; common leafspot. Brazil.
Dasyspora gregaria (Kunze) P. Henn. (synonyms: D. foveolata Berk. & Curt. , Puccinia foveolata (Berk. & Curt.) P. Henn. , P. gregaria Kunze); the common rust. Guatemala, Peru, Surinam, Brazil, Argentina, Uruguay.
Meliola xylopiae Stev. ; black mildew. Probably widespread.
Phyllachora xylopiae (P. Henn.) Th. & Syd. ; tar spotting. Peru.
Phyllosticta xylopiae Sacc. ; leafspot. Mexico.
Puccinia winteria Paz. ; rust. S. America.
Scolecopeltis xylopiae Bat. & Gayão; black mildew. Brazil.

YAM see DIOSCOREA

YUCCA ALOIFOLIA L. , Y. ELEPHANTIPES L. , Y. FILAMENTOSA L. Native in the approximate region of southern North America and northern Central America, the perennial tough and short woody-trunked plants crowned with stiff spine-tipped leaves and flowering at the top. Grown for fences, ornament, tying fiber, soapy root juices, and edible blooms; "iuca, " "yucca, " "a guija, " "bayoneta, " "izote" Sp. Ind.

Anthostomella nigro-annulata Berk. & Curt. ; severe leafspot that on coalescence may kill leaf. Subtropical USA, Cuba, Bermuda. A. sphaeroidea Speg. ; a semiparasite on decaying leaf. Argentina.
Botryosphaeria agave (P. Henn.) Syd. = Diplodia circinans Berk. & Br. ; leafspot. Texas, Mexico, C. America.
Cercospora concentrica Cke. & Ell. ; leafspot. Texas, Puerto Rico, El Salvador, Bolivia. C. yuccae Cke. ; leafspot. Argentina.
Colletotrichum gloeosporoides Penz. , synonyms: Gleosporium yuccae Politis, G. yuccae Poll, G. yuccogenum Ell. & Ev. (asco-state is Glomerella cingulata (Stonem.) Sp. & Schr. given also as G. yuccae Gutner); common as target-zoned leafspot, flower stem lesion. C. America and probably much more widespread.
Coniothyrium concentricum (Desm.) Sacc. , apparently same as C. yuccae Speg. ; a leaf spot. W. Indies, C. America, Argentina.
Cryptovalsa yuccae Speg. ; infection on stem of inflorescence. Argentina.
Cylindrosporium angustifolium Ell. & Kell. ; leafspot. Subtropical USA, Mexico.
Diaporthe arctii (Lasch) Nits. ; leaf and inflorescence lesion. Subtropical USA, Virgin Islands.

Didymosphaeria clementsii Sacc. & Sacc., and D. yuccogena
 (Cke.) Sacc.; lesions on senescent leaves. El Salvador,
 California. (A Didymosphaeria of undescribed species dif-
 ferent from the two named, was collected in El Salvador.)
Diplodia yuccae Speg.; on dead leaves. Argentina (see also
 Botryosphaeria above).
Gloeosporium (see Colletotrichum)
Kellermannia anomala (Cke.) Hoehn., and K. major Dearn. &
 Barth.; leafblights. Subtropical USA.
Leptosphaeria filamentosa Ell. & Ev.; a leaf blotch in Sub-
 tropical USA, El Salvador, Mexico and other countries.
 L. obtusispora Speg.; leaf blotch in El Salvador and Argen-
 tina.
Neottiospora yuccifolia Hall; leafspot. Subtropical USA, Guate-
 mala.
Nigrospora sp.; identification from spore scrapings of leaf-
 edge disease. El Salvador.
Pestalotia sp.; leaf lesion. Widespread in moist localities.
Pestalozziella yuccae Karst.; leaf rot and injury. Texas,
 Virgin Islands.
Phomatospora argyrostigma (Berk.) Sacc.; on old leaf. Sub-
 tropical USA.
Phyllosticta yuccogena Ell. & Ev.; leafspotting. Subtropical
 USA.
Physalospora sp.; disease of stem. Subtropical USA, C.
 America.
Pleospora parvula Speg. = Pyrenophora parvula (Speg.) Sacc.;
 leaf lesions. Argentina. P. thuemeniana Sacc.; lesions
 on old leaves. Subtropical USA.
Plowrightia circumscissa Tracy & Earle; stem disease.
 Florida, Virgin Islands.
Septoria sp.; leafspot. Texas.
Sphaerodothis pringlei (Pk.) Th. & Syd.; tarspot. Subtropical
 USA, El Salvador.
Stagonospora gigantea Heald & Wolf; leafspot. California.
Torula maculans Cke.; leafmold. Subtropical USA.

ZAMIA spp. Native very short-trunked trees with palm-like crown
 of shining dark green pinnate leaves, they are cultivated in small
 plantings and protected in the wild for the handsome leaves used
 by florists and for religious purposes; "palma de funebre,"
 "palmita," "marunguey" Sp.
 Ainsworthia xanthoxyli Bat. & Costa; leaflet blemish. Brazil.
 Anabaena cycadeae Reinke; internal symbiotic alga in corra-
 loid roots. Widespread.
 Aschersonia turbinata Berk.; scab deposit from growth of
 parasite on scale insects that gives appearance of leaf infection. Widespread.
 Cephaleuros virescens Kunze; algal leaf-scurf. Costa Rica,
 Panama and probably widespread.
 Cephalosporium lecanii Zimm.; fungus parasite of insect. S.
 America.

Cladosporium herbarum (Pers.) Lk. ; leaf blemish on old
   leaves in markets. Costa Rica, El Salvador.
Epiphytic fungus mycelium on leaf epidermis. No observable
   disease but such leaflets lose shiny green color. Costa
   Rica.
Leafspots similar to a condition called weak-spot on coffee
   and some other hosts. Spots sunken, with reddish edges
   (believed from fungi but possible result of insect punctures).
   Costa Rica.
Meloidogyne sp. ; rootknot nematode. Florida.
Mendoziopeltis ilicifolia Bat. & Peres; leaflet spot. W.
   Indies, Brazil.
Microxyphium aciculiforme Cif. , Bat. , & Nasc. ; sootymold.
   W. Indies, Brazil.
Pestalotia cycadis Allesch. ; leafspots on leaflets of old leaves.
   Florida, C. America.
Podoxyphium azevedoi Bat. , Nasc. , & Cif. ; leaflet mold
   blemish. Cuba, Puerto Rico, Brazil. P. citricola Bat. ,
   Vit. & Cif. (at times associated with Microxyphium aciculi-
   forme); sooty mold. Cuba, Colombia, Brazil, probably
   Widespread.
Reticularia lycoperdon Bull. ; wood decay of old stump and
   trunk. Scattered.
Rhizoctonia sp. ; mycelium on surface of old leaf still living
   but at ground level (cultured fungus was not R. solani).
   Costa Rica.
Strigula complanata Feé; white lichen-spot of leaflet. Costa
   Rica.
Trichomerium didymopanacis Bit. & Cif. (believed asco-state
   is Microxyphium aciculiforme); common sootymold. Spe-
   cially noted in Puerto Rico, Subtropical USA, Cuba, Brazil,
   but is probably Neotropical.
Triposporium stelligerum Speg. ; disease of old leaves. Puerto
   Rico, St. Thomas, Argentina.

ZANTEDESCHIA AETHIOPICA Spreng. Origin in Africa, grown in
   the American tropics for its so-called "lilly" flowers that arise
   from rhizomes, the crop is important economically for florists
   and ornamental gardeners; "lirio de egipto, " "calla" Sp. , "lirio
   de cala, " "cala" Po. , "calla" En.
   Alternaria sp. (small spored organism); a few spots on aging
      leaves. Subtropical USA, Puerto Rico, El Salvador.
   Armillaria mellea = Armilariella mellea (Fr.) Karst. ; root-
      rot. California.
   Botrytis sp. ; flower browning and collapse in market. Guate-
      mala.
   Cercospora callae Pk. & Clint. ; leafspot. Colombia. C.
      richardiaecola Atk. ; leafspot. Subtropical USA, Puerto Rico,
      Guatemala.
   Colletotrichum gloeosporioides Penz. = Gloeosporium callae
      Oud. (asco-state is Glomerella cingulata (Stonem.) Sp. &
      Schr.); common but seldom severe infections on flowers,
      leaves, stems, rhizomes. Widespread.

Coniothecium richardiae (Merc.) Jauch; leafspot and fumago-
   like growth.   Argentina, Brazil, Peru, California, Florida.
   (Has been reported incorrectly as a Phyllosticta.)
Discomycete, undetermined but with typical sporocarps.   On
   stem and exposed rhizome of old plant in Cathedral patio.
   Guatemala.
Erwinia spp.; bacterial soft rots of cut flower stalk and leaf
   ends in flower markets, and stalk and rhizome rot in
   gardens.   Widespread.
Gloeosporium (see Colletotrichum)
Meloidogyne sp.; rootknot.   Subtropical USA, Puerto Rico.
Phoma sp.; rhizome dryrot.   California, Costa Rica, Hon-
   duras, Guatemala.
? Phyllosticta report (see Coniothecium)
Phytophthora cryptogea Pethyb. & Laff. var. richardiae
   (Buis.) Ashby; rootrot.   Subtropical USA, W. Indies, C.
   America.   P. erythroseptica Pethyb.; rhizomerot.   Cali-
   fornia.
Rhizoctonia solani Kuehn (its known spore-producing stage is
   Thanatephorous cucumeris (Frank) Donk); rootrot, plant
   base disease.   California, Costa Rica.   Rhizoctonia sp.
   (aerial type); causing large target-board leafspot in Costa
   Rica.
Rosellinia bunodes (Berk. & Br.) Sacc.; black rootrot.   El
   Salvador.
Sclerotium rolfsii (Curzi) Sacc. = perfect state is Pellicularia
   rolfsii (Sacc.) West; basal rot.   Subtropical USA.
Viruses: Dasheen mosaic virus in Florida, W. Indies.   A
   Mosaic mottle of leaves and green streak and blemish of
   flowers in Argentina, Guatemala, Honduras.   Tomato
   Spotted Wilt causing ringspotting and malformed leaves
   and flowers in Argentina, Subtropical USA.

ZANTHOXYLUM spp. (XANTHOXYLUM spp.).   A few species of
   trees unusual for the aromatic properties of the wood and its
   centuries of persistence buried in various types of soils in wet
   and dry climates; "acetillo," "naranjillo," "cenizo," "espinosa,"
   "palo de hueso" Sp.
   Acanthotecium mirabile Speg.; leaf disease.   Brazil.
   Antennaria scoriadea Berk.; sootymold.   Chile, probably
      well spread.
   Catacauma zanthoxyli Stev. (See also Phyllachora).   Black
      patch on leaf.   Panama.
   Cercospora xanthoxyli Pk.; leafspot.   Texas.
   Diplodia sp. is most common report = Botryodiplodia sp.;
      dieback.   Widespread.
   Epiphytic growths (non-parasitizing bacteria and fungi, lichens,
      mosses, miniature liverworts, orchids, bromeliads, etc.),
      most common in moist areas, causes smothering injury.
      C. America.
   Higher Basidiomycetes attack on sapwood but not heartwood of
      fallen logs.   Reported as occasional in C. America by farm
      owners and Indians.

Hypoxylon anthracnodes (Fr.) Mont. and H. deustum (Hoffm.)
   Grev. (both involved in branch-wood disease). Chile.
Lasiosphaeria brasiliensis Speg. (may be the ascogenous state
   of Phyllachora brasiliensis Speg.); leafspot, defoliation.
   Widespread.
Limacinia fernandeziana Neger; sootymold. Chile.
Loranthaceae, various genera and species (observed in tree
   tops many not collected or identified, see Oryctanthus and
   Phorodendron below); tree branch parasites. Widespread.
Melanconium phyllostictoides Speg. ; leaf disease. Brazil.
Meliola pilocarpi Stev. ; black mildew on leaves. Panama.
Oryctanthus occidentalis (L.) Eichl. ; parasitic woody bush on
   host branches. Identified in Costa Rica, widely distributed
   in C. America.
Ovulariopsis farinosa Syd. ; powdery mildew. Florida.
Phorodendron flavescens (Pursh) Nutt. ; mistletoe-like tree
   parasite. Florida.
Phyllachora spp. all leafspots, some shiny tarspot symptom:
   P. applanata Wint. in Brazil, P. selenospora Speg. (=
   Catacauma selenospora (Speg.) Chard.) in Brazil and other
   countries of S. America, P. tijucensis (Rehm) Th. & Syd.
   in Brazil, P. winteri Sacc. & Syd. in Brazil, P. zanthoxyli
   (Lév.) Cke. in Subtropical USA, W. Indies, Brazil, Vene-
   zuela.
Phyllactinia guttata (Fr.) Lév. (compare with Ovulariopsis);
   powdery mildew. Widespread.
Phyllosticta brasiliensis Speg. ; leaf spotting. Brazil and
   probably other countries in S. America.
Puccinia andropogonis Schw. var. xanthoxyli (Pk.) Arth. ;
   leaf-rust with grass alternate host. Subtropical USA.
Septoria pachyspora Ell. & Holw. ; leafspot. Subtropical USA.
Seynesia olivascens Speg. and S. australis Speg. ; foliage
   diseases. Argentina, Brazil, Chile.
Sorokinia uleana Rehm; leaf disease. Brazil.
Sphaeria deusta Hoffm. = Hypoxylon denstum see above.
Trabutia spp. , scab-like black leafspots: T. brasiliensis
   (Speg.) Chard. = Phyllachora brasiliensis (Speg.) Chard.
   and T. xanthoxyli Chard. = Phyllachora xanthoxyli Wint.
   Puerto Rico, Virgin Islands, Costa Rica, Dominican Re-
   public, Venezuela, Brazil.

ZEA MAYS L.   An annual giant grass grain plant that originated
   in subtropical and tropical America, the single most important
   crop of the Neotropics, and is used for food and drinks, bedding,
   animal feeding, woven materials, foot gear, wall panels, temporary fences, fuel for cooking and warmth, toys, and in Indian
   religious ceremonies.   It produces under the widest range of
   ecological conditions of any major world cereal and furnishes 80%
   or more of the daily diet of many in tropical America; "maíz"
   Sp. , "milho" Po. , "maize, " "Indian corn, " "corn" En. , "ixiim, "
   "tizate, " "cacapo, " "malacajo, " "mah-híz, " "pupo, " "capulín"
   Ind. (Peasant growers from cool mountain locations and hot

jungle clearings have asked about and described to me diseases,
but without samples, that no one I know has determined.) [Notes
on environmental effects that are known about some diseases are
included in brackets.]

Acremoniella verucosa Togn.; earmold.   Puerto Rico.

Aerobacter sp.; decay of stalks (standing and fallen).   [Moist
    country conditions.]   Widespread.

Algal (green) growth, smothering effect on basal leaves.
    [Cloud-forest ecology.]   Costa Rica, Guatemala.

Alternaria alternata (Fr.) Keiss = A. tenuis Nees; seed mold,
    rare leaf edge infection, rare mold on ear husk.   El
    Salvador, Guatemala.   [Seen in dry season.]

Angiospora (see Physopella)

Anthostomella punctulata (Rob. & Desm.) Sacc.; leafspot.
    Scattered in S. America.

Apiospora phomatopsis Speg.; culm disease.   Argentina, nearby
    countries.

Ascochyta zeae Stout; leaf-and-stalk spot.   Dominican Republic.

Aspergillus spp. (includes at least A. flavus Lk., A. niger
    v. Tiegh., and A. ruber (Spieck. & Brem.) Thom &
    Church); grain molds, reduce seed germination.   Wide-
    spread.   [Especially in the warm-moist lowlands of W.
    Indies, C. America.]

Bacterium sp.; a rare rot in the core of cob.   Jamaica.

Basisporium (see Nigrospora)

Beauveria sp.; confusing presence of fungus parasite of in-
    sects.   S. America.

Belonolaimus gracilis Stein. and B. longicaudatus Rau; root
    nematodes in Florida.

Botryodiplodia (see Diplodia)

Botryosphaeria (see Diplodia)

Botryosporium pulchrum Cda.; saprophyte to weak parasite of
    senescent leaf.   Dominican Republic, S. America, C.
    America.   [Most notable in C. America at close of wet
    season and first part of dry season.]

Botrytis cinerea Pers.; grain graymold on mature unshelled
    ear.   El Salvador, Ecuador.   [Rainy season disease.]

Camarosporium sp.; small scattered perithecia on dying stalks.
    Mexico, El Salvador.

Capnodiaceae (unidentified as to genera or species), sooty-
    molds.   Widespread.   (Note: Nearly harmless, surface
    inhabiting, black to filmy growths.   Sometimes with sclero-
    tium-like structures.   [Most spectacular in dryseaons and
    definitely associated with insect honeydew secretions; in
    El Salvador mountains such sootymolds along with their
    insects were also on wild grasses and may have spread
    from them to the cultivated maize.]

Cassytha filiformis L.; greenish pungent-smelling dodder.   St.
    Croix, St. Thomas, Puerto Rico.   [Occurrence near sea-
    shore.]

Cephalosporium acremonium Cda.; black bundle, kernel rot.
    Subtropical USA.   Especially noted in Nicaragua and Guate-
    mala, possibly widespread.

Ceratostomella paradoxa (Dade) Mor.; stalk and root disease. Venezuela.

Cercospora sorghi Ell. & Ev.; most common leafspot, the spots small, may make streaks on leaves and leafsheaths. [In moist tropics of C. America considered lowland disease.] Subtropical USA, C. America, Mexico, W. Indies, Colombia, Peru, Brazil, perhaps widespread. C. zeae-maydis Tehon & Dan.; large leafspot, brown to gray in color. W. Indies, Guatemala, El Salvador, Costa Rica, Panama, Colombia, Ecuador, Venezuela, Brazil.

Cercosporella sp.; leafspot. Costa Rica.

Chaetomium melioloides Cke. & Pk.; saprophyte on discolored husks. El Salvador, Guatemala, Costa Rica. [In Guatemala seen in rainyseason in field, with ears doubled down on old standing stalk, found also on husks incubated in moist chambers.]

Chaetophoma maydis Speg. = Lasiophoma maydis Speg.; leaf disease. Paraguay.

Cladosporium herbarum Pers. = Hormodendrum sp.; on senescent leaf, ear bract, black mold of grain. Puerto Rico, El Salvador, Costa Rica, Colombia, Nicaragua, Panama, Ecuador, Bolivia, Brazil, probably widespread. (C. zeae Pk. also reported occasionally, ? synonym of C. herbarum.)

Claviceps (see Ustilagopsis)

Colletotrichum graminicola (Ces.) Wils. (a few synonyms: C. zeae Lobik, C. falcatum Wint., C. bromi Jenn., C. pamparum Speg., Vermicularia culmigena Cke.) its ascostate is Glomerella tucumanensis (Speg.) Arx & Mull. = Physalospora tucumanensis Speg.; the common anthracnose of leaf and stalk, leaf crumple, ear peduncle disease. [Anthracnose develops under very wide temperature range, but not cold condition.] Colletotrichum sp. (large spores, it did not infect orange, maize leaf or stalk, apple); growth in exudates had smothering effect on aphids on stalks. El Salvador.

Cordyceps sp.; on ear worm larvae. Colombia.

Corticium (see Rhizoctonia grisea)

Corynespora cassiicola (Berk. & Curt.) Wei; leaf speck-spot. El Salvador. Not commonly reported but no doubt in several countries.

Curvularia spp. all cause more or less blackened leaf lesions, sometimes elongated and on edge tissues: C. geniculata (Tracy & Earle) Boed. in Puerto Rico, El Salvador, Costa Rica, C. (? inaequalis) in Colombia, C. lunata (Wakk.) Boed. in Nicaragua, Panama and C. maculans (Bancr.) Boed. in Bolivia, Brazil.

Cuscuta americana L.; yellow fast growing dodder. Guatemala, Costa Rica, Puerto Rico. Cuscuta spp.; unidentified field dodders. C. America. [In warm, often inland areas.]

Cylindrocladium sp.; isolated from decayed weak roots of plants that had lodged in wind-free area. Costa Rica.

Darluca filum (Biv. & Fr.) Cast. = Eudarluca caricis (Fr.)
    Eriks.; can be mistaken for a primary disease but is
    actually a hyperparasite of rusts.   Puerto Rico, El Salva-
    dor, Costa Rica, Panama, Guatemala, Colombia.
Diaporthe zeina Speg. = D. austro-americana Speg.; culm
    disease.   Argentina.
Diplodia sp. (fungus name often given for all four of these
    different parasites on maize): 1. D. frumenti Ell. & Ev.;
    ear rot from disease starting in old silk.   S. America.
    2. D. macrospora Earle; stalk dry rot, ear disease.   Sub-
    tropical USA, Costa Rica, Panama, Brazil, Argentina.
    3. D. theobromae (Pat.) Now. (a few synonyms: D. natalen-
    sis P. Evans, D. gossypina Cke., Botryodiplodia theo-
    bromae Pat.) with its perfect state Botryosphaeria quercuum
    (Schw.) Sacc. = Physalospora rhodina Berk. & Curt.; com-
    mon stalk black rot, ear rot at peduncle end.   Neotropical
    in distribution.   4. D. zeae (Schw.) Lev. = D. maydis
    (Berk.) Sacc. with its perfect state of Phaeostagonosporop-
    sis zeae (Schw.) Wr.; dry rot in stalk and ear.   Rare in
    Subtropical USA, more common in Brazil, Argentina, and
    Peru.   [Field observations and reports suggest D. theo-
    bromae occurs over especially wide range of environmental
    conditions.]
Ephelis sp.; rare thin graymold.   W. Indies, S. America.
Epicoccum neglectum Desm.; leafspot, lives as a saprophyte
    on ear husk.   Mexico.   E. nigrum Lk.; infection of senes-
    cent-leaf.   Brazil.
Erwinia dissolvens (Rosen) Burkh.; bacterial disease of root
    and stalk.   Subtropical USA, Puerto Rico.
Eudarluca (see Darluca)
Fusarium spp. (are variable and often confusing as to species
    in maize): 1. F. culmorum (W. G. Sm.) Sacc.; root and
    stalk injury, isolated from lesions on husks of unharvested
    ears.   Florida, Jamaica, Chile.   2. F. graminearum
    Schwabe its perfect stage is Gibberella zeae (Schw.) Petch;
    causes stalk canker and ear disease.   [In warm land fields
    and somewhat restricted in spread, does not flourish in
    deserts but is found in dryer areas.]   Texas and Florida,
    dry sides of Virgin Islands, Puerto Rico, El Salvador,
    Costa Rica, Nicaragua, Surinam, Ecuador, Colombia,
    Argentina, Brazil.   3. F. moniliforme Sheld. its perfect
    stage is given the name Gibberella moniliforme (Sheld.)
    Winel., or G. fujikuroi (Saw.) Wr.; common pink disease
    of ear, "pokka boeng," root-rot and occasional wilt, stalk
    lesion.   [Warm-moist fields in slightly acid soils.]   Wide-
    spread.   4. An isolation from discrete root lesions from
    a mountain field in Costa Rica yielded pure cultures of
    typical F. oxysporum Schl.; and it caused wilt of maize
    seedling in warm greenhouse. 5. F. roseum Lk.; ear
    disease (not pink colored), grain lesion, seedling blight.
    Mexico, C. America, Chile, Venezuela, Colombia, Argen-
    tina, Uruguay.

Gibberella (see Fusarium)

Gloeocercospora sorghi Bain & Edg.; zonate leafspot. Subtropical USA, Panama.

Glomerella (see Colletotrichum)

Helminthosporium spp.: H. carbonum Ullst., asco-state is
    Cochliobolus carbonus Nelson; blackened ear and/or stalk,
    seedling blast, spreading spot on leaf, blotch. Mexico,
    Costa Rica, El Salvador, Guatemala, Colombia. H.
    maydis Nisik. & Miy., asco-state is C. heterostrophus
    Drechs.; blotch, seedling blight. Subtropical USA, W.
    Indies, C. America, Surinam, Ecuador, Bolivia, Mexico,
    Panama. [Found mostly in areas where conditions are
    warm and moist.] H. penicillus Speg.; on leaves. S.
    America. H. ravenellii Berk. & Curt.; leaf disease. S.
    America. H. sativum Pam., King & Bakke; stalk and root
    rot. Mexico, Panama. H. turcicum Pass.; leaf blast,
    spots may become stripes, large leafspotting. [In C.
    America apparently low hot areas tend to be almost free,
    seemingly most common in fields with ecology where
    temperatures are moderate and there is medium rainfall.]
    Probably the most widely distributed maize parasite of this
    genus.

Hemicycliophora sp.; nematode collected in rootzone. Peru.

Himantia stellifera Johnst.; white mycelial stalk and root
    disease. Scattered occurrence in Puerto Rico, Virgin
    Islands, Guiana.

Hormodendrum (see Cladosporium)

Iron deficiency, chlorosis and weak growth. Puerto Rico,
    Costa Rica, El Salvador.

Lasiophoma (see Chaetophoma)

Leaf fleck from arrested infections from fungus spores.
    Occasional.

Leptosphaeria arundinacea Sow. = L. orthogramma (Berk. &
    Curt.) Sacc.; weak parasite following other primary infec-
    tions by such strong fungi as Phyllachora or Septoria,
    result is enlarged leaf and stalk lesions. Subtropical USA,
    St. Thomas.

Lightning strike, impact point discernible from split-open
    maize stalks, plants killed on all sides of point of strike.
    (In one example, dead area was 12 paces in diameter.)
    Nicaragua.

Lodging, certainly seasonal during wind storm periods, es-
    pecially if stalks are tall and bear heavy ears. Mexico,
    Honduras, Guatemala, Panama, and no doubt elsewhere in
    storm prone areas.

Lodging, physiological root weakness noted in an introduced
    experimental maize variety. Costa Rica.

Macrophomina phaseoli (Maubl.) Ashby with its imperfect
    state Sclerotium bataticola Taub.; charcoal stalk rot.
    Scattered in Subtropical USA, W. Indies and Antilles, C.
    America, Argentina, perhaps widespread. [Often in hot
    and dry country.]

Magnesium deficiency, white and light yellow streaks between leaf veins. Puerto Rico, Florida.

Manganese deficiency, whitish to light yellow chlorosis at leaf tip accompanied by necrotic-end effect. Puerto Rico, Panama.

Marasmius graminicola Speg.; not spectacular but sometimes common on old decaying stubble. Mexico, Guatemala, El Salvador, Nicaragua, Costa Rica, Puerto Rico, St. Thomas, Panama, possibly widespread.

Melanconium sacchari Mass., leaf disease. Bolivia.

Melanospora pampeana Speg. with its conidial state Spicaria mucoricola Speg.; a culm disease. Argentina.

Meloidogyne sp.; rootknot nematode. Subtropical USA, Brazil.

Metasphaeria sp.; leafspot. S. America.

Mucor olivacellus Speg. = Rhizopus olivacellus (Speg.) Naum.; tassel rot. El Salvador, Argentina, Guatemala [in fields in turned down ears dried on stalk, disease appeared at close of dry season and start of rains]. M. plumbeus Bon.; seed rot. C. America.

Mycena citricola (Berk. & Br.) Sacc. (the gemma state used for inoculum) caused leaf spot in laboratory. Costa Rica.

Myriogenospora aciculisporae Viz.; stromatic masses on leaf disease. Brazil.

Nigrospora oryzae (Berk. & Br.) Petch; part of leafspot and ear husk-spot complex, black kernel, cobrot. [In medium cool, semi-dry conditions, neither wet nor hot.] W. Indies in mountain-side fields, Mexico, El Salvador, Guatemala, Peru, Brazil. N. sphaeria (Sacc.) Mason = Basisporium gallorum Moell.; in El Salvador, Costa Rica, Brazil.

Nitrogen deficiency was especially noted in certain varieties imported from "corn belt" in U.S. temperate zone, result is premature yellowing of old leaves with the chlorosis following midrib. Not in local varieties. Costa Rica.

Oidium sp.; light powdery mildew. Colombia.

Oospora heteromera Speg.; seed-grain with tightly adherent fungus growth on surface that reduces germination. Argentina.

Ophiobolus acuminata (Sow.) Duby; plant footrot. Brazil.

Patellaria atrata (Hedw.) Fr.; weak parasitic stalk invasion, saprophytic involvement of dying and dead stalks. Puerto Rico, Virgin Islands, probably further spread.

Penicillium spp.; musty grain, musty cattle fodder, seedmold in moist storage. El Salvador, Costa Rica, Guatemala, Panama, Colombia, Ecuador, Chile, Brazil. P. viridicatum Westl.; of special note as it causes blue-eye-spot on grain. Uruguay, Argentina.

Periconia sp.; part of leaf rot and a leaf spot complex. Costa Rica.

Peronospora sp.; downy-type mildew. Mexico.

Phaeomarssonia sp., in association with Phyllosticta; increases size and severity of leafspot. Brazil.

Phaeostagnosporopsis (see Diplodia)

Phoma zeicola Ell. & Ev. ; on leaves.   Rare in Subtropical
    USA, and Peru, found in Brazil.
Phosphorus deficiency occasional on acid soils, symptoms are
    purplish leaves, late maturity, poor ears.   El Salvador,
    Costa Rica.
Phyllachora maydis Maubl. ; common tarspot, leaf collapse
    [noted especially at close of wet season, has caused com-
    plete crop failure at times in plots in some areas].   Wide-
    spread.
Phyllosticta hispida Ell. & Dearn. ; leafspot in Brazil.   P.
    zeae Bat. ; leafspot.   Brazil, Costa Rica, believed seen
    for first time in Florida in 1932.
Phymatotrichum omnivorum (Shear) Dug. ; quiescent fungus re-
    covered from roots that showed no signs of tissue invasions.
    Subtropical USA [in dry land areas].
Physoderma maydis Miy. = P. zeae-maydis Shaw; brownspot,
    broken node.   [In El Salvador most severe in maize crop
    grown in that part of year when wet season was interrupted
    by medium short rain-free periods.]   Subtropical USA, El
    Salvador, Mexico, Nicaragua, Guatemala, Costa Rica,
    Colombia, Venezuela, Ecuador, Brazil.
Physopella zeae (Mains) Cumm. & Ram. (with synonyms:
    Angiospora zeae Mains, A. pallescens (Arth.) Mains,
    Puccinia pallescens Arth., Uredo pallida (Diet. & Holw.)
    Arth., Dicaeoma pallescens (Arth.) Arth. & Fromme); rust
    pustules develop under leaf tissue retaining spores until
    host tissue splits or disintegrates.   Texas, Louisiana,
    Mexico, Guatemala, Honduras, El Salvador, Costa Rica,
    Nicaragua, Dominican Republic, Puerto Rico, Jamaica,
    Islands in the Antilles, Panama, Trinidad, Colombia,
    Ecuador, Venezuela, Peru, Brazil.   [This rust somewhat
    restricted in occurrence, in C. America is called Guate-
    mala rust, most readily found in cool country or mountain
    side fields.]
Phytophthora sp. ; rootlet rot.   [Isolation from diseased roots
    in high mountain clearing in wet jungle.]   Guatemala.
Poria cocos Wolf; root disease accompanied by formation of
    giant subterranean sclerotium.   Subtropical USA.
Potassium deficiency, result is leaf-margin necrosis, weak
    and dwarfed stalks, poorly developed ears.   Costa Rica,
    Panama, Florida.
Pratylenchus spp., nematodes feeding on maize roots: P.
    brachyurus (Godf.) Filip. in Brazil, P. pratensis (De Man)
    Filip. in Subtropical USA, P. zeae Graham in Subtropical
    USA, Jamaica, Puerto Rico, Brazil, and a Pratylenchus
    sp. was collected in Chile and in Peru.
Pseudomonas alboprecipitans Rosen; bacterial leaf and stalk
    disease.   Subtropical USA, Puerto Rico, Guatemala.   P.
    syringae v. Hall; bacterial leafspot.   Subtropical USA, El
    Salvador.
Puccinia polysora Underw. = Dicaeoma polysorum Arth. ;
    brown leafrust, spores borne in open sori, spots relatively

small and rounded.   Neotropics.   Specially noted in Sub-
tropical USA, Cuba, Puerto Rico, Mexico, Guatemala, El
Salvador, Honduras, Costa Rica, Nicaragua, St. Lucia,
Dominican Republic, Jamaica, Panama, British Honduras,
Trinidad, Venezuela, Colombia, Peru, Brazil. [Most
severe in warmer-moist areas, in C. American mountains
especially at low elevations long feared by Indians there.]
P. sorghi Schw. = P. maydis Bereng.; common yellow to
light brown leafrust, large spots, occasionally on stalk.
Neotropics.
Pyricularia grisea (Cke.) Sacc.; [Severe at medium high ele-
vations in C. American mountains.] young seedling disease.
Fungus identified by isolation and microscopic means.
Costa Rica and Panama, both in areas near diseased rice
fields.
Pythium spp. (about seven species involved); seedling damping-
off, older plant rootrot. [In wet soils, most seen in wet
seasons in fields with shallow hardpan.] Subtropical USA,
Costa Rica, Mexico, Guatemala, Panama, Colombia, Ecua-
dor, Peru, Venezuela, Brazil.
Radopholus similis (Cobb) Thorne; nematode at roots. C.
America.
Rhizoctonia grisea (J. A. Stev.) Matz. = Sclerotium griseum
J. A. Stev.; leaf rot, plant base-rot. Puerto Rico, Mexico,
Cuba, Haiti, Ecuador. R. solani Kuehn (its spore bearing
state is Thanatephorous cucumeris (Frank) Donk); footrot,
root disease, wilt and death. Subtropical USA, Cuba, El
Salvador, Costa Rica, Puerto Rico, Mexico, Colombia,
Panama, Peru. R. zeae Voorh.; ear rot. Florida, Guate-
mala, Panama, Costa Rica (aerial type of the genus).
Rhizopus (see Mucor, but see immediately below)
Rhizopus nigricans Ehr.; reduces seed germination, causes
mustyness of grain, preemergence damping-off. Costa
Rica, Guatemala, Panama, Ecuador, Argentina.
Rhopographus zeae Pat. = Clasterosporium longisporum
Voorh.; a stalk disease. Subtropical USA, W. Indies.
Rhynchosporium oryzae Hashi & Yok.; zonal leaf spot. Costa
Rica.
Rofillardia sp.; part of leaf disease complex, associated with
other primary disease producing fungi. Brazil.
(Rusts. Rusts of maize, especially the mixed-bred and folk-
selected varieties, are often managed in such a way that
the rusts seldom cause extreme losses. Small farmers
may follow a system of rapid cropping and shifting of fields.
When farms are very large where broad stretches are
planted to pure stands of highly selected maize with narrow
range of genetic composition, or where a big area is altogether given over to growing this extremely important crop
and no other, there is much more danger from severe
losses.)
Sclerospora graminicola (Sacc.) Schr.; downy mildew often
causing witches' brooming. Mexico, Guatemala, Brazil and

spreading. S. macrospora Sacc.; leafscald in Florida and
Peru and may be spreading. S. sorghi Weston & Uppal
in Texas and in some S. American countries.
Sclerotium rolfsii (Curzi) Sacc., its sporuliferous state is
   Pellicularia rolfsii (Sacc.) West; basal rot, plant-drop.
   Subtropical USA, Panama, Surinam.
Scolecotrichum graminus Fckl.; leafspotting. Puerto Rico,
   Subtropical USA.
Septoria zeae Stout = S. zeicola Stout; leafspot. Subtropical
   USA, Mexico.
Sirococcus maydis Speg.; stalk infection. Chile.
Spegazzinia tessarthra (Berk. & Curt.) Sacc. = Sporodesmium
   tessarthra Berk. & Curt. (collected in Cuba in 1829); semi-
   parasitic small black lesions on weakened leaf and stem
   tissues. Cuba, Puerto Rico (seen on hand woven bed mats
   in Guatemala).
Sphacelotheca reiliana (Kuehn) Clint. = S. reilanum (Kuehn)
   McAlp.; tassel smut, occasional multiple tassel. Sub-
   tropical USA, Mexico, Guatemala, El Salvador, Costa Rica,
   Jamaica, Colombia, Uruguay, Brazil, Argentina. S.
   paranensis Hirsch. and S. platensis Hirsch.; tassel smuts
   in S. America.
Sphaerulina maydis P. Henn.; angular and circular leafspots.
   Brazil.
Spicaria sp.; a weak parasite associated with leafspots and
   stalk diseases started by other fungi. Texas, Argentina,
   scattered in Honduras, El Salvador (also see Melanospora).
Stemphylium botryosum Wallr.; small dark lesions, scattered
   in occurrence. Costa Rica, Panama.
Stictis maydis P. Henn.; on dead leaves and husk. Brazil,
   Argentina.
Stilbum sp.; possible saprophyte. Probably scattered in
   most countries on standing dead stalks. Identified in
   Panama and Honduras.
Striga sp.; phanerogamic herb active as a root parasite.
   Subtropical USA, Dominican Republic, apparently will
   spread.
Trichoderma lignorum Tode; accompanying root disease.
   Colombia. T. viride Pers.; seeding blight from virulent-
   type of parasite, and there is a non-virulent-type of the
   fungus wholly saprophytic. El Salvador.
Trichodorus christiei Allen; a root-invading nematode. Florida.
Trichothecium roseum Lk. = Cephalotecium roseum Cda.;
   cobrot, semiparasite often on edges of leaf tips. [Rainy
   season disease of cobs of grain still on stalks standing in
   the field.] W. Indies, S. America, Costa Rica, Trinidad.
Tylenchorhynchus sp.; nematode inhabitant of rhizosphere.
   Chile.
Unknown disease, possibly a Spirochaete, top collapse and
   plant death. Puerto Rico, Dominican Republic.
Urocystis agropyri (Preuss.) Schr.; rare maize smut, reported
   in Costa Rica.

Ustilaginoidea virens (Cke.) Tak.; false smut of leaf.   Subtropical USA, Mexico, Panama.
Ustilago fischeri Pass.; smut.   Argentina and the U. maydis
   (DC.) Cda. (synonyms: U. maydis (DC.) Tul., U. zeae
   (Beckm.) Ung., U. zeae Berk., U. abortifera Speg.
   (1898)); the common earsmut, occasional attack of leaf
   and stalk.   Neotropics.   [Occurs over wide range of conditions in C. America from warm lowlands to cool mountain-
   sides, is most severe in somewhat warm but not too dry
   areas, and the coolest-drizzly highest mountain lands are
   almost free.]
Ustilagopsis deliquescens Speg. (synonyms: Claviceps deliquescens (Speg.) Haum., C. paspali Stev. & Hall, Sphacelia
   deliquescens (Speg.) Haum.); parasite of flower parts.
   Argentina, Mexico, Guatemala.
Viruses (apparently at least ten): 1. Cucumber Mosaic virus
   transmitted by aphid, causes mottling and mild chlorosis
   in Florida, Costa Rica.   2. Cucumber Mosaic strain
   Celery-Commelina, aphid transmission, is a more severe
   chlorosis and mottle and early infection results in severe
   dwarfing in Florida, Puerto Rico, Costa Rica, Guatemala,
   Cuba.   3. Maize Bird's-eye-spot in Guatemala.   4. Maize
   Fleck in Florida, Puerto Rico, Costa Rica.   5. Maize
   Mosaic is apparently the most common mottling with some
   malformation in W. Indies, Subtropical USA, C. America,
   Panama, Dominican Republic, Trinidad, Jamaica, Bolivia,
   Venezuela, Peru, probably widespread but often not recognized as it does not kill.   6. Maize Necrotic Streak virus
   in Brazil.   7. Maize Stripe (the virus particle is a round
   ended short rod) causes stripe and shortening of internodes
   but no dwarfing of leaf, in Florida, Cuba, Puerto Rico,
   Nicaragua, Mexico, Costa Rica, Trinidad, Panama, Venezuela, Colombia, and may be in more countries.   8. Maize
   Stunt also called Maize Dwarf (with particles found to be
   both long flexuous thread-like bodies and mycoplasma-like
   bodies) cause of stunting of whole plant with leaves not only
   shortened but also streaked and reddish yellow in Subtropical USA, Mexico, Guatemala, El Salvador, Honduras,
   Costa Rica (symptoms seen by me in 1941 published in
   1949), Nicaragua, Dominican Republic, Panama, Ecuador,
   Venezuela, and spreading.   9. Maize White-streak (?
   possible strain of common mottling mosaic, but transmitted
   by leafhopper) in Cuba, Florida (rarely found there but has
   appeared after a few years of unseasonbly warm winters),
   Nicaragua, Venezuela.   10. Sugarcane Mosaic transmitted
   from disease sugarcane by aphids in Subtropical USA, El
   Salvador, Costa Rica, Jamaica, Haiti, Dominican Republic,
   Peru, Venezuela, Brazil, Argentina.
Water-spot or grease-spot.   Cause unknown, considered physio-
   logical.   Not believed to be damaging.   Widespread.
Xanthomonas albilineans (Ashby) Dows.; leafscald, wilt.
   Guianas, Uruguay, Brazil.   X. stewartii (E. F. Sm.) Dows.;

bacterial wilt. [Specially noted in lowland and warm fields.] W. Indies, Guatemala, Costa Rica, Mexico, Guyana, Brazil. X. vasculorum (Cobb) Dows.; bacterial gumming, wilting. W. Indies and Antilles, Mexico, Costa Rica, Trinidad, Colombia, Brazil.

Xiphinema sp.; nematode in rhizosphere. Chile, Peru.

Zinc deficiency, results in apical chlorosis of leaf, causes white bud of stalk tip, yellowish leaf streak, if severe no ears develop, where mild ears poorly formed. Florida, Texas, Costa Rica, Panama.

ZINGIBER OFFICINALE Roscoe. A perennial herb native to the Far East and known in the ancient world since prehistoric times, it has handsome leaves on a reed-like more than knee-high stem but is specially grown for its irregular rootstalks that are dug and processed as a spice for world commerce in candied root, flavor for cooking and in drinks, and medicinal purposes and is becoming a crop of importance in certain West Indies islands and a few areas in other tropical American countries; "jengibre," "ajengibre" Sp., "gengibre" Po., "ginger" En.

Algae, green scum at plant base reducing growth of rootstalk. Puerto Rico.

Botryodiplodia theobromae Pat. = Diplodia theobromae (Pat.) Now. or D. natalensis P. Evans, with asco-state Botryosphaeria quercuum (Schw.) Sacc.; = Physalospora rhodina Berk. & Curt.; causes stem lesion, leafspot. Puerto Rico, Jamaica.

Cassytha filiformis L.; the slender olive-green dodder infection by laboratory inoculation. Puerto Rico.

Cephaleuros virescens Kunze; algal scurfy leafspot. Puerto Rico, Jamaica.

Cercospora sp.; leafspot and blotch. Costa Rica.

Cladosporium sp.; on necrotic leafedge. Puerto Rico.

Colletotrichum graminicola (Ces.) Wils. = Gloeosporium tucumanensis (Speg.) Arx. & Moell. and the asco-state is Glomerella tucumanensis (Speg.) Arx. & Moell.; anthracnose. Widespread.

Cuscuta americana L.; the golden-yellow fast growing dodder infected by laboratory inoculation. Puerto Rico.

Erwinia sp.; bacterial softrot of flower stem. Costa Rica.

Fusarium roseum Lk.; stem and rootstalk dryrot. Puerto Rico, Jamaica. F. solani (Mart.) Sacc.; large discrete rounded lesions on market root. Florida.

Infection-body (gemma, but not Mycena, oblong in shape and grayish salmon-pink in color, nearly size of a pinhead) ? from untouched jungle close by; leafspot. Costa Rica.

Leaf spots, cause unidentified, color red and rounded in shape. Puerto Rico.

Mycena citricolor (Berk. & Br.) Sacc. = Omphalia flavida (Cke.) Maubl. & Rang.; leafspot from infection by round lemon-yellow gemma. Laboratory inoculation and natural occurrence in field. Costa Rica.

<u>Penicillium</u> sp. isolated from rotted ends of dried roots in
market.   Guatemala.
<u>Phycopeltis</u> sp. ; alga on leaf and stem.   Guatemala.
<u>Phyllosticta</u> sp. ; leafspot.   Puerto Rico, Costa Rica.
<u>Phytophthora-Pythium</u> (in culture unable to distinguish which);
rootrot of fibrous roots.   Puerto Rico.
<u>Pseudomonas</u> <u>solanacearum</u> E. F. Sm. ; bacterial wilt.
Jamaica.
Ragged leaf, both young and old leaves, undetermined cause.
Puerto Rico.
<u>Rhizoctonia</u> <u>microsclerotia</u> Matz; web-blight.   Guatemala.
<u>R.</u> <u>solani</u> Kuehn; root-rot.   Costa Rica.
<u>Rosellina</u> <u>bunodes</u> (Berk. & Br.) Sacc. ; root disease.   Costa
Rica.
<u>Strigula</u> <u>complanata</u> Fée; white milk-spot from lichen on older
leaves and stems.   Widespread.
Virus.   Mottle and curiously contorted rootstalks for sale in
market.   Ecuador.
Yellow leaf-scurf (lichen) in Venezuela.

ZINNIA PERUVIANA L. = Z. ELEGANS Jacq. , and Z. MULTIFLORA
L.   With apparently some chance interspecific crosses, native to
tropical America from Mexico south to below Peru, these are
popular and profusely flowering annuals to semi-perennial somewhat woody shrubs, often grown in sunny gardens for massed
colors, also commercially important to countless small market
florists and growers; "copetón, " "rascamoños, " "clavél" Sp. ,
"velhas, " "zabumba, " "moças" Po. , "zinnia" En.
<u>Alternaria</u> <u>sonchi</u> Davis and <u>Z.</u> <u>zinniae</u> Pape; leafspot and
collapse.   Argentina, rare spot in Costa Rica and Guate-
mala in cool dry localities.
<u>Botrytis</u> sp. ; disease of unopened flower.   C. America in
warm-moist locations.
<u>Briarea</u> <u>gigantea</u> Speg. ; discoloring disease of flower-head.
Argentina.
<u>Capnodium</u> sp. ; sootymold.   W. Indies, C. America.
<u>Cassytha</u> <u>filiformis</u> L. ; dodder on self sown plants in field.
Puerto Rico.
<u>Cercospora</u> zinniae Ell. & Mart. = <u>C.</u> <u>atricinta</u> Heald &
Wolf; leafspot, petal spot, stem lesions.   Noted especially
in subtropical USA, Trinidad, Puerto Rico, Costa Rica, El
Salvador, Jamaica, Venezuela, Brazil.
<u>Choanephora</u> sp. ; blossom rot.   Florida, Puerto Rico, Costa
Rica.
<u>Cicinnobolus</u> sp. ; parasite of mildew on leaf.   Southern S.
America.
<u>Erysiphe</u> (see Oidium)
<u>Fusarium</u> spp. ; stem disease, basal rot, wilt.   Scattered.
<u>Meloidogyne</u> sp. ; rootknot.   Florida, Puerto Rico.
<u>Mycena</u> <u>citricolor</u> (Berk. & Br.) Sacc. ; leafspot, from both
inoculation in laboratory and natural infection in garden ad-
jacent to leafspot diseased coffee.   Costa Rica, El Salvador.

Oidium sp., asco-state is Erysiphe cichoracearum DC.,
latter is seldom collected; common white powdery mildew,
disease increases whiteness as it accumulates on older
plants. Widespread.
Phoma zinnae Speg.; stem canker. Argentina.
Phymatotrichum omnivorum (Shear) Dug.; dry rootrot. Texas.
Pseudomonas (see Xanthomonas)
Puccinia melampodis Diet. & Holw. = P. zinniae P. & H.
Syd.; common brown leafrust. Widespread.
Pythium sp.; damping-off. Widespread.
Rhizoctonia solani Kuehn is the parasitic mycelial state while
the nonparasitic spore-bearing growth is Thanatephorous
cucumeris (Frank) Donk; damping-off, stem canker, rot of
plant base. California, Mexico, Costa Rica (found in long
cultivated medium dry, neutral soils, not in rain-forest
in newly cleared areas). Puerto Rico, Panama, probably
nearly Widespread.
Sclerotinia sclerotiorum (Lib.) dBy. = Whetzeliana sclerotiorum
(Lib.) Korf & Dumont; bud disease. Costa Rica, Chile.
Sclerotium rolfsii (Curzi) Sacc. its asco-state is Pellicularia
rolfsii (Sacc.) West; bottom rot. Florida, Puerto Rico.
Septoria zinniae Speg.; small leafspots. Argentina, Bolivia.
Sootymold (not Capnodium), in Panama.
Stereum purpureum (Pers.) Fr.; on old plant infection. Argen-
tina.
Viruses: Beet (Argentine) Curly Top in Argentina. Crucifer
Mosaic = Turnip Mosaic (by inoculation) in Trinidad. Cu-
cumber Mosaic = Zinnia Mosaic in Subtropical USA, Puerto
Rico, El Salvador, Panama. Tobacco Streak (by inoculation) in Brazil. Tomato Spotted Wilt scattered in Subtropi-
cal USA, Brazil, Argentina.
Xanthomonas solanacearum (E. F. Sm.) Dows. = Pseudomonas
solanacearum E. F. Sm.; bacterial vascular wilt. Widespread.

ZIZYPHUS JUJUBA Mill. From the Oriental tropics, a tough small
tree that grows often under conditions of extreme heat, and pro-
duces fruit much liked in the Eastern tropics and of slowly in-
creasing popularity where tested in the Occident; "jujube, "
"aprín." (Very little disease study on this tree, thus far, in
the American tropics.)
Ascochyta ziziphi Putt.; leafspot. Brazil.
Bud failure (undetermined fungus). Guatemala, El Salvador.
Cercospora sp.; large gray leafspot. Costa Rica.
Dieback (unidentified fungus). Honduras.
Phakopsora zizyphi-vulgaris (P. Henn.) Diet.; rust. Florida.
Phymatotrichum omnivorum (Shear) Dug.; dry rootrot. Texas.
Root disease (unidentified cause). Costa Rica.
Schizophyllum commune Fr.; old-branch disease, following
dieback. Honduras.
Strigula complanata Fée; white lichen leafspot. Honduras.

Tryblidiella sp.; secondary attack of old diseased branch tip.
  Panama.
Uredo zizyphi Pat.; leafrust. Brazil.

ZOYSIA spp. Originating in the Oriental tropics, these are short,
  creeping perennial grasses for pastures, lawns and playing fields,
  of apparent special value in some tropical areas of the Americas where other lawn grasses are difficult to maintain.   Tests
  still in progress; "hierbafino, " "zoizia" Sp. , "zoisia, " "graneado
  oriental" Po. , "zoysia, " "velvet grass" En.
    Colletotrichum graminicola (Ces.) Wils. (its perfect state not
      collected by me on this host); necrotic streak-spots on
      leaves and the short stalks, plants killed. Puerto Rico,
      Costa Rica.
    Curvularia lunata (Wakk.) Boed.; leafspot, lesion at soil line.
      Nicaragua.
    Fusarium spp. (a long-spored species and a short-spored
      type) no isolations made and species not determined.   Plant
      failure. Costa Rica.
    Helminthosporium sp.; killing of lawn edge along walk in
      Puerto Rico, also reported in Florida.
    Nigrospora sp.; spores seen in microscopic examinations of
      old plants killed by insects. Costa Rica.
    Penicillium sp.; from dead plants of turf edge sent to Guate-
      mala from unknown source.   (The Penicillium disappeared
      on plants grown a year in new location.)
    Pythium sp.; from decaying plants in warm lowland moist
      area. El Salvador.
    Rhizoctonia microsclerotia Matz which is the mycelial and
      parasitic stage, (its sporuliferous state was seen, that is
      the non-parasitic Pellicularia microsclerotia (Matz) Weber);
      severe as web-blight. Costa Rica, Puerto Rico. R.
      solani Kuehn the common mycelial parasite (its spore-bear-
      ing but not infective fungus stage Thanatephorous cucumeris
      (Frank) Donk was not found); large spot injury of lawn
      studied in Puerto Rico, also reported in Florida, several
      countries in C. America and W. Indies.
    Sclerotinia sp.; small discrete dead spots in lawn. Florida,
      Costa Rica.

AFTERWORD

In my work on plant diseases I have been perpetually impressed with the interest and dedication to our science of many a freely helpful colleague and friend. In human relationships where so many live on the efforts of others, the plant pathologist belongs to one of the least parasitic of professions. In certain parts of the tropics in the struggle for survival, I have seen that sometimes the pathologist is one of the less understood and less appreciated of the very important members of his community.

INDEX OF HOST NAMES*

abacá 250
abacate (abacateiro) 304
abacaxi 14
abele 337
abelmoscho 184
Abelmoschus moschatus 184
abió 220
abóbora 106
aboboreira 439
abricatiero 340
abutilon (abutiloniera) 1
Abutilon sp. 1
acácia 2, 214
Acacia bursario 2
Acacia caven 2
Acacia julibrissin 6, 7
Acacia lebek 7
Acacia riparia 7
acacu 192
acae 148
acafrá 38
acaju 413
acalífa 3
acalypha 3
Acalypha tricolor 3
Acalypha wilkesiana 3
aceituno 271
acelga 36
acetillo 459
achiera 50
achiote 38
achoccha 109
Achras sapota 220, 383
Achras zapota 3
acid mamey 356
Acrocomia sp. 285
açucena 13

açucena 193
adeira 50
adelfa 259
african daisy 163
african marigold 417
african tulip tree 408
african violet 377
agave 4
agave, giant type 160
Agave americana 4
Agave fourcroydes 4
Agave letonae 4
ageratum 5
Ageratum conyzoides 5
Aglaonema simplex 5
agropiro 6
Agropyron sp. 6
agróstide 6
Agrostis sp. 6
aguacate 304
aguija 456
ahuacatl 304
aipim 238
ajengibre 470
ají 52
ajipa 285
ají pimentão 52
ají pimentón 52
ajo 8
ajonjoli 389
akee 39
ala 124
ala de pajaro 27
alamo 337
albaricoque 340
Albizzia julibrissin 6, 7
Albizzia lebek 7

*Scientific (Latin) names are underlined; others are the common or
vernacular names in various languages.